Thermoelasticity with Finite Wave Speeds

Józef Ignaczak

Institute of Fundamental Technological Research
Polish Academy of Sciences
Warsaw, Poland

Martin Ostoja-Starzewski

Department of Mechanical Science and Engineering,
Institute for Condensed Matter Theory,
and Beckman Institute
University of Illinois at Urbana-Champaign
Urbana, USA

OXFORD
UNIVERSITY PRESS

OXFORD
UNIVERSITY PRESS

Great Clarendon Street, Oxford OX2 6DP

Oxford University Press is a department of the University of Oxford.
It furthers the University's objective of excellence in research, scholarship,
and education by publishing worldwide in

Oxford New York

Auckland Cape Town Dar es Salaam Hong Kong Karachi
Kuala Lumpur Madrid Melbourne Mexico City Nairobi
New Delhi Shanghai Taipei Toronto

With offices in

Argentina Austria Brazil Chile Czech Republic France Greece
Guatemala Hungary Italy Japan Poland Portugal Singapore
South Korea Switzerland Thailand Turkey Ukraine Vietnam

Oxford is a registered trade mark of Oxford University Press
in the UK and in certain other countries

Published in the United States
by Oxford University Press Inc., New York

First published 2010

British Library Cataloguing in Publication Data
Data available

Library of Congress Cataloging in Publication Data

Data available

Typeset by SPI Publisher Services, Pondicherry, India
Printed in Great Britain
on acid-free paper by
CPI Antony Rowe, Chippenham, Wiltshire

ISBN 978–0–19–954164–5

1 3 5 7 9 10 8 6 4 2

CONTENTS

Józef Ignaczak dedicates the book to Krystyna on the 45th anniversary of their marriage.
Martin Ostoja-Starzewski dedicates the book to his wife Iwona.

PREFACE

This book focuses on a generalized dynamic coupled thermoelasticity theory of solid materials, free of the classical paradox of infinite propagation speeds of thermal signals. As is well known, that paradox is caused by the Fourier model of heat conduction. Dating back to Maxwell (late nineteenth century) and Cattaneo (mid-twentieth century), several models have been developed and intensively studied over the past four decades, and it is now time to write an up to date monograph on this subject. Besides a few monographs on extended rational continuum mechanics and thermodynamics, which only touch upon the dynamic thermoelasticity of solid materials per se, let us mention a book on generalized dynamic thermoelasticity by Podstrigach and Kolano (in Russian) and a book on classical and generalized dynamic thermoelasticity by Dhaliwal and Singh, both published in the late 1970s. That is, our book focuses on dynamic thermoelasticity governed by hyperbolic equations, rather than on a wide range of continuum theories. Hence the title: *Thermoelasticity with Finite Wave Speeds.*

Besides the paradox of infinite propagation speeds, the classical dynamic thermoelasticity theory offers either unsatisfactory or poor descriptions of a solid's response to a fast transient loading (say, due to short laser pulses) and at low temperatures. Such drawbacks have led many researchers to advance various generalized thermoelasticity theories. Following in the steps of Maxwell and Cattaneo, they proposed thermoelastic models with one or two relaxation times, focused on low temperatures, absence of energy dissipation, a dual-phase-lag theory, or even anomalous heat conduction described by fractional calculus.

The present book concentrates on the two leading theories of hyperbolic thermoelasticity: that of Lord–Shulman (with one relaxation time), and that of Green–Lindsay (with two relaxation times). They are both set in small strains, and so, the resulting field equations are linear partial differential ones. The complexity of theories is due to the coupling of mechanical with thermal fields. The book is concerned with the mathematical aspects of both theories – existence and uniqueness theorems, domain of influence theorems, convolutional variational principles – as well as with the methods of dealing with a range of initial/boundary value problems. In the latter respect, following the establishment of the central equation of thermoelasticity with finite wave speeds, we consider: the exact, aperiodic-in-time solutions of the Green–Lindsay theory; Kirchhoff-type formulas and integral equations in the Green–Lindsay theory; thermoelastic polynomials; moving discontinuity surfaces; and time-periodic solutions. This is followed by a chapter on the physical and microstructural aspects of generalized thermoelasticity, where we review other models and theories. We conclude with a chapter on a non-linear hyperbolic theory of a rigid heat conductor for which a number of asymptotic solutions are obtained using a method of weakly non-linear geometric optics.

The book we present is a monograph. It may be used to augment graduate-level courses on advanced continuum mechanics, elasticity, and thermoelasticity. It also offers a basis for lab tests, as well as a basis for further research in the area, indeed an area that is actively being developed in several research centers in Europe, America and Asia.

Acknowledgements

We warmly acknowledge the support of Professor Richard B. Hetnarski who prompted the authors to write this book, and thank Professor Bruno A. Boley for comments on an earlier version of the manuscript. Through the years the work of both authors was partly facilitated by grants from the National Science Foundation, the Polish Academy of Sciences and the Natural Sciences and Engineering Research Council of Canada.

INTRODUCTION

This book focuses on a generalized dynamic coupled thermoelasticity theory of solid materials, free of the classical paradox of infinite propagation speeds of thermal signals. As is well known, that paradox is caused by the Fourier model of heat conduction, first observed in rigid isotropic conductors,

$$q_i = -k\frac{\partial \theta}{\partial x_i}, \tag{1}$$

which leads to a diffusion (i.e. parabolic-type) equation

$$\frac{\partial \theta}{\partial t} = \frac{k}{\rho c_p}\nabla^2\theta. \tag{2}$$

In eqns (1) and (2), q_i is the heat flux, θ is the temperature, k is the thermal conductivity, ρ is the mass density and c_p is the specific heat at constant pressure. The first equation above also introduces the index notation, which we follow throughout the book when dealing with tensors.

To remove the said paradox, following the proposal dating back to Maxwell (1867) and Cattaneo (1948), one should replace eqn (1) by

$$\left(1 + t_0\frac{\partial}{\partial t}\right) q_i = -k\frac{\partial \theta}{\partial x_i}, \tag{3}$$

whereupon, instead of eqn (2), there holds a telegraph (i.e. hyperbolic type) equation

$$t_0\frac{\partial^2 \theta}{\partial t^2} + \frac{\partial \theta}{\partial t} = \frac{k}{\rho c_p}\nabla^2\theta. \tag{4}$$

Clearly, this models heat conduction as a wave, often called a *second sound.* In eqns (3) and (4) t_0 is a relaxation time, and $c_T = (k/\rho c_p t_0)^{1/2}$ represents the speed of the second sound.

Properties of this model as well as a wealth of other, more complex models, all in the setting of rigid conductors, were developed and intensively studied over the past half a century (Joseph and Preziosi, 1989, 1990), and the trend continues. Over the same time period, parallel to this activity, there has been a major growth of thermoelasticity accounting for non-Fourier-type heat conduction in elastic bodies.

Besides the paradox of infinite propagation speeds, the classical dynamic thermoelasticity theory offers either unsatisfactory or poor descriptions of a solid's response to a fast transient loading (say, due to short laser pulses) and at low temperatures. Such drawbacks have led many researchers to advance various generalized thermoelasticity theories. Following in the steps of Maxwell and Cattaneo, they proposed thermoelastic models with one or two relaxation times, models focused on low temperatures, models with absence of energy dissipation, a dual-phase-lag theory, or even anomalous heat conduction described by fractional calculus. A number of reviews on the subject, already sometimes in a book form, have appeared over the past two decades (Chandrasekharaiah, 1986, 1998; Ieşan, 2004; Ignaczak, 1980b, 1989a,b, 1991; Hetnarski and Ignaczak, 1999, 2000). One should also mention here a book on generalized thermoelasticity (Podstrigach and Kolano, 1976) and a book on classical and generalized dynamic coupled thermoelasticity (Dhaliwal and Singh, 1980).

In the past two decades there has been an increased research activity in generalized thermoelasticity – among others, in Germany, India, Iran, Japan, Poland, Spain, UK, USA, and several Arabic countries, to name a few. Besides numerous technological applications we mention here the interest in modelling heat ransfer in living tissues, e.g. (Dai *et al.*, 2008).

Overall, it is now time to write an up-to-date monograph on the subject of hyperbolic thermoelasticity, or dynamic thermoelasticity governed by hyperbolic equations, which represents a subset of generalized thermoelasticity, the latter being part of a wide range of continuum theories. A contact with such theories occurs in a handful of monographs on extended rational continuum mechanics and thermodynamics (Müller and Ruggeri, 1993, 1998; Wilmański, 1998). Our book's title, *Thermoelasticity with Finite Wave Speeds*, reflects the defining concept of a *domain of influence*, $\mathcal{D}(t)$, in dealing with the paradox of infinite propagation speeds of thermal signals. Dating back to Zaremba (1915), $\mathcal{D}(t)$ is a generalization of the concept of support of a function $f(t)$, which is a set of all the points in the body that may be reached by the thermomechanical disturbances propagating from the locus of the disturbance with a speed not greater than a finite v, see Fig. I.

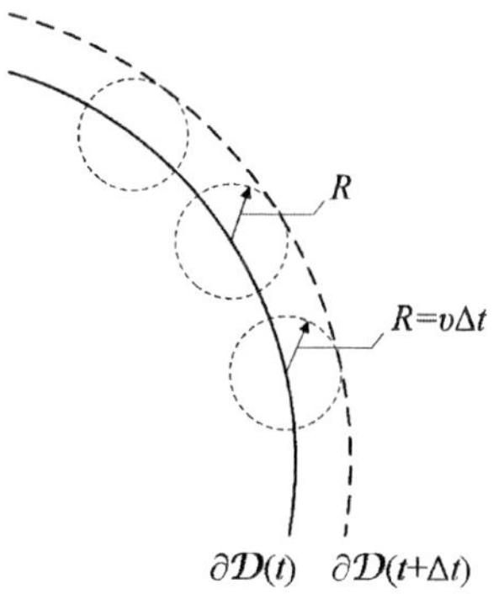

Figure I A schematic showing the evolution of the domain of influence via Huygens' principle from time t to $t + \Delta t$.

The following example illustrates the concept of a domain of influence (DOI).

Example 0.1 Cauchy problem for a hyperbolic equation.

Consider the following one-dimensional (1D) initial-boundary value problem. Find a solution of the wave equation

$$\left(\frac{\partial^2}{\partial x^2} - \frac{1}{c^2}\frac{\partial^2}{\partial t^2}\right) u(x,t) = 0, \quad |x| < \infty, \quad t \geq 0, \quad c > 0, \tag{5}$$

subject to the initial conditions

$$\begin{aligned} u(x,0) \equiv \eta(x) = u_0 \left[H(x+x_0) - H(x-x_0)\right] \quad |x| < \infty, \quad x_0 > 0 \\ \frac{\partial}{\partial t} u(x,0) = 0 \quad |x| < \infty, \end{aligned} \tag{6}$$

where $H = H(t)$ is the Heaviside function

$$H(t) = \begin{cases} 1 & \text{for} \quad x > 0, \\ 0 & \text{for} \quad x < 0. \end{cases} \tag{7}$$

and u_0 is a constant. A unique solution to the problem (5)–(7) takes the form

$$\begin{aligned} u(x,t) &= \frac{1}{2}\left[\eta(x+ct) + \eta(x-ct)\right] \\ &= \frac{1}{2} u_0 \left[H(x+x_0+ct) - H(x-x_0+ct)\right. \\ &\quad \left. + H(x+x_0-ct) - H(x-x_0-ct)\right], \end{aligned} \tag{8}$$

and it follows from eqn (8) that

$$u(x,t) = 0 \quad \text{for} \quad |x| > x_0 + ct. \tag{9}$$

A motion produced by the rectangular initial "disturbance" over the interval $[-x_0, x_0]$ (see eqn (6)) is a sum of two rectangular waves, each moving with velocity $c > 0$, one to the right and the other to the left, and in such a way that for fixed $t > 0$ the points $x > x_0 + ct$ and $x < -x_0 - ct$ are undisturbed. Therefore, the interval $[-x_0 - ct, x_0 + ct]$ represents a domain of influence (DOI) of the initial data at time $t > 0$ for the problem (5)–(7). And this example, related to a Cauchy problem for classical wave equation, clearly illustrates the concept of DOI. The concept is extended in the book to include the hyperbolic thermoelastic L–S and G–L theories.

To illustrate the concept of "infinite velocity" of a temperature field governed by a parabolic equation consider another example.

Example 0.2 Cauchy problem for a parabolic equation.

Find a solution of the temperature equation

$$\left(\frac{\partial^2}{\partial x^2} - \frac{1}{\kappa}\frac{\partial}{\partial t}\right) T(x,t) = 0, \quad |x| < \infty, \quad t \geq 0, \tag{10}$$

subject to the initial condition

$$T(x,0) \equiv T_0(x) = T_0^* \left[H(x+x_0) - H(x-x_0)\right], \qquad |x| < \infty, \qquad x_0 > 0. \tag{11}$$

In eqns (10) and (11) $\kappa > 0$, $T_0^* > 0$.

To find a solution of the problem (10) and (11) we use the Laplace transform method, and obtain

$$\frac{\partial^2}{\partial x^2}\bar{T} - \frac{1}{\kappa}(p\bar{T} - T_0) = 0, \tag{12}$$

where the bar indicates the Laplace transform and p is the transform parameter. An equivalent form of eqn (12) reads

$$\left(D^2 - \frac{p}{\kappa}\right)\bar{T} = -\frac{T_0}{\kappa}, \qquad D = \frac{\mathrm{d}}{\mathrm{dx}}. \tag{13}$$

Consider now a Green's function $\bar{f}$ that satisfies the equation

$$\left(D^2 - \frac{p}{\kappa}\right)\bar{f}(x,p) = -\delta(x-\xi) = -\frac{1}{2\pi}\int_{-\infty}^{\infty} \cos\alpha(x-\xi)\mathrm{d}\alpha. \tag{14}$$

We look for $\bar{f}$ in the form

$$\bar{f}(x,p) = \int_{-\infty}^{\infty} A(\alpha,p)\cos\alpha(x-\xi)\mathrm{d}\alpha, \tag{15}$$

and substituting eqn (15) into eqn (14) we obtain

$$\bar{f}(x,p) = \frac{1}{\pi}\int_0^{\infty} \frac{\cos\alpha(x-\xi)}{\alpha^2 + p/\kappa}\mathrm{d}\alpha = \frac{\sqrt{\kappa}}{2}\frac{\exp[-|x-\xi|\sqrt{p/\kappa}]}{\sqrt{p}}. \tag{16}$$

Taking the inverse Laplace transform of eqn (16) we obtain

$$f(x,t) = \frac{\sqrt{\kappa}}{2\sqrt{\pi t}}\exp\left[\frac{-(x-\xi)^2}{4\kappa t}\right]. \tag{17}$$

As a result, the only solution of the Cauchy problem (10) and (11) takes the form

$$T(x,t) = \frac{T_0^*}{\sqrt{4\kappa\pi t}}\int_{-x_0}^{x_0} \exp\left[\frac{-(x-\xi)^2}{4\kappa t}\right]\mathrm{d}\xi. \tag{18}$$

The formula (18) implies that, for every $t > 0$ and $|x| < \infty$, the temperature $T = T(x,t)$ takes positive values, meaning it propagates with an infinite velocity.

Remark 0.1 The field equations in Examples 1 and 2 can be treated as particular cases of the wave equation with dissipation

$$\left(\frac{\partial^2}{\partial x^2} - \frac{1}{c^2}\frac{\partial^2}{\partial t^2} - 2h\frac{\partial}{\partial t}\right)u(x,t) = 0 \quad |x| < \infty, \quad t \geq 0, \tag{19}$$

where $h > 0$. An asymptotic analysis of a solution to a Cauchy problem for eqn (19) [when $c \to \infty$, $h > 0$, or $c > 0$, $h \to 0$] provides the concept of an *asymptotic domain of influence* (ADOI) for a classical wave equation and of a *temperature speed tending to infinity* for a parabolic equation. The concept of ADOI is a natural one when formulating a Saint Venant's Principle [Sternberg, 1954; Boley, 1955; Boley and Weiner, 1960; Chirita, 1995, 2007; Ignaczak, 1998, 2002; Quintanilla, 2001], or identifying the role of inertia terms by means of a so-called Boley's number [Boley, 1972; Boley and Barber, 1957].

The present book concentrates on the two leading theories of hyperbolic thermoelasticity: that of Lord–Shulman (1967) – also called a *theory with one relaxation time* – and that of Green–Lindsay (1972) – also called a *theory with two relaxation times.* They are both set in small strains and small departures from equilibrium temperatures, and so, the resulting field equations are linear partial differential ones. The complexity as well as the richness – and, therefore, the attractiveness of these theories – is due to the coupling of mechanical with thermal fields. The book is concerned with the mathematical aspects of both theories – existence and uniqueness theorems, domain of influence theorems, convolutional variational principles – as well as with the methods of dealing with a range of initial-boundary value problems. In the following we give an overview of the book's contents.

The first chapter reviews the basic equations of classical thermoelasticity and, then, gives the corresponding equations of the thermoelasticity with one relaxation time (or the L–S theory), followed by an analogous set of equations of thermoelasticity with two relaxation times (or the G–L theory). In each case the global balance laws are stated in terms of a displacement–temperature pair or, alternatively, a stress–heat-flux pair, or in terms of another pair of thermomechanical variables. This provides an Ansatz for the entire book.

The second chapter first presents a conventional and non-conventional characterization of a thermoelastic process, giving the mixed initial-boundary value problems: displacement–temperature and stress–heat–flux of the L–S theory, and displacement–temperature and stress–temperature of the G–L theory. This is followed by a discussion of the relations among descriptions of a thermoelastic process in terms of various pairs of thermomechanical variables.

The third chapter gives the existence and uniqueness theorems for conventional and non-conventional thermoelastic processes. On this basis, the fourth chapter expounds the domain of influence theorems characteristic of both theories for: the potential–temperature problem, the natural stress–heat–flux problem, the natural stress-temperature problem, and the displacement–temperature problem.

The fifth chapter is devoted to convolutional variational principles in the L–S and G–L theories. It first presents the alternative descriptions of a conventional thermoelastic process in the Green–Lindsay theory, and then gives the variational principles for a conventional and non-conventional thermoelastic processes in the L–S and G–L theories.

The sixth chapter is focused on a central equation of thermoelasticity with finite wave speeds, i.e. a partial differential equation that, most remarkably, has a similar form for both theories. We then develop a decomposition theorem for a central equation of the Green–Lindsay theory, the wave-like equations with a convolution, and the speed and attenuation of thermoelastic disturbances. We end with an analysis of the convolution coefficient and kernel.

The seventh chapter opens with fundamental solutions for a three-dimensional (3D) bounded domain, and then develops solutions for several key problems of thermoelasticity: the potential–temperature problem for a 3D bounded domain, a thermoelastic layer, the Nowacki-type solution, Danilovskaya-type solution, and a thermoelastic response of a half-space to laser irradiation.

The eighth chapter deals with integral representations and integral equations for fundamental solutions, integral representation of a solution to a central system of equations, and integral equations for a potential–temperature problem.

In the ninth chapter we start from the observation that the fundamental solutions of the G–L theory may be determined with the help of polynomial sequences on the time axis, the so-called polynomials of thermoelasticity. Here, we give a number of recurrence relations describing these polynomials and then show that a pair of thermoelastic polynomials can be identified with an element of the null space of a linear ordinary differential operator. From this are developed an integral relation, and associated thermoelastic polynomials.

The tenth chapter focuses on singular surfaces propagating in a thermoelastic medium, and studies the propagation of a plane shock wave in a thermoelastic half-space with one relaxation time, as well as the propagation of a plane acceleration wave in a thermoelastic half-space with two relaxation times.

On the other hand, the eleventh chapter studies time-periodic plane, spherical, and cylindrical waves as well as the fundamental solutions and for the potential–temperature solution, all in the setting of the G–L theory.

The twelfth chapter first provides a brief review of several other theories, all classified as generalized thermoelasticity and due to Green and Naghdi. Next, follows a justification of the presence of the material time derivative rather than the partial time derivative in the Maxwell–Cattaneo equation (3); this is done following the recent work by Christov and Jordan (2005). We thus see that a partial derivative may be employed in a theory focused on solid mechanics in infinitesimal strains. Another way to see how the material time derivative in eqn (3) arises naturally is to take the thermodynamics with internal variables as a starting point for the derivation of constitutive laws – this is done in the subsequent section of Chapter 12. What follows next is an account of some applications of the L–S and G–L theories: to helices and chiral media, both with homogeneous as well as composite structures; to surface waves; and to thermoelastic damping in nanomechanical resonators. The chapter culminates with a thermoelasticity with anomalous heat conduction treated via fractional calculus, and a formulation of thermoelasticity of fractal media in the vein of dimensional regularization.

While Chapters 1 through 12 are devoted to linear hyperbolic theories of thermoelasticity, the thirteenth chapter concerns a rigid but non-linear hyperbolic heat conductor due to Coleman *et al.* (1982, 1983, 1986). This particular material model obeys the law of conservation of energy, the dissipation inequality, Cattaneo's equation, and a generalized energy–entropy relation with a parabolic variation of the energy and entropy along the heat-flux axis. Following a review of the field equations for a 1D case, a number of closed-form solutions to the non-linear governing equations are obtained, and then a method of weakly non-linear geometric optics is applied to obtain an asymptotic solution to the Cauchy problem with a weakly perturbed initial condition associated with the non-linear model.

Given the book's focus on two special theories in the first 11 chapters, accompanied by an account of only some further models in Chapters 12 and 13, a Reference Supplement is given at the book's end. It collects much of the existing literature on hyperbolic linear and non-linear thermoelastic models and their various generalizations.

Finally, we would like to point out the well-known analogy of classical thermoelasticity to mass transfer in strained elastic solids, which is based on the even more basic analogy of Fourier-type heat conduction to Fick-type diffusion. Thus, a paradox of infinite propagation speeds arises also in those classical mass-transfer and diffusion problems, so that its resolution may be offered by a range of models already developed in theories of generalized thermoelasticity. Such a resolution has been proposed recently by Sherief *et al.* (2004), Aouadi (2007, 2008) and Sharma *et al.* (2008).

1

FUNDAMENTALS OF LINEAR THERMOELASTICITY WITH FINITE WAVE SPEEDS

1.1 Fundamentals of classical thermoelasticity

1.1.1 *Basic considerations*

The theory of classical (linear) thermoelasticity is the starting point for a number of various generalizations including: visco-thermoelasticity, thermoelasticity with diffusion, electro-magneto-thermoelasticity, or thermoelasticity with finite wave speeds. Thus, before proceeding with the exposition of the latter of these theories, it is beneficial to review the derivation of fundamental equations of classical thermoelasticity (Carlson, 1972).

Given the fact that the concept of a thermoelastic body may be defined in various ways,[1] let us begin by stating that in this book it is reserved for a body B in which there occurs a coupled dynamical process of an exchange of mechanical energy into thermal energy under the action of externally applied thermomechanical loadings. This process is accompanied by strains and temperature changes within the body, all of which vanish upon the removal of the said thermomechanical loadings. Notably, this process is described in terms of the field variables[2] by the set of relations:

the strain–displacement relations

$$E_{ij} = \frac{1}{2}\left(u_{i,j} + u_{j,i}\right), \tag{1.1.1}$$

the dynamic equilibrium equations

$$S_{ij,j} + b_i = \rho \ddot{u}_i, \qquad S_{ij} = S_{ji}, \tag{1.1.2}$$

the energy conservation law

$$\dot{e} = S_{ij}\dot{E}_{ij} - q_{i,i} + r, \tag{1.1.3}$$

[1] One can deal with a static thermoelasticity, uncoupled quasi-static thermoelasticity, coupled quasi-static thermoelasticity, uncoupled dynamic thermoelasticity, and coupled dynamic thermoelasticity; see e.g. (Nowacki, 1975).

[2] By the field variables we mean the physical fields that characterize a thermoelastic process: the displacement u_i, strain tensor E_{ij}, temperature change ϑ, and others.

the dissipation inequality

$$\dot{\eta} \geqslant -\left(\frac{q_i}{\theta}\right),_i + \frac{r}{\theta} \qquad (\theta > 0)\,, \tag{1.1.4}$$

and the (yet to be specified) constitutive relations. Employing the standard tensor notation in the above, u_i, E_{ij}, S_{ij}, q_i, θ, η and e denote, respectively, the displacements, strains, stresses, heat flux, absolute temperature, entropy and internal energy of the body. Moreover, b_i, r and ρ stand for the body forces, external heat sources and mass density of the body, respectively; here we use the standard notations: $\dot{\eta} = \partial\eta/\partial t$, $(\cdot),_i = \partial/\partial x_i(\cdot)$.[3] As is well known, the relations (1.1.3) and (1.1.4) are also called the *first and second laws of thermodynamics*, respectively. Introducing the *free energy* ψ through

$$\psi = e - \eta\theta, \tag{1.1.5}$$

and combining eqn (1.1.3) with eqn (1.1.4), we arrive at the dissipation inequality involving ψ:

$$\dot{\psi} + \eta\dot{\theta} - S_{ij}\dot{E}_{ij} + \frac{q_i\theta,_i}{\theta} \leqslant 0. \tag{1.1.6}$$

A *classical thermoelastic body* is a body in which, besides relations (1.1.1)–(1.1.6),

$$S_{ij} = \frac{\partial\psi}{\partial E_{ij}}, \tag{1.1.7}$$

$$\eta = -\frac{\partial\psi}{\partial\theta}, \tag{1.1.8}$$

where $\psi = \psi(E_{ij}, \theta)$ is a function given a priori. With the thus chosen free energy, we have

$$\dot{\psi} = \frac{\partial\psi}{\partial E_{ij}}\dot{E}_{ij} + \frac{\partial\psi}{\partial\theta}\dot{\theta} = S_{ij}\dot{E}_{ij} - \eta\dot{\theta} \tag{1.1.9}$$

and eqn (1.1.6) takes the form

$$q_i\theta,_i \leqslant 0. \tag{1.1.10}$$

Moreover, the relation (1.1.5) implies

$$\dot{e} = S_{ij}\dot{E}_{ij} + \dot{\eta}\theta, \tag{1.1.11}$$

and the energy conservation law (1.1.3) may be written as

$$\theta\dot{\eta} = -q_i,_i + r. \tag{1.1.12}$$

For a *linear thermoelastic body* we assume that

$$|E_{ij}|, |\dot{E}_{ij}|, |\theta - \theta_0|, |\dot{\theta}|, |\theta,_i| \leqslant \hat{\epsilon}, \tag{1.1.13}$$

[3] The relations (1.1.1)–(1.1.4) constitute a linear version of the basic laws of mechanics and thermodynamics of a continuum written in index notation.

where $\hat{\epsilon}$ is a small number, while $\theta_0 > 0$ is a *constant reference temperature*, such that[4]

$$S_{ij}(0, \theta_0) = 0, \quad \eta(0, \theta_0) = 0. \tag{1.1.14}$$

A *classical linear thermoelastic anisotropic body* is the body in which relations (1.1.1)–(1.1.14) hold along with this choice of free energy

$$\psi(E_{ij}, \theta) = \frac{1}{2} C_{ijkl} E_{ij} E_{kl} + M_{ij} E_{ij} \vartheta - \frac{C_E}{2\theta_0} \vartheta^2, \tag{1.1.15}$$

whereby

$$\vartheta = \theta - \theta_0. \tag{1.1.16}$$

Here, C_{ijkl}, M_{ij} and C_E denote the *elasticity tensor*, the *stress–temperature tensor* and the *specific heat at zero strain*, respectively. These quantities satisfy the following relations

$$C_{ijkl} = C_{jikl} = C_{ijlk} = C_{klij}, \tag{1.1.17}$$

$$M_{ij} = M_{ji}, \quad C_E > 0, \tag{1.1.18}$$

$$C_{ijkl} E_{ij} E_{kl} > 0. \tag{1.1.19}$$

Evidently, the dissipation inequality (1.1.10) is to be satisfied by an appropriate heat flux q_i, independent of the choice of ψ. We do so by choosing

$$q_i = -k_{ij} \vartheta,_j, \tag{1.1.20}$$

where k_{ij} is the *thermal conductivity tensor*, such that

$$k_{ij} = k_{ji}, \quad k_{ij} \vartheta,_i \vartheta,_j > 0. \tag{1.1.21}$$

It follows from the relations (1.1.7), (1.1.8) and (1.1.15)–(1.1.18) that the constitutive relations for a linear thermoelastic body are

$$S_{ij} = C_{ijkl} E_{kl} + M_{ij} \vartheta, \tag{1.1.22}$$

$$\eta = -M_{ij} E_{ij} + \frac{C_E}{\theta_0} \vartheta, \tag{1.1.23}$$

while the energy equation for such a body is obtained upon replacing $\theta\, \dot{\eta}$ in eqn (1.1.12) by $\theta_0\, \dot{\eta}$. We thus arrive at the following alternative definition of a *linear thermoelastic body*: it is a body in which the process of exchange of

[4] Since the function ψ depends on (E_{ij}, θ), the functions S_{ij} and η in eqns (1.1.7), (1.1.8) and (1.1.14) are also viewed as functions of (E_{ij}, θ). In the following, when we deal with a thermoelastic process taking place in a body B, the functions S_{ij} and η are also treated as fields on $\mathrm{B} \times [0, \infty)$, where B is a domain occupied by the body, and $[0, \infty)$ is the time interval.

mechanical and thermal energies is described by the equations

$$E_{ij} = \frac{1}{2}\left(u_{i,j} + u_{j,i}\right), \tag{1.1.24}$$

$$S_{ij,j} + b_i = \varrho \ddot{u}_i, \tag{1.1.25}$$

$$\theta_0 \dot{\eta} = -q_{i,i} + r, \tag{1.1.26}$$

$$S_{ij} = C_{ijkl} E_{kl} + M_{ij} \vartheta, \tag{1.1.27}$$

$$\theta_0 \eta = -\theta_0 M_{ij} E_{ij} + C_E \vartheta, \tag{1.1.28}$$

$$q_i = -k_{ij} \vartheta_{,j}. \tag{1.1.29}$$

The system of functions $[u_i, E_{ij}, S_{ij}, \vartheta, \eta, q_i]$ satisfying eqns (1.1.24)–(1.1.29) on $\mathrm{B} \times [0, \infty)$, where B is a domain occupied by the body and $[0, \infty)$ is a time interval, is called a *classical thermoelastic process* corresponding to a body force b_i and a heat source r.

When the tensors C_{ijkl} and k_{ij} are invertible, then, using notations

$$K_{ijkl} = \left(C_{ijkl}\right)^{-1}, \; \lambda_{ij} = \left(k_{ij}\right)^{-1}, \tag{1.1.30}$$

$$A_{ij} = -K_{ijkl} M_{kl}, \; C_S = C_E - \theta_0 M_{ij} A_{ij}, \tag{1.1.31}$$

we arrive at an alternative form of eqns (1.1.24)–(1.1.29):

$$E_{ij} = \frac{1}{2}\left(u_{i,j} + u_{j,i}\right), \tag{1.1.32}$$

$$S_{ij,j} + b_i = \varrho \ddot{u}_i, \tag{1.1.33}$$

$$\theta_0 \dot{\eta} = -q_{i,i} + r, \tag{1.1.34}$$

$$E_{ij} = K_{ijkl} S_{kl} + A_{ij} \vartheta, \tag{1.1.35}$$

$$\theta_0 \eta = \theta_0 A_{ij} S_{ij} + C_S \vartheta, \tag{1.1.36}$$

$$\vartheta_{,i} = -\lambda_{ij} q_j. \tag{1.1.37}$$

Here, K_{ijkl} is the *elastic compliance tensor*, A_{ij} is the *thermal expansion tensor*, λ_{ij} is the *thermal resistivity tensor*, and C_S is the *specific heat at zero stress.* These quantities satisfy the following relations

$$K_{ijkl} = K_{jikl} = K_{ijlk} = K_{klij}, \tag{1.1.38}$$

$$A_{ij} = A_{ji}, \; C_S > 0, \tag{1.1.39}$$

$$K_{ijkl} S_{ij} S_{kl} > 0, \tag{1.1.40}$$

$$\lambda_{ij} = \lambda_{ji}, \; \lambda_{ij} q_i q_j > 0. \tag{1.1.41}$$

In the case of an *inhomogeneous anisotropic body*, the quantities ϱ, C_E, k_{ij}, M_{ij}, and C_{ijkl} as well as their counterparts ϱ^{-1}, C_S^{-1}, λ_{ij}, A_{ij}, and K_{ijkl} depend on the location in the body, but not on time. They describe the physical properties of the body, and are therefore called *material functions of the thermoelastic medium.*

Clearly, a thermoelastic process corresponding to the loading (b_i, r) may be described by eqns (1.1.24–1.1.29) or (1.1.32–37). Since both equation systems are quite complex, we usually reduce them to simpler ones involving a minimum of unknown fields. For instance, by eliminating the entropy η from eqns (1.1.24)–(1.1.29) or eqns (1.1.32)–(1.1.37), we find that the process $[u_i, E_{ij}, S_{ij}, \vartheta, q_i]$ is described by the equations

$$E_{ij} = \frac{1}{2}\left(u_{i,j} + u_{j,i}\right), \tag{1.1.42}$$

$$S_{ij,j} + b_i = \varrho \ddot{u}_i, \tag{1.1.43}$$

$$-q_{i,i} + r = C_E \dot{\vartheta} - \theta_0 M_{ij} \dot{E}_{ij}, \tag{1.1.44}$$

$$S_{ij} = C_{ijkl} E_{kl} + M_{ij} \vartheta, \tag{1.1.45}$$

$$q_i = -k_{ij} \vartheta_{,j}, \tag{1.1.46}$$

or by the equations

$$E_{ij} = \frac{1}{2}\left(u_{i,j} + u_{j,i}\right), \tag{1.1.47}$$

$$S_{ij,j} + b_i = \rho \ddot{u}_i, \tag{1.1.48}$$

$$-q_{i,i} + r = C_S \dot{\vartheta} + \theta_0 A_{ij} \dot{S}_{ij}, \tag{1.1.49}$$

$$E_{ij} = K_{ijkl} S_{kl} + A_{ij} \vartheta, \tag{1.1.50}$$

$$\vartheta_{,i} = -\lambda_{ij} q_j. \tag{1.1.51}$$

For an *isotropic thermoelastic body* we have

$$C_{ijkl} E_{kl} = 2\mu E_{ij} + \lambda E_{kk} \delta_{ij}, \tag{1.1.52}$$

$$M_{ij} = -\left(3\lambda + 2\mu\right) \alpha \delta_{ij}, \; k_{ij} = k \delta_{ij}, \tag{1.1.53}$$

and

$$K_{ijkl} S_{kl} = \frac{1}{2\mu}\left(S_{ij} - \frac{\lambda}{3\lambda + 2\mu} S_{kk} \delta_{ij}\right), \tag{1.1.54}$$

$$A_{ij} = \alpha \delta_{ij}, \quad \lambda_{ij} = \frac{1}{k} \delta_{ij}, \tag{1.1.55}$$

$$C_S = C_E + 3\theta_0 \left(3\lambda + 2\mu\right) \alpha^2, \tag{1.1.56}$$

where λ and μ are the *Lamé moduli*, α is the *coefficient of thermal expansion*, and k is the *thermal conductivity coefficient.* In this particular case, in place of the system (1.1.42–46) we obtain

$$E_{ij} = \frac{1}{2}\left(u_{i,j} + u_{j,i}\right), \tag{1.1.57}$$

$$S_{ij,j} + b_i = \rho \ddot{u}_i, \tag{1.1.58}$$

$$-q_{i,i} + r = C_E\dot{\vartheta} + (3\lambda + 2\mu)\,\alpha\theta_0\dot{E}_{kk}, \tag{1.1.59}$$

$$S_{ij} = 2\mu E_{ij} + \lambda E_{kk}\delta_{ij} - (3\lambda + 2\mu)\,\alpha\vartheta\delta_{ij}, \tag{1.1.60}$$

$$q_i = -k\vartheta_{,i}, \tag{1.1.61}$$

while in place of the system (1.1.47–51) we obtain

$$E_{ij} = \frac{1}{2}\left(u_{i,j} + u_{j,i}\right), \tag{1.1.62}$$

$$S_{ij,j} + b_i = \rho \ddot{u}_i, \tag{1.1.63}$$

$$-q_{i,i} + r = C_S\dot{\vartheta} + \theta_0\alpha\dot{S}_{kk}, \tag{1.1.64}$$

$$E_{ij} = \frac{1}{2\mu}\left(S_{ij} - \frac{\lambda}{3\lambda + 2\mu}S_{kk}\delta_{ij}\right) + a\vartheta\delta_{ij}, \tag{1.1.65}$$

$$\vartheta_{,i} = -\frac{1}{k}q_i. \tag{1.1.66}$$

The restrictions on the material functions of an isotropic body are

$$\begin{aligned} &\rho > 0, \ \ \mu > 0, \ \ 3\lambda + 2\mu > 0, \\ &k > 0, \ C_E > 0, \qquad \alpha \neq 0. \end{aligned} \tag{1.1.67}$$

The field equations (1.1.42–1.1.46) or (1.1.47–1.1.51) for an anisotropic body, as well as eqns (1.1.57–1.1.61) or eqns (1.1.62–1.1.66) for an isotropic body, still represent complex systems of equations. However, a number of field variables may be eliminated out of each of these systems in such a way as to obtain a description of the thermoelastic process in terms of a pair. For example, the system (1.1.42–46) may be reduced to that involving a pair (u_i, ϑ), while the system (1.1.47–51) may be reduced to that involving a pair (S_{ij}, q_i). Other possible pairs are either (u_i, q_i) or (S_{ij}, ϑ).

When working with the pair (u_i, ϑ), we employ a so-called *displacement–temperature description*, which is obtained by eliminating E_{ij}, S_{ij}, and q_i from eqns (1.1.42–1.1.46)

$$\begin{aligned} (C_{ijkl}u_{k,l})_{,j} - \rho\ddot{u}_i + (M_{ij}\vartheta)_{,j} &= -b_i, \\ (k_{ij}\vartheta_{,j})_{,i} - C_E\dot{\vartheta} + \theta_0 M_{ij}\dot{u}_{i,j} &= -r. \end{aligned} \tag{1.1.68}$$

Analogously, when working with the pair (S_{ij}, q_i), we employ a so-called *stress–heat-flux description*, which is obtained by eliminating u_i, E_{ij}, and ϑ from eqns (1.1.47–1.1.51)[5]

$$\begin{aligned}(\rho^{-1}S_{(ik,k)},_{j)} - K'_{ijkl}\ddot{S}_{kl} + C_S^{-1}A_{ij}\dot{q}_{k,k} &= -(\rho^{-1}b_{(i)},_{j)} + C_S^{-1}A_{ij}\dot{r},\\ (C_S^{-1}q_{k,k}),_i - \lambda_{ij}\dot{q}_j + \theta_0(C_S^{-1}A_{pq}\dot{S}_{pq}),_i &= (C_S^{-1}r),_i,\end{aligned} \tag{1.1.69}$$

where

$$K'_{ijkl} = K_{ijkl} - \theta_0 C_S^{-1}A_{ij}A_{kl}. \tag{1.1.70}$$

The field equations described in terms of other pairs – such as (S_{ij}, ϑ) – will be discussed later in this book.

Let us note that the eqns (1.1.68) and (1.1.69) represent necessary conditions for a thermoelastic process to be satisfied on $B \times (0, \infty)$. A question arises whether these conditions are also sufficient. The answer to this question is simple relative to the system (1.1.68). Indeed, if (u_i, ϑ) is a solution of eqn (1.1.68), then specifying the dependent variables E_{ij}, S_{ij}, and q_i with the help of formulas (1.1.42), (1.1.45) and (1.1.46), we conclude that the system $[u_i, E_{ij}, S_{ij}, \vartheta, q_i]$ satisfies eqns (1.1.42–1.1.46), i.e. it is a thermoelastic process. Thus, we obtain sufficiency.

The related issue with respect to eqn (1.1.69) is more complex. A solution was given (Nickell and Sackman, 1968) based on the assumption that the initial conditions for the pair (S_{ij}, q_i) are generated from the conventional initial values of the pair (u_i, ϑ). This solution consists in writing down the relations specifying u_i, E_{ij} and ϑ in terms of the pair (S_{ij}, q_i) in such a way that the system $[u_i, E_{ij}, S_{ij}, \vartheta, q_i]$ is a thermoelastic process. Since a general thermoelastic process does not need to be consistent with conventional initial values of the pair (u_i, ϑ), we shall return to the issue of sufficiency of field equations in terms of one pair of thermomechanical variables when formulating the initial-boundary value problems. At this point, we derive the basis for such problems, i.e. the global balance laws in terms of (u_i, ϑ) and (S_{ij}, q_i), corresponding, respectively, to eqns (1.1.68) and (1.1.69).

1.1.2 *Global balance law in terms of* (u_i, ϑ)

Multiplying eqn (1.1.43) through by $\dot{u}_i$, we obtain

$$\dot{u}_i S_{ij},_j + \dot{u}_i b_i = \rho \dot{u}_i \ddot{u}_i, \tag{1.1.71}$$

from which, in view of eqn (1.1.42),

$$(\dot{u}_i S_{ij}),_j - \dot{E}_{ij}S_{ij} + \dot{u}_i b_i = \frac{1}{2}\rho\frac{\partial}{\partial t}\dot{u}_i\dot{u}_i. \tag{1.1.72}$$

[5] In eqn (1.1.69) the parentheses on the index level indicate the symmetric part of a second-order tensor; e.g. $b_{(i,j)} = (b_{i,j} + b_{j,i})/2$, $(\rho^{-1}S_{(ik,k)},_{j)} = [(\rho^{-1}S_{ik,k}),_j + (\rho^{-1}S_{jk,k}),_i]/2$; see (Hetnarski and Ignaczak, 2004).

Next, integrating over B, and employing the divergence theorem, we find

$$\int_{\partial B} \dot{u}_i S_{ij} n_j \mathrm{d}a + \int_{B} \dot{u}_i b_i \mathrm{d}v = \int_{B} \dot{E}_{ij} S_{ij} \mathrm{d}v + \frac{1}{2}\frac{\mathrm{d}}{\mathrm{d}t}\int_{B} \rho \dot{u}_i \dot{u}_i \mathrm{d}v. \tag{1.1.73}$$

Here, n_i is the outer normal to the boundary ∂B. Noting from eqns (1.1.45) and (1.1.17) that

$$\dot{E}_{ij} S_{ij} = C_{ijkl} \dot{E}_{ij} E_{kl} + M_{ij} \dot{E}_{ij} \vartheta = \frac{1}{2}\frac{\partial}{\partial t}\left(C_{ijkl} E_{ij} E_{kl}\right) + M_{ij} \dot{E}_{ij} \vartheta, \tag{1.1.74}$$

eqn (1.1.73) may be written as

$$\int_{\partial B} \dot{u}_i S_{ij} n_j \mathrm{d}a + \int_{B} \dot{u}_i b_i \mathrm{d}v = \frac{1}{2}\frac{\mathrm{d}}{\mathrm{d}t}\int_{B} (C_{ijkl} E_{ij} E_{kl} + \rho \dot{u}_i \dot{u}_i) \mathrm{d}v + \int_{B} M_{ij} \dot{E}_{ij} \vartheta \mathrm{d}v. \tag{1.1.75}$$

Introducing q_i from eqn (1.1.46) into eqn (1.1.44), we obtain

$$C_E \dot{\vartheta} - \theta_0 M_{ij} \dot{E}_{ij} = \left(k_{ij} \vartheta,_j\right),_i + r, \tag{1.1.76}$$

from which

$$\begin{aligned} M_{ij} \dot{E}_{ij} \vartheta &= \frac{1}{\theta_0}\left[C_E \vartheta \dot{\vartheta} - \left(k_{ij} \vartheta,_j\right),_i \vartheta - r\vartheta\right] \\ &= \frac{1}{2\theta_0} C_E \frac{\partial}{\partial t}\vartheta^2 - \frac{1}{\theta_0}\left(k_{ij}\vartheta,_j \vartheta\right),_i + \frac{1}{\theta_0} k_{ij} \vartheta_{,j} \vartheta,_i - \frac{r\vartheta}{\theta_0}. \end{aligned} \tag{1.1.77}$$

Integration of this equation over B gives

$$\begin{aligned} \int_{B} M_{ij} \dot{E}_{ij} \vartheta \mathrm{d}v = \frac{1}{2\theta_0}\frac{\mathrm{d}}{\mathrm{d}t}\int_{B} C_E \vartheta^2 \mathrm{d}v + \frac{1}{\theta_0}\int_{B} k_{ij} \vartheta,_i \vartheta,_j \mathrm{d}v \\ -\frac{1}{\theta_0}\int_{B} r\vartheta \mathrm{d}v - \frac{1}{\theta_0}\int_{\partial B} n_i k_{ij} \vartheta,_j \vartheta \mathrm{d}a. \end{aligned} \tag{1.1.78}$$

Replacing the last integral on the right-hand side of eqn (1.1.75) by the right-hand side of eqn (1.1.78), we obtain the global form of the balance law in terms of the pair (u_i, ϑ)[6]

$$\frac{\mathrm{d}}{\mathrm{d}t}\left[\frac{1}{2}\int_{\mathrm{B}}\rho\dot{u}_i\dot{u}_i\mathrm{d}v + \frac{1}{2}\int_{\mathrm{B}}C_{ijkl}E_{ij}E_{kl}\mathrm{d}v + \frac{1}{2\theta_0}\int_{\mathrm{B}}C_E\vartheta^2\mathrm{d}v\right]$$
$$+\frac{1}{\theta_0}\int_{\mathrm{B}}k_{ij}\vartheta,_i\vartheta,_j\mathrm{d}v = \int_{\partial\mathrm{B}}\dot{u}_iS_{ij}n_j\mathrm{d}a + \int_{\mathrm{B}}\dot{u}_ib_i\mathrm{d}v \tag{1.1.79}$$
$$+\frac{1}{\theta_0}\int_{\partial\mathrm{B}}n_ik_{ij}\vartheta,_j\vartheta\mathrm{d}a + \frac{1}{\theta_0}\int_{\mathrm{B}}r\vartheta\mathrm{d}v.$$

Note that eqn (1.1.79) was obtained only from the field eqns (1.1.42–1.1.46), similar to the derivation of field eqns (1.1.68) in terms of (u_i, ϑ). Thus, eqn (1.1.79) represents an integral of thermoelastic energy associated with eqn (1.1.68).

1.1.3 *Global balance law in terms of* (S_{ij}, q_i)

In order to obtain an integral of energy associated with the thermoelastic process in terms of (S_{ij}, q_i), we multiply eqn $(1.1.69)_1$ by $\dot{S}_{ij}$ and eqn $(1.1.69)_2$ by $\dot{q}_i$ so as to obtain[7]

$$(\rho^{-1}S_{ik},_k),_j\dot{S}_{ij} - K'_{ijkl}\,\dot{S}_{ij}\ddot{S}_{kl} + C_S^{-1}A_{ij}\dot{S}_{ij}\dot{q}_k,_k$$
$$= -(\rho^{-1}b_i),_j\dot{S}_{ij} + C_S^{-1}A_{ij}\dot{S}_{ij}\dot{r} \tag{1.1.80}$$

and

$$\left(C_S^{-1}q_{k,k}\right),_i\dot{q}_i - \lambda_{ij}\dot{q}_iq_j + \theta_0(C_S^{-1}A_{pq}\dot{S}_{pq}),_i\dot{q}_i = (C_S^{-1}r),_i\dot{q}_i. \tag{1.1.81}$$

Given that

$$(\rho^{-1}S_{ik,k}),_j\dot{S}_{ij} = (\rho^{-1}S_{ik},_k\dot{S}_{ij}),_j - \rho^{-1}S_{ik},_k\dot{S}_{ij},_j, \tag{1.1.82}$$

$$(C_S^{-1}q_k,_k),_i\dot{q}_i = (C_S^{-1}q_k,_k\dot{q}_i),_i - C_S^{-1}q_k,_k\dot{q}_i,_i, \tag{1.1.83}$$

$$(C_S^{-1}A_{pq}\dot{S}_{pq}),_i\dot{q}_i = (C_S^{-1}A_{pq}\dot{S}_{pq}\dot{q}_i),_i - C_S^{-1}A_{pq}\dot{S}_{pq}\dot{q}_i,_i, \tag{1.1.84}$$

[6] In eqn (1.1.79), the fields E_{ij} and S_{ij} are defined in terms of the pair (u_i, ϑ) by eqns (1.1.42) and (1.1.45), respectively. They have been left in the global law for simplicity.

[7] We use the following result: if S_{ij} is a symmetric tensor, and T_{ij} is an arbitrary second-order tensor, then $T_{(ij)}S_{ij} = T_{ij}S_{ij}$.

by integrating eqns (1.1.80) and (1.1.81), and using the divergence theorem, we find

$$\begin{aligned}
&\int_{\partial B} \rho^{-1} S_{ik},_k \dot{S}_{ij} n_j \mathrm{d}a - \int_{B} \rho^{-1} S_{ik},_k \dot{S}_{ij},_j \mathrm{d}v \\
&- \int_{B} K'_{ijkl} \dot{S}_{ij} \ddot{S}_{kl} \mathrm{d}v + \int_{B} C_S^{-1} A_{ij} \dot{S}_{ij} \dot{q}_k,_k \mathrm{d}v \\
&= - \int_{B} \left(\rho^{-1} b_i\right),_j \dot{S}_{ij} \mathrm{d}v + \int_{B} C_S^{-1} A_{ij} \dot{S}_{ij} \dot{r} \mathrm{d}v
\end{aligned} \tag{1.1.85}$$

and

$$\begin{aligned}
&\frac{1}{\theta_0} \int_{\partial B} C_S^{-1} q_k,_k \dot{q}_i n_i \mathrm{d}a - \frac{1}{\theta_0} \int_{B} C_S^{-1} q_k,_k \dot{q}_i,_i \mathrm{d}v \\
&- \frac{1}{\theta_0} \int_{B} \lambda_{ij} \dot{q}_i \dot{q}_j \mathrm{d}v + \int_{\partial B} C_S^{-1} A_{pq} \dot{S}_{pq} \dot{q}_i n_i \mathrm{d}a \\
&- \int_{B} C_S^{-1} A_{pq} \dot{S}_{pq} \dot{q}_i,_i \mathrm{d}v = \frac{1}{\theta_0} \int_{B} \left(C_S^{-1} r\right),_i \dot{q}_i \mathrm{d}v.
\end{aligned} \tag{1.1.86}$$

Upon adding eqns (1.1.85) and (1.1.86), after rearranging the terms, we arrive at

$$\begin{aligned}
&\frac{\mathrm{d}}{dt} \left\{ \frac{1}{2} \int_{B} \rho^{-1} S_{ik},_k S_{ij,j} \mathrm{d}v + \frac{1}{2} \int_{B} K'_{ijkl} \dot{S}_{ij} \dot{S}_{kl} \mathrm{d}v + \frac{1}{2\theta_0} \int_{B} C_S^{-1} (q_k,_k)^2 \mathrm{d}v \right\} \\
&+ \frac{1}{\theta_0} \int_{B} \lambda_{ij} \dot{q}_i \dot{q}_j \mathrm{d}v = \int_{B} \left\{ (\rho^{-1} b_i),_j \dot{S}_{ij} - C_S^{-1} A_{ij} \dot{S}_{ij} \dot{r} - \theta_0^{-1} (C_S^{-1} r),_i \dot{q}_i \right\} \mathrm{d}v \\
&+ \int_{\partial B} \left\{ \rho^{-1} S_{ik},_k \dot{S}_{ij} n_j + C_S^{-1} (A_{pq} \dot{S}_{pq} + \theta_0^{-1} q_k,_k) \dot{q}_i n_i \right\} \mathrm{d}a.
\end{aligned} \tag{1.1.87}$$

This is the sought for global balance law in terms of the pair (S_{ij}, q_i).[8] Overall, the balance laws (1.1.79) and (1.1.87) pertain to a general anisotropic thermoelastic body. Their counterparts for an isotropic body can be readily obtained using the relations (1.1.52–56) and (1.1.70). A general analysis of local equations (1.1.42–46) or (1.1.47–51) indicates that these equations are of a hyperbolic-parabolic type (Kupradze *et al.*, 1979), that is, they describe thermoelastic disturbances "propagating" with infinite speeds. This result is also confirmed by inspection of explicit solutions of a number of typical problems of classical thermoelasticity (Nowacki, 1975). Let us note that this is at odds with a fundamental fact in physics: for a finite time interval $[0, t]$, a disturbance of

[8] The global stress–heat-flux law of classical thermoelasticity (1.1.87) is obtained here for the first time.

a bounded support may only generate the response of a bounded support. In order to eliminate this paradox (or inconsistency) a number of modifications of classical thermoelasticity have appeared in the literature over the past one hundred years or so. The next part of this chapter will treat one such modified theory: *thermoelasticity with one relaxation time.*

1.2 Fundamentals of thermoelasticity with one relaxation time

1.2.1 *Basic considerations*

The theory of thermoelasticity with one relaxation time arose as a result of a modification of the equation of heat conduction (1.1.29) in Section 1.1, originally proposed by Maxwell (1867) in the context of theory of gases, and later by Cattaneo (1948) in the context of heat conduction in rigid bodies. Accounting for such a change in the description of a thermoelastic process in a deformable body – as proposed among others by Lord and Shulman (1967)[9] – leads to the following system of equations, corresponding to eqns (1.1.24–1.1.29):

$$E_{ij} = \frac{1}{2}\left(u_{i,\,j} + u_{j,\,i}\right), \tag{1.2.1}$$

$$S_{ij,\,j} + b_i = \rho \ddot{u}_i, \tag{1.2.2}$$

$$\theta_0 \dot{\eta} = -q_{i,\,i} + r, \tag{1.2.3}$$

$$S_{ij} = C_{ijkl}E_{kl} + M_{ij}\vartheta, \tag{1.2.4}$$

$$\theta_0 \eta = -\theta_0 M_{ij}E_{ij} + C_E\vartheta, \tag{1.2.5}$$

$$Lq_i = -k_{ij}\vartheta_{,\,j}, \tag{1.2.6}$$

where L is an operator of the form

$$L = 1 + t_0 \partial/\partial t, \tag{1.2.7}$$

and t_0 is the so-called *relaxation time*, satisfying the condition

$$t_0 > 0. \tag{1.2.8}$$

The L–S theory described by the relations (1.2.1)–(1.2.8) is confined to the case when a second-order tensor of relaxation times is reduced to a spherical part; in this case eqn (7) of (Lord and Shulman, 1967) can readily be reduced to eqn (1.2.6).

In eqns (1.2.1–1.2.6) all the symbols have the meaning similar to those of the system (1.1.24–29). In particular, we assume that all the material functions are restricted by the same constitutive inequalities as in the classical theory. It follows from the analysis of Section 1.1 that a thermoelastic process described by eqns (1.2.1–1.2.6) satisfies identical strain–displacement relations, dynamic equilibrium equations, energy conservation law, and dissipation inequality just

[9] A survey of results related to the L–S theory can be found in (Ignaczak, 1981, 1987, 1989a) (see also (Lebon, 1982; Pao and Banerjee, 1978; Kaliski, 1965)).

as the classical thermoelastic process. This holds with one caveat: The dissipation inequality $q_i\vartheta,_i \leq 0$ is satisfied so long as the term $t_0\dot{q}_i\vartheta,_i$ is "small" relative to the term $k_{ij}\vartheta,_i\vartheta,_j$, and this is true provided t_0 is a small parameter.

Remark 1.1 The L–S theory obeys the governing equations of a *linear extended thermoelasticity* in which the first and second laws of thermodynamics are satisfied identically in the following sense. We extend a domain of the free energy ψ to include the heat flux vector q_i into a set of independent variables. In other words, we postulate that

$$\psi = \psi(E_{ij}, \theta, q_i). \tag{R1.1.1}$$

Then, the internal energy e is also a function of these variables

$$e = e(E_{ij}, \theta, q_i), \tag{R1.1.2}$$

and proceeding along the lines of obtaining the field equations of classical thermoelasticity [see eqns (1.1.1)–(1.1.12)] we find that the *extended second law of thermodynamics* takes the form [see eqn (1.1.10)]

$$\frac{\partial \psi}{\partial q_i}\dot{q}_i + \frac{q_i\theta,_i}{\theta} \leq 0, \tag{R1.1.3}$$

while the *extended energy conservation law* is written as [see eqn (1.1.12)]

$$\theta\dot{\eta} = -q_i,_i + r - \frac{\partial \psi}{\partial q_i}\dot{q}_i. \tag{R1.1.4}$$

Next, by letting $\psi = \psi(E_{ij}, \theta, q_i)$ in the form [see eqn (1.1.15)]

$$\psi(E_{ij}, \theta, q_i) = \frac{1}{2}C_{ijkl}E_{ij}E_{kl} + M_{ij}E_{ij}\vartheta - \frac{C_E}{2\theta_0}\vartheta^2 + \frac{t_0}{2\theta_0}\lambda_{ij}q_iq_j, \tag{R1.1.5}$$

we recover eqns (1.2.1), (1.2.2), (1.2.4), and (1.2.5). To recover the linear energy conservation law (1.2.3) we replace the term $\theta\dot{\eta}$ by $\theta_0\dot{\eta}$, and ignore a second-order term on the RHS of eqn (R1.1.4).

Finally, to show that the second law of thermodynamics [eqn (R1.1.3)] is satisfied we use eqns (R1.1.3) and (R1.1.5), and obtain

$$\frac{t_0}{\theta_0}\lambda_{ij}\, q_j\, \dot{q}_i + \frac{q_i\theta,_i}{\theta} \leq 0, \tag{R1.1.6}$$

or

$$q_i\left(\frac{t_0}{\theta_0}\lambda_{j\,i}\,\dot{q}_j + \frac{\theta_{,i}}{\theta}\right) \leq 0. \tag{R1.1.7}$$

Since [see eqn (1.2.6)]

$$q_i + t_0\dot{q}_i = -k_{ij}\theta,_j, \tag{R1.1.8}$$

therefore, multiplying eqn (R1.1.8) by $\lambda_{ia} = \lambda_{ai} = (k_{ai})^{-1}$, we obtain

$$\lambda_{ai}\, q_i + t_0\lambda_{ai}\, \dot{q}_i = -\theta,_a, \tag{R1.1.9}$$

or

$$\frac{t_0 \lambda_{ij} \dot{q}_j}{\theta_0} + \frac{\theta_{,\,i}}{\theta} = -\frac{1}{\theta_0}\lambda_{ij} q_j - \left(\frac{1}{\theta_0} - \frac{1}{\theta}\right) \theta_{,\,i}. \tag{R1.1.10}$$

Next, multiplying eqn (R1.1.10) by q_i and letting $\theta \approx \theta_0$ in the second term on the RHS of the resulting equation, we obtain

$$q_i \left(\frac{t_0 \lambda_{ij} \dot{q}_j}{\theta_0} + \frac{\theta_{,i}}{\theta}\right) = -\frac{1}{\theta_0}\lambda_{ij} q_i q_j. \tag{R1.1.11}$$

Equation (R1.1.11) together with the positive-definiteness of the tensor λ_{ij} implies that the extended second law of thermodynamics in the form (R1.1.7) is identically satisfied. As a result, the L–S theory represents a linear extended thermoelasticity for which the extended 1st and 2nd laws of thermodynamics are satisfied identically.

Equation (1.2.6) is often called the *Maxwell–Cattaneo equation*, and the theory defined by eqns (1.2.1–1.2.6) the *thermoelasticity with one relaxation time*. An alternative form of that system of equations is the following system (recall eqns (1.1.32–37)):

$$E_{ij} = \frac{1}{2}(u_{i,\,j} + u_{j,\,i}), \tag{1.2.9}$$

$$S_{ij,\,j} + b_i = \rho \ddot{u}_i, \tag{1.2.10}$$

$$\theta_0 \dot{\eta} = -q_{i,\,i} + r, \tag{1.2.11}$$

$$E_{ij} = K_{ijkl} S_{kl} + A_{ij} \vartheta, \tag{1.2.12}$$

$$\theta_0 \eta = \theta_0 A_{ij} S_{ij} + C_S \vartheta, \tag{1.2.13}$$

$$\vartheta_{,\,i} = -\lambda_{ij}(q_j + t_0 \dot{q}_j). \tag{1.2.14}$$

Just like in Section 1.1, we introduce here the concept of a *thermoelastic body with one relaxation time* and a *thermoelastic process with one relaxation time.*

Now, by eliminating the variables E_{ij}, S_{ij}, η and q_i from eqns (1.2.1–1.2.6), we obtain the the displacement–temperature field equations

$$\begin{aligned} (C_{ijkl} u_{k,\,l})_{,\,j} - \rho \ddot{u}_i + (M_{ij} \vartheta)_{,\,j} &= -b_i, \\ (k_{ij} \vartheta_{,\,j})_{,\,i} - C_E \dot{\hat{\vartheta}} + \theta_0 M_{ij} \dot{\hat{u}}_{i,\,j} &= -\hat{r}, \end{aligned} \tag{1.2.15}$$

where a hut denotes action of the operator L, that is, for a function f on $\mathrm{B} \times (0, \infty)$,

$$\hat{f} = Lf. \tag{1.2.16}$$

Furthermore, upon eliminating u_i, E_{ij}, η and ϑ from the system (1.2.9–14), we obtain the field equations in terms of the pair (S_{ij}, q_i)

$$\begin{aligned}(\rho^{-1}S_{(ik},_{k}),_{j)} - K'_{ijkl}\ddot{S}_{kl} + C_S^{-1}A_{ij}\dot{q}_{k},_{k} &= -(\rho^{-1}b_{(i}),_{j)} + C_S^{-1}A_{ij}\dot{r}, \\ (C_S^{-1}q_{k},_{k}),_{i} - \lambda_{ij}\dot{\hat{q}}_j + \theta_0(C_S^{-1}A_{pq}\dot{S}_{pq}),_{i} &= (C_S^{-1}r),_{i}.\end{aligned} \tag{1.2.17}$$

Comparing eqns (1.1.68) and (1.1.69) with eqns (1.2.15) and (1.2.17) we observe a formal similarity of these two sets of equations, whereby eqn (1.2.15) appears to introduce a "larger" modification than eqn (1.2.17) relative to eqns (1.1.68) and (1.1.69), respectively.

1.2.2 *Global balance law in terms of* (u_i, ϑ)

This balance law is associated with the systems (1.2.1–6) and (1.2.15). Operating with L on (1.2.1–5) and employing notation (1.2.16), we arrive at[10]

$$\hat{E}_{ij} = \frac{1}{2}(\hat{u}_{i},_{j} + \hat{u}_{j},_{i}), \tag{1.2.18}$$

$$\hat{S}_{ij},_{j} + \hat{b}_i = \rho\ddot{\hat{u}}_i, \tag{1.2.19}$$

$$\theta_0\dot{\hat{\eta}} = -\hat{q}_{i},_{i} + \hat{r}, \tag{1.2.20}$$

$$\hat{S}_{ij} = C_{ijkl}\hat{E}_{kl} + M_{ij}\hat{\vartheta}, \tag{1.2.21}$$

$$\theta_0\hat{\eta} = -\theta_0 M_{ij}\hat{E}_{ij} + C_E\hat{\vartheta}, \tag{1.2.22}$$

$$\hat{q}_i = -k_{ij}\vartheta,_{j}. \tag{1.2.23}$$

Proceeding in a way analogous to that involving eqns (1.1.71–1.1.75) – that is, employing eqns (1.2.18), (1.2.19) and (1.2.21) – we obtain

$$\begin{aligned}&\int_{\partial B}\dot{\hat{u}}_i\hat{S}_{ij}n_j \mathrm{d}a + \int_B \hat{b}_i\dot{\hat{u}}_i \mathrm{d}v \\ &= \frac{1}{2}\frac{\mathrm{d}}{\mathrm{d}t}\int_B (C_{ijkl}\hat{E}_{ij}\hat{E}_{kl} + \rho\dot{\hat{u}}_i\dot{\hat{u}}_i)\mathrm{d}v + \int_B M_{ij}\dot{\hat{E}}_{ij}\hat{\vartheta}\mathrm{d}v.\end{aligned} \tag{1.2.24}$$

From eqns (1.2.20), (1.2.22) and (1.2.23) we next find

$$C_E\dot{\hat{\vartheta}} - \theta_0 M_{ij}\dot{\hat{E}}_{ij} = -\hat{q}_{i},_{i} + \hat{r}, \tag{1.2.25}$$

and

$$C_E\dot{\hat{\vartheta}} - \theta_0 M_{ij}\dot{\hat{E}}_{ij} = (k_{ij}\vartheta,_{j}),_{i} + \hat{r}. \tag{1.2.26}$$

[10] To obtain eqns (1.2.18)–(1.2.22) we also use the hypothesis that the material functions in eqns (1.2.1)–(1.2.6) do not depend on time, and L commutes with the spatial partial derivatives.

Hence,

$$M_{ij}\dot{\hat{E}}_{ij}\hat{\vartheta} = \frac{C_E}{\theta_0}\dot{\hat{\vartheta}}\hat{\vartheta} - \frac{(k_{ij}\vartheta_{,j}),_i\hat{\vartheta}}{\theta_0} - \frac{\hat{r}\hat{\vartheta}}{\theta_0}, \tag{1.2.27}$$

or

$$\begin{aligned} M_{ij}\dot{\hat{E}}_{ij}\hat{\vartheta} &= \frac{1}{2\theta_0}\frac{\partial}{\partial t}(C_E\hat{\vartheta}^2) \\ +\frac{1}{\theta_0}k_{ij}\vartheta,_i\vartheta,_j &+ \frac{t_0}{2\theta_0}\frac{\partial}{\partial t}(k_{ij}\vartheta,_i\vartheta,_j) - \frac{(k_{ij}\vartheta,_j\hat{\vartheta}),_i}{\theta_0} - \frac{\hat{r}\hat{\vartheta}}{\theta_0}. \end{aligned} \tag{1.2.28}$$

Integration of eqn (1.2.28) over B, and use of the divergence theorem, yields

$$\begin{aligned} \int_B M_{ij}\dot{\hat{E}}_{ij}\hat{\vartheta}\mathrm{d}v &= \frac{1}{2\theta_0}\frac{\mathrm{d}}{\mathrm{d}t}\int_B C_E\hat{\vartheta}^2\mathrm{d}v + \frac{t_0}{2\theta_0}\frac{\mathrm{d}}{\mathrm{d}t}\int_B k_{ij}\vartheta,_i\vartheta,_j\mathrm{d}v \\ &+\frac{1}{\theta_0}\int_B k_{ij}\vartheta,_i\vartheta,_j\mathrm{d}v - \frac{1}{\theta_0}\int_{\partial B}\hat{\vartheta}k_{ij}\vartheta,_jn_i\mathrm{d}a - \frac{1}{\theta_0}\int_B\hat{r}\hat{\vartheta}\mathrm{d}v. \end{aligned} \tag{1.2.29}$$

Finally, combining eqn (1.2.24) with eqn (1.2.29), we arrive at

$$\begin{aligned} &\frac{\mathrm{d}}{\mathrm{d}t}\left\{\frac{1}{2}\int_B\rho\dot{\hat{u}}_i\dot{\hat{u}}_i\mathrm{d}v + \frac{1}{2}\int_B C_{ijkl}\hat{E}_{ij}\hat{E}_{kl}\mathrm{d}v\right. \\ &\left.+\frac{1}{2\theta_0}\int_B C_E\hat{\vartheta}^2\mathrm{d}v + \frac{t_0}{2\theta_0}\int_B k_{ij}\vartheta,_i\vartheta,_j\mathrm{d}v\right\} + \tfrac{1}{\theta_0}\int_B k_{ij}\vartheta,_i\vartheta,_j\mathrm{d}v \\ &= \int_{\partial B}\dot{\hat{u}}_i\hat{S}_{ij}n_j\mathrm{d}a + \int_B\dot{\hat{u}}_i\hat{b}_i\mathrm{d}v + \frac{1}{\theta_0}\int_{\partial B}\hat{\vartheta}k_{ij}\vartheta,_jn_i\mathrm{d}a + \frac{1}{\theta_0}\int_B\hat{r}\hat{\vartheta}\mathrm{d}v. \end{aligned} \tag{1.2.30}$$

In comparison with the analogous relation (1.1.79) of classical thermoelasticity, the relation (1.2.30) is the global balance law containing the *thermoelastic energy of higher order in time.*[11]

1.2.3 *Global balance law in terms of* (S_{ij}, q_i)

This law is associated with eqns (1.2.17). Given the fact that eqns (1.2.17) constitute only a small modification of eqns (1.1.69), a global energy integral for eqns (1.2.17) is obtained by a small modification of the method leading to

[11] The global displacement–temperature law (1.2.30) is obtained here for the first time. A particular form of eqn (1.2.30) related to a homogeneous isotropic thermoelastic body can be found in (Ignaczak, 1982).

the result (1.1.87). Proceeding in that way, we obtain the global balance law in terms of (S_{ij}, q_i) for thermoelasticity with one relaxation time[12]

$$\frac{\mathrm{d}}{\mathrm{d}t}\left\{\frac{1}{2}\int_{B}\rho^{-1}S_{ik,k}S_{ij,j}\mathrm{d}v + \frac{1}{2}\int_{B}K'_{ijkl}\dot{S}_{ij}\dot{S}_{kl}\mathrm{d}v + \frac{1}{2\theta_0}\int_{B}C_S^{-1}(q_{k,k})^2\mathrm{d}v \right.$$
$$\left. + \frac{t_0}{2\theta_0}\int_{B}\lambda_{ij}\dot{q}_i\dot{q}_j\mathrm{d}v\right\} + \frac{1}{\theta_0}\int_{B}\lambda_{ij}\dot{q}_i\dot{q}_j\mathrm{d}v \tag{1.2.31}$$
$$= \int_{B}\{(\rho^{-1}b_i)_{,j}\dot{S}_{ij} - C_S^{-1}A_{ij}\dot{S}_{ij}\dot{r} - \theta_0^{-1}(C_S^{-1}r)_{,i}\dot{q}_i\}\mathrm{d}v$$
$$+ \int_{\partial B}\{\rho^{-1}S_{ik,k}\dot{S}_{ij}n_j + C_S^{-1}(A_{pq}\dot{S}_{pq} + \theta_0^{-1}q_{k,k})\dot{q}_i n_i\}\mathrm{d}a.$$

Although there exists a formal similarity between the local and global balance laws of classical thermoelasticity and thermoelasticity with one relaxation time, those laws describe qualitatively different processes so long as the relaxation time is positive. Indeed, one may show that the field equations (1.2.15) are of hyperbolic type – that is, they describe a thermoelastic disturbance propagating with finite speeds – in contradistinction to eqns (1.1.68).

Remark 1.2 In an attempt to model ultrafast processes of thermoelasticity Tzou (1997) proposed a *dual-phase-lag model* (DPLM) of thermoelasticity in which the Maxwell–Cattaneo equation (1.2.6) is replaced by the relation

$$q_i + \tau_q\dot{q}_i = -k_{ij}(\vartheta + \tau_T\dot{\vartheta})_{,j} \tag{R1.2.1}$$

while eqns (1.2.1)–(1.2.5) remain the same; in eqn (R1.2.1) τ_q and τ_T stand for *the heat flux and temperature gradient phase-lags*, respectively, and $\tau_q \geq 0$, $\tau_T \geq 0$. Clearly, a DPLM covers the hyperbolic L–S model when $\tau_q = t_0 > 0$, and $\tau_T = 0$. And, by letting $k_{ij} = k\delta_{ij}$, $\tau_T = \tau_q\xi$ in eqn (R1.2.1), where $k > 0$ and $0 < \xi < 1$; and, confining to a rigid body, a *Jeffrey's isotropic rigid heat conductor* is obtained.

Also, one can show that a DPLM, restricted to an isotropic rigid heat conductor, reduces to: (i) a *hyperbolic heat conductor* if $\tau = \tau_q - \tau_T > 0$, (ii) a *parabolic heat conductor* if $\tau = \tau_q - \tau_T = 0$, and (iii) an *elliptic heat conductor* if $\tau = \tau_q - \tau_T < 0$. To show this, introduce the phase shift

$$t^* = t + \tau_T, \tag{R1.2.2}$$

[12] The global stress–heat-flux law (1.2.31) is obtained here for the first time. For a homogeneous isotropic thermoelastic body the law is obtained in (Ignaczak, 1979).

and for any function $f = f(x, t^*)$ define the notations $[x \in \mathrm{B},\ t^* \geq \tau_T]$

$$\hat{f} = \left(1 + \tau \frac{\partial}{\partial t^*}\right) f, \tag{R1.2.3}$$

$$\tilde{f} = \left(1 - \tau_T \frac{\partial}{\partial t^*}\right) f, \tag{R1.2.4}$$

where

$$\tau = \tau_q - \tau_T. \tag{R1.2.5}$$

Then, the equation

$$q_i(x, t + \tau_q) = -kT,_i(x, t + \tau_T) \tag{R1.2.6}$$

takes the form

$$q_i(x, t^* + \tau) = -kT,_i(x, t^*), \tag{R1.2.7}$$

while the energy equation

$$C_p \dot{T}(x, t) = -q_i,_i(x, t) \tag{R1.2.8}$$

is reduced to

$$C_p \dot{T}(x, t^* - \tau_T) = -q_i,_i(x, t^* - \tau_T). \tag{R1.2.9}$$

Expanding equations (R1.2.7) and (R1.2.9) in the Taylor's series in a neighborhood of $\tau \approx 0$ and $\tau_T \approx 0$, and using the notations (R1.2.3) and (R1.2.4) we obtain

$$\hat{q}_i(x, t^*) = -kT,_i(x, t^*) \tag{R1.2.10}$$

and

$$-\tilde{q}_i,_i(x, t^*) = C_p \tilde{\dot{T}}(x, t^*). \tag{R1.2.11}$$

Applying the "hut" operator [see eqn (R1.2.3)] to (R1.2.11) we obtain

$$-\tilde{\hat{q}}_i,_i(x, t^*) = C_p \tilde{\hat{\dot{T}}}(x, t^*). \tag{R1.2.12}$$

Also, applying the "div" operator to eqn (R1.2.10) and substituting the result into eqn (R1.2.12) we obtain

$$k\tilde{T},_{ii}(x, t^*) = C_p \tilde{\hat{\dot{T}}}(x, t^*), \tag{R1.2.13}$$

or dividing by k and using the definition of "$\sim$" operator [see eqn (R1.2.4)] we obtain

$$\left(1 - \tau_T \frac{\partial}{\partial t^*}\right) \left[\nabla^2 T - \frac{1}{\kappa} \frac{\partial}{\partial t^*} \left(1 + \tau \frac{\partial}{\partial t^*}\right) T\right] = 0, \quad \kappa = k/C_p. \tag{R1.2.14}$$

In addition, for a model with a quiescent past for which

$$T(x, \tau_T) = \dot{T}(x, \tau_T) = \ddot{T}(x, \tau_T) = 0 \quad (\cdot = \partial/\partial t^*) \tag{R1.2.14}$$

eqn (R1.2.14) is equivalent to

$$\nabla^2 T - \frac{1}{\kappa}\frac{\partial}{\partial t^*}\left(1 + \tau\frac{\partial}{\partial t^*}\right) T = 0. \tag{R1.2.14}$$

As a result, the following theorem holds true

Theorem 1.1 *In the time frame $t^* = t + \tau_T$ $(t \geq 0)$ the temperature of a dual-phase lag model with a quiescent past is governed by a hyperbolic equation if $\tau = \tau_q - \tau_T > 0$, by an elliptic equation if $\tau = \tau_q - \tau_T < 0$, and by a parabolic equation if $\tau = \tau_q - \tau_T = 0$.*

Since a type of the equation is invariant with respect to a phase shift, the dual-phase lag model is hyperbolic, elliptic, and parabolic for $\tau > 0$, $\tau < 0$, and $\tau = 0$, respectively.

Finally, note that the case of $\tau < 0$ describes an ultrafast phonon–electron interaction model on a microscale where the values of τ_T and τ_q are of the order of picoseconds to femtoseconds. For a gold film subjected to a short-pulse laser heating, according to Table 5.1 in (Tzou, 1997) we obtain: $\tau_T = 89.28\,\text{ps}$, $\tau_q = 0.74\,\text{ps}$ and $\tau = -88.54\,\text{ps}$.

At this point we shall move to the setup of another modification of classical thermoelasticity, which also eliminates the paradox of infinite speeds of propagation. This theory is based on a generalized dissipation inequality and introduces two relaxation times in the description of a thermoelastic process. Its foundations were given by Green and Lindsay (1972),[13] and, in the following, we shall refer to it as the *G–L theory* or a *thermoelasticity with two relaxation times*.

1.3 Fundamentals of thermoelasticity with two relaxation times

1.3.1 *Basic considerations*

In the place of eqns (1.1.1–1.1.4) we now have the relations

$$E_{ij} = \frac{1}{2}\left(u_{i,j} + u_{j,i}\right), \tag{1.3.1}$$

$$S_{ij,j} + b_i = \rho \ddot{u}_i, \tag{1.3.2}$$

$$\dot{e} = S_{ij}\dot{E}_{ij} - q_{i,i} + r, \tag{1.3.3}$$

$$\dot{\eta} \geq -\left(\frac{q_i}{\phi}\right)_{,i} + \frac{r}{\phi}, \tag{1.3.4}$$

[13] See also (Müller, 1971; Green and Laws, 1972; Şuhubi, 1975; and Erbay and Şuhubi, 1986).

where

$$\phi = \phi(\theta, \dot{\theta}) > 0 \tag{1.3.5}$$

is a scalar function of two independent variables to be specified in the following. Clearly, the laws (1.3.1–3) are identical with eqns (1.1.1–1.1.3), whereas the inequality (1.3.4) reduces to eqn (1.1.4), providing ϕ is set equal to θ. In that sense, eqn (1.3.4) is a generalization of the classical dissipation inequality.

Besides the foregoing, we introduce a *generalized free energy*

$$\psi = e - \eta\phi, \tag{1.3.6}$$

from which

$$\dot{\psi} = \dot{e} - \dot{\eta}\phi - \eta\dot{\phi}, \tag{1.3.7}$$

and the relations (1.3.3–4), written in terms of functions ϕ and ψ, take the forms

$$\dot{\psi} + \dot{\eta}\phi + \eta\dot{\phi} = S_{ij}\dot{E}_{ij} - q_{i,\,i} + r, \tag{1.3.8}$$

$$\dot{\psi} + \eta\dot{\phi} - S_{ij}\dot{E}_{ij} + q_i\phi_{,\,i}\phi^{-1} \leq 0. \tag{1.3.9}$$

In comparison with the classical thermoelasticity, we also extend the domain of ψ by taking

$$\psi = \psi(E_{ij}, \theta, \dot{\theta}, \theta_{,\,i}). \tag{1.3.10}$$

From this, we obtain

$$\dot{\psi} = \frac{\partial\psi}{\partial E_{ij}}\dot{E}_{ij} + \frac{\partial\psi}{\partial\theta}\dot{\theta} + \frac{\partial\psi}{\partial\dot{\theta}}\ddot{\theta} + \frac{\partial\psi}{\partial\theta_{,\,k}}\dot{\theta}_{,\,k}. \tag{1.3.11}$$

Moreover, the derivatives of ϕ are expressed as

$$\dot{\phi} = \frac{\partial\phi}{\partial\theta}\dot{\theta} + \frac{\partial\phi}{\partial\dot{\theta}}\ddot{\theta}, \tag{1.3.12}$$

$$\phi_{,\,i} = \frac{\partial\phi}{\partial\theta}\theta_{,\,i} + \frac{\partial\phi}{\partial\dot{\theta}}\dot{\theta}_{,\,i}. \tag{1.3.13}$$

Employing eqns (1.3.11–1.3.13) we can now reduce eqns (1.3.8) and (1.3.9) to the forms

$$\begin{aligned} &\left(\frac{\partial\psi}{\partial E_{ij}} - S_{ij}\right)\dot{E}_{ij} + \left(\frac{\partial\psi}{\partial\theta} + \eta\frac{\partial\phi}{\partial\theta}\right)\dot{\theta} \\ &+ \left(\frac{\partial\psi}{\partial\dot{\theta}} + \eta\frac{\partial\phi}{\partial\dot{\theta}}\right)\ddot{\theta} + \frac{\partial\psi}{\partial\theta_{,\,k}}\dot{\theta}_{,\,k} + \dot{\eta}\phi = -q_{i,\,i} + r, \end{aligned} \tag{1.3.14}$$

and

$$\begin{gathered}\left(\frac{\partial\psi}{\partial E_{ij}}-S_{ij}\right)\dot{E}_{ij}+\left(\frac{\partial\psi}{\partial\theta}+\eta\frac{\partial\phi}{\partial\theta}\right)\dot{\theta}+\left(\frac{\partial\psi}{\partial\dot{\theta}}+\eta\frac{\partial\phi}{\partial\dot{\theta}}\right)\ddot{\theta}\\+\left(\frac{\partial\psi}{\partial\theta,_k}+\frac{q_k}{\phi}\frac{\partial\phi}{\partial\dot{\theta}}\right)\dot{\theta},_k+\frac{q_k}{\phi}\frac{\partial\phi}{\partial\theta}\theta,_k\leq 0.\end{gathered}\tag{1.3.15}$$

Let us now postulate the constitutive equations in the forms

$$S_{ij}=\frac{\partial\psi}{\partial E_{ij}},\tag{1.3.16}$$

$$\eta=-\left(\frac{\partial\psi}{\partial\dot{\theta}}\right)\left(\frac{\partial\phi}{\partial\dot{\theta}}\right)^{-1},\tag{1.3.17}$$

$$q_k=-\left(\frac{\partial\psi}{\partial\theta,_k}\right)\left(\frac{1}{\phi}\frac{\partial\phi}{\partial\dot{\theta}}\right)^{-1},\tag{1.3.18}$$

and reduce eqns (1.3.14) and (1.3.15) to

$$\left(\frac{\partial\psi}{\partial\theta}+\eta\frac{\partial\phi}{\partial\theta}\right)\dot{\theta}+\frac{\partial\psi}{\partial\theta,_k}\dot{\theta},_k+\dot{\eta}\phi=-q_i,_i+r,\tag{1.3.19}$$

$$\left(\frac{\partial\psi}{\partial\theta}+\eta\frac{\partial\phi}{\partial\theta}\right)\dot{\theta}+\frac{q_k}{\phi}\frac{\partial\phi}{\partial\theta}\theta,_k\leq 0.\tag{1.3.20}$$

We will now prove that, for a linear thermoelastic body, relations (1.3.19) and (1.3.20) are satisfied, if[14]

$$\phi=\theta_0+\vartheta+t_1\dot{\vartheta},\tag{1.3.21}$$

$$\begin{gathered}\psi=\frac{1}{2}C_{ijkl}E_{ij}E_{kl}+M_{ij}E_{ij}(\vartheta+t_1\dot{\vartheta})\\-\frac{C_E}{2\theta_0}\vartheta^2-\frac{C_E}{\theta_0}t_1\vartheta\dot{\vartheta}-\frac{C_E}{2\theta_0}t_0t_1\dot{\vartheta}^2+\frac{t_1}{2\theta_0}k_{ij}\vartheta,_i\vartheta,_j,\end{gathered}\tag{1.3.22}$$

where t_0 and t_1 are parameters with dimension of time, satisfying the inequalities[15]

$$t_1\geq t_0>0,\tag{1.3.23}$$

whereby the remaining symbols appearing in eqns (1.3.21) and (1.3.22) have the same meaning as in the classical thermoelasticity. Note that, as $t_1\to 0$, $\phi\to\theta$ and ψ tends to the free energy of the classical thermoelastic body, recall eqn (1.1.15). Considering that

$$\vartheta=\theta-\theta_0,\tag{1.3.24}$$

[14] Functions (1.3.21) and (1.3.22) are restrictions of those proposed by Green and Lindsay (1972).

[15] The relaxation time t_0 in eqn (1.3.23) is, in general, different from that of Section 1.2.

and using eqns (1.3.21) and (1.3.22), we find that

$$\frac{\partial \phi}{\partial \theta} = 1, \quad \frac{\partial \phi}{\partial \dot{\theta}} = t_1, \tag{1.3.25}$$

$$\frac{\partial \psi}{\partial \theta} = M_{ij} E_{ij} - \frac{C_E}{\theta_0}(\vartheta + t_1 \dot{\vartheta}), \tag{1.3.26}$$

$$\frac{\partial \psi}{\partial \dot{\theta}} = t_1 [M_{ij} E_{ij} - \frac{C_E}{\theta_0}(\vartheta + t_0 \dot{\vartheta})], \tag{1.3.27}$$

$$\frac{\partial \psi}{\partial \theta,_k} = \frac{t_1}{\theta_0} k_{ki} \vartheta,_i, \tag{1.3.28}$$

$$\frac{\partial \psi}{\partial E_{ij}} = C_{ijkl} E_{kl} + M_{ij}(\vartheta + t_1 \dot{\vartheta}). \tag{1.3.29}$$

From the relations (1.3.16) and (1.3.29) we obtain

$$S_{ij} = C_{ijkl} E_{kl} + M_{ij}(\vartheta + t_1 \dot{\vartheta}), \tag{1.3.30}$$

whereas, in view of eqns (1.3.17), $(1.3.25)_2$ and (1.3.27), there is

$$\theta_0 \eta = -\theta_0 M_{ij} E_{ij} + C_E(\vartheta + t_0 \dot{\vartheta}). \tag{1.3.31}$$

Furthermore, eqns (1.3.18), (1.3.21), $(1.3.25)_2$ and (1.3.28) lead to

$$q_k t_1 = -(\theta_0 + \vartheta + t_1 \dot{\vartheta}) \frac{t_1}{\theta_0} k_{ki} \vartheta,_i. \tag{1.3.32}$$

Since for a linear thermoelastic body the terms $\vartheta \vartheta,_i$ and $\dot{\vartheta} \vartheta,_i$ may be neglected relative to $\vartheta,_i$, the formula (1.3.32) results in[16]

$$q_k = -k_{kj} \vartheta,_j. \tag{1.3.33}$$

In order to demonstrate the validity of eqns (1.3.19) and (1.3.20), we now calculate the term

$$\left(\frac{\partial \psi}{\partial \theta} + \eta \frac{\partial \phi}{\partial \theta} \right) = \frac{\partial \psi}{\partial \theta} + \eta. \tag{1.3.34}$$

In view of eqns (1.3.26) and (1.3.31), we get

$$\frac{\partial \psi}{\partial \theta} + \eta \frac{\partial \phi}{\partial \theta} = -\frac{C_E}{\theta_0} (t_1 - t_0) \dot{\vartheta}, \tag{1.3.35}$$

from which

$$\left(\frac{\partial \psi}{\partial \theta} + \eta \frac{\partial \phi}{\partial \theta} \right) \dot{\theta} = -\frac{C_E}{\theta_0} (t_1 - t_0) \dot{\vartheta}^2, \tag{1.3.36}$$

[16] Equation (1.3.33) represents the classical Fourier law of heat conduction [see eqn (1.1.20)].

so that the energy balance equation (1.3.19) may be written as

$$-\frac{C_E}{\theta_0}(t_1 - t_0)\dot{\vartheta}^2 + \frac{t_1}{\theta_0}k_{kj}\vartheta,_j\dot{\vartheta},_k + \dot{\eta}(\theta_0 + \vartheta + t_1\dot{\vartheta}) = -q_i,_i + r. \tag{1.3.37}$$

Neglecting on the left-hand side of the above all the terms proportional to $\dot{\vartheta}^2$, $\vartheta,_j\dot{\vartheta},_k$, $\vartheta\dot{E}_{ij}$ and $\vartheta\dot{\vartheta}$ as small relative to those proportional to $\dot{E}_{ij}$, $\dot{\vartheta}$ and $\ddot{\vartheta}$, we find the following linear form of energy balance

$$\theta_0\dot{\eta} = -q_i,_i + r. \tag{1.3.38}$$

In order to prove the validity of the dissipation inequality (1.3.20), we first note that

$$\frac{1}{\phi}\frac{\partial\phi}{\partial\theta} \sim \frac{1}{\theta_0}, \tag{1.3.39}$$

which, together with eqn (1.3.33), leads to

$$\frac{q_k}{\phi}\frac{\partial\phi}{\partial\theta}\theta,_k = -\frac{1}{\theta_0}k_{ij}\vartheta,_i\vartheta,_j. \tag{1.3.40}$$

Adding eqn (1.3.36) and eqn (1.3.40) gives

$$\left(\frac{\partial\psi}{\partial\theta} + \eta\frac{\partial\phi}{\partial\theta}\right)\dot{\theta} + \frac{q_k}{\phi}\frac{\partial\phi}{\partial\theta}\theta,_k = -\frac{C_E}{\theta_0}(t_1 - t_0)\dot{\vartheta}^2 - \frac{1}{\theta_0}k_{ij}\vartheta,_i\vartheta,_j. \tag{1.3.41}$$

Observe now that, the left-hand side of eqn (1.3.41) is identical with the left-hand side of inequality (1.3.20). Thus, noting eqn (1.3.23) and the inequalities postulated for C_E, θ_0 and k_{ij} in Section 1.1, the right-hand side of eqn (1.3.41) is non-positive. This then implies that the dissipation inequality (1.3.20) is satisfied. As a result, a *linear thermoelastic body with two relaxation times* can be defined as the one in which a thermoelastic process $[u_i, E_{ij}, S_{ij}, \vartheta, \eta, q_i]$ satisfies the field equations [see also eqns (1.3.1), (1.3.2), (1.3.30), (1.3.31), (1.3.33) and (1.3.38)]

$$E_{ij} = \frac{1}{2}(u_i,_j + u_j,_i), \tag{1.3.42}$$

$$S_{ij},_j + b_i = \rho\ddot{u}_i, \tag{1.3.43}$$

$$\theta_0\dot{\eta} = -q_i,_i + r, \tag{1.3.44}$$

$$S_{ij} = C_{ijkl}E_{kl} + M_{ij}(\vartheta + t_1\dot{\vartheta}), \tag{1.3.45}$$

$$\theta_0\eta = -\theta_0 M_{ij}E_{ij} + C_E(\vartheta + t_0\dot{\vartheta}), \tag{1.3.46}$$

$$q_i = -k_{ij}\vartheta,_j. \tag{1.3.47}$$

The following is an alternative system of equations describing a thermoelastic body with two relaxation times

$$E_{ij} = \frac{1}{2}(u_{i,j} + u_{j,i}), \tag{1.3.48}$$

$$S_{ij,j} + b_i = \rho \ddot{u}_i, \tag{1.3.49}$$

$$\theta_0 \dot{\eta} = -q_{i,i} + r, \tag{1.3.50}$$

$$E_{ij} = K_{ijkl} S_{kl} + A_{ij}(\vartheta + t_1 \dot{\vartheta}), \tag{1.3.51}$$

$$\theta_0 \eta = \theta_0 A_{pq} S_{pq} + C_S(\vartheta + t_{(0)} \dot{\vartheta}), \tag{1.3.52}$$

$$\vartheta_{,i} = -\lambda_{ij} q_j, \tag{1.3.53}$$

where $t_{(0)}$ is a *reduced relaxation time*, defined through

$$t_{(0)} = \left(1 - \frac{C_E}{C_S}\right) t_1 + \frac{C_E}{C_S} t_0, \tag{1.3.54}$$

and all the remaining symbols in eqns (1.3.48–1.3.53) have the same meaning as those in eqns (1.1.32–1.1.37). Given that

$$C_S \geq C_E > 0, \tag{1.3.55}$$

and that the inequality (1.3.23) is satisfied, we have

$$t_1 \geq t_{(0)} \geq \left(1 - \frac{C_E}{C_S}\right) t_1 > 0. \tag{1.3.56}$$

Of course, in the case of an anisotropic inhomogeneous body, the scalars C_S and C_E are functions of position, and so $t_{(0)}$ is also a function of position. The system of functions $[u_i, E_{ij}, S_{ij}, \vartheta, \eta, q_i]$ satisfying either eqns (1.3.42–1.3.47) or eqns (1.3.48–1.3.53) on $\mathrm{B} \times (0, \infty)$ will be called a *thermoelastic process with two relaxation times*.

Upon elimination of E_{ij}, S_{ij}, η and q_i from eqns (1.3.42–1.3.47), we obtain the system of field equations in terms of (u_i, ϑ):

$$\begin{aligned} (C_{ijkl} u_{k,l})_{,j} - \rho \ddot{u}_i + [M_{ij}(\vartheta + t_1 \dot{\vartheta})]_{,j} &= -b_i, \\ (k_{ij} \vartheta_{,j})_{,i} - C_E(\dot{\vartheta} + t_0 \ddot{\vartheta}) + \theta_0 M_{ij} \dot{u}_{i,j} &= -r. \end{aligned} \tag{1.3.57}$$

In order to obtain a description of a thermoelastic process with two relaxation times in terms of a pair (S_{ij}, ϑ), we consider the system of eqns (1.3.48–1.3.53), in which eqn (1.3.53) is replaced by the equivalent eqn (1.3.47). Eliminating the variables u_i, E_{ij}, η and q_i, we arrive at

$$\begin{gathered}(\rho^{-1}S_{(ik,k)},_{j)} - K_{ijkl}\ddot{S}_{kl} + A_{ij}L_1\ddot{\vartheta} = -(\rho^{-1}b_{(i)},_{j)}, \\ (k_{ij}\vartheta,_j),_i - C_S L_{(0)}\dot{\vartheta} - \theta_0 A_{pq}\dot{S}_{pq} = -r,\end{gathered} \tag{1.3.58}$$

where

$$L_1 = 1 + t_1\partial/\partial t, \tag{1.3.59}$$

$$L_{(0)} = 1 + t_{(0)}\partial/\partial t. \tag{1.3.60}$$

Clearly, the tensorial eqn $(1.3.58)_1$ contains $\dddot{\vartheta}$. We will now show that the system (1.3.58) may be replaced by one without $\dddot{\vartheta}$. To that end, let us differentiate eqn $(1.3.58)_2$ with respect to time, so as to get

$$\dddot{\vartheta} = -t_{(0)}^{-1}\left\{\ddot{\vartheta} + C_S^{-1}\left[\theta_0 A_{pq}\ddot{S}_{pq} - (k_{ij}\dot{\vartheta},_j),_i - \dot{r}\right]\right\}, \tag{1.3.61}$$

and

$$\begin{gathered}L_1\ddot{\vartheta} = \ddot{\vartheta} + t_1\dddot{\vartheta} = -t_{(0)}^{-1}(t_1 - t_0)\ddot{\vartheta} \\ -t_1 t_{(0)}^{-1} C_S^{-1}\left[\theta_0 A_{pq}\ddot{S}_{pq} - (k_{ij}\dot{\vartheta},_j),_i - \dot{r}\right].\end{gathered} \tag{1.3.62}$$

Substituting eqn (1.3.62) into eqn $(1.3.58)_1$, we now find the following form of field equations in terms of (S_{ij}, ϑ)[17]

$$\begin{gathered}(\rho^{-1}S_{(ik,k)},_{j)} - \widetilde{K}_{ijkl}\ddot{S}_{kl} - A_{ij}t_{(0)}^{-1}[t_1 C_S^{-1}(k_{pq}\dot{\vartheta},_q),_p - (t_1 - t_{(0)})\,\ddot{\vartheta}] = -\tilde{b}_{(ij)}, \\ C_S^{-1}(k_{pq}\vartheta,_q),_p - (\dot{\vartheta} + t_{(0)}\ddot{\vartheta}) - \theta_0 C_S^{-1}A_{pq}\dot{S}_{pq} = -C_S^{-1}r,\end{gathered} \tag{1.3.63}$$

where

$$\widetilde{K}_{ijkl} = K_{ijkl} - \frac{t_1}{t_{(0)}}\frac{\theta_0}{C_S}A_{ij}A_{kl}, \tag{1.3.64}$$

$$\tilde{b}_{(ij)} = (\rho^{-1}b_{(i)},_{j)} - \frac{t_1}{t_{(0)}}\frac{\dot{r}}{C_S}A_{ij}. \tag{1.3.65}$$

Since we are aiming at the formulation of initial-boundary value problems in the G–L theory, we shall now write down the global balance laws associated with the field equations (1.3.57) and (1.3.63).

[17] A restriction of eqns (1.3.63) to a homogeneous isotropic body is obtained in (Ignaczak, 1978c).

1.3.2 *Global balance law in terms of* (u_i, ϑ)[18]

Multiplying eqn (1.3.43) through by $\dot{u}_i$, and integrating over B, as well as using eqns (1.3.42) and (1.3.45), we arrive at

$$\frac{\mathrm{d}}{\mathrm{d}t}\left\{\int_{B}\frac{1}{2}\rho\dot{u}_i\dot{u}_i\mathrm{d}v+\frac{1}{2}\int_{B}C_{ijkl}E_{ij}E_{kl}\mathrm{d}v\right\}+\int_{B}M_{ij}\dot{E}_{ij}(\vartheta+t_1\dot{\vartheta})\mathrm{d}v \\ =\int_{\partial B}\dot{u}_iS_{ij}n_j\mathrm{d}a+\int_{B}\dot{u}_ib_i\mathrm{d}v. \tag{1.3.66}$$

Equations (1.3.44) and (1.3.46–47) imply that

$$C_E\overline{(\vartheta+t_0\dot{\vartheta})}^{\cdot}-\theta_0M_{ij}\dot{E}_{ij}=(k_{ij}\vartheta,_j),_i+r. \tag{1.3.67}$$

Next, multiplying eqn (1.3.67) through by $(\vartheta+t_0\dot{\vartheta})$, we obtain

$$C_E(\vartheta+t_0\dot{\vartheta})\overline{(\vartheta+t_0\dot{\vartheta})}^{\cdot}-\theta_0M_{ij}\dot{E}_{ij}\overline{(\vartheta+t_0\dot{\vartheta})}^{\cdot} \\ =[(k_{ij}\vartheta_{,j}),_i+r](\vartheta+t_0\dot{\vartheta}), \tag{1.3.68}$$

or

$$\frac{C_E}{2}\frac{\partial}{\partial t}(\vartheta+t_0\dot{\vartheta})^2-\theta_0M_{ij}\dot{E}_{ij}(\vartheta+t_1\dot{\vartheta}) \\ +\theta_0M_{ij}\dot{E}_{ij}(t_1-t_0)\dot{\vartheta}=\left[(k_{ij}\vartheta_{,j}),_i+r\right](\vartheta+t_0\dot{\vartheta}). \tag{1.3.69}$$

On the other hand, multiplying eqn (1.3.67) through by $(t_1-t_0)\dot{\vartheta}$, we obtain

$$C_E(t_1-t_0)\dot{\vartheta}\overline{(\vartheta+t_0\dot{\vartheta})}^{\cdot}-\theta_0M_{ij}\dot{E}_{ij}(t_1-t_0)\dot{\vartheta} \\ =\left[(k_{ij}\vartheta_{,j}),_i+r\right](t_1-t_0)\dot{\vartheta}. \tag{1.3.70}$$

Now, adding eqn (1.3.69) to eqn (1.3.70), we find

$$\frac{C_E}{2}\frac{\partial}{\partial t}(\vartheta+t_0\dot{\vartheta})^2+\frac{C_E}{2}t_0(t_1-t_0)\frac{\partial}{\partial t}\dot{\vartheta}^2 \\ +(t_1-t_0)C_E\dot{\vartheta}^2+k_{ij}\vartheta,_j(\vartheta+t_1\dot{\vartheta}),_i-\theta_0M_{ij}\dot{E}_{ij}(\vartheta+t_0\dot{\vartheta}) \\ =[k_{ij}\vartheta,_j(\vartheta+t_1\dot{\vartheta})],_i+r(\vartheta+t_1\dot{\vartheta}). \tag{1.3.71}$$

[18] The law in terms of (u_i, ϑ) was obtained for the first time in (Green and Lindsay, 1972).

Integrating over B and using the divergence theorem, we convert this to

$$\int_{\mathrm{B}} M_{ij}\dot{E}_{ij}(\vartheta + t_1\dot{\vartheta})\mathrm{d}v = -\frac{1}{\theta_0}\int_{\partial\mathrm{B}}(\vartheta + t_1\dot{\vartheta})n_i k_{ij}\vartheta,_j\mathrm{d}a - \frac{1}{\theta_0}\int_{\mathrm{B}}(\vartheta + t_1\dot{\vartheta})r\mathrm{d}v$$
$$+\frac{1}{2\theta_0}\frac{\mathrm{d}}{\mathrm{d}t}\left\{\int_{\mathrm{B}} C_E(\vartheta + t_1\dot{\vartheta})^2\mathrm{d}v + t_0(t_1 - t_0)\int_{\mathrm{B}} C_E\dot{\vartheta}^2\mathrm{d}v + t_1\int_{\mathrm{B}} k_{ij}\vartheta,_i\vartheta,_j\mathrm{d}v\right\}$$
$$+\frac{t_1 - t_0}{\theta_0}\int_{\mathrm{B}} C_E\dot{\vartheta}^2\mathrm{d}v + \frac{1}{\theta_0}\int_{\mathrm{B}} k_{ij}\vartheta,_i\vartheta,_j\mathrm{d}v. \tag{1.3.72}$$

Finally, by eliminating from eqns (1.3.66) and (1.3.72) the volume integral containing the tensor M_{ij}, we obtain the following global balance law in terms of (u_i, ϑ)

$$\frac{\mathrm{d}}{\mathrm{d}t}\left\{\frac{1}{2}\int_{\mathrm{B}}\rho\dot{u}_i\dot{u}_i\mathrm{d}v + \frac{1}{2}\int_{\mathrm{B}} C_{ijkl}E_{ij}E_{kl}\mathrm{d}v\right.$$
$$\left.+\frac{1}{2\theta_0}\int_{\mathrm{B}} C_E[(\vartheta + t_0\dot{\vartheta})^2 + t_0\,(t_1 - t_0)\,\dot{\vartheta}^2]\mathrm{d}v + \frac{t_1}{2\theta_0}\int_{\mathrm{B}} k_{ij}\vartheta,_i\vartheta,_j\mathrm{d}v\right\}$$
$$+\frac{1}{\theta_0}\int_{\mathrm{B}} k_{ij}\vartheta,_i\vartheta,_j\mathrm{d}v + \frac{t_1 - t_0}{\theta_0}\int_{\mathrm{B}} C_E\dot{\vartheta}^2\mathrm{d}v$$
$$= \int_{\partial\mathrm{B}}\dot{u}_i S_{ij}n_j\mathrm{d}a + \int_{\mathrm{B}}\dot{u}_i b_i\mathrm{d}v + \frac{1}{\theta_0}\int_{\partial\mathrm{B}} n_i k_{ij}\vartheta,_j(\vartheta + t_1\dot{\vartheta})\mathrm{d}a + \frac{1}{\theta_0}\int_{\mathrm{B}}(\vartheta + t_1\dot{\vartheta})r\mathrm{d}v. \tag{1.3.73}$$

1.3.3 *Global balance law in terms of* (S_{ij}, ϑ)[19]

This law represents the energy integral associated with the field equations (1.3.63). It will be derived assuming that $t_{(0)}$ does not depend on position. This hypothesis is satisfied either when the quotient C_E/C_S does not depend on position or when $t_1 = t_0$; in the latter case $t_{(0)} = t_1$. Multiplying eqn $(1.3.63)_1$ through by $\dot{S}_{ij}$, we arrive at

$$\frac{1}{2}\frac{\partial}{\partial t}\left(\rho^{-1}S_{ik},{}_k S_{ij},{}_j\right) + \frac{1}{2}\frac{\partial}{\partial t}\widetilde{K}_{ijkl}\dot{S}_{ij}\dot{S}_{kl} + A_{ij}\dot{S}_{ij}t_{(0)}^{-1}\Big[t_1 C_S^{-1}(k_{pq}\dot{\vartheta},{}_q),{}_p$$
$$-\left(t_1 - t_{(0)}\right)\ddot{\vartheta}\Big] = b_{ij}\dot{S}_{ij} + (\rho^{-1}S_{ik},{}_k\dot{S}_{ij}),{}_j. \tag{1.3.74}$$

[19] The law is presented here for the first time. A restriction of the law to a homogeneous isotropic thermoelastic body was obtained in (Biały, 1983).

Next, taking the gradient of eqn $(1.3.63)_2$ and multiplying through by $k_{is}\dot{\vartheta},_s$, we find

$$C_S^{-1}\left[(k_{pq}\vartheta,_q),_p - \theta_0 A_{pq}\dot{S}_{pq}\right],_i k_{is}\dot{\vartheta},_s - (\dot{\vartheta} + t_{(0)}\ddot{\vartheta}),_i k_{is}\dot{\vartheta},_s$$
$$= -(C_S^{-1}r),_i k_{is}\dot{\vartheta},_s. \qquad (1.3.75)$$

Also, employing eqn $(1.3.63)_2$, we obtain

$$-C_S^{-1}\left[(k_{pq}\vartheta,_q),_p - \theta_0 A_{pq}\dot{S}_{pq}\right](k_{is}\dot{\vartheta},_s),_i - k_{is}\dot{\vartheta},_i\dot{\vartheta},_s - t_{(0)}k_{is}\ddot{\vartheta},_i\dot{\vartheta},_s$$
$$= -\left[(\dot{\vartheta} + t_{(0)}\ddot{\vartheta} - C_S^{-1}r)k_{is}\dot{\vartheta},_s\right],_i - (C_S^{-1}r),_i k_{is}\dot{\vartheta},_s. \qquad (1.3.76)$$

From this follows

$$t_1 t_{(0)}^{-1} C_S^{-1} A_{pq}\dot{S}_{pq}(k_{is}\dot{\vartheta},_s),_i = t_1 t_{(0)}^{-1}\theta_0^{-1}\left\{\frac{1}{2}\frac{\partial}{\partial t}C_S^{-1}[(k_{pq}\vartheta,_q),_p]^2\right.$$
$$\left. + \frac{1}{2}t_{(0)}\frac{\partial}{\partial t}\left(k_{is}\dot{\vartheta},_i\dot{\vartheta},_s\right) + k_{is}\dot{\vartheta},_i\dot{\vartheta},_s\right\} \qquad (1.3.77)$$
$$-t_1 t_{(0)}^{-1}\theta_0^{-1}\left\{\left[(\dot{\vartheta} + t_{(0)}\ddot{\vartheta} - C_S^{-1}r)k_{is}\dot{\vartheta},_s\right],_i + (C_S^{-1}r),_i k_{is}\dot{\vartheta},_s\right\}.$$

Clearly, the left-hand side of eqn (1.3.77) is identical with the third term on the left-hand side of eqn (1.3.74). We shall now use eqn $(1.3.63)_2$ to determine the fourth term on the left-hand side of eqn (1.3.74): multiplying eqn $(1.3.63)_2$ by $C_S^{-1}\ddot{\vartheta}$, we obtain

$$(k_{pq}\vartheta,_q),_p\ddot{\vartheta} - \theta_0 A_{pq}\dot{S}_{pq}\ddot{\vartheta} - C_S(\dot{\vartheta}\ddot{\vartheta} + t_{(0)}\ddot{\vartheta}^2) = r\ddot{\vartheta}. \qquad (1.3.78)$$

Since

$$(k_{pq}\vartheta,_q),_p\ddot{\vartheta} = \frac{\partial}{\partial t}\left[(k_{pq}\vartheta,_q),_p\dot{\vartheta}\right] - (k_{pq}\dot{\vartheta},_q),_p\dot{\vartheta}$$
$$= \frac{\partial}{\partial t}\left[(k_{pq}\vartheta,_q),_p\dot{\vartheta}\right] - (k_{pq}\dot{\vartheta},_q\dot{\vartheta}),_p + k_{pq}\dot{\vartheta},_q\dot{\vartheta},_p, \qquad (1.3.79)$$

in view of eqn (1.3.78), the fourth term on the left-hand side of eqn (1.3.74) becomes

$$-t_{(0)}^{-1}(t_1 - t_{(0)})A_{pq}\dot{S}_{pq}\ddot{\vartheta} = -\theta_0^{-1}t_{(0)}^{-1}(t_1 - t_{(0)})\left\{\frac{\partial}{\partial t}\left[(k_{pq}\vartheta,_q),_p\dot{\vartheta}\right]\right.$$
$$\left. -(k_{pq}\dot{\vartheta},_q\dot{\vartheta}),_p + k_{pq}\dot{\vartheta},_q\dot{\vartheta},_p + r\ddot{\vartheta} - C_S(\dot{\vartheta}\ddot{\vartheta} + t_{(0)}\ddot{\vartheta}^2)\right\}. \qquad (1.3.80)$$

Adding the sides of eqns (1.3.77) and (1.3.80), we obtain

$$A_{pq}\dot{S}_{pq}t_{(0)}^{-1}\left[t_1 C_S^{-1}(k_{pq}\dot{\vartheta},_q),_p - (t_1 - t_{(0)})\ddot{\vartheta}\right]$$

$$= \frac{1}{2\theta_0}\frac{\partial}{\partial t}\left\{\frac{t_1}{t_{(0)}}\frac{[(k_{pq}\vartheta,_q),_p]^2}{C_S} + t_1 k_{is}\dot{\vartheta},_i\dot{\vartheta},_s - 2\frac{t_1 - t_{(0)}}{t_{(0)}}\left[(k_{pq}\vartheta,_q),_p\dot{\vartheta}\right]\right.$$

$$\left. + C_S\frac{t_1 - t_{(0)}}{t_{(0)}}\dot{\vartheta}^2\right\} + \frac{1}{\theta_0}k_{pq}\dot{\vartheta},_p\dot{\vartheta},_q + \frac{C_S}{\theta_0}(t_1 - t_{(0)})\ddot{\vartheta}^2$$

$$-\frac{1}{\theta_0}\left[\frac{t_1 - t_{(0)}}{t_{(0)}}\ddot{\vartheta} + \frac{t_1}{t_{(0)}}(C_S^{-1}r),_i k_{is}\dot{\vartheta},_s\right] - \frac{1}{\theta_0}\left[\left(\dot{\vartheta} + t_1\ddot{\vartheta} - \frac{t_1}{t_{(0)}}\frac{r}{C_S}\right)k_{is}\dot{\vartheta},_s\right],_i. \tag{1.3.81}$$

Now, since

$$\frac{t_1}{t_{(0)}}\frac{[(k_{pq}\vartheta,_q),_p]^2}{C_S} - 2\left(\frac{t_1}{t_{(0)}} - 1\right)[(k_{pq}\vartheta,_q),_p]\,\dot{\vartheta} + C_S\frac{t_1 - t_{(0)}}{t_{(0)}}\dot{\vartheta}^2$$
$$= \frac{t_1 - t_{(0)}}{t_{(0)}}\left[\frac{(k_{pq}\vartheta,_q),_p}{\sqrt{C_S}} - \sqrt{C_S}\dot{\vartheta}\right]^2 + \frac{[(k_{pq}\vartheta,_q),_p]^2}{C_S}, \tag{1.3.82}$$

the relation (1.3.81) may be written as

$$A_{pq}\dot{S}_{pq}t_{(0)}^{-1}\left[t_1 C_S^{-1}(k_{ij}\dot{\vartheta}_{,j})_{,i} - (t_1 - t_{(0)})\ddot{\vartheta}\right]$$

$$= \frac{1}{2\theta_0}\frac{\partial}{\partial t}\left\{\left(\frac{t_1}{t_{(0)}} - 1\right)\left[\frac{(k_{pq}\vartheta,_q),_p}{\sqrt{C_S}} - \sqrt{C_S}\dot{\vartheta}\right]^2\right.$$

$$\left. + \frac{\left[(k_{pq}\vartheta_{,q})_{,p}\right]^2}{C_S} + t_1 k_{pq}\dot{\vartheta},_p\dot{\vartheta},_q\right\} + \frac{1}{\theta_0}\left[k_{pq}\dot{\vartheta},_p\dot{\vartheta},_q + C_S\left(t_1 - t_{(0)}\right)\ddot{\vartheta}^2\right]$$

$$-\frac{1}{\theta_0}\left[\frac{t_1 - t_{(0)}}{t_{(0)}}r\ddot{\vartheta} + \frac{t_1}{t_{(0)}}(C_S^{-1}r),_i k_{is}\dot{\vartheta},_s\right] - \frac{1}{\theta_0}\left[\left(\dot{\vartheta} + t_1\ddot{\vartheta} - \frac{t_1}{t_{(0)}}\frac{r}{C_S}\right)k_{is}\dot{\vartheta},_s\right],_i. \tag{1.3.83}$$

Finally, integrating eqn (1.3.74) and (1.3.83) over B, employing the divergence theorem, and eliminating from these equations the volume integral containing the tensor A_{ij}, we obtain the following global balance law in terms of (S_{ij}, ϑ)

$$\frac{\mathrm{d}}{dt}\left\{\frac{1}{2}\int\limits_{B}\rho^{-1}S_{ik},{}_{k}S_{ij},{}_{j}\mathrm{d}v+\frac{1}{2}\int\limits_{B}\widetilde{K}_{ijkl}\dot{S}_{ij}\dot{S}_{kl}\mathrm{d}v\right.$$

$$\left.+\frac{1}{2\theta_0}\int\limits_{B}\left\{\frac{t_1-t_{(0)}}{t_{(0)}}\left[\frac{(k_{pq}\vartheta,{}_{q}),{}_{p}}{\sqrt{C_S}}-\sqrt{C_S}\dot{\vartheta}\right]^2+\frac{\left[(k_{pq}\vartheta,{}_{q}),{}_{p}\right]^2}{C_S}+t_1k_{pq}\dot{\vartheta},{}_{p}\dot{\vartheta},{}_{q}\right\}\mathrm{d}v\right\}$$

$$+\frac{1}{\theta_0}\int\limits_{B}\left[k_{pq}\dot{\vartheta},{}_{p}\dot{\vartheta},{}_{q}+C_S\left(t_1-t_{(0)}\right)\ddot{\vartheta}^2\right]\mathrm{d}v$$

$$=\int\limits_{\partial B}\rho^{-1}n_i\dot{S}_{ij}S_{jk},{}_{k}\mathrm{d}a+\int\limits_{B}\dot{S}_{ij}\left[(\rho^{-1}b_{(i)},{}_{j})-\frac{t_1}{t_{(0)}}\frac{\dot{r}}{C_S}A_{ij}\right]\mathrm{d}v$$

$$+\frac{1}{\theta_0}\int\limits_{\partial B}n_ik_{is}\dot{\vartheta},{}_{s}\left(\dot{\vartheta}+t_1\ddot{\vartheta}-\frac{t_1}{t_{(0)}}\frac{r}{C_S}\right)\mathrm{d}a$$

$$+\frac{1}{\theta_0}\int\limits_{B}\left[\left(\frac{t_1-t_{(0)}}{t_{(0)}}\right)r\ddot{\vartheta}+\frac{t_1}{t_{(0)}}\left(\frac{r}{C_S}\right),{}_{i}k_{is}\dot{\vartheta},{}_{s}\right]\mathrm{d}v. \tag{1.3.84}$$

Noting that

$$\frac{t_1}{t_{(0)}}\rightarrow 1+0 \text{ for } t_1\rightarrow t_0+0, \tag{1.3.85}$$

one may then obtain the global law for a body in which $t_1 = t_0 > 0$, as well as for the special case of a classical thermoelastic body.

2

FORMULATIONS OF INITIAL-BOUNDARY VALUE PROBLEMS

2.1 Conventional and non-conventional characterization of a thermoelastic process

In Section 1.1, focused on the classical theory, we defined the thermoelastic process corresponding to a loading (b_i, r) as a system of functions $[u_i, E_{ij}, S_{ij}, \vartheta, \eta, q_i]$ satisfying field equations (1.1.24–29) or (1.1.32–37). In Sections 1.2 and 1.3 we have similarly defined thermoelastic processes with relaxation times. It was noted there that, by eliminating four out of six field variables describing the process, one can write field equations in terms of various thermomechanical pairs such as (u_i, ϑ), (u_i, q_i), (S_{ij}, ϑ), and (S_{ij}, q_i).

The thermomechanical pair created from the variables describing a given process will be called a pair corresponding to that process, provided all other field variables may be recovered from it.[1] For instance, a pair (u_i, ϑ) satisfying eqns (1.1.68) is such a pair because it generates fields $E_{ij}, S_{ij}, \vartheta$ and η, so that $[u_i, E_{ij}, S_{ij}, \vartheta, \eta, q_i]$ is a classical thermoelastic process corresponding to the loading (b_i, r).

Clearly, the definition of a thermoelastic process introduced in Chapter 1 does not uniquely determine that process, because in the case of a specific bounded body B, this process depends not only on the body forces b_i and heat sources r, but also on the initial thermomechanical load and the boundary thermomechanical load. It is common in the classical thermoelasticity (Carlson, 1972; Nowacki, 1975) to describe an initial state of the body in terms of a pair (u_i, ϑ), assuming that the fields $u_i(\cdot, 0)$, $\dot{u}_i(\cdot, 0)$ and $\vartheta(\cdot, 0)$ are known at time $t = 0$. These assumptions are analogous to those appearing in classical (Newtonian) mechanics, in which the motion of a material particle is determined by its initial position and velocity, as well as in classical heat-conduction theory for a rigid conductor with an initial temperature. Also, in classical thermoelasticity it became an accepted practice to specify a boundary loading in terms of the pair (u_i, ϑ). Thus, a thermoelastic process corresponding to the loading (b_i, r), and the initial and boundary conditions in terms of the pair (u_i, ϑ) will be called

[1] The definition is an extension of that involving a pair corresponding to solution of a mixed conventional problem of classical thermoelasticity [see Theorem 1 on p. 355 in (Carlson, 1972); see p. 18 in (Nickell and Sackman, 1968)].

a *conventional thermoelastic process*, while the formulation of the associated initial-boundary value problem will be called a *conventional description of a thermoelastic process.*

A thermoelastic process corresponding to the loading (b_i, r), and the initial and boundary conditions not in terms of the pair (u_i, ϑ) will be called a *non-conventional thermoelastic process*, while the formulation of the associated initial-boundary value problem will be called a *non-conventional description of a thermoelastic process.*

In classical thermoelasticity the conventional thermoelastic process may be described in terms of various pairs of thermomechanical variables. Most commonly, this is done with the help of (u_i, ϑ), see page 335 in (Carlson, 1972). Its description in terms of (u_i, q_i) and (S_{ij}, ϑ) may be found in (Ieşan, 1966), and in terms of (S_{ij}, q_i) in (Nickell and Sackman, 1968).

The non-conventional thermoelastic process remains so far little known even in the realm of classical thermoelasticity. The known cases involve some results for an associated initial-boundary value problem in terms of the pair describing the initial thermomechanical state of the body. However, the recovery from the pair of all other field variables remains an open problem. Also, alternative descriptions of this non-conventional process in terms of various thermomechanical pairs are unknown. In this chapter we shall formulate two conventional and two non-conventional initial-boundary value problems, corresponding to the global conservation laws of Sections 1.2 and 1.3, as well as analyze some relations among these formulations.

2.1.1 *Two mixed initial-boundary value problems in the L–S theory*

In order to formulate an initial-boundary value problem, we divide the boundary $\partial\mathrm{B}$ into two parts in two different ways:

$$\partial\mathrm{B} = \partial\mathrm{B}_1 \cup \partial\mathrm{B}_2 = \partial\mathrm{B}_3 \cup \partial\mathrm{B}_4, \tag{2.1.1}$$

where

$$\partial\mathrm{B}_1 \cap \partial\mathrm{B}_2 = \partial\mathrm{B}_3 \cap \partial\mathrm{B}_4 = \varnothing. \tag{2.1.2}$$

Mixed displacement–temperature problem in the L–S theory

Find a pair (u_i, ϑ) satisfying the field equations[2]

$$\left.\begin{aligned} (C_{ijkl}u_{k,l})_{,j} - \rho\ddot{u}_i + (M_{ij}\vartheta)_{,j} &= -b_i \\ (k_{ij}\vartheta_{,j})_{,i} - C_E\hat{\dot{\vartheta}} + \theta_0 M_{ij}\hat{\dot{u}}_{i,j} &= -\hat{r} \end{aligned}\right\} \quad \text{on } \mathrm{B}\times[0,\infty), \tag{2.1.3}$$

the initial conditions

$$\left.\begin{aligned} u_i(\cdot,0) = u_{i0},\ \dot{u}_i(\cdot,0) &= \dot{u}_{i0} \\ \vartheta(\cdot,0) = \vartheta_0,\ \ \dot{\vartheta}(\cdot,0) &= \dot{\vartheta}_0 \end{aligned}\right\} \quad \text{on } \mathrm{B}, \tag{2.1.4}$$

[2] See eqns (1.2.15).

and the boundary conditions

$$
\begin{aligned}
u_i &= u_i' && \text{on} \quad \partial B_1 \times [0,\infty), \\
(C_{ijkl} u_{k,l} + M_{ij}\vartheta)\, n_j &= s_i' && \text{on} \quad \partial B_2 \times [0,\infty), \\
\vartheta &= \vartheta' && \text{on} \quad \partial B_3 \times [0,\infty), \\
-k_{ij}\vartheta_{,j}\, n_i &= q' && \text{on} \quad \partial B_4 \times [0,\infty).
\end{aligned}
\tag{2.1.5}
$$

Here, $(u_{i0}, \dot{u}_{i0}, \vartheta_0, \dot{\vartheta}_0)$ and $(u_i', s_i', \vartheta', q')$ are prescribed functions determining the initial and boundary loadings of the body, respectively. Thus, the thermoelastic process corresponding to the problem (2.1.3–5) is caused by the thermomechanical loading represented by a system of functions

$$
(b_i, r, u_{i0}, \dot{u}_{i0}, \vartheta_0, \dot{\vartheta}_0, u_i', s_i', \vartheta', q'). \tag{2.1.6}
$$

One can show that a pair (u_i, ϑ) that satisfies (2.1.3–5) corresponds to the conventional thermoelastic process defined in Section 1.2. Also, one can demonstrate that the problem (2.1.3–5) is associated with the global balance law (1.2.30).

Mixed stress–heat-flux problem in the L–S theory

Find a pair (S_{ij}, q_i) satisfying the field equations[3]

$$
\begin{aligned}
(\rho^{-1} S_{(ik,k)})_{,j)} - K'_{ijkl}\ddot{S}_{kl} + C_S^{-1} A_{ij}\dot{q}_{k,k} &= -(\rho^{-1} b_{(i)})_{,j)} + C_S^{-1}\dot{r} A_{ij} \\
(C_S^{-1} q_{k,k})_{,i} - \lambda_{ij}\dot{\hat{q}}_j + \theta_0 (C_S^{-1} A_{pq}\dot{S}_{pq})_{,i} &= (C_S^{-1} r)_{,i} \\
\text{on} \quad B \times [0,\infty),
\end{aligned}
\tag{2.1.7}
$$

the initial conditions

$$
\begin{aligned}
S_{ij}(\cdot,0) = S_{ij}^{(0)}, \quad \dot{S}_{ij}(\cdot,0) = \dot{S}_{ij}^{(0)} \\
q_i(\cdot,0) = q_i^{(0)}, \quad \dot{q}_i(\cdot,0) = \dot{q}_i^{(0)}
\end{aligned}
\quad \text{on} \quad B,
\tag{2.1.8}
$$

and the boundary conditions

$$
\begin{aligned}
S_{ik,k} &= p_i' && \text{on} \quad \partial B_1 \times [0,\infty), \\
S_{ij} n_j &= s_i' && \text{on} \quad \partial B_2 \times [0,\infty), \\
q_{k,k} + \theta_0 A_{pq}\dot{S}_{pq} &= \vartheta'' && \text{on} \quad \partial B_3 \times [0,\infty), \\
q_i n_i &= q' && \text{on} \quad \partial B_4 \times [0,\infty).
\end{aligned}
\tag{2.1.9}
$$

Here, $(S_{ij}^{(0)}, \dot{S}_{ij}^{(0)}, q_i^{(0)}, \dot{q}_i^{(0)})$ and $(p_i', s_i', \vartheta'', q')$ are prescribed functions determining the initial and boundary loadings of the body, respectively. Thus, the thermoelastic process corresponding to the problem (2.1.7–9) is caused by the thermomechanical loading represented by a system of functions

$$
(b_i, r, S_{ij}^{(0)}, \dot{S}_{ij}^{(0)}, q_i^{(0)}, \dot{q}_i^{(0)}, p_i', s_i', \vartheta', q'). \tag{2.1.10}
$$

[3] See eqns (1.2.17).

Given the non-conventional initial conditions (2.1.8) and boundary conditions $(2.1.9)_1$ and $(2.1.9)_3$, this thermoelastic process is a non-conventional one. These conditions are dictated by the global conservation law (1.2.31) of Section 1.2.

2.1.2 *Two mixed initial-boundary value problems in the G–L theory*

Mixed displacement–temperature problem in the G–L theory

Find a pair (u_i, ϑ) satisfying the field equations[4]

$$\left.\begin{aligned}(C_{ijkl}u_{k,l})_{,j} - \rho\ddot{u}_i + [M_{ij}(\vartheta + t_1\dot{\vartheta})]_{,j} &= -b_i\\ (k_{ij}\vartheta_{,j})_{,i} - C_E(\dot{\vartheta} + t_0\ddot{\vartheta}) + \theta_0 M_{ij}\dot{u}_{i,j} &= -r\end{aligned}\right\} \text{ on } \mathrm{B}\times[0,\infty), \tag{2.1.11}$$

the initial conditions

$$\left.\begin{aligned}u_i(\cdot,0) = u_{i0},\ \dot{u}_i(\cdot,0) = \dot{u}_{i0}\\ \vartheta(\cdot,0) = \vartheta_0,\ \ \dot{\vartheta}(\cdot,0) = \dot{\vartheta}_0\end{aligned}\right\} \text{ on } \mathrm{B}, \tag{2.1.12}$$

and the boundary conditions

$$\begin{aligned}u_i &= u_i' && \text{on } \partial\mathrm{B}_1\times[0,\infty),\\ [C_{ijkl}u_{k,l} + M_{ij}(\vartheta + t_1\dot{\vartheta})]n_j &= s_i' && \text{on } \partial\mathrm{B}_2\times[0,\infty),\\ \vartheta &= \vartheta' && \text{on } \partial\mathrm{B}_3\times[0,\infty),\\ -k_{ij}\vartheta_{,j}\,n_i &= q' && \text{on } \partial\mathrm{B}_4\times[0,\infty).\end{aligned} \tag{2.1.13}$$

Here, the external thermomechanical loading causing this thermoelastic process, in accordance with eqns (2.1.11–13), is represented by a system of functions

$$(b_i, r, u_{i0}, \dot{u}_{i0}, \vartheta_0, \dot{\vartheta}_0, u_i', s_i', \vartheta', q'). \tag{2.1.14}$$

One can show that a pair (u_i, ϑ) satisfying eqns (2.1.11–13) corresponds to a conventional thermoelastic process with two relaxation times. Also, the problem (2.1.11–13) is associated with the global balance law (1.3.73) of Section 1.3.

Mixed stress–temperature problem in the G–L theory

Find a pair (S_{ij}, ϑ) satisfying the field equations[5]

$$\left.\begin{aligned}(\rho^{-1}S_{(ik,k})_{,j)} - \widetilde{K}_{ijkl}\ddot{S}_{kl} - A_{ij}t_{(0)}^{-1}[t_1 C_S^{-1}(k_{pq}\dot{\vartheta}_{,q})_{,p} - (t_1 - t_{(0)})\ddot{\vartheta}]\\ = -[(\rho^{-1}b_{(i})_{,j)} - t_{(0)}^{-1}t_1 C_S^{-1}\dot{r}A_{ij}]\\ (k_{pq}\vartheta_{,q})_{,p} - C_S(\dot{\vartheta} + t_{(0)}\ddot{\vartheta}) - \theta_0 A_{pq}\dot{S}_{pq} = -r\end{aligned}\right\} \text{ on } \mathrm{B}\times[0,\infty), \tag{2.1.15}$$

the initial conditions

$$\left.\begin{aligned}S_{ij}(\cdot,0) = S_{ij}^{(0)},\ \dot{S}_{ij}(\cdot,0) = \dot{S}_{ij}^{(0)}\\ \vartheta(\cdot,0) = \vartheta_0,\qquad \dot{\vartheta}(\cdot,0) = \dot{\vartheta}_0\end{aligned}\right\} \text{ on } \mathrm{B}, \tag{2.1.16}$$

4 See eqns (1.3.57).
5 See eqns (1.3.63).

and the boundary conditions

$$\begin{aligned} S_{ik},_{k} &= p_i' \quad \text{on } \partial\mathrm{B}_1 \times [0,\infty), \\ S_{ij}n_j &= s_i' \quad \text{on } \partial\mathrm{B}_2 \times [0,\infty), \\ \vartheta &= \vartheta' \quad \text{on } \partial\mathrm{B}_3 \times [0,\infty), \\ -k_{ij}\vartheta,_j n_i &= q' \text{ on } \partial\mathrm{B}_4 \times [0,\infty). \end{aligned} \tag{2.1.17}$$

Here, the external thermomechanical loading causing this thermoelastic process, in accordance with eqns (2.1.15–17), is represented by a system of functions

$$(b_i, r, S_{ij}^{(0)}, \dot{S}_{ij}^{(0)}, \vartheta_0, \dot{\vartheta}_0, p_i', s_i', \vartheta', q'). \tag{2.1.18}$$

We note that this process also belongs to non-conventional processes – this is due to the non-conventional initial conditions $(2.1.16)_1$ and the boundary condition $(2.1.17)_1$. The latter has no physical interpretation if not referred to a displacement vector. Furthermore, we observe that the problem described by eqns (2.1.15–17) is associated with the global balance law (1.3.84) of Section 1.3.

The formulated four mixed initial-boundary value problems contain as special cases the so-called *natural problems*. If the mixed displacement–temperature problem is denoted by MDTP, the following terminology is introduced:

- A *natural displacement–temperature problem* (NDTP) in the L–S or G–L theory is defined as a limiting case of MDTP when $\partial\mathrm{B}_2 = \partial\mathrm{B}_4 = \varnothing$.
- A *natural stress–heat flux problem* (NSHFP) in the L–S theory is defined as a limiting case of MSHFP when $\partial\mathrm{B}_1 = \partial\mathrm{B}_3 = \varnothing$.
- A *natural stress–temperature problem* (NSTP) in the G–L theory is defined as a limiting case of MSTP when $\partial\mathrm{B}_1 = \partial\mathrm{B}_4 = \varnothing$.

Clearly, in the case of NSHFP in the L–S theory, as well as in the case of NSTP in the G–L theory, the thermomechanical loading possesses a natural physical interpretation, even though the initial data are non-conventional.

2.2 Relations among descriptions of a thermoelastic process in terms of various pairs of thermomechanical variables

Proceeding similarly as in the case of classical thermoelasticity (e.g. Nickell and Sackman, 1968; Ieşan, 1966), one can prove the following theorem:

Theorem 2.1 *A conventional thermoelastic process in the L–S as well as G–L theory, described in terms of* (u_i, ϑ) *through eqns (2.1.3–5) and (2.1.11–13), respectively, may also be described in terms of the pairs* (S_{ij}, ϑ), (S_{ij}, q_i), *or* (u_i, q_i); *each of these alternative descriptions contains data that are uniquely determined by those of the conventional process.*

An analogous theorem for a non-conventional process is not known as yet. Note here that a non-conventional process is associated with the initial-boundary value problem for the pair of variables that belongs to a certain vector space of a higher dimension than that of the conventional process.[6] For example, a process associated with NSHFP in the L–S theory is consistent with the pair (S_{ij}, q_i) that belongs to a vector space of dimension 9, whereas a process associated with NSTP in the G–L theory is consistent with the pair (S_{ij}, ϑ) that belongs to a vector space of dimension 7.

Furthermore, a non-conventional thermoelastic process is generated by a thermomechanical loading that is more general than that generating a conventional process. With reference to the four mixed initial-boundary value problems of Section 2.1, we have these two theorems:

Theorem 2.2 *MSHFP in the L–S theory is more general than MDTP in that theory, i.e.* $MDTP \subseteq MSHFP$.

Theorem 2.3 *MSTP in the G–L theory is more general than MDTP in that theory, i.e.* $MDTP \subseteq MSTP$.

A proof of these theorems is based on the two observations:

(i) the thermomechanical loading in MSHFP (or MSTP) is more general than that in MDTP;
(ii) there is equivalence of the two problems in the case of a particular loading in MSHFP (or MSTP).

Observation (i) follows from the fact that the thermomechanical loading in MDTP in both theories belongs to a vector space of dimension 20, whereas the thermomechanical loading in a MSHFP of the L–S theory belongs to a vector space of dimension 30, and the thermomechanical loading in a MSTP of the G–L theory belongs to a vector space of dimension 26.

In order to justify the second observation, we first show that a MSHFP, for an appropriately restricted thermomechanical loading of the L–S theory, is equivalent to a MDTP of that theory. To this end, we multiply the field equations (2.1.7) by t in the sense of convolution to include the initial conditions (2.1.8). Next, restricting the non-conventional initial and boundary conditions[7] to those that appear in a MSHFP equivalent to a MDTP of the L–S theory, we conclude that MSHFP = MDTP.

[6] A set of all pairs (u_i, ϑ) can be identified with a vector space of dimension 4.

[7] The non-conventional initial and boundary conditions are defined by eqns (2.1.8), $(2.1.9)_1$ and $(2.1.9)_3$. The initial conditions (2.1.8) are replaced by those consistent with eqns (2.1.4) and (1.2.1)–(1.2.6) at $t = 0$. The boundary conditions $(2.1.9)_1$ and $(2.1.9)_3$ are replaced by those equivalent to eqns $(2.1.5)_1$ and $(2.1.5)_3$. And to define the displacement and temperature in terms of (S_{ij}, q_i) we use eqns (1.2.1)–(1.2.6).

The proof that a MSTP for a thermomechanical loading in the G–L theory is equivalent to a MDTP in that theory is completely analogous. This completes the proof of Theorems 2.2 and 2.3. On this basis we formulate:

Corollary 2.1 *Each non-conventional thermoelastic process in the L–S (or G–L) theory is more universal than the conventional thermoelastic process in the L–S (or G–L) theory.*

Corollary 2.2 *Each natural non-conventional thermoelastic process in the L–S (or G–L) theory, restricted to a pair of thermomechanical variables that describe it, possesses an equally good physical interpretation as that of a conventional thermoelastic process in the L–S (or G–L) theory.*

3

EXISTENCE AND UNIQUENESS THEOREMS

3.1 Uniqueness theorems for conventional and non-conventional thermoelastic processes

We now state four uniqueness theorems corresponding to the four mixed initial-boundary value problems of Chapter 2.

Theorem 3.1 *MDTP in the L–S theory has at the most one solution.*

Proof. MDTP in the L–S theory is described by eqns (2.1.3–5) of Section 2.1. Now, assume there exist two solutions of this problem. Their difference (u_i, ϑ) corresponds to the null initial conditions

$$\left.\begin{aligned} u_i(\cdot,0) = 0,\ \dot{u}_i(\cdot,0) = 0 \\ \vartheta(\cdot,0) = 0,\ \ \dot{\vartheta}(\cdot,0) = 0 \end{aligned}\right\} \text{ on B,} \tag{3.1.1}$$

the null boundary conditions

$$\begin{aligned} u_i &= 0 \quad \text{on } \partial \mathrm{B}_1 \times [0,\infty), \\ S_{ij} n_j &= 0 \quad \text{on } \partial \mathrm{B}_2 \times [0,\infty), \\ \vartheta &= 0 \quad \text{on } \partial \mathrm{B}_3 \times [0,\infty), \\ -k_{ij}\vartheta_{,j}\, n_i &= 0 \quad \text{on } \partial \mathrm{B}_4 \times [0,\infty), \end{aligned} \tag{3.1.2}$$

and the null body force and heat source fields

$$b_i = 0,\ r = 0 \text{ on } \overline{\mathrm{B}} \times [0,\infty). \tag{3.1.3}$$

It follows from eqns (1.2.1), (1.2.2) and (1.2.4), taken for the difference of solutions and for $t = 0$ that

$$\ddot{u}_i(\cdot,0) = \rho^{-1}\left[C_{ijkl}u_{(k,l)} + M_{ij}\vartheta\right]_{,j}(\cdot,0) \quad \text{on B,} \tag{3.1.4}$$

so that, in view of the definition of the hut operator (recall eqn (1.2.16)) and conditions (3.1.1–4),

$$\begin{aligned} \hat{u}_i(\cdot,0) = 0,\ \dot{\hat{u}}_i(\cdot,0) = 0 \\ \vartheta(\cdot,0) = 0,\ \ \hat{\vartheta}(\cdot,0) = 0 \end{aligned} \quad \text{on B,} \tag{3.1.5}$$

$$\begin{aligned} \dot{\hat{u}}_i &= 0 \text{ on } \partial \mathrm{B}_1 \times [0,\infty), \\ \hat{S}_{ij} n_j &= 0 \text{ on } \partial \mathrm{B}_2 \times [0,\infty), \\ \hat{\vartheta} &= 0 \text{ on } \partial \mathrm{B}_3 \times [0,\infty), \\ -k_{ij}\vartheta,_j n_i &= 0 \text{ on } \partial \mathrm{B}_4 \times [0,\infty), \end{aligned} \tag{3.1.6}$$

and

$$\hat{b}_i = 0,\ \hat{r} = 0 \text{ on B} \times [0,\infty). \tag{3.1.7}$$

Given eqns (3.1.6) and (3.1.7), the global conservation law (1.2.30), associated with MDTP of Section 2.1 and applied to the difference (u_i, ϑ), takes on the form

$$\frac{\mathrm{d}}{\mathrm{dt}}\left\{\frac{1}{2}\int_{\mathrm{B}} \rho \dot{\hat{u}}_i \dot{\hat{u}}_i \mathrm{d}v + \frac{1}{2}\int_{\mathrm{B}} C_{ijkl}\hat{u}_{i},_j\, \hat{u}_{k},_l\, \mathrm{d}v + \frac{1}{2\theta_0}\int_{\mathrm{B}} C_E \hat{\vartheta}^2 \mathrm{d}v + \frac{t_0}{2\theta_0}\int_{\mathrm{B}} k_{ij}\vartheta,_i \vartheta,_j\, \mathrm{d}v\right\}$$
$$+\frac{1}{\theta_0}\int_{\mathrm{B}} k_{ij}\vartheta,_i \vartheta,_j\, \mathrm{d}v = 0. \tag{3.1.8}$$

Now, integrating over time, and using conditions (3.1.5), we find

$$\frac{1}{2}\int_{\mathrm{B}} \rho \dot{\hat{u}}_i \dot{\hat{u}}_i \mathrm{d}v + \frac{1}{2}\int_{\mathrm{B}} C_{ijkl}\hat{u}_{i},_j\, \hat{u}_{k},_l\, \mathrm{d}v + \frac{1}{2\theta_0}\int_{\mathrm{B}} C_E \hat{\vartheta}^2 \mathrm{d}v + \frac{t_0}{2\theta_0}\int_{\mathrm{B}} k_{ij}\vartheta,_i \vartheta,_j\, \mathrm{d}v$$
$$+\frac{1}{\theta_0}\int_0^t \int_{\mathrm{B}} k_{ij}\vartheta,_i \vartheta,_j\, \mathrm{d}v\mathrm{d}\tau = 0. \tag{3.1.9}$$

Since

$$\rho > 0,\ C_E > 0 \text{ on B}, \tag{3.1.10}$$

$$\theta_0 > 0,\ t_0 > 0 \text{ on B}, \tag{3.1.11}$$

$$C_{ijkl}E_{ij}E_{kl} > 0\ \forall E_{ij} = E_{(ij)} \text{ on } \bar{\mathrm{B}} \times [0,\infty), \tag{3.1.12}$$

$$k_{ij}t_i t_j > 0\ \forall t_i \text{ on } \bar{\mathrm{B}} \times [0,\infty), \tag{3.1.13}$$

each integral appearing on the left-hand side of eqn (3.1.9) is non-negative. Thus, in view of eqns (3.1.9–13), we get

$$\dot{\hat{u}}_i = 0,\ \hat{\vartheta} = 0 \text{ on } \bar{\mathrm{B}} \times [0,\infty), \tag{3.1.14}$$

or, on the basis of eqn (3.1.5),

$$\hat{u}_i = 0,\ \vartheta = 0 \text{ on } \bar{\mathrm{B}} \times [0, \infty). \tag{3.1.15}$$

Thus, using eqn (3.1.1), we arrive at

$$(u_i, \vartheta) = (0, 0) \text{ on } \bar{\mathrm{B}} \times [0, \infty), \tag{3.1.16}$$

which completes the proof of Theorem 3.1.[1] □

Prior to formulating a uniqueness theorem for MSHFP in the L–S theory, in the place of constitutive inequalities (3.1.10–13), we assume[2]

$$\rho^{-1} > 0,\ C_S^{-1} > 0 \text{ on B}, \tag{3.1.17}$$

$$\theta_0 > 0,\ t_0 > 0, \tag{3.1.18}$$

$$K'_{ijkl} S_{ij} S_{kl} > 0\ \forall S_{ij} = S_{(ij)} \text{ on } \bar{\mathrm{B}} \times [0, \infty), \tag{3.1.19}$$

$$\lambda_{ij} t_i t_j > 0\ \forall t_i \text{ on } \bar{\mathrm{B}} \times [0, \infty). \tag{3.1.20}$$

Observe that the inequalities (3.1.17–20) are an alternative to eqns (3.1.10–13).

Theorem 3.2 *MSHFP in the L–S theory has at the most one solution.*

Proof. MSHFP in the L–S theory is described by eqns (2.1.7–9) of Section 2.1. Now, assume there exist two solutions of this problem. Their difference (S_{ij}, q_i) corresponds to the null initial data

$$\left.\begin{aligned} S_{ij}(\cdot, 0) = 0,\ \dot{S}_{ij}(\cdot, 0) = 0 \\ q_i(\cdot, 0) = 0,\quad \dot{q}_i(\cdot, 0) = 0 \end{aligned}\right. \quad \text{on B}, \tag{3.1.21}$$

the null boundary conditions

$$\begin{aligned} S_{ik},_k &= 0 \text{ on } \partial\mathrm{B}_1 \times (0, \infty), \\ S_{ij} n_j &= 0 \text{ on } \partial\mathrm{B}_2 \times (0, \infty), \\ q_k,_k + \theta_0 A_{pq} \dot{S}_{pq} &= 0 \text{ on } \partial\mathrm{B}_3 \times (0, \infty), \\ q_i n_i &= 0 \text{ on } \partial\mathrm{B}_4 \times (0, \infty), \end{aligned} \tag{3.1.22}$$

and

$$b_i = 0,\ r = 0 \text{ on } \mathrm{B} \times (0, \infty). \tag{3.1.23}$$

On the basis of eqns (3.1.22) and (3.1.23), the global conservation law (1.2.31), associated with the MSHFP described by eqns (2.1.7–9) of Section 2.1, applied

[1] Theorem 3.1 is proved here for the first time. The proof is similar to that of a homogeneous isotropic thermoelasticity with one relaxation time presented in (Ignaczak, 1982).

[2] In the second part of this chapter we show that the hypothesis (3.1.19) restricted to a homogeneous isotropic thermoelasticity is implied by the inequalities (1.1.67).

to the difference (S_{ij}, q_i) takes the form

$$\frac{\mathrm{d}}{\mathrm{d}t}\left\{\frac{1}{2}\int_B \rho^{-1} S_{ik},_k S_{ij},_j \,\mathrm{d}v + \frac{1}{2}\int_B K'_{ijkl}\dot{S}_{ij}\dot{S}_{kl}\mathrm{d}v\right.$$
$$\left.+\frac{1}{2\theta_0}\int_B C_S^{-1}(q_k,_k)^2\mathrm{d}v + \frac{t_0}{2\theta_0}\int_B \lambda_{ij}\dot{q}_i\dot{q}_j\mathrm{d}v\right\} + \frac{1}{\theta_0}\int_B \lambda_{ij}\dot{q}_i\dot{q}_j\mathrm{d}v = 0. \quad (3.1.24)$$

Henceforth, integrating eqn (3.1.24) over the time interval $(0,t)$ and using eqns (3.1.21), we obtain

$$\frac{1}{2}\int_B \rho^{-1} S_{ik},_k S_{ij},_j \,\mathrm{d}v + \frac{1}{2}\int_B K'_{ijkl}\dot{S}_{ij}\dot{S}_{kl}\mathrm{d}v$$
$$+\frac{1}{2\theta_0}\int_B C_S^{-1}(q_k,_k)^2\,\mathrm{d}v + \frac{t_0}{2\theta_0}\int_B \lambda_{ij}\dot{q}_i\dot{q}_j\mathrm{d}v + \frac{1}{\theta_0}\int_0^t\int_B \lambda_{ij}\dot{q}_i\dot{q}_j\mathrm{d}v\mathrm{d}\tau = 0. \quad (3.1.25)$$

The last relation and the inequalities (3.1.17–20) imply that

$$\dot{S}_{ij} = 0,\ \dot{q}_i = 0 \text{ on } \bar{\mathrm{B}} \times [0,\infty). \quad (3.1.26)$$

Thus, using eqns (3.1.21), we arrive at

$$(S_{ij}, q_i) = (0,0) \text{ on } \bar{\mathrm{B}} \times [0,\infty), \quad (3.1.27)$$

which completes the proof of Theorem 3.2.[3] □

The next two uniqueness theorems pertain to thermoelasticity with two relaxation times.

Theorem 3.3 *MDTP in the G–L theory has at the most one solution.*

Proof. This problem is described by eqns (2.1.11–13). Let us assume there exist two solutions of this problem. Their difference (u_i, ϑ) corresponds to the null data

$$\left.\begin{aligned} u_i(\cdot,0) = 0,\ \dot{u}_i(\cdot,0) = 0 \\ \vartheta(\cdot,0) = 0,\ \dot{\vartheta}(\cdot,0) = 0 \end{aligned}\right\} \text{ on B}, \quad (3.1.28)$$

$$\begin{aligned} u_i &= 0 \quad \text{on } \partial\mathrm{B}_1 \times (0,\infty), \\ S_{ij}n_j &= 0 \quad \text{on } \partial\mathrm{B}_2 \times (0,\infty), \\ \vartheta &= 0 \quad \text{on } \partial\mathrm{B}_3 \times (0,\infty), \\ -k_{ij}\vartheta,_j n_i &= 0 \quad \text{on } \partial\mathrm{B}_4 \times (0,\infty), \end{aligned} \quad (3.1.29)$$

[3] Theorem 3.2, restricted to an isotropic thermoelastic body with one relaxation time, was proved in (Ignaczak, 1979).

and

$$b_i = 0,\ r = 0 \text{ on } \bar{\mathrm{B}} \times [0, \infty). \tag{3.1.30}$$

On the basis of eqns (3.1.29) and (3.1.30), the global conservation law (1.3.73), associated with the problem at hand, and applied to the difference (u_i, ϑ) takes the form

$$\frac{\mathrm{d}}{\mathrm{d}t}\left\{\frac{1}{2}\int_{\mathrm{B}} \rho \dot{u}_i \dot{u}_i \mathrm{d}v + \frac{1}{2}\int_{\mathrm{B}} C_{ijkl} u_{i,j}\, u_{k,l}\, \mathrm{d}v\right.$$
$$\left.+\frac{1}{2\theta_0}\int_{\mathrm{B}} C_E[(\vartheta + t_0\dot{\vartheta})^2 + t_0(t_1 - t_0)\dot{\vartheta}^2]\mathrm{d}v + \frac{t_1}{2\theta_0}\int_{\mathrm{B}} k_{ij}\vartheta_{,i}\,\vartheta_{,j}\,\mathrm{d}v\right\} \tag{3.1.31}$$
$$+\frac{1}{\theta_0}\int_{\mathrm{B}} k_{ij}\vartheta_{,i}\,\vartheta_{,j}\,\mathrm{d}v + \frac{t_1 - t_0}{\theta_0}\int_{\mathrm{B}} C_E \dot{\vartheta}^2 \mathrm{d}v = 0.$$

Henceforth, integrating this equation over the time interval $(0, t)$ and using eqns (3.1.28), we obtain

$$\frac{1}{2}\int_{\mathrm{B}} \rho \dot{u}_i \dot{u}_i \mathrm{d}v + \frac{1}{2}\int_{\mathrm{B}} C_{ijkl} u_{i,j}\, u_{k,l}\, \mathrm{d}v$$
$$+\frac{1}{2\theta_0}\int_{\mathrm{B}} C_E[(\vartheta + t_0\dot{\vartheta})^2 + t_0\,(t_1 - t_0)\,\dot{\vartheta}^2]\mathrm{d}v + \frac{t_1}{2\theta_0}\int_{\mathrm{B}} k_{ij}\vartheta_{,i}\,\vartheta_{,j}\,\mathrm{d}v \tag{3.1.32}$$
$$+\frac{1}{\theta_0}\int_{\mathrm{B}}\int_0^t k_{ij}\vartheta_{,i}\,\vartheta_{,j}\,\mathrm{d}\tau\mathrm{d}v + \frac{t_1 - t_0}{\theta_0}\int_{\mathrm{B}}\int_0^t C_E\dot{\vartheta}^2\mathrm{d}\tau\mathrm{d}v = 0.$$

The left-hand side of eqn (3.1.32) represents the total energy of the thermoelastic body with two relaxation times, expressed in terms of the pair (u_i, ϑ). Since

$$\rho > 0,\ C_E > 0 \text{ on B}, \tag{3.1.33}$$

$$\theta_0 > 0,\ t_1 > \ t_0 > 0 \text{ on B}, \tag{3.1.34}$$

$$C_{ijkl}E_{ij}E_{kl} > 0\ \forall E_{ij} = E_{(ij)} \text{ on } \bar{\mathrm{B}} \times [0, \infty), \tag{3.1.35}$$

$$k_{ij}p_ip_j > 0\ \forall p_i \text{ on } \ \bar{\mathrm{B}} \times [0, \infty), \tag{3.1.36}$$

each term appearing on the left-hand side of eqn (3.1.32) is non-negative. Thus, we conclude that

$$\dot{u}_i = 0,\ \dot{\vartheta} = 0 \text{ on } \bar{\mathrm{B}} \times [0, \infty). \tag{3.1.37}$$

Relations (3.1.37) and (3.1.28) imply

$$(u_i, \vartheta) = (0, 0) \text{ on } \bar{\mathrm{B}} \times [0, \infty), \tag{3.1.38}$$

which completes the proof of Theorem 3.3. □

Theorem 3.4 *MSTP in the G–L theory has at the most one solution.*

Proof. MSTP is described by eqns (2.1.15–17). Similar to what was done in the proofs of Theorems 3.1–3.3, let us assume there exist two solutions of this problem. Their difference (S_{ij}, ϑ) corresponds to the null data

$$\begin{aligned} S_{ij}(\cdot,0)=0,\ \dot{S}_{ij}(\cdot,0)=0 \\ \vartheta(\cdot,0)=0,\ \ \dot{\vartheta}(\cdot,0)=0 \end{aligned} \quad \text{on B,} \tag{3.1.39}$$

$$\begin{aligned} S_{ik,k}&=0 \ \text{ on } \partial \mathrm{B}_1\times(0,\infty),\\ S_{ij}n_j&=0 \ \text{ on } \partial \mathrm{B}_2\times(0,\infty),\\ \vartheta&=0 \ \text{ on } \partial \mathrm{B}_3\times(0,\infty),\\ -k_{ij}\vartheta_{,j}\,n_i&=0 \ \text{ on } \partial \mathrm{B}_4\times(0,\infty), \end{aligned} \tag{3.1.40}$$

and

$$b_i=0,\ r=0 \text{ on } \bar{\mathrm{B}}\times[0,\infty). \tag{3.1.41}$$

On the basis of eqns (3.1.40 and 3.1.41), the global conservation law (1.3.84), associated with the problem at hand, and applied to the difference (S_{ij}, ϑ) takes the form

$$\begin{aligned} &\frac{\mathrm{d}}{\mathrm{d}t}\left\{\frac{1}{2}\int_{\mathrm{B}}\rho^{-1}S_{ik},_k\,S_{ij},_j\,\mathrm{d}v+\frac{1}{2}\int_{\mathrm{B}}\widetilde{K}_{ijkl}\dot{S}_{ij}\dot{S}_{kl}\mathrm{d}v\right.\\ &\left.+\frac{1}{2\theta_0}\int_{\mathrm{B}}\left\{\left(\frac{t_1}{t_{(0)}}-1\right)\left[\frac{(k_{pq}\vartheta,_q),_p}{\sqrt{C_S}}-\sqrt{C_S}\dot{\vartheta}\right]^2+\frac{[(k_{pq}\vartheta,_q),_p]^2}{C_S}+t_1k_{ij}\dot{\vartheta},_i\,\dot{\vartheta},_j\right\}\mathrm{d}v\right\}\\ &+\frac{1}{\theta_0}\int_{\mathrm{B}}\left[k_{ij}\dot{\vartheta},_i\,\dot{\vartheta},_j+C_S(t_1-t_{(0)})\ddot{\vartheta}^2\right]\mathrm{d}v=0. \end{aligned} \tag{3.1.42}$$

Henceforth, integrating eqn (3.1.42) over the time interval $(0, t)$ and using eqns (3.1.39), we obtain

$$\begin{aligned} &\frac{1}{2}\int_{\mathrm{B}}\rho^{-1}S_{ik},_k\,S_{ij},_j\,\mathrm{d}v+\frac{1}{2}\int_{\mathrm{B}}\widetilde{K}_{ijkl}\dot{S}_{ij}\dot{S}_{kl}\mathrm{d}v\\ &+\frac{1}{2\theta_0}\int_{\mathrm{B}}\left\{\left(\frac{t_1}{t_{(0)}}-1\right)\left[\frac{(k_{pq}\vartheta,_q),_p}{\sqrt{C_S}}-\sqrt{C_S}\dot{\vartheta}\right]^2+\frac{[(k_{pq}\vartheta,_q),_p]^2}{C_S}+t_1k_{ij}\dot{\vartheta},_i\,\dot{\vartheta},_j\right\}\mathrm{d}v\\ &+\frac{1}{\theta_0}\int_{\mathrm{B}}\int_0^t\left[k_{pq}\dot{\vartheta},_p\,\dot{\vartheta},_q+C_S\left(t_1-t_{(0)}\right)\ddot{\vartheta}^2\right]\mathrm{d}\tau\mathrm{d}v=0. \end{aligned} \tag{3.1.43}$$

The left-hand side of eqn (3.1.43) represents the total energy of the thermoelastic body with two relaxation times, expressed in terms of the pair (S_{ij}, ϑ). Because[4]

$$\rho^{-1} > 0,\, C_S^{-1} > 0 \text{ on } \mathrm{B}, \tag{3.1.44}$$

$$\theta_0 > 0,\, t_1 > t_{(0)} > 0 \text{ on } \mathrm{B}, \tag{3.1.45}$$

$$\widetilde{K}_{ijkl} S_{ij} S_{kl} > 0 \;\forall S_{ij} = S_{(ij)} \text{ on } \bar{\mathrm{B}} \times [0, \infty), \tag{3.1.46}$$

$$k_{ij} p_i p_j > 0 \;\forall p_i \text{ on } \bar{\mathrm{B}} \times [0, \infty), \tag{3.1.47}$$

each term appearing on the left-hand side of eqn (3.1.43) is non-negative, so that

$$\dot{S}_{ij} = 0,\, \dot{\vartheta} = 0 \text{ on } \bar{\mathrm{B}} \times [0, \infty). \tag{3.1.48}$$

These relations together with eqns (3.1.39) imply

$$(S_{ij}, \vartheta) = (0, 0) \text{ on } \bar{\mathrm{B}} \times [0, \infty), \tag{3.1.49}$$

which completes the proof of Theorem 3.4.[5] □

3.2 Existence theorem for a non-conventional thermoelastic process

In this section we formulate and prove an existence theorem for a generalized solution to NSHFP of Section 2.1 under the following hypotheses: (i) there is a classical solution of the problem; (ii) the body is homogeneous and isotropic; and (iii) the thermomechanical boundary loading vanishes.

In the classical formulation this problem hinges on finding a pair (S_{ij}, q_i) satisfying the field equations[6]

$$\begin{aligned} \rho^{-1} S_{(ik,kj)} - \left[\frac{1}{2\mu}\left(\ddot{S}_{ij} - \frac{\lambda}{3\lambda + 2\mu} \ddot{S}_{kk}\delta_{ij} \right) - \frac{\alpha^2\theta_0}{C_S} \ddot{S}_{kk}\delta_{ij} \right] & \\ + \frac{\alpha}{C_S} \dot{q}_{k,k}\, \delta_{ij} = -F_{(ij)} & \quad \text{on } \mathrm{B} \times [0, \infty), \\ \frac{1}{C_S}(q_{k,k} + \alpha\theta_0 \dot{S}_{kk})_{,i} - \frac{1}{k}(\dot{q}_i + t_0 \ddot{q}_i) = -g_i & \end{aligned} \tag{3.2.1}$$

[4] The inequality $(3.1.45)_2$ is equivalent to eqn (1.3.56). The inequality (3.1.46), restricted to an isotropic thermoelastic body, is implied by the conventional constitutive inequalities of the G–L theory.

[5] For a particular natural stress–temperature problem of homogeneous isotropic thermoelasticity with two relaxation times, Theorem 3.4 was proved in (Ignaczak, 1978). Theorem 3.4 in the general form is formulated here for the first time. Other uniqueness theorems of the L–S and G–L theories can be found in (Wojnar, 1984, 1985a, 1985b, 1985c; Chandrasekharaiah, 1984).

[6] See eqns (2.1.7–9) in which $\partial\mathrm{B}_1 = \partial\mathrm{B}_3 = \varnothing$, $s_i' = 0$, $q' = 0$, $A_{ij} = \alpha\delta_{ij}$, $\lambda_{ij} = k^{-1}\delta_{ij}$,

$$K'_{ijkl} S_{kl} = \frac{1}{2\mu}\left(S_{ij} - \frac{\lambda}{3\lambda + 2\mu} S_{kk}\delta_{ij} \right) - \frac{\alpha^2\theta_0}{C_S} S_{kk}\delta_{ij},$$

and the material functions are constant.

the initial conditions

$$\left.\begin{array}{l} S_{ij}(\cdot,0)=S_{ij}^{(0)},\ \dot{S}_{ij}(\cdot,0)=\dot{S}_{ij}^{(0)} \\ q_i(\cdot,0)=q_i^{(0)},\quad \dot{q}_i(\cdot,0)=\dot{q}_i^{(0)} \end{array}\right\} \text{ on B,} \tag{3.2.2}$$

and the boundary conditions

$$\begin{aligned} S_{ij}n_j &= 0 \text{ on } \partial\mathrm{B}\times(0,\infty), \\ q_in_i &= 0 \ \text{ on } \partial\mathrm{B}\times(0,\infty). \end{aligned} \tag{3.2.3}$$

Here,

$$\left.\begin{array}{l} F_{(ij)}=\rho^{-1}b_{(i,j)}-C_S^{-1}\alpha\dot{r}\delta_{ij} \\ g_i=-C_S^{-1}r_{,i} \\ C_S=C_E+3(3\lambda+2\mu)\alpha^2\theta_0 \end{array}\right\} \text{ on } \overline{\mathrm{B}}\times[0,\infty). \tag{3.2.4}$$

The material constants appearing in eqns (3.2.1) satisfy the inequalities

$$\begin{array}{ll} \rho>0, & C_E>0, \\ \theta_0>0, & t_0>0, \\ \mu>0, & 3\lambda+2\mu>0, \\ k>0, & |\alpha|>0. \end{array} \tag{3.2.5}$$

Observe that the inequalities (3.2.5) jointly imply the inequality[7]

$$\begin{aligned} &\left[\frac{1}{2\mu}\left(S_{ij}-\frac{\lambda}{3\lambda+2\mu}S_{kk}\delta_{ij}\right)-\frac{\alpha^2\theta_0}{C_S}S_{kk}\delta_{ij}\right]S_{ij} \\ &\quad=\frac{1}{2\mu}\left(S_{ij}-\frac{1}{3}S_{kk}\delta_{ij}\right)\left(S_{ij}-\frac{1}{3}S_{kk}\delta_{ij}\right) \qquad \forall S_{ij} \text{ on } \overline{\mathrm{B}}\times[0,\infty). \\ &\quad+\frac{1}{3(3\lambda+2\mu)}\frac{C_E}{C_S}S_{kk}^2\geq 0 \end{aligned} \tag{3.2.6}$$

For further convenience we transform eqns (3.2.1–4) to a non-dimensional form. Thus, let x_0, ρ_0, μ_0, and θ_0 denote, respectively, the units of length, density, stress, and temperature. Next, introduce the notations

$$\begin{array}{ll} \overset{\triangle}{t}_0=x_0\rho_0^{\frac{1}{2}}\mu_0^{-\frac{1}{2}}, & \overset{\triangle}{q}_0=\mu_0x_0\overset{\triangle}{t}_0^{-1}, \\ \overset{\triangle}{b}_0=\mu_0x_0^{-1}, & \overset{\triangle}{r}_0=\mu_0\overset{\triangle}{t}_0^{-1}, \\ \overset{\triangle}{C}_{S0}=\mu_0\theta_0^{-1}, & \overset{\triangle}{k}_0=x_0^2\overset{\triangle}{C}_{S0}\overset{\triangle}{t}_0^{-1}. \end{array} \tag{3.2.7}$$

[7] The inequality (3.2.6) can also be written in the form $K'_{ijkl}S_{ij}S_{kl}\geq 0\ \forall S_{ij}$ on $\bar{\mathrm{B}}\times[0,\infty)$, that is equivalent to the inequality (3.1.19).

Clearly, the parameters $\overset{\triangle}{t}_0$, $\overset{\triangle}{q}_0$, $\overset{\triangle}{b}_0$, $\overset{\triangle}{r}_0$, $\overset{\triangle}{C}_{S0}$, and $\overset{\triangle}{k}_0$ have, respectively, the dimensions of time, heat flux, body force, heat source, specific heat, and thermal conductivity. Assuming those parameters as reference units for respective quantities appearing in eqns (3.2.1–4), and keeping the same notations as in eqns (3.2.1–4), we pass to the following non-dimensional form of that problem[8]

$$\rho^{-1} S_{(ik,kj)} - \left[\frac{1}{2\mu} \left(\ddot{S}_{ij} - \frac{\lambda}{3\lambda + 2\mu} \ddot{S}_{kk} \delta_{ij} \right) - \frac{\alpha^2}{C_S} \ddot{S}_{kk} \delta_{ij} \right] + \frac{\alpha}{C_S} \dot{q}_{k,k}\, \delta_{ij} = -F_{(ij)}$$

$$\frac{1}{C_S} (q_{k,k} + \alpha \dot{S}_{kk})_{,i} - \frac{1}{k} (\dot{q}_i + t_0 \ddot{q}_i) = -g_i \quad \text{on B} \times [0, \infty), \tag{3.2.8}$$

$$\left.\begin{aligned} S_{ij}(\cdot, 0) &= S_{ij}^{(0)}, \; \dot{S}_{ij}(\cdot, 0) = \dot{S}_{ij}^{(0)} \\ q_i(\cdot, 0) &= q_i^{(0)}, \quad \dot{q}_i(\cdot, 0) = \dot{q}_i^{(0)} \end{aligned}\right. \quad \text{on B}, \tag{3.2.9}$$

$$S_{ij} n_j = 0, \; q_i n_i = 0 \text{ on } \partial\text{B} \times (0, \infty), \tag{3.2.10}$$

where

$$\begin{aligned} F_{(ij)} &= \rho^{-1} b_{(i,j)} - C_S^{-1} \alpha \dot{r} \delta_{ij} \\ g_i &= -C_S^{-1} r_{,i} \qquad \text{on } \bar{\text{B}} \times [0, \infty). \\ C_S &= C_E + 3\,(3\lambda + 2\mu)\, \alpha^2 \end{aligned} \tag{3.2.11}$$

In the following, we restrict a domain of the pair (S_{ij}, q_i) to the Cartesian product $\bar{\text{B}} \times [0, T]$, where T is a non-dimensional finite time. Let $L_2(t)$, $t \in [0, T]$, denote a linear space of the dimensionless pairs $p = \left(U_{(ij)}, v_i\right)$ defined over $\bar{\text{B}} \times [0, T]$, whose norm is[9]

$$\|p\|_{L_2(t)} = \left[\int_{\text{B}} (U_{ij} U_{ij} + v_i v_i)\, \mathrm{d}v \right]^{\frac{1}{2}} \forall t \in [0, T]. \tag{3.2.12}$$

Let $E_2(t)$ denote a linear space of the dimensionless pairs $p = \left(U_{(ij)}, v_i\right)$ defined over $\bar{\text{B}} \times [0, T]$, whose norm is

$$\begin{aligned} \|p\|_{E_2(t)} = \Bigg\{ \int_{\text{B}} \Big[& U_{ij} U_{ij} + v_i v_i + \rho^{-1} U_{ik,k}\, U_{ij,j} \\ & + \frac{1}{2\mu} \left(\dot{U}_{ij} - \frac{1}{3} \dot{U}_{kk} \delta_{ij} \right) \left(\dot{U}_{ij} - \frac{1}{3} \dot{U}_{kk} \delta_{ij} \right) + \frac{1}{3(3\lambda + 2\mu)} \frac{C_E}{C_S} \dot{U}_{kk}^2 \quad \forall t \in [0, T]. \\ & + \frac{1}{C_S} (v_{k,k})^2 + \frac{t_0}{k} \dot{v}_i \dot{v}_i + \frac{2}{k} \int_0^t \dot{v}_i \dot{v}_i \mathrm{d}\tau \Big] \mathrm{d}v \Bigg\}^{1/2} \end{aligned} \tag{3.2.13}$$

[8] Equations (3.2.8–11) are obtained formally by letting $\theta_0 = 1$ in eqns (3.2.1–4). The coefficient α in eqns (3.2.8–11) is equal to $\alpha\theta_0$.

[9] In the following we write $p = (U_{ij}, v_i)$ since $U_{ij} = U_{(ij)}$.

Thus, $L_2(t)$ and $E_2(t)$ are Banach spaces parameterized by $t \in [0,T]$, whereby the time differentiation and integration in eqn (3.2.13) is understood in a classical sense, whereas the spatial differentiation is understood in a generalized sense. One can show that if $\mathcal{E}(t)$ denotes the total thermoelastic energy of B at time t associated with eqns (3.2.8–11) and $p = (S_{ij}, q_i)$, then

$$\mathcal{E}(t) = \frac{1}{2}\|p\|^2_{E_2(t)} - \frac{1}{2}\|p\|^2_{L_2(t)}. \tag{3.2.14}$$

We now introduce the notations

$$\begin{aligned} p &= (S_{ij}, q_i), & p^{(n)} &= (S^{(n)}_{ij}, q^{(n)}_i),\\ p_1 &= (F_{(ij)}, g_i), & p^{(n)}_1 &= (F^{(n)}_{(ij)}, g^{(n)}_i),\\ p_0 &= (S^{(0)}_{ij}, q^{(0)}_i), & p^{(n)}_0 &= (S^{(0)(n)}_{ij}, q^{(0)(n)}_i),\\ \dot{p}_0 &= (\dot{S}^{(0)}_{ij}, \dot{q}^{(0)}_i), & \dot{p}^{(n)}_0 &= (\dot{S}^{(0)(n)}_{ij}, \dot{q}^{(0)(n)}_i), \end{aligned} \tag{3.2.15}$$

where the pairs p, p_1, p_0 and $\dot{p}_0$ are constructed from the tensor and vector fields appearing in the problem (3.2.8–11), while the pairs $p^{(n)}$, $p^{(n)}_1$, $p^{(n)}_0$ and $\dot{p}^{(n)}_0$ are constructed from the analogous tensor and vector fields for $n = 1, 2, 3, \ldots$ In the following, we identify a thermoelastic process associated with the problem (3.2.8–11) with a pair (S_{ij}, q_i).

Definition 3.1 *A thermoelastic process p corresponding to the thermomechanical loading $(p_1, p_0, \dot{p}_0)$ and satisfying eqns (3.2.8–3.2.11) pointwise will be called a classical one.*

Definition 3.2 *A thermoelastic process p corresponding to the thermoelastic loading $(p_1, p_0, \dot{p}_0)$ will be called a generalized one provided it is a limit in $E_2(t)$ of a sequence of classical processes $p^{(n)}$ $(n = 1, 2, 3, \ldots)$ corresponding to the thermomechanical loading $(p^{(n)}_1, p^{(n)}_0, \dot{p}^{(n)}_0)$ such that*

$$\lim_{n\to\infty} \|p^{(n)}_1 - p_1\|_{L_2(t)} = 0 \quad \forall t \in [0,T], \tag{3.2.16}$$

$$\lim_{n\to\infty} \|p^{(n)} - p\|_{E_2(0)} = 0. \tag{3.2.17}$$

Clearly, the generalized thermoelastic process corresponds to a situation in which the thermomechanical loading is generally discontinuous (non-differentiable in a classical sense) on certain surfaces within the domain $\overline{\mathrm{B}} \times [0,T]$.

Theorem 3.5 *(On a continuous dependence of a classical thermoelastic process on the thermomechanical loading)*[10] *If $p^{(1)}$ is a classical thermoelastic process corresponding to a thermomechanical loading $(p^{(1)}_1, p^{(1)}_0, \dot{p}^{(1)}_0)$ and $p^{(2)}$ is another*

[10] See (Bem, 1982).

classical thermoelastic process corresponding to a thermomechanical loading $(p_1^{(2)}, p_0^{(2)}, \dot{p}_0^{(2)})$*, and if*

$$\|p^{(1)} - p^{(2)}\|_{E_2(0)} \leq \epsilon_1, \tag{3.2.18}$$

$$\|p_1^{(1)} - p_1^{(2)}\|_{L_2(t)} \leq \epsilon_2 \qquad \forall t \in [0, T], \tag{3.2.19}$$

where ϵ_1 *and* ϵ_2 *are small positive numbers, then*

$$\|p^{(1)} - p^{(2)}\|_{E_2(t)} \leq \epsilon \qquad \forall t \in [0, T], \tag{3.2.20}$$

where ϵ *is a small positive number.*

Proof. Let $p = (S_{ij}, q_i)$ be a classical thermoelastic process corresponding to the thermomechanical loading $(p_1, p_0, \dot{p}_0)$. On the basis of eqn (1.2.31), specialized to eqns (3.2.8–11), we obtain

$$\frac{\mathrm{d}}{\mathrm{d}t}\mathcal{E}(t) = \int_{\mathrm{B}} (F_{ij}\dot{S}_{ij} + g_i\dot{q}_i)\mathrm{d}v, \tag{3.2.21}$$

where $\mathcal{E}(t)$ is the total thermoelastic energy associated with p at time t. On the other hand, differentiating eqn (3.2.14) with respect to time, we obtain

$$\frac{\mathrm{d}}{\mathrm{d}t}\mathcal{E}(t) = \frac{1}{2}\frac{\mathrm{d}}{\mathrm{d}t}\|p\|^2_{E_2(t)} - \frac{1}{2}\frac{\mathrm{d}}{\mathrm{d}t}\|p\|^2_{L_2(t)}. \tag{3.2.22}$$

Using the formula (3.2.12) with $p = (S_{ij}, q_i)$, on the basis of eqns (3.2.21) and (3.2.22), we obtain

$$\frac{\mathrm{d}}{\mathrm{d}t}\|p\|^2_{E_2(t)} = 2\int_{\mathrm{B}} (F_{ij}\dot{S}_{ij} + g_i\dot{q}_i + S_{ij}\dot{S}_{ij} + q_i\dot{q}_i)\mathrm{d}v. \tag{3.2.23}$$

From this, with the inequalities

$$\begin{aligned} 2U_{ij}V_{ij} &\leq U_{ij}U_{ij} + V_{ij}V_{ij} \quad \forall\, U_{ij}, V_{ij}, \\ 2p_iq_i &\leq p_ip_i + q_iq_i \qquad\quad \forall\, p_i, q_i, \end{aligned} \tag{3.2.24}$$

we arrive at the estimate

$$\frac{\mathrm{d}}{\mathrm{d}t}\|p\|^2_{E_2(t)} \leq \|p\|^2_{L_2(t)} + 2\|\dot{p}\|^2_{L_2(t)} + \|p_1\|^2_{L_2(t)}, \tag{3.2.25}$$

where

$$\dot{p} = (\dot{S}_{ij}, \dot{q}_i). \tag{3.2.26}$$

Since

$$\dot{S}_{ij}\dot{S}_{ij} \leq a\left[\frac{1}{2\mu}\left(\dot{S}_{ij} - \frac{1}{3}\dot{S}_{kk}\delta_{ij}\right)\left(\dot{S}_{ij} - \frac{1}{3}\dot{S}_{kk}\delta_{ij}\right) + \frac{1}{3(3\lambda + 2\mu)}\frac{C_E}{C_S}\dot{S}^2_{ij}\right], \tag{3.2.27}$$

where

$$a = \max\left\{2\mu, (3\lambda + 2\mu)\frac{C_E}{C_S}\right\}, \tag{3.2.28}$$

and

$$\int\limits_{B} \dot{q}_i \dot{q}_i \mathrm{d}v \leq \frac{k}{t_0} \|p\|^2_{E_2(t)}, \tag{3.2.29}$$

therefore, in view of eqn (3.2.25), we get

$$\frac{\mathrm{d}}{\mathrm{d}t}\|p\|^2_{E_2(t)} - m\|p\|^2_{E_2(t)} \leq \|p_1\|^2_{L_2(t)}, \tag{3.2.30}$$

where

$$m = 2\left(1 + a + kt_0^{-1}\right). \tag{3.2.31}$$

Integrating the inequality (3.2.30) over $(0, t)$, we obtain

$$\|p\|^2_{E_2(t)} \leq \mathrm{e}^{mt}\|p\|^2_{E_2(0)} + \int_0^t \mathrm{e}^{m(t-\tau)}\|p_1\|^2_{L_2(\tau)}\mathrm{d}\tau. \tag{3.2.32}$$

From the definition of processes $p^{(1)}$ and $p^{(2)}$ appearing in Theorem 3.5, it follows that $p^{(1)} - p^{(2)}$ is a process corresponding to the loading $(p_1^{(1)} - p_1^{(2)}, p_0^{(1)} - p_0^{(2)}, \dot{p}_0^{(1)} - \dot{p}_0^{(2)})$. Thus, setting $p = p_1^{(1)} - p_1^{(2)}$ in eqn (3.2.32), we find

$$\|p^{(1)} - p^{(2)}\|^2_{E_2(t)} \leq \mathrm{e}^{mt}\|p^{(1)} - p^{(2)}\|^2_{E_2(0)} + \int_0^t \mathrm{e}^{m(t-\tau)}\|p_1^{(1)} - p_1^{(2)}\|^2_{L_2(\tau)}\mathrm{d}\tau. \tag{3.2.33}$$

From this, in view of eqns (3.2.18 and 3.2.19), we get

$$\|p^{(1)} - p^{(2)}\|^2_{E_2(t)} \leq \epsilon \quad \forall t \in [0, T], \tag{3.2.34}$$

where

$$\epsilon = \left[\left(\epsilon_1^2 + T\epsilon_2^2\right)\exp\left(mT\right)\right]^{\frac{1}{2}}. \tag{3.2.35}$$

□

Theorem 3.6 *(On the existence of a generalized solution to problem (3.2.8–11))*[11] *The problem (3.2.8–11) possesses a generalized solution in the space* $E_2(t)$ *provided that*

$$\|p_1\|_{L_2(t)} < \infty, \tag{3.2.36}$$

and

$$\|p\|_{E_2(0)} < \infty. \tag{3.2.37}$$

[11] See (Bem, 1982).

Proof. Given that $E_2(t)$ is a Banach space $\forall t \in [0, T]$, it suffices to show that the sequence $p^{(n)}$ appearing in the Definition 3.2 is a fundamental sequence in $E_2(t)$. To this end let $p^{(n)}$ be a sequence from the Defnition 3.2. Then it follows that $p^{(n)} - p^{(k)}$ is a classical solution of the problem (3.2.8–11) corresponding to the loading $\left(p_1^{(n)} - p_1^{(k)}, p_0^{(n)} - p_0^{(k)}, \dot{p}_0^{(n)} - \dot{p}_0^{(k)}\right)$ for $n, k = 1, 2, 3 \ldots$ Furthermore, in view of eqn (3.2.36)

$$\left\| p_1^{(n)} - p^{(k)} \right\|_{L_2(t)} = \left\| (p_1^{(n)} - p_1) + (p_1 - p_1^{(k)}) \right\|_{L_2(t)}. \tag{3.2.38}$$

From this, and on the basis of the triangle inequality in $L_2(t)$, and using eqn (3.2.16), we obtain

$$\left\| p_1^{(n)} - p_1^{(k)} \right\|_{L_2(t)} \leq \hat{\epsilon}_1 \quad \forall t \in [0, T], \tag{3.2.39}$$

where $\hat{\epsilon}_1$ is a small positive number. Also, in view of eqn (3.2.37),

$$\left\| p^{(n)} - p^{(k)} \right\|_{E_2(0)} = \left\| \left(p^{(n)} - p\right) + \left(p - p^{(k)}\right) \right\|_{E_2(0)}. \tag{3.2.40}$$

The latter relation, the triangle inequality, and the condition (3.2.17) jointly imply that

$$\left\| p^{(n)} - p^{(k)} \right\|_{E_2(0)} \leq \hat{\epsilon}_2, \tag{3.2.41}$$

where $\hat{\epsilon}_2$ is a small positive number. As a result, the hypothesis of Theorem 3.5 in which $p^{(1)} = p^{(n)}$ and $p^{(2)} = p^{(k)}$ are satisfied, and on account of eqn (3.2.20), we obtain

$$\left\| p^{(n)} - p^{(k)} \right\|_{E_2(t)} \leq \hat{\epsilon} \quad \forall t \in [0, T], \tag{3.2.42}$$

where $\hat{\epsilon}$ is a small positive number. The inequality (3.2.42) means that $p^{(n)}$ is a fundamental sequence in $E_2(t)$[12]. □

Clearly, Theorem 3.6 concerns the existence of a generalized solution to eqns (3.2.8–11), which is defined as a limit of a sequence of classical solutions to this problem in a functional space. Analogous existence theorems in the G–L theory are given in (Bem, 1982, 1983).

Remark 3.1 The uniqueness results of Section 3.1 are based on the hypothesis that there is a classical solution to an initial-boundary value problem. Also, the existence theorem of Section 3.2, related to a generalized solution of an initial-boundary value problem, is proved under the hypothesis that there is a classical solution to the problem. In this sense, Sections 3.1 and 3.2 should be viewed as introductory to a vast literature on the existence and uniqueness in the L–S

[12] The proof of Theorem 3.6 is based on the Theorem 3.5 in which $m < \infty$, that is, for $k > 0$, $a > 0$, and $t_0 > 0$. If $t_0 \to 0 + 0$, then $m \to \infty$ and Theorem 3.6 is no longer true, i.e. Theorem 3.6 does not imply an existence theorem of classical thermoelasticity when $t_0 \to 0 + 0$.

and G–L theories, and in their extensions, in which the refined results have been obtained. See, for example, the existence and uniqueness results obtained in (Gawinecki, 1987; Burchuladze, 1997; Bem, 1988; De Cicco and Diaco, 2002; Karakostas and Massalas, 1991; Chirita, 1988; Sherief, 1987; Quintanilla and Straughan, 2000; Wang and Dhaliwal, 1993; Chandrasekharaiah, 1996a; Ieşan, 2004; Ezzat and El-Karamany, 2002).

4

DOMAIN OF INFLUENCE THEOREMS

In what follows, a solution to an initial-boundary value problem associated with a thermoelastic process will also be called a *thermoelastic disturbance.* In particular, a solution to NDTP of the L–S (or G–L) theory is to be called a *displacement–temperature disturbance* of the L–S (or G–L) theory, and a solution to NSHFP of the L–S theory is to be called a *stress–heat-flux disturbance* of this theory. In this chapter we shall formulate a number of theorems that imply that the thermoelastic disturbances described by the L–S and G–L theories have a character of waves propagating in B with finite wave speeds. Such theorems are called the *domain of influence theorems.*[1]

The first part of this chapter concerns a domain of influence theorem for a *potential–temperature disturbance*, which is a particular form of a displacement–temperature disturbance in the L–S theory. The second part of this chapter concerns an analogous theorem for the G–L theory. In the third part we formulate a domain of influence theorem for a *natural stress–heat flux disturbance in the L–S theory.* In the fourth part we present a domain of influence theorem for a *natural stress–temperature disturbance in the G–L theory.* Parts 1–4 are restricted to the setting of a homogeneous and isotropic thermoelastic body. Finally, in the fifth part we formulate a number of domain of influence theorems for a non-homogeneous anisotropic thermoelastic solid in the L–S and G–L theories.

4.1 The potential–temperature problem in the Lord–Shulman theory

That problem represents a restriction of a NDTP of the L–S theory in which the displacement is taken as a gradient of a scalar field. Let us recall the formulation of this problem for a homogeneous and isotropic medium.[2] Find a pair (u_i, ϑ) satisfying the field equations

$$\begin{aligned} \mu u_{i,kk} + (\lambda+\mu)u_{k,ki} - \rho\ddot{u}_i - (3\lambda+2\mu)\alpha\vartheta_{,i} &= -b_i \\ k\vartheta_{,ii} - C_E(\dot{\vartheta} + t_0\ddot{\vartheta}) & \\ -(3\lambda+2\mu)\alpha\theta_0(\dot{u}_{k,k} + t_0\ddot{u}_{k,k}) &= -(r + t_0\dot{r}) \end{aligned} \quad \text{on } B\times[0,\infty), \quad (4.1.1)$$

[1] The domain of influence theorems motivates the concept of "thermoelasticity with finite wave speeds" appearing in the title of the book.

[2] See eqns (2.1.3–5) reduced to a homogeneous isotropic thermoelastic solid for which $\partial B_2 = \partial B_4 = \varnothing$.

the initial conditions

$$\begin{matrix} u_i(\cdot,0) = u_{i0}, & \dot{u}_i(\cdot,0) = \dot{u}_{i0} \\ \vartheta(\cdot,0) = \vartheta_0, & \dot{\vartheta}(\cdot,0) = \dot{\vartheta}_0 \end{matrix} \quad \text{on B}, \tag{4.1.2}$$

and the boundary conditions

$$u_i = u_i', \ \vartheta = \vartheta' \text{ on } \partial \mathrm{B} \times (0, \infty). \tag{4.1.3}$$

Now, if we take $b_i = 0$ on $\overline{\mathrm{B}} \times [0, \infty)$ and let

$$u_i = \phi_{,i} \text{ on } \overline{\mathrm{B}} \times [0, \infty), \tag{4.1.4}$$

where ϕ is a scalar field defined on $\overline{\mathrm{B}} \times [0, \infty)$, then the system (4.1.1) is satisfied as long as the pair (ϕ, ϑ) satisfies the system of equations

$$\begin{aligned} \nabla^2 \phi - \frac{\rho}{\lambda + 2\mu}\ddot{\phi} - \frac{3\lambda + 2\mu}{\lambda + 2\mu}\alpha\vartheta = 0 \\ \nabla^2 \vartheta - \frac{C_E}{k}(\dot{\vartheta} + t_0\ddot{\vartheta}) \qquad \text{on B} \times [0, \infty). \\ -\frac{3\lambda + 2\mu}{k}\alpha\theta_0 \nabla^2(\dot{\phi} + t_0\ddot{\phi}) = -\frac{1}{k}(r + t_0\dot{r}) \end{aligned} \tag{4.1.5}$$

At this point, let us introduce the notations[3]

$$\hat{x}_0 = \frac{k}{C_E C_1}, \quad \hat{t}_0 = \frac{k}{C_E C_1^2}, \tag{4.1.6}$$

where

$$\frac{1}{C_1^2} = \frac{\rho}{\lambda + 2\mu},$$

and

$$\hat{\phi}_0 = \frac{(3\lambda + 2\mu)\,\alpha\theta_0}{\lambda + 2\mu}\hat{x}_0^2, \quad \hat{\vartheta}_0 = \theta_0, \quad \hat{r}_0 = \frac{k\theta_0}{\hat{x}_0^2}. \tag{4.1.7}$$

Clearly, $\hat{x}_0$ and $\hat{t}_0$ have the dimensions of length and time, respectively, while $\hat{\phi}_0$, $\hat{\vartheta}_0$ and $\hat{r}_0$ have the dimensions of potential, temperature and heat source. Taking these parameters as units of reference for respective quantities appearing in eqns (4.1.5) and keeping there the same notations for dimensionless quantities, we pass to the dimensionless form of eqns (4.1.5)

$$\begin{aligned} \nabla^2 \phi - \ddot{\phi} - \vartheta = 0 \\ \nabla^2 \vartheta - L\dot{\vartheta} - \epsilon \nabla^2 L\dot{\phi} = -(r + t_0\dot{r}) \end{aligned} \quad \text{on B} \times [0, \infty), \tag{4.1.8}$$

[3] The huts over the symbols x_0 and t_0 in eqn (4.1.6) must not be confused with the operator $\widehat{\ }$ in eqn (1.2.16).

where

$$L = 1 + t_0 \frac{\partial}{\partial t}, \tag{4.1.9}$$

$$\epsilon = \frac{(3\lambda + 2\mu)^2 \alpha^2 \theta_0}{(\lambda + 2\mu) C_E}. \tag{4.1.10}$$

We now formulate the *potential–temperature problem (PTP) in the L–S theory*, with no heat sources ($r = 0$) and body forces ($b_i = 0$). Find a pair (ϕ, ϑ) satisfying the field equations

$$\begin{aligned} \nabla^2 \phi - \ddot{\phi} - \vartheta = 0 \\ \nabla^2 \vartheta - L\dot{\vartheta} - \epsilon \nabla^2 L\dot{\phi} = 0 \end{aligned} \quad \text{on} \;\; \mathrm{B} \times [0, \infty), \tag{4.1.11}$$

the initial conditions

$$\begin{aligned} \phi(\cdot, 0) = \phi_0, \quad \dot{\phi}(\cdot, 0) = \dot{\phi}_0 \\ \vartheta(\cdot, 0) = \vartheta_0, \quad \dot{\vartheta}(\cdot, 0) = \dot{\vartheta}_0 \end{aligned} \quad \text{on} \;\; \mathrm{B}, \tag{4.1.12}$$

and the boundary conditions

$$\phi_{,k} n_k = f, \; \vartheta = g \text{ on } \; \partial\mathrm{B} \times (0, \infty). \tag{4.1.13}$$

The pair (ϕ, ϑ) describes a thermoelastic process caused by a thermomechanical loading $(\phi_0, \dot{\phi}_0, \vartheta_0, \dot{\vartheta}_0, f, g)$. It is evident that the initial conditions (4.1.12) are consistent with the conditions (4.1.2) and the hypothesis (4.1.4) for $u_{i0} = \phi_{0,i}$ and $\dot{u}_{i0} = \dot{\phi}_{0,i}$. Moreover, the boundary conditions (4.1.3) and (4.1.13) are identical if eqn $(4.1.3)_1$ is replaced by the conditions for the normal component of the displacement vector, and $g = \vartheta'$.

The system of equations (4.1.11) is called a *central system of equations* in the L–S theory, while the problem described by eqns (4.1.11–13) is called a *central problem* of that theory.[4] The role of the central problem of the L–S theory in solving a general problem of the theory is similar to that of an initial-boundary value problem for the classical wave equation in solving a general problem of linear isothermal elastodynamics.

Definition 4.1 *Let $t \in (0, \infty)$ be a fixed time. The set*

$$\begin{aligned} \mathcal{D}_0(t) = \{x \in \bar{B} : &\text{(1) if } x \in B, \text{ then } \phi_0 \neq 0 \text{ or } \dot{\phi}_0 \neq 0 \text{ or } \vartheta_0 \neq 0 \text{ or } \dot{\vartheta}_0 \neq 0, \\ &\text{(2) if } (x, \tau) \in \partial B \times [0, t], \text{ then } f(x, \tau) \neq 0 \text{ or } g(x, \tau) \neq 0\} \end{aligned} \tag{4.1.14}$$

is called the support of a thermomechanical load of PTP in the L–S theory.

[4] Another central problem of the L–S theory is obtained if the boundary conditions (4.1.13) are replaced by the conditions $\phi = f$ and $\vartheta = g$ on $\partial\mathrm{B} \times [0, \infty)$.

This definition is a generalization of the concept of the support of a function.[5] It is apparent that, if the domain B fills the entire E^3 space, then in the central problem (4.1.11–13) there are no functions f and g. In such a case the PTP of the L–S theory is a Cauchy problem for an unbounded space and the set $\mathcal{D}_0(t)$ does not depend on time.

Definition 4.2 *Let $\upsilon > 0$ be a number satisfying the inequality*

$$\upsilon \geq \max\left(2, 1+\epsilon, t_0^{-1}\right), \tag{4.1.15}$$

and let $\Sigma_{\upsilon t}(x)$ be an open ball of radius υt, centered at x. The domain of influence for the thermomechanical load at time t for the PTP (4.1.11–13) is the set

$$\mathcal{D}(t) = \{x \in \bar{B} : \mathcal{D}_0(t) \cap \overline{\Sigma_{\upsilon t}(x)} \neq \varnothing\}. \tag{4.1.16}$$

Clearly, $\mathcal{D}(t)$ is a set of all the points of $\overline{\mathrm{B}}$ that may be reached by the thermomechanical disturbances propagating from $\mathcal{D}_0(t)$ with a speed not greater than υ.

We shall now formulate a theorem stating that the thermomechanical load restricted to the interval $[0,t]$ does not influence the points outside the domain $\mathcal{D}(t)$.

Theorem 4.1 *(On the domain of influence for a PTP in the L–S theory[6]) If the pair (ϕ, ϑ) is a smooth solution of PTP (4.1.11–13) and if $\mathcal{D}(t)$ is the domain of influence for the thermomechanical load at time t, then*

$$\phi = \vartheta = 0 \quad on \quad \{\bar{B} - \mathcal{D}(t)\} \times [0,t]. \tag{4.1.17}$$

The proof of this theorem is based on the following lemma:

Lemma 4.1 *Let (ϕ, ϑ) be a solution to eqns (4.1.11–13), and let $p \in C^1(\bar{B})$ denote a scalar field such that the set*

$$E_0 = \{x \in \bar{B} : p(x) > 0\} \tag{4.1.18}$$

is bounded. Then

$$\begin{aligned} &\frac{1}{2}\int_B \left[P(x,p(x)) - P(x,0)\right] \mathrm{d}\upsilon + \int_B \int_0^{p(x)} Q(x,t)\mathrm{d}t\mathrm{d}\upsilon \\ &+ \int_B R_i(x,p(x))\, p_{,i}(x)\mathrm{d}\upsilon = \int_{\partial B} \int_0^{p(x)} R_i(x,t)n_i(x)\mathrm{d}t\mathrm{d}a, \end{aligned} \tag{4.1.19}$$

[5] The support of a function $f(x)$ is a set of points on which f does not vanish, i.e. supp $f = \{x : f(x) \neq 0\}$.

[6] See (Ignaczak and Biały, 1980). Theorem 4.1 generalizes a domain of influence result for a classical scalar wave equation obtained by Zaremba (1915). An analogous extension of Zaremba's result to include a displacement problem of classical isothermal elastodynamics is due to Gurtin (1972). See also (Eringen and Şuhubi, 1975).

where

$$P(x,t) = \epsilon(\nabla^2\hat{\phi})^2 + \epsilon(\dot{\hat{\phi}},_i)^2 + \hat{\vartheta}^2 + t_0(\vartheta,_i)^2, \tag{4.1.20}$$

$$Q(x,t) = (\vartheta,_i)^2, \tag{4.1.21}$$

$$R_i(x,t) = \epsilon(\nabla^2\hat{\phi})\,\dot{\hat{\phi}},_i + \hat{\vartheta}\,(\vartheta,_i - \epsilon\dot{\hat{\phi}},_i)\ \forall (x,t) \in \bar{B}\times[0,\infty), \tag{4.1.22}$$

and the hut denotes the operator L[7], *that is*

$$\hat{\phi} = L\phi, \quad \hat{\vartheta} = L\vartheta \ on\ \bar{B}\times[0,\infty). \tag{4.1.23}$$

Proof of Lemma 4.1 Applying the operator L to both sides of eqn $(4.1.11)_1$, and taking the gradient, we infer that the pair (ϕ,ϑ) satisfies the equations

$$\left.\begin{aligned} &(\nabla^2\hat{\phi}),_i - \ddot{\hat{\phi}},_i - \hat{\vartheta},_i = 0 \\ &\nabla^2\vartheta - \dot{\hat{\vartheta}} - \epsilon\nabla^2\dot{\hat{\phi}} = 0 \end{aligned}\right\} \text{ on } \bar{\mathrm{B}}\times[0,\infty). \tag{4.1.24}$$

Multiplying eqn $(4.1.24)_1$ through by $\epsilon\dot{\hat{\phi}},_i$ and using the identities

$$\dot{\hat{\phi}},_i(\nabla^2\hat{\phi}),_i = (\dot{\hat{\phi}},_i\nabla^2\hat{\phi}),_i - \frac{1}{2}\frac{\partial}{\partial t}(\nabla^2\hat{\phi})^2, \tag{4.1.25}$$

$$\dot{\hat{\phi}},_i\ddot{\hat{\phi}},_i = \frac{1}{2}\frac{\partial}{\partial t}(\dot{\hat{\phi}},_i)^2, \tag{4.1.26}$$

we get

$$\frac{\epsilon}{2}\frac{\partial}{\partial t}\{(\nabla^2\hat{\phi})^2 + (\dot{\hat{\phi}},_i)^2\} + \epsilon\dot{\hat{\phi}},_i\hat{\vartheta},_i = \epsilon(\dot{\hat{\phi}},_i\nabla^2\hat{\phi}),_i\,. \tag{4.1.27}$$

Next, multiplying eqn $(4.1.24)_2$ through by $\hat{\vartheta}$ and using the relations

$$\hat{\vartheta}\nabla^2\vartheta = (\hat{\vartheta}\,\vartheta,_k),_k - \hat{\vartheta},_k\vartheta,_k\,, \tag{4.1.28}$$

$$\hat{\vartheta}\nabla^2\dot{\hat{\phi}} = (\hat{\vartheta}\,\dot{\hat{\phi}},_k),_k - \hat{\vartheta},_k\dot{\hat{\phi}},_k\,, \tag{4.1.29}$$

we arrive at

$$\frac{1}{2}\frac{\partial}{\partial t}\hat{\vartheta}^2 + \hat{\vartheta},_k\vartheta,_k - \epsilon\dot{\hat{\phi}},_k\hat{\vartheta},_k = [\hat{\vartheta}(\vartheta,_k - \epsilon\dot{\hat{\phi}},_k)],_k\,. \tag{4.1.30}$$

Now, adding eqns (4.1.27) and (4.1.30), and recalling the definition of the hut operator, we obtain

$$\frac{1}{2}\frac{\partial}{\partial t}P(x,t) + Q(x,t) = R_{i,i}(x,t), \tag{4.1.31}$$

[7] See eqn (1.2.16).

where P, Q and R_i are defined by eqns (4.1.20), (4.1.21) and (4.1.22), respectively. Since

$$\int_0^{p(x)} R_{i,i}(x,t)\,\mathrm{d}t = \left[\int_0^{p(x)} R_i(x,t)\mathrm{d}t\right]_{,i} - R_i(x,p(x))p_{,i}(x)\,, \qquad (4.1.32)$$

the integration of eqn (4.1.31) from $t=0$ up to $t=p(x)$, will result in

$$\frac{1}{2}\left[P(x,p(x)) - P(x,0)\right] + \int_0^{p(x)} Q(x,t)dt$$

$$+ R_i(x,p(x))\,p_{,i}(x) = \left[\int_0^{p(x)} R_i(x,t)\mathrm{d}t\right]_{,i}. \qquad (4.1.33)$$

Since the set E_0 defined by eqn (4.1.18) is bounded, each term in eqn (4.1.33) has a bounded support. Therefore, integrating eqn (4.1.33) over B and using the divergence theorem, we obtain eqn (4.1.19). □

Remark 4.1 The relation (4.1.19) is called a *generalized energy identity* for the problem (4.1.11–13). Setting $p(x)=t$, we obtain the classical energy identity for that problem.[8]

Proof of Theorem 4.1 Let $(z,\lambda) \in \{\mathrm{B} - \mathcal{D}(t)\} \times (0,t)$ be a fixed point. Let

$$\Omega = \overline{\mathrm{B}} \cap \overline{\Sigma_{\upsilon\lambda}(z)} \qquad (4.1.34)$$

and let

$$p_\lambda(x) = \begin{cases} \lambda - \dfrac{1}{\upsilon}\mid x - z \mid & \text{for } x \in \Omega, \\ 0 & \text{for } x \notin \Omega, \end{cases} \qquad (4.1.35)$$

where υ is a parameter defined by the inequality (4.1.15). Since $\lambda < t$, it follows from the definitions of $\mathcal{D}(t)$ and Ω (recall eqns (4.1.16) and (4.1.34)) that the sets $\mathcal{D}_0(t)$ and Ω are disjoint:

$$\Omega \cap \mathcal{D}_0(t) = \varnothing. \qquad (4.1.36)$$

Hence,

$$\phi_{,i}\, n_i = 0, \quad \vartheta = 0 \text{ on } (\Omega \cap \partial\mathrm{B}) \times [0,t], \qquad (4.1.37)$$

and

$$\dot{\hat{\phi}}_{,k}\, n_k = 0, \quad \hat{\vartheta} = 0 \text{ on } (\Omega \cap \partial\mathrm{B}) \times [0,t]. \qquad (4.1.38)$$

[8] The classical energy identity for the problem described by eqns (4.1.11–13) implies that the problem has at most one solution.

Furthermore,

$$\phi(\cdot,0)=\dot{\phi}(\cdot,0)=\vartheta(\cdot,0)=\dot{\vartheta}(\cdot,0)=0 \text{ on } \Omega. \tag{4.1.39}$$

Thus, in view of eqns (4.1.35), (4.1.38), and (4.1.22), we obtain

$$\int_{\partial B}\int_0^{p_\lambda(x)} R_i(x,t)\,n_i(x)\,\mathrm{d}t\mathrm{d}a=0. \tag{4.1.40}$$

Also, eqns (4.1.39) and $(4.1.11)_1$ imply that

$$\nabla^2\hat{\phi}(\cdot,0)=\hat{\vartheta}(\cdot,0)=0 \text{ on } \Omega, \tag{4.1.41}$$

$$\dot{\hat{\phi}}_{,i}(\cdot,0)=\vartheta_{,i}(\cdot,0)=0 \text{ on } \Omega. \tag{4.1.42}$$

Thus, from the definitions of $P(x,t)$ and $p_\lambda(x)$ we get

$$P(x,p_\lambda(x))-P(x,0)=\begin{cases} P(x,p_\lambda(x)) & \text{for } x\in\Omega,\\ 0 & \text{for } x\notin\Omega. \end{cases} \tag{4.1.43}$$

Upon substitution of $p_\lambda(x)$ into eqn (4.1.19), and using eqns (4.1.40) and (4.1.43), we find

$$\frac{1}{2}\int_\Omega P(x,p_\lambda(x))\,\mathrm{d}v+\int_\Omega\int_0^{p_\lambda(x)} Q(x,t)\,\mathrm{d}t\mathrm{d}v=-\int_\Omega R_i(x,p_\lambda(x))\,p_{\lambda,i}(x)\,\mathrm{d}v. \tag{4.1.44}$$

Since $Q\geq 0$ on Ω, from eqns (4.1.35) and (4.1.44) we obtain

$$\frac{1}{2}\int_\Omega P(x,p_\lambda(x))\,\mathrm{d}v\leq\frac{1}{\upsilon}\int_\Omega |R_i(x,p_\lambda(x))|\,\mathrm{d}v. \tag{4.1.45}$$

From the definition of R_i (recall eqn (4.1.22)) we obtain

$$\begin{aligned} |R_i(x,p_\lambda(x))| &\leq \epsilon|\nabla^2\hat{\phi}||\dot{\hat{\phi}}_{,i}|+|\hat{\vartheta}|(|\vartheta_{,i}|+\epsilon|\dot{\hat{\phi}}_{,i}|)\\ &\leq \frac{\epsilon}{2}\left\{\left(\nabla^2\hat{\phi}\right)^2+2\left(\dot{\hat{\phi}}_{,i}\right)^2+\hat{\vartheta}^2\right\}+\frac{1}{2}\left\{\hat{\vartheta}^2+(\vartheta_{,i})^2\right\}. \end{aligned} \tag{4.1.46}$$

Therefore, from the definition of $P(x,t)$, and in view of the inequalities (4.1.45) and (4.1.46), we arrive at

$$\begin{aligned} &\int_\Omega\left\{\frac{\epsilon}{2}\left(1-\frac{1}{\upsilon}\right)\left(\nabla^2\hat{\phi}\right)^2+\frac{\epsilon}{2}\left(1-\frac{2}{\upsilon}\right)\left(\dot{\hat{\phi}}_{,i}\right)^2\right.\\ &\left.+\frac{1}{2}\left(1-\frac{1+\epsilon}{\upsilon}\right)\hat{\vartheta}^2+\frac{1}{2}\left(t_0-\frac{1}{\upsilon}\right)(\vartheta_{,i})^2\right\}\mathrm{d}v\leq 0. \end{aligned} \tag{4.1.47}$$

The definition of υ (recall eqn (4.1.15)) implies that the integrand of eqn (4.1.47) is a sum of non-negative terms. Thus, the inequality (4.1.47) implies that each of those terms must vanish in Ω. In particular, we have

$$\nabla^2 \hat{\phi}(x, p_\lambda(x)) = 0, \quad \hat{\vartheta}(x, p_\lambda(x)) = 0 \quad \text{on } \Omega. \tag{4.1.48}$$

In view of the definition of $p_\lambda(x)$ (recall eqn (4.1.35)), and since the pair (ϕ, ϑ) is sufficiently smooth, we also have

$$\begin{aligned} \nabla^2 \hat{\phi}(x, p_\lambda(x)) &\longrightarrow \nabla^2 \hat{\phi}(z, \lambda) \\ \hat{\vartheta}(x, p_\lambda(x)) &\longrightarrow \hat{\vartheta}(z, \lambda) \end{aligned} \quad \text{as } x \to z. \tag{4.1.49}$$

Hence, taking the limits in eqns (4.1.48) as $x \to z$, we find

$$\nabla^2 \hat{\phi}(z, \lambda) = \hat{\vartheta}(z, \lambda) = 0. \tag{4.1.50}$$

Since (z, λ) is an arbitrary point in $\{B - \mathcal{D}(t)\} \times (0, t)$, and since (ϕ, ϑ) is sufficiently smooth, hence

$$\nabla^2 \hat{\phi} = \hat{\vartheta} = 0 \quad \text{on } \{\bar{B} - \mathcal{D}(t)\} \times (0, t). \tag{4.1.51}$$

Thus, eqns (4.1.51) and (4.1.11)$_1$ imply that

$$\ddot{\hat{\phi}} = \hat{\vartheta} = 0 \quad \text{on } \{\bar{B} - \mathcal{D}(t)\} \times (0, t), \tag{4.1.52}$$

from which

$$\begin{aligned} \ddot{\phi}(x, \tau) &= \ddot{\phi}(\cdot, 0) \exp(-\tau t_0^{-1}) \\ \vartheta(x, \tau) &= \vartheta(\cdot, 0) \exp(-\tau t_0^{-1}) \end{aligned} \quad \text{for } (x, \tau) \in \{\bar{B} - \mathcal{D}(t)\} \times (0, t). \tag{4.1.53}$$

In view of the definition of domain $\mathcal{D}(t)$ and eqn (4.1.11)$_1$,

$$\ddot{\phi}(\cdot, 0) = \vartheta(\cdot, 0) = 0 \quad \text{on } \bar{B} - \mathcal{D}(t), \tag{4.1.54}$$

so that, from eqns (4.1.53) we obtain

$$\ddot{\phi} = \vartheta = 0 \quad \text{on } \{\bar{B} - \mathcal{D}(t)\} \times [0, t]. \tag{4.1.55}$$

Finally, recalling the definition of $\mathcal{D}(t)$ once again, and using eqn (4.1.55), we obtain

$$\phi = \vartheta = 0 \quad \text{on } \{\bar{B} - \mathcal{D}(t)\} \times [0, t]. \tag{4.1.56}$$

$\square$

This theorem implies that, for a finite time t and a bounded support of the thermomechanical loading (that is, for a bounded set $\mathcal{D}_0(t)$, recall eqn (4.1.14)), a thermoelastic disturbance generated by the pair (ϕ, ϑ) satisfying the system (4.1.11–13) vanishes outside the bounded set $\mathcal{D}(t)$, which depends on the support of the load, the material constants, and the relaxation time.

In other words, the said disturbance is propagated with a finite speed, bounded from above by the speed υ. If $t_0 \to 0$, then in view of the definition of υ, it follows that $\upsilon \to \infty$. Thus, for a vanishing relaxation time, the thermoelastic disturbance described by the pair (ϕ, ϑ) attains an infinite speed, a fact that may well have been expected since the PTP of the L–S theory reduces to a potential–temperature problem of classical thermoelasticity.

4.2 The potential–temperature problem in the Green–Lindsay theory

That problem is an analog of the PTP in the L–S theory (recall eqns (4.1.11–13)) and the related theorem on the domain of influence of Section 4.1. Hence, both the formulation and proof of the theorem are similar to what was given in Section 4.1.

For a homogeneous isotropic thermoelastic body, the NDTP in the G–L theory is formulated as follows:[9] Find a pair (u_i, ϑ) satisfying the field equations

$$\mu u_{i,kk} + (\lambda + \mu) u_{k,ki} - \rho \ddot{u}_i - (3\lambda + 2\mu) \alpha(\vartheta + t_1\dot{\vartheta})_{,i} = -b_i$$
$$k\vartheta_{,ii} - C_E(\dot{\vartheta} + t_0\ddot{\vartheta}) - (3\lambda + 2\mu) \alpha\theta_0 \dot{u}_{k,k} = -r$$

on $B \times 0, \infty)$,

(4.2.1)

the initial conditions

$$u_i(\cdot, 0) = u_{i0}, \quad \dot{u}_i(\cdot, 0) = \dot{u}_{i0}$$
$$\vartheta(\cdot, 0) = \vartheta_0, \quad \dot{\vartheta}(\cdot, 0) = \dot{\vartheta}_0$$

on B,

(4.2.2)

and the boundary conditions

$$u_i = u_i^{'}, \quad \vartheta = \vartheta^{'} \text{ on } \partial B \times (0, \infty).$$

(4.2.3)

Setting $b_i = 0$ on $\overline{B} \times 0, \infty)$, and

$$u_i = \phi_{,i} \text{ on } \overline{B} \times [0, \infty),$$

(4.2.4)

where ϕ is a potential on $\overline{B} \times [0, \infty)$, we conclude that eqns (4.2.1) are satisfied so long as the pair (ϕ, ϑ) satisfies the equations

$$\nabla^2 \phi - \frac{\rho}{\lambda + 2\mu}\ddot{\phi} - \frac{3\lambda + 2\mu}{\lambda + 2\mu}\alpha(\vartheta + t_1\dot{\vartheta}) = 0$$
$$\nabla^2 \vartheta - \frac{C_E}{k}(\dot{\vartheta} + t_0\ddot{\vartheta}) - \frac{3\lambda + 2\mu}{k}\alpha\theta_0\nabla^2\dot{\phi} = -\frac{r}{k}$$

on $B \times [0, \infty)$. (4.2.5)

Transforming eqns (4.2.5) into a dimensionless form, in a way similar as in Section 4.1, and keeping the same notations for dimensionless quantities, we obtain

$$\nabla^2 \phi - \ddot{\phi} - (\vartheta + t_1\dot{\vartheta}) = 0$$
$$\nabla^2 \vartheta - (\dot{\vartheta} + t_0\ddot{\vartheta}) - \epsilon\nabla^2\dot{\phi} = -r$$

on $B \times [0, \infty)$.

(4.2.6)

---

[9] See eqns (2.1.11)–(2.1.13) restricted to a homogeneous isotropic thermoelastic body with $\partial B_2 = \partial B_4 = \varnothing$.

The potential–temperature problem in the G–L theory with null body forces and null heat sources is now formulated as follows: Find a pair (ϕ, ϑ) satisfying the field equations

$$\left.\begin{aligned} \nabla^2\phi - \ddot{\phi} - (\vartheta + t_1\dot{\vartheta}) = 0 \\ \nabla^2\vartheta - (\dot{\vartheta} + t_0\ddot{\vartheta}) - \epsilon\nabla^2\dot{\phi} = 0 \end{aligned}\right\} \quad \text{on} \quad \mathrm{B} \times [0, \infty), \tag{4.2.7}$$

the initial conditions

$$\left.\begin{aligned} \phi(\cdot, 0) = \phi_0, \quad \dot{\phi}(\cdot, 0) = \dot{\phi}_0 \\ \vartheta(\cdot, 0) = \vartheta_0, \quad \dot{\vartheta}(\cdot, 0) = \dot{\vartheta}_0 \end{aligned}\right\} \quad \text{on B}, \tag{4.2.8}$$

and the boundary conditions

$$\phi_{,k}\, n_k = f,\ \vartheta = g \text{ on } \partial\mathrm{B} \times (0, \infty). \tag{4.2.9}$$

All the symbols here have the analogous meaning as in Section 4.1.

Relations (4.2.7) represent a central system of equations of the G–L theory, while a PTP described by the eqns (4.2.7–9) is a central problem of that theory. Clearly, the set $\mathcal{D}_0(t)$ of Section 4.1 is a support of the thermomechanical load for that problem (see eqn (4.1.14)), while the concept of a domain of influence for that problem is contained in the following definition.

Definition 4.3 *Let $\upsilon > 0$ be a number satisfying the inequality*

$$\upsilon \geq \max\left\{2, (1+\epsilon)\frac{t_1}{t_0}, \frac{1}{t_1}\right\}, \tag{4.2.10}$$

and let $\Sigma_{\upsilon t}(x)$ be a ball of radius υt and center at point x. The domain of influence of a thermomechanical loading at time t for the central problem (4.2.7–9) is the set

$$\mathcal{D}(t) = \left\{x \in \overline{B} : \mathcal{D}_0(t) \cap \overline{\Sigma_{\upsilon t}(x)} \neq \varnothing\right\}. \tag{4.2.11}$$

Here, $\mathcal{D}_0(t)$ is given by the formula (4.1.14). Using this definition, we shall now prove

Theorem 4.2 *(On the domain of influence for a PTP in the G–L theory)*[10] *If the pair (ϕ, ϑ) is a smooth solution of PTP (4.2.7–9) and if $\mathcal{D}(t)$ is the domain of influence for the thermomechanical load at time t, then*

$$\phi = \vartheta = 0 \quad on \quad \left\{\overline{B} - \mathcal{D}(t)\right\} \times [0, t]. \tag{4.2.12}$$

The proof of this theorem is analogous to the proof of Theorem 4.1 of Section 4.1, and is based on the following lemma:

[10] This theorem is a modification of that from (Ignaczak, 1978b).

Lemma 4.2 *Let (ϕ, ϑ) be a solution of (4.2.7–9) and let p be a scalar field of Lemma 4.1. Then, the following generalized energy identity holds true*

$$\begin{aligned} &\frac{1}{2}\int_B \left[P\left(x, p\left(x\right)\right) - P\left(x, 0\right)\right] \mathrm{d}v + \int_B \int_0^{p(x)} Q\left(x, t\right) \mathrm{d}t\mathrm{d}v \\ &+ \int_B R_i\left(x, p\left(x\right)\right) p_{,i}\left(x\right) \mathrm{d}v = \int_{\partial B} \int_0^{p(x)} R_i\left(x, t\right) n_i(x) \mathrm{d}t\mathrm{d}a, \end{aligned} \tag{4.2.13}$$

where

$$\begin{aligned} P\left(x, t\right) &= \epsilon(\nabla^2\phi)^2 + \epsilon(\dot{\phi}_{,i})^2 \\ +t_1\left(\vartheta_{,i}\right)^2 &+ \frac{t_0}{t_1}\left[(\vartheta + t_1\dot{\vartheta})^2 + \left(\frac{t_1}{t_0} - 1\right)\dot{\vartheta}^2\right] \quad \forall\left(x, t\right) \in B \times [0, \infty), \end{aligned} \tag{4.2.14}$$

$$Q\left(x, t\right) = \left(\vartheta_{,i}\right)^2 + \left(t_1 - t_0\right)\dot{\vartheta}^2 \quad \forall\left(x, t\right) \in B \times [0, \infty), \tag{4.2.15}$$

$$R_i\left(x, t\right) = (\vartheta + t_1\dot{\vartheta})\vartheta_{,i} + \epsilon[\nabla^2\phi - (\vartheta + t_1\dot{\vartheta})]\dot{\phi}_{,i} \ \forall\left(x, t\right) \in B \times [0, \infty). \tag{4.2.16}$$

Similar to what was done in Section 4.1, for convenience we have dropped (x, t) in the right-hand sides of eqns (4.2.14–16).

Proof. By assumption, the pair (ϕ, ϑ) satisfies the system (4.2.7). Taking the gradient on both sides of eqn $(4.2.7)_1$ and multiplying through by $\dot{\phi}_{,i}$, we obtain

$$\dot{\phi}_{,i}\nabla^2\phi_{,i} - \dot{\phi}_{,i}\ddot{\phi}_{,i} - \dot{\phi}_{,i}(\vartheta + t_1\dot{\vartheta})_{,i} = 0, \tag{4.2.17}$$

from which

$$\frac{1}{2}\frac{\partial}{\partial t}(\dot{\phi}_{,i})^2 + \frac{1}{2}\frac{\partial}{\partial t}(\nabla^2\phi)^2 + \dot{\phi}_{,i}(\vartheta + t_1\dot{\vartheta})_{,i} + (t_1 - t_0)\dot{\phi}_{,i}\dot{\vartheta}_{,i} = (\dot{\phi}_{,i}\nabla^2\phi)_{,i}\,. \tag{4.2.18}$$

On the other hand, multiplying eqn $(4.2.7)_2$ through by $(\vartheta + t_0\dot{\vartheta})$, we obtain

$$(\vartheta + t_0\dot{\vartheta})\left[(\vartheta - \epsilon\dot{\phi})_{,ii} - (\dot{\vartheta} + t_0\ddot{\vartheta})\right] = 0, \tag{4.2.19}$$

so that

$$\frac{1}{2}\frac{\partial}{\partial t}(\vartheta + t_0\dot{\vartheta})^2 + (\vartheta + t_0\dot{\vartheta})_{,i}(\vartheta - \epsilon\dot{\phi})_{,i} = [(\vartheta + t_0\dot{\vartheta})(\vartheta - \epsilon\dot{\phi})_{,i}]_{,i}\,, \tag{4.2.20}$$

or

$$\begin{aligned} &\frac{1}{2}\frac{\partial}{\partial t}(\vartheta + t_0\dot{\vartheta})^2 + \frac{t_0}{2}\frac{\partial}{\partial t}(\vartheta_{,i})^2 + (\vartheta_{,i})^2 \\ &-\epsilon\dot{\phi}_{,i}(\vartheta + t_0\dot{\vartheta})_{,i} = [(\vartheta + t_0\dot{\vartheta})(\vartheta - \epsilon\dot{\phi})_{,i}]_{,i}\,. \end{aligned} \tag{4.2.21}$$

Then, multiplying eqn $(4.2.7)_2$ through by $\dot{\vartheta}$, we get

$$\dot{\vartheta}[(\vartheta - \epsilon\dot{\phi}),_{ii} - (\dot{\vartheta} + t_0\ddot{\vartheta})] = 0, \tag{4.2.22}$$

so that

$$\frac{t_0}{2}\frac{\partial}{\partial t}\dot{\vartheta}^2 + \frac{1}{2}\frac{\partial}{\partial t}(\vartheta,_i)^2 + \dot{\vartheta}^2 - \epsilon\phi,_i\dot{\vartheta},_i = [\dot{\vartheta}(\vartheta - \epsilon\dot{\phi}),_i],_i\,. \tag{4.2.23}$$

Now, adding eqn (4.2.18) multiplied through by ϵ with eqns (4.2.21) and (4.2.23) multiplied through by $(t_1 - t_0)$, we obtain

$$\frac{1}{2}\frac{\partial}{\partial t}\{\epsilon[(\dot{\phi},_i)^2 + (\nabla^2\phi)^2] + (\vartheta + t_0\dot{\vartheta})^2 + t_0(\vartheta,_i)^2 + (t_1 - t_0)[t_0\dot{\vartheta}^2 + (\vartheta,_i)^2]\}$$
$$+(\vartheta,_i)^2 + (t_1 - t_0)\dot{\vartheta}^2 = \{(\vartheta + t_1\dot{\vartheta})\vartheta,_i + \epsilon[\nabla^2\phi - (\vartheta + t_1\dot{\vartheta})]\dot{\phi},_i\},_i\,. \tag{4.2.24}$$

The latter relation may also be written in the form

$$\frac{1}{2}\frac{\partial}{\partial t}P(x,t) + Q(x,t) = [R,_i(x,t)],_i\,, \tag{4.2.25}$$

where P, Q and R are defined by eqns (4.2.14–16), respectively. Now, integrating eqn (4.2.25) from $t = 0$ to $t = p(x)$, and using eqn (4.1.32), we arrive at

$$\frac{1}{2}[P(x, p(x)) - P(x,0)] + \int_0^{p(x)} Q(x,t)\,\mathrm{d}t$$
$$= \left[\int_0^{p(x)} R_i(x,t)\,\mathrm{d}t\right],_i - R_i(x, p(x))p,_i(x)\,. \tag{4.2.26}$$

Finally, integrating eqn (4.2.26) over B and using the divergence theorem, we obtain eqn (4.2.13). □

Proof of Theorem 4.2 Similarly to Section 4.1, we fix a point $(z,\lambda) \in \{\mathrm{B} - \mathcal{D}(t)\} \times (0,t)$ and introduce the set

$$\Omega = \overline{\mathrm{B}} \cap \overline{\Sigma_{\upsilon\lambda}(z)}. \tag{4.2.27}$$

Moreover, we define a scalar function $p_\lambda(x)$ using the formula

$$p_\lambda(x) = \begin{cases} \lambda - \upsilon^{-1}|x - z| & \text{for } x \in \Omega, \\ 0 & \text{for } x \notin \Omega. \end{cases} \tag{4.2.28}$$

Then, $p_\lambda(x) > 0$ on Ω and

$$|p_\lambda,_i(x)| = \begin{cases} \upsilon^{-1} & \text{on } \Omega, \\ 0 & \text{on } \overline{\mathrm{B}} - \Omega. \end{cases} \tag{4.2.29}$$

Using the definitions of domains $\mathcal{D}(t)$ and Ω, and the inequality $\lambda < t$, we conclude that $\mathcal{D}_0(t)$ and Ω are disjoint, that is

$$\Omega \cap \mathcal{D}_0(t) = \varnothing, \tag{4.2.30}$$

so that

$$\dot{\phi}_{,i} n_i = 0, \; \vartheta + t_1\dot{\vartheta} = 0 \text{ on } (\Omega \cap \partial\mathrm{B}) \times [0,t], \tag{4.2.31}$$

$$\phi(\cdot,0) = \dot{\phi}(\cdot,0) = \vartheta(\cdot,0) = \dot{\vartheta}(\cdot,0) = 0 \text{ on } \Omega. \tag{4.2.32}$$

From the definition of R_i (recall eqn (4.2.16)), and from the formulas (4.2.28) and (4.2.31), we find

$$\int_{\partial\mathrm{B}} \int_0^{p_\lambda(x)} R_i(x,t)\, n_i(x)\, \mathrm{d}t\mathrm{d}a = 0. \tag{4.2.33}$$

Also, from the definition of P (recall eqn (4.2.14)), and from the formulas (4.2.28) and (4.2.32), we find

$$P(x,p_\lambda(x)) - P(x,0) = \begin{cases} P(x,p_\lambda(x)) & \text{for } x \in \Omega, \\ 0 & \text{for } x \notin \Omega. \end{cases} \tag{4.2.34}$$

Clearly, $p_\lambda(x)$ satisfies the assumptions of Lemma 4.2, and so, substituting $p(x) \equiv p_\lambda(x)$ into eqn (4.2.13), as well as making use of eqns (4.2.33–34) and $(4.2.29)_2$, we obtain

$$\begin{aligned} \frac{1}{2}\int_\Omega P(x,p_\lambda(x))\,\mathrm{d}v &+ \int_\Omega \int_0^{p_\lambda(x)} Q(x,t)\,\mathrm{d}t\mathrm{d}v \\ &= -\int_\Omega R_i(x,p_\lambda(x))\, p_{\lambda,i}(x)\,\mathrm{d}v. \end{aligned} \tag{4.2.35}$$

Since $Q \geq 0$ $(t_1 > t_0)$, the relations $(4.2.29)_1$ and (4.2.35) imply the inequality

$$\frac{1}{2}\int_\Omega P(x,p_\lambda(x)) \leq \frac{1}{v}\int_\Omega |R_i(x,p_\lambda(x))|\,\mathrm{d}v. \tag{4.2.36}$$

From the definition of R_i (recall eqn (4.2.16)) we get

$$\begin{aligned} |R_{,i}| &\leq |\vartheta + t_1\dot{\vartheta}||\vartheta_{,i}| + \epsilon|\nabla^2\phi - (\vartheta + t_1\dot{\vartheta})||\dot{\phi}_{,i}| \\ &\leq |\vartheta + t_1\dot{\vartheta}||\vartheta_{,i}| + \epsilon|\dot{\phi}_{,i}|(|\nabla^2\phi| + |\vartheta + t_1\dot{\vartheta}|) \\ &= |\vartheta + t_1\dot{\vartheta}|(|\vartheta_{,i}| + \epsilon|\dot{\phi}_{,i}|) + \epsilon|\dot{\phi}_{,i}||\nabla^2\phi| \\ &\leq \frac{1}{2}\{(1+\epsilon)(\vartheta + t_1\dot{\vartheta})^2 + (\vartheta_{,i})^2 + 2\epsilon(\dot{\phi}_{,i})^2 + 2\epsilon(\nabla^2\phi)^2\}. \end{aligned} \tag{4.2.37}$$

Thus, the definition of P (recall eqn (4.2.14)) and the inequalities (4.2.36) and (4.2.37) lead to the relation

$$\begin{aligned}&\frac{\epsilon}{2}\left(1-\frac{2}{\upsilon}\right)\int_{\Omega}\left[(\nabla^2\phi)^2+(\dot{\phi}_{,i})^2\right]\mathrm{d}\upsilon+\frac{1}{2}\left(t_1-\frac{1}{\upsilon}\right)\int_{\Omega}(\vartheta_{,i})^2\mathrm{d}\upsilon\\&+\frac{1}{2}\left(\frac{t_0}{t_1}-\frac{1+\epsilon}{\upsilon}\right)\int_{\Omega}(\vartheta+t_1\dot{\vartheta})^2\mathrm{d}\upsilon+\frac{1}{2}\left(1-\frac{t_0}{t_1}\right)\int_{\Omega}\vartheta^2\mathrm{d}\upsilon\leq 0.\end{aligned} \tag{4.2.38}$$

The definition of the parameter υ (recall eqn (4.2.10)) implies that the coefficients in front of the integrals in the inequality (4.2.38) are non-negative. With these integrals being non-negative as well, the inequality (4.2.38) implies

$$\nabla^2\phi\left(x,p_\lambda\left(x\right)\right)=0,\ \vartheta\left(x,p_\lambda\left(x\right)\right)=0\ x\in\Omega. \tag{4.2.39}$$

Hence, in view of the definition of $p_\lambda\left(x\right)$ and the continuity of $\nabla^2\phi$ and ϑ,

$$\left.\begin{aligned}\nabla^2\phi\left(x,p_\lambda\left(x\right)\right)&\to\nabla^2\phi\left(z,\lambda\right)\\ \vartheta\left(x,p_\lambda\left(x\right)\right)&\to\vartheta\left(z,\lambda\right)\end{aligned}\right.\quad \text{as } x\to z. \tag{4.2.40}$$

Thus, taking the limit $x\to z$ in eqns (4.2.39), we obtain

$$\nabla^2\phi\left(z,\lambda\right)=0,\ \vartheta\left(z,\lambda\right)=0. \tag{4.2.41}$$

Since (z,λ) is an arbitrary point of the set $\{\mathrm{B}-\mathcal{D}\left(t\right)\}\times(0,t)$ and the pair (ϕ,ϑ) is sufficiently smooth on $\overline{\mathrm{B}}\times[0,\infty)$, the relations (4.2.41) imply

$$\vartheta=\nabla^2\phi=0\ \text{ on }\ \left\{\overline{\mathrm{B}}-\mathcal{D}\left(t\right)\right\}\times\left[0,t\right]. \tag{4.2.42}$$

From this and eqn $(4.2.7)_1$ we find

$$\ddot{\phi}=0\ \text{ on }\ \left\{\overline{\mathrm{B}}-\mathcal{D}\left(t\right)\right\}\times\left[0,t\right]. \tag{4.2.43}$$

Now, since

$$\phi\left(\cdot,0\right)=\dot{\phi}\left(\cdot,0\right)=0\ \text{ on } \overline{\mathrm{B}}-\mathcal{D}\left(t\right), \tag{4.2.44}$$

the relation (4.2.43) yields

$$\phi=0\ \text{ on }\ \left\{\overline{\mathrm{B}}-\mathcal{D}\left(t\right)\right\}\times\left[0,t\right], \tag{4.2.45}$$

which, in view of eqns $(4.2.42)_1$ and (4.2.45) gives eqn (4.2.12), thus completing the proof of Theorem 4.2. □

Theorem 4.2 implies that a potential–temperature disturbance described by eqns (4.2.7–9) propagates with a speed not greater than υ specified by eqn (4.2.10). The maximum speed of the disturbance propagating out of the domain $\mathcal{D}_0\left(t\right)$ depends on both relaxation times t_0 and t_1, and on the parameter of thermoelastic coupling ϵ. The speed υ becomes unbounded in two cases: (a) for $t_1\to 0$; (b) for $t_1>0$ with $t_0\to 0$.

Clearly, the thermoelastic disturbances governed by eqns (4.2.7–9) are generally different from those governed by eqns (4.1.11–13). However, these disturbances have a number of common characteristics. For instance, for $t_1 = t_0 > 0$ the potential–temperature disturbances of the G–L theory possess the same domain of influence as the potential–temperature disturbances of the L–S theory; compare here the Definition 4.3 of Section 4.2 with the Definitions 4.1 and 4.2 of Section 4.1. Also note that, proceeding in the same manner as in Sections 4.1 and 4.2, we can formulate a number of general theorems on the domain of influence for the conventional and non-conventional thermoelastic processes both in the L–S and the G–L theories. In particular, these general theorems may be formulated for the mixed displacement–temperature problems of Section 2.1, see (Ignaczak *et al.*, 1986).

4.3 The natural stress–heat-flux problem in the Lord–Shulman theory

First, we note that a NSHFP for a homogeneous isotropic body with one relaxation time involves finding a pair (S_{ij}, q_i) satisfying the field equations[11]

$$\rho^{-1}S_{(ik,kj)} - \left[\frac{1}{2\mu}\left(\ddot{S}_{ij} - \frac{\lambda}{3\lambda+2\mu}\ddot{S}_{kk}\delta_{ij}\right) - \frac{\alpha^2\theta_0}{C_S}\ddot{S}_{kk}\delta_{ij}\right] + \frac{\alpha}{C_S}\dot{q}_{k,k}\,\delta_{ij} = -F_{(ij)} \qquad \text{on B} \times [0,\infty),$$

$$\frac{1}{C_S}(q_{k,k} + \alpha\theta_0\dot{S}_{k,k})_{,i} - \frac{1}{k}(\dot{q}_i + t_0\ddot{q}_i) = -g_i \tag{4.3.1}$$

the initial conditions

$$\begin{aligned} S_{ij}(\cdot,0) &= S_{ij}^{(0)}, & \dot{S}_{ij}(\cdot,0) &= \dot{S}_{ij}^{(0)} \\ q_i(\cdot,0) &= q_i^{(0)}, & \dot{q}_i(\cdot,0) &= \dot{q}_i^{(0)} \end{aligned} \quad \text{on B}, \tag{4.3.2}$$

and the boundary conditions

$$\begin{aligned} S_{ij}n_j &= s_i' \\ q_i n_i &= q' \end{aligned} \quad \text{on } \partial\text{B} \times (0,\infty). \tag{4.3.3}$$

In eqns (4.3.1) we have set[12]

$$\begin{aligned} F_{(ij)} &= \rho^{-1}b_{(i,j)} - C_S^{-1}\alpha\dot{r}\delta_{ij} \\ g_i &= -C_S^{-1}r_{,i} \end{aligned} \quad \text{on } \overline{\text{B}} \times [0,\infty). \tag{4.3.4}$$

Certainly, the stress–heat-flux thermoelastic disturbances described by the eqns (4.3.1–4) are more general than those studied in Sections 4.1 and 4.2.

[11] See eqns (2.1.6–8) particularized for a homogeneous isotropic body for which $\partial\text{B}_1 = \partial\text{B}_3 = \varnothing$.
[12] See eqn (3.2.4).

Therefore, the domain of influence is also more general here than the previous ones.

The set

$$\mathcal{D}_0(t) = \{x \in \overline{\mathrm{B}} : (1) \text{ If } x \in \mathrm{B} \text{ then } S_{ij}^{(0)} \neq 0 \text{ or } \dot{S}_{ij}^{(0)} \neq 0 \text{ or } q_i^{(0)} \neq 0 \text{ or } \dot{q}_i^{(0)} \neq 0;$$
$$(2) \text{ If } (x, \tau) \in \mathrm{B} \times [0, t], \text{ then } F_{(ij)} \neq 0 \text{ or } g_i \neq 0;$$
$$(3) \text{ If } (x, \tau) \in \partial\mathrm{B} \times [0, t], \text{ then } s_i' \neq 0 \text{ or } q' \neq 0\} \tag{4.3.5}$$

is called a support of the thermomechanical loading at time t for the problem (4.3.1–3).

The domain of influence of the thermomechanical loading at time t for the problem (4.3.1–3) is the set

$$\mathcal{D}(t) = \left\{x \in \overline{\mathrm{B}} : \mathcal{D}_0(t) \cap \overline{\Sigma_{vt}(x)} \neq \varnothing\right\}, \tag{4.3.6}$$

where v is a parameter with dimension of velocity, satisfying the inequality[13]

$$v \geq \max(v_1, v_2, v_3), \tag{4.3.7}$$

where

$$v_1 = \left(\frac{2\mu}{\rho}\right)^{\frac{1}{2}}, \tag{4.3.8}$$

$$v_2 = \left\{\frac{3\lambda + 2\mu}{\rho}\frac{C_S}{C_E}\left[1 - \left(1 - \frac{C_E}{C_S}\right)^{\frac{1}{2}}\right]^{-1}\right\}^{\frac{1}{2}}, \tag{4.3.9}$$

$$v_3 = \left\{\frac{k}{t_0}\frac{1}{C_S}\left[1 + \frac{C_S}{C_E}\left(1 - \frac{C_E}{C_S}\right)^{\frac{1}{2}}\right]\right\}^{\frac{1}{2}}. \tag{4.3.10}$$

The following theorem holds true:

Theorem 4.3 *(On the domain of influence for a NSHFP in the L–S theory)*[14] *If the pair (S_{ij}, q_i) is a smooth solution of NSHFP described by eqns (4.3.1–3) and if $\mathcal{D}(t)$ is given by the formula (4.3.6), then*

$$S_{ij} = 0, \quad q_i = 0 \text{ on } \left\{\overline{B} - \mathcal{D}(t)\right\} \times [0, t]. \tag{4.3.11}$$

The proof of the theorem is based on the following lemma of Zaremba type:

[13] Contrary to the dimensionless velocity v of Sections 4.1 and 4.2, the formulation of problem (4.3.1–3) is dimensional; in particular, v in eqn (4.3.7) and v_1, v_2, and v_3 in eqns (4.3.8–10) have the dimensions of m/s.

[14] The theorem was proved in (Bialy, 1983). Here, we present another proof of the theorem.

Lemma 4.3 *Let (S_{ij}, q_i) be a solution of eqns (4.3.1–3) and let p be a scalar field of Lemma 4.1 of Section 4.1. Then, the following generalized energy identity holds true for the problem (4.1.1–3)*[15]

$$\frac{1}{2}\int_B \left[P\left(x,p\left(x\right)\right)-P\left(x,0\right)\right]\mathrm{d}v+\int_B\int_0^{p(x)} Q\left(x,t\right)\mathrm{d}t\mathrm{d}v+\int_B R_i\left(x,p\left(x\right)\right)p_{,i}\left(x\right)\mathrm{d}v$$
$$=\int_B\int_0^{p(x)} S\left(x,t\right)\mathrm{d}t\mathrm{d}v+\int_{\partial B}\int_0^{p(x)} R_i\left(x,t\right)n_i(x)\mathrm{d}t\mathrm{d}a, \tag{4.3.12}$$

where

$$P\left(x,t\right)=\rho^{-1}S_{ik,k}\,S_{ij,j}+\frac{1}{2\mu}\left(\dot{S}_{ij}\dot{S}_{ij}-\frac{\lambda}{3\lambda+2\mu}\,\dot{S}_{kk}^2\right) - \frac{\alpha^2\theta_0}{C_S}\,\dot{S}_{kk}^2+\frac{1}{\theta_0 C_S}(q_{k,k})^2+\frac{t_0}{\theta_0 k}(\dot{q}_i)^2 \quad on\ \overline{B}\times[0,\infty), \tag{4.3.13}$$

$$Q\left(x,t\right)=\frac{1}{\theta_0 k}(\dot{q}_i)^2 \ on\ \overline{B}\times[0,\infty), \tag{4.3.14}$$

$$R_i\left(x,t\right)=\rho^{-1}\dot{S}_{ij}S_{jk,k}+C_S^{-1}(\theta_0^{-1}q_{k,k}+\alpha\dot{S}_{kk})\dot{q}_i \ on\ \overline{B}\times[0,\infty), \tag{4.3.15}$$

$$S\left(x,t\right)=F_{(ij)}\dot{S}_{ij}+\theta_0^{-1}g_i\dot{q}_i \ on\ \overline{B}\times[0,\infty). \tag{4.3.16}$$

Proof of Lemma 4.3 Multiplying eqn $(4.3.1)_1$ through by $\dot{S}_{ij}$and eqn $(4.3.1)_2$ through by $\theta_0^{-1}\dot{q}_i$, and adding the results, we obtain

$$\frac{1}{2}\frac{\partial}{\partial t}P\left(x,t\right)+Q\left(x,t\right)=\left[R_i\left(x,t\right)\right]_{,i}+S\left(x,t\right), \tag{4.3.17}$$

where P, Q, R_i, S are given by formulas (4.3.13–16). Next, we integrate eqn (4.3.17) from $t=0$ to $t=p(x)$ with the use of formula (4.1.32). Finally, integrating the result over B and using the divergence theorem, we obtain the required identity (4.3.12). □

Proof of Theorem 4.3 Proceeding in a way similar to that employed in the proof of Theorem 4.1 of Section 4.1, we fix a point $(z,\lambda)\in\{\mathrm{B}-\mathcal{D}(t)\}\times(0,t)$, and introduce the set

$$\Omega=\overline{\mathrm{B}}\cap\overline{\Sigma_{\upsilon\lambda}\left(z\right)}. \tag{4.3.18}$$

Moreover, we define a scalar function $p_\lambda\left(x\right)$ using the formula

$$p_\lambda\left(x\right)=\begin{cases}\lambda-\upsilon^{-1}\left|x-z\right| & \text{for } x\in\Omega,\\ 0 & \text{for } x\notin\Omega.\end{cases} \tag{4.3.19}$$

[15] The generalized energy identity (4.3.12) is similar to eqn (4.1.19) or eqn (4.2.13).

where υ is specified by eqn (4.3.7). Then, $p_\lambda(x) > 0$ on Ω and

$$\Omega \cap \mathcal{D}_0(t) = \varnothing. \tag{4.3.20}$$

From this,

$$\dot{S}_{ij} n_i = 0,\ \dot{q}_i n_i = 0 \text{ on } (\Omega \cap \partial \mathrm{B}) \times (0, t), \tag{4.3.21}$$

$$F_{(ij)} = 0,\ g_i = 0 \text{ on } \Omega \times (0, t), \tag{4.3.22}$$

$$S_{ij}(\cdot, 0) = \dot{S}_{ij}(\cdot, 0) = q_i(\cdot, 0) = \dot{q}_i(\cdot, 0) = 0 \ \text{ on } \Omega, \tag{4.3.23}$$

as well as

$$\int_{\partial \mathrm{B}} \int_0^{p_\lambda(x)} R_i(x, t)\, n_i(x)\, \mathrm{d}t \mathrm{d}a = 0, \tag{4.3.24}$$

$$\int_{\mathrm{B}} \int_0^{p_\lambda(x)} S(x, t)\, \mathrm{d}t \mathrm{d}v = 0. \tag{4.3.25}$$

Next, using the definitions of $P(x,t)$ and $p_\lambda(x)$, we obtain

$$P(x, p_\lambda(x)) - P(x, 0) = \begin{cases} P(x, p_\lambda(x)) & \text{for } x \in \Omega, \\ 0 & \text{for } x \notin \Omega. \end{cases} \tag{4.3.26}$$

Thus, putting $p_\lambda(x)$ into the identity (4.3.12), and exploiting eqns (4.3.24–26) along with the non-negative character of Q (recall eqn (4.3.14)), we arrive at the estimate

$$\frac{1}{2} \int_\Omega P(x, p_\lambda(x))\, \mathrm{d}v \leq \frac{1}{\upsilon} \int_\Omega |R_i(x, p_\lambda(x))|\, \mathrm{d}v. \tag{4.3.27}$$

From the definition of R_i (recall eqn (4.3.15)) there follows

$$\begin{aligned} \frac{1}{\upsilon} |R_i| &\leq \rho^{-1} \left| \frac{\dot{S}_{ij}}{\upsilon} S_{jk},_k \right| + C_S^{-1} \left| \theta_0^{-1} q_k,_k + \alpha \dot{S}_{kk} \right| \left| \frac{\dot{q}_i}{\upsilon} \right| \\ &\leq \rho^{-1} \left| \frac{\dot{S}_{ij}}{\upsilon} \right| |S_{jk},_k| + C_S^{-1} \theta_0^{-1} |q_k,_k| \left| \frac{\dot{q}_i}{\upsilon} \right| + C_S^{-1} |\alpha| \left| \dot{S}_{kk} \right| \left| \frac{\dot{q}_i}{\upsilon} \right|. \end{aligned} \tag{4.3.28}$$

The first two terms on the right-hand side of eqn (4.3.28) are estimated from above using the relations

$$\left| \frac{\dot{S}_{ij}}{\upsilon} \right| |S_{jk},_k| \leq \frac{1}{2} (S_{ij},_j S_{ik},_k + \frac{1}{\upsilon^2} \dot{S}_{ij} \dot{S}_{ij}), \tag{4.3.29}$$

$$\left| \frac{\dot{q}_i}{\upsilon} \right| |q_k,_k| \leq \frac{1}{2} \left[(q_k,_k)^2 + \frac{1}{\upsilon^2} (\dot{q}_i)^2 \right]. \tag{4.3.30}$$

In order to estimate the last term of the inequality (4.3.28), we use the relation

$$\sqrt{ab} \le \frac{1}{2}\left(\hat{\varepsilon}a + \hat{\varepsilon}^{-1}b\right), \tag{4.3.31}$$

which is true for arbitrary non-negative (physical) fields a and b having the same dimension, as well as for a non-dimensional positive parameter $\hat{\varepsilon}$. Setting

$$a = (\dot{S}_{kk})^2,\ b = \frac{1}{\upsilon^2\left(\alpha\theta_0\right)^2}(\dot{q}_i)^2, \tag{4.3.32}$$

and

$$\hat{\varepsilon} = \frac{C_E}{C_S}\left(1 - \frac{C_E}{C_S}\right)^{-\frac{1}{2}} \tag{4.3.33}$$

in eqn (4.3.31), we obtain

$$\left|\dot{S}_{kk}\right| \frac{1}{\upsilon\left|\alpha\right|\theta_0}\left|\dot{q}_i\right| \le \frac{1}{2}\left\{\frac{C_E}{C_S}\left(1 - \frac{C_E}{C_S}\right)^{-\frac{1}{2}}\dot{S}_{kk}^2 + \frac{C_S}{C_E}\left(1 - \frac{C_E}{C_S}\right)^{\frac{1}{2}}\frac{1}{\upsilon^2\alpha^2\theta_0^2}\dot{q}_i{}^2\right\}. \tag{4.3.34}$$

Thus, in view of eqns (4.3.28–30) and eqn (4.3.34), we obtain

$$\frac{1}{\upsilon}\left|R_i\right| \le \frac{1}{2}\left\{\rho^{-1}\left(S_{ij,j}S_{ik,k} + \frac{1}{\upsilon^2}\dot{S}_{ij}\dot{S}_{ij}\right) + C_S^{-1}\theta_0^{-1}\left[\left(q_{k,k}\right)^2 + \dot{q}_i^2/\upsilon^2\right]\right.$$
$$\left. + C_S^{-1}\alpha^2\theta_0^{-1}\left[\frac{C_E}{C_S}\left(1 - \frac{C_E}{C_S}\right)^{-\frac{1}{2}}\dot{S}_{kk}^2 + \frac{C_S}{C_E}\left(1 - \frac{C_E}{C_S}\right)^{\frac{1}{2}}\frac{1}{\upsilon^2\alpha^2\theta_0^2}\dot{q}_i{}^2\right]\right\}. \tag{4.3.35}$$

Finally, from the definition of $P(x,t)$ and the inequalities (4.3.27) and (4.3.35), we find

$$\begin{aligned}
&\left(\frac{1}{2\mu} - \frac{1}{\rho\upsilon^2}\right)\int_\Omega\left(\dot{S}_{ij} - \frac{1}{3}\dot{S}_{kk}\delta_{ij}\right)\left(\dot{S}_{ij} - \frac{1}{3}\dot{S}_{kk}\delta_{ij}\right)\mathrm{d}\upsilon \\
&+\frac{1}{3}\left\{\frac{1}{3\lambda + 2\mu}\frac{C_E}{C_S}\left[1 - \left(1 - \frac{C_E}{C_S}\right)^{\frac{1}{2}}\right] - \frac{1}{\rho\upsilon^2}\right\}\int_\Omega\dot{S}_{kk}^2\mathrm{d}\upsilon \\
&+\frac{1}{\theta_0}\left\{\frac{t_0}{k} - \frac{1}{C_S\upsilon^2}\left[1 + \frac{C_S}{C_E}\left(1 - \frac{C_E}{C_S}\right)^{\frac{1}{2}}\right]\right\}\int_\Omega\dot{q}_i{}^2\mathrm{d}\upsilon \le 0.
\end{aligned} \tag{4.3.36}$$

It follows from eqn (4.3.7) that the coefficients in front of the integrals in eqn (4.3.36) are non–negative. Since these integrals are also non-negative, the inequality (4.3.36) is satisfied iff

$$\dot{S}_{ij}\left(x, p_\lambda(x)\right) = 0,\ \dot{q}_i\left(x, p_\lambda(x)\right) = 0 \text{ on } \Omega. \tag{4.3.37}$$

Using the definition (4.3.19) of $p_\lambda(x)$ and passing in eqn (4.3.37) to the limit $x \to z$, we obtain

$$\dot{S}_{ij}(z,\lambda) = 0, \quad \dot{q}_i(z,\lambda) = 0. \tag{4.3.38}$$

Since (z,λ) is an arbitrary point of the Cartesian product $\{\mathrm{B} - \mathcal{D}(t)\} \times (0,t)$, and since the pair (S_{ij}, q_i) is smooth in $\overline{\mathrm{B}} \times [0,\infty)$, from eqn (4.3.38) we infer that

$$\dot{S}_{ij} = 0, \quad \dot{q}_i = 0 \text{ on } \left\{\overline{\mathrm{B}} - \mathcal{D}(t)\right\} \times (0,t). \tag{4.3.39}$$

Finally, given the definition of $\mathcal{D}(t)$,

$$S_{ij}(\cdot,0) = 0, \quad q_i(\cdot,0) = 0 \text{ on } \overline{\mathrm{B}} - \mathcal{D}(t), \tag{4.3.40}$$

so that, from eqns (4.3.39 and 40) we obtain

$$S_{ij} = 0, \quad q_i = 0 \text{ on } \left\{\overline{\mathrm{B}} - \mathcal{D}(t)\right\} \times [0,t]. \tag{4.3.41}$$

This completes the proof of Theorem 4.3. □

This theorem implies that the stress–heat-flux disturbances described by eqns (4.3.1–3) propagate in the thermoelastic body with speeds not greater than v specified by eqn (4.3.7). In order to illustrate the order of magnitude of v, let us consider an aluminum alloy [p. 36 in (Biały, 1983)], for which

$$\theta_0 = 25\,[^o\mathrm{C}], \quad t_0 = 8 \cdot 10^{-12}\mathrm{s}, \tag{4.3.42}$$

$$\rho = \frac{2.7}{981} \cdot 10^{-3}\left[\frac{\mathrm{kG} \cdot s^2}{\mathrm{cm}^4}\right], \quad \lambda = 5.707 \cdot 10^5 \left[\frac{\mathrm{kG}}{\mathrm{cm}^2}\right], \mu = 2.686 \cdot 10^5 \left[\frac{\mathrm{kG}}{\mathrm{cm}^2}\right], \tag{4.3.43}$$

$$k = 23.884 \left[\frac{\mathrm{kG}}{^o\mathrm{C} \cdot \mathrm{s}}\right], \quad \alpha = 23.86 \cdot 10^{-6}\left[^o\mathrm{C}^{-1}\right], \tag{4.3.44}$$

$$C_E = 24.643 \left[\frac{\mathrm{kG}}{^o\mathrm{C} \cdot \mathrm{cm}^2}\right], \quad C_S = 24.739 \left[\frac{\mathrm{kG}}{^o\mathrm{C} \cdot \mathrm{cm}^2}\right]. \tag{4.3.45}$$

Substituting these material parameters in the formulas (4.3.8–10), we find

$$v_1 = 4.418 \cdot 10^5 \left[\frac{\mathrm{cm}}{\mathrm{s}}\right], v_2 = 9.358 \cdot 10^5 \left[\frac{\mathrm{cm}}{\mathrm{s}}\right], v_3 = 3.581 \cdot 10^5 \left[\frac{\mathrm{cm}}{\mathrm{s}}\right]. \tag{4.3.46}$$

Thus, v_2 is an upper bound of speeds of the stress–heat-flux disturbances. In general, that bound depends on the speeds $v_1^{(0)} = (2\mu/\rho)^{1/2}$, $v_2^{(0)} = [(3\lambda + 2\mu)\ /\rho]^{1/2}$, $v_3^{(0)} = (k/t_0 C_S)^{1/2}$, and on the ratio C_E/C_S. For $t_0 \to 0$ that bound tends to infinity. For $\alpha = 0$ $(C_E = C_S)$, the NSHFP described by eqns (4.3.1–4) separates into two problems: (a) the problem of propagation of isothermal stress disturbances described by eqns $(4.3.1)_1$, $(4.3.2)_1$, $(4.3.3)_1$, and $(4.3.4)_1$; at $\alpha = 0$; (b) the problem of propagation of heat flux in a rigid conductor described by eqns $(4.3.1)_2$, $(4.3.2)_2$, $(4.3.3)_2$, and $(4.3.4)_2$; at $\alpha = 0$. In the context of the first problem we have a theorem on the domain of influence with a speed $v \geq \max\left(v_1^{(0)}, v_2^{(0)}\right)$,[16] while for the second problem we have a theorem on the domain of influence with a speed $v \geq v_3^{(0)}$.[17]

4.4 The natural stress–temperature problem in the Green–Lindsay theory

Let us assume, similarly to what was done in Sections 4.1–3, that the thermoelastic body is homogeneous and isotropic. In that case, a NSTP of the G–L theory hinges on finding a pair (S_{ij}, ϑ) satisfying the field equations[18]

$$\begin{gathered}
\rho^{-1} S_{(ik,kj)} - \left[\frac{1}{2\mu}\left(\ddot{S}_{ij} - \frac{\lambda}{3\lambda + 2\mu}\ddot{S}_{kk}\delta_{ij}\right) - \frac{\alpha^2\theta_0}{C_S}\frac{t_1}{t_{(0)}}\ddot{S}_{kk}\delta_{ij}\right] \\
-\alpha t_{(0)}^{-1}\left[t_1 C_S^{-1} k\dot{\vartheta}_{,pp} - \left(t_1 - t_{(0)}\right)\ddot{\vartheta}\right]\delta_{ij} = -b_{(ij)} \qquad \text{on } \mathrm{B} \times [0, \infty), \\
C_S^{-1}(k\vartheta_{,ii} - \alpha\theta_0 \dot{S}_{kk}) - (\dot{\vartheta} + t_{(0)}\ddot{\vartheta}) = -C_S^{-1} r
\end{gathered} \tag{4.4.1}$$

the initial conditions

$$\begin{aligned}
S_{ij}(\cdot, 0) &= S_{ij}^{(0)}, \quad & \dot{S}_{ij}(\cdot, 0) &= \dot{S}_{ij}^{(0)} \\
\vartheta(\cdot, 0) &= \vartheta_0, & \dot{\vartheta}(\cdot, 0) &= \dot{\vartheta}_0
\end{aligned} \quad \text{on B}, \tag{4.4.2}$$

and the boundary conditions

$$S_{ij} n_j = s'_i, \quad \vartheta = \vartheta' \quad \text{on } \partial\mathrm{B} \times (0, \infty). \tag{4.4.3}$$

Here,

$$b_{(ij)} = \rho^{-1} b_{(i,j)} - t_{(0)}^{-1} t_1 C_S^{-1} \alpha \dot{r} \delta_{ij} \quad \text{on } \overline{\mathrm{B}} \times [0, \infty). \tag{4.4.4}$$

The set

$$\begin{aligned}
\mathcal{D}_0(t) = \{x \in \overline{\mathrm{B}} : &(1) \text{ If } x \in \mathrm{B}, \text{ then } S_{ij}^{(0)} \neq 0 \text{ or } \dot{S}_{ij}^{(0)} \neq 0 \text{ or } \vartheta_0 \neq 0 \text{ or } \dot{\vartheta}_0 \neq 0; \\
&(2) \text{ If } (x, \tau) \in \mathrm{B} \times [0, t], \text{ then } b_{(ij)} \neq 0 \text{ or } r \neq 0; \\
&(3) \text{ If } (x, \tau) \in \partial\mathrm{B} \times [0, t], \text{ then } s'_i \neq 0 \text{ or } \vartheta' \neq 0\}
\end{aligned} \tag{4.4.5}$$

is called a support of the thermomechanical loading at time t for the problem (4.4.1–3).

The domain of influence of the thermomechanical loading at time t for the problem (4.4.1–3) is the set

$$\mathcal{D}(t) = \left\{x \in \overline{\mathrm{B}} : \mathcal{D}_0(t) \cap \overline{\Sigma_{vt}(x)} \neq \varnothing\right\}, \tag{4.4.6}$$

[16] This theorem was proved in (Ignaczak, 1974).
[17] See pp. 37–38 in (Biały, 1983).
[18] Equations (4.4.1–4) are obtained by restricting eqns (2.1.15–17) to a homogeneous isotropic body with $\partial\mathrm{B}_1 = \partial\mathrm{B}_4 = \varnothing$.

where υ is a constant speed satisfying the inequality

$$\upsilon \geq \max(\upsilon_1, \upsilon_2, \upsilon_3), \tag{4.4.7}$$

while

$$\upsilon_1 = \left(\frac{2\mu}{\rho}\right)^{\frac{1}{2}}, \tag{4.4.8}$$

$$\upsilon_2 = \left(\frac{3\lambda+2\mu}{\rho}\right)^{\frac{1}{2}} \left\{\left[1-\left(1-\frac{C_E}{C_S}\right)^{\frac{1}{2}}\right]\left[1-\left(1-\frac{C_E}{C_S}\right)\frac{t_1}{t_{(0)}}\right]\right\}^{-\frac{1}{2}}, \tag{4.4.9}$$

$$\upsilon_3 = \left(\frac{k}{t_{(0)}C_S}\right)^{\frac{1}{2}} \left(\frac{t_1}{t_{(0)}}\right)^{\frac{3}{2}} \left\{\frac{t_1}{t_{(0)}}\frac{C_S}{C_E} + \left(1-\frac{C_E}{C_S}\right)^{\frac{1}{2}}\left[1-\left(1-\frac{C_E}{C_S}\right)\frac{t_1}{t_{(0)}}\right]^{-1}\right\}^{\frac{1}{2}}. \tag{4.4.10}$$

The following theorem holds true:

Theorem 4.4 *(On the domain of influence for NSTP in the G–L theory)*[19] *If the pair (S_{ij}, ϑ) is a smooth solution of NSTP described by eqns (4.4.1–3) and if $\mathcal{D}(t)$ is given by the formula (4.4.6), then*

$$S_{ij} = 0,\ \vartheta = 0 \ \text{on} \ \left\{\overline{B} - \mathcal{D}(t)\right\} \times [0, t]. \tag{4.4.11}$$

The proof of this theorem is based on the following:

Lemma 4.4 *Let (S_{ij}, ϑ) be a solution of eqns (4.4.1–3) and let $p = p(x)$ be a scalar field of Lemma 4.1. Then*

$$\frac{1}{2}\int_B [P(x,p(x)) - P(x,0)]\,\mathrm{d}\upsilon + \int_B \int_0^{p(x)} Q(x,t)\,\mathrm{d}t\mathrm{d}\upsilon + \int_B R_i(x,p(x))\,p_{,i}(x)\,\mathrm{d}\upsilon$$
$$= \int_B \int_0^{p(x)} S(x,t)\,\mathrm{d}t\mathrm{d}\upsilon + \int_{\partial B} \int_0^{p(x)} R_i(x,t)\,n_i(x)\mathrm{d}t\mathrm{d}a, \tag{4.4.12}$$

where the functions P, Q, R_i and S are given by the formulas

$$P(x,t) = \rho^{-1} S_{ik,k}\, S_{ij,j} + \left[\frac{1}{2\mu}\left(\dot{S}_{ij}\dot{S}_{ij} - \frac{\lambda}{3\lambda+2\mu}\dot{S}_{kk}^2\right) - \frac{\alpha^2\theta_0}{C_S}\frac{t_1}{t_{(0)}}\dot{S}_{kk}^2\right]$$
$$+\frac{1}{\theta_0}\left\{\left(\frac{t_1}{t_{(0)}} - 1\right)\left[\frac{k\vartheta_{,ii}}{\sqrt{C_S}} - \sqrt{C_S}\dot{\vartheta}\right]^2 + \frac{(k\vartheta_{,ii})^2}{C_S} + t_1 k(\dot{\vartheta}_{,i})^2\right\}, \tag{4.4.13}$$

[19] This theorem was proved in (Bialy, 1983).

$$Q\left(x,t\right) = \theta_0^{-1}\left[k(\dot{\vartheta}_{,i})^2 + C_S(t_1 - t_{(0)})\ddot{\vartheta}^2\right], \tag{4.4.14}$$

$$R_i\left(x,t\right) = \rho^{-1}\dot{S}_{ij}S_{jk,k} + k\theta_0^{-1}\dot{\vartheta}_{,i}\left(\dot{\vartheta} + t_1\ddot{\vartheta}\right), \tag{4.4.15}$$

$$S\left(x,t\right) = b_{(ij)}\dot{S}_{ij} + \frac{r}{\theta_0}\left[\left(\frac{t_1}{t_{(0)}} - 1\right)\ddot{\vartheta} - \frac{t_1}{t_{(0)}}\frac{k}{C_S}\dot{\vartheta}_{,ii}\right]. \tag{4.4.16}$$

Proof of Lemma 4.4 Proceeding in a manner similar to that used when deriving eqn (4.3.17), for the field equations (4.4.1) we construct the local conservation law

$$\frac{1}{2}\frac{\partial}{\partial t}P\left(x,t\right) + Q\left(x,t\right) = \left[R_i\left(x,t\right)\right]_{,i} + S\left(x,t\right), \tag{4.4.17}$$

where P, Q, R_i and S are given by formulas (4.4.13–16). Next, we integrate eqn (4.4.17) from $t = 0$ to $t = p\left(x\right)$ with the use of formula (4.1.32), so as to get

$$\frac{1}{2}\left[P\left(x,p\left(x\right)\right) - P\left(x,0\right)\right] + \int_0^{p(x)} Q\left(x,t\right)\mathrm{d}t + R_i\left(x,p\left(x\right)\right)p_{,i}\left(x\right)$$

$$= \left[\int_0^{p(x)} R_i\left(x,t\right)\mathrm{d}t\right]_{,i} + \int_0^{p(x)} S\left(x,t\right)\mathrm{d}t. \tag{4.4.18}$$

Finally, integrating eqn (4.4.18) over B and using the divergence theorem, we obtain the required identity (4.4.12); this is a generalized energy identity for the problem (4.4.1–3). $\qquad\square$

Proof of Theorem 4.4 For a fixed point $(z,\lambda) \in \{B-\mathcal{D}\left(t\right)\} \times (0,t)$, we introduce the set

$$\Omega = \overline{B} \cap \overline{\Sigma_{\upsilon\lambda}\left(z\right)}, \tag{4.4.19}$$

where υ is determined by the inequality (4.4.7). Furthermore, we introduce a function $p_\lambda\left(x\right)$ using this formula

$$p_\lambda\left(x\right) = \begin{cases} \lambda - \upsilon^{-1}\left|x - z\right| & \text{for } x \in \Omega, \\ 0 & \text{for } x \notin \Omega. \end{cases} \tag{4.4.20}$$

Then, $p_\lambda\left(x\right) > 0$ in Ω and

$$\Omega \cap \mathcal{D}_0\left(t\right) = \varnothing. \tag{4.4.21}$$

From this and the definitions of R_i and S, we get

$$\int_{\partial B} \int_0^{p_\lambda(x)} R_i\left(x,t\right)n_i\left(x\right)\mathrm{d}t\mathrm{d}a = 0, \tag{4.4.22}$$

and

$$\int_B \int_0^{p_\lambda(x)} S(x,t)\,\mathrm{d}t\mathrm{d}v = 0. \tag{4.4.23}$$

Next, using the definitions of $P(x,t)$ and $p_\lambda(x)$, we obtain

$$P(x,p_\lambda(x)) - P(x,0) = \begin{cases} P(x,p_\lambda(x)) & \text{for } x \in \Omega, \\ 0 & \text{for } x \notin \Omega. \end{cases} \tag{4.4.24}$$

Thus, putting $p_\lambda(x)$ into the identity (4.4.12), and using eqns (4.4.22–24) along with the non-negative character of Q, we arrive at the estimate

$$\frac{1}{2}\int_\Omega P(x,p_\lambda(x))\,\mathrm{d}v \le \frac{1}{\upsilon}\int_\Omega |R_i(x,p_\lambda(x))|\,\mathrm{d}v. \tag{4.4.25}$$

From the field equation $(4.4.1)_2$, there follows

$$\dot{\vartheta} + t_{(0)}\ddot{\vartheta} = C_S^{-1}(k\vartheta_{,ii} - \alpha\theta_0\dot{S}_{kk}) \text{ on } \Omega \times (0,t), \tag{4.4.26}$$

whence

$$\dot{\vartheta} + t_1\ddot{\vartheta} = \frac{t_1}{t_{(0)}}C_S^{-1}(k\vartheta_{,ii} - \alpha\theta_0\dot{S}_{kk}) - \left(\frac{t_1}{t_{(0)}} - 1\right)\dot{\vartheta} \text{ on } \Omega \times (0,t). \tag{4.4.27}$$

Now, in view of eqns (4.4.15) and (4.4.27) an alternative form of the integrand on the right-hand side of the inequality (4.4.25) is

$$|R_i(x,p_\lambda(x))| = \\ |\rho^{-1}\dot{S}_{ij}S_{jk,k} + k\dot{\vartheta}_{,i}\theta_0^{-1}\left[\frac{t_1}{t_{(0)}}C_S^{-1}(k\vartheta_{,kk} - \alpha\theta_0\dot{S}_{kk}) + \left(\frac{t_1}{t_{(0)}} - 1\right)\dot{\vartheta}\right]|, \tag{4.4.28}$$

which leads to the estimate

$$\upsilon^{-1}|R_i(x,p_\lambda(x))| \le \rho^{-1}\left|\frac{\dot{S}_{ij}}{\upsilon}\right||S_{jk,k}| + \frac{t_1}{t_{(0)}}\frac{k^2}{\theta_0 C_S}\left|\frac{\dot{\vartheta}_{,i}}{\upsilon}\right||\vartheta_{,kk}| \\ + \frac{t_1}{t_{(0)}}\frac{1}{\theta_0 C_S}\left|\frac{k\dot{\vartheta}_{,i}}{\upsilon}\right|\left|\alpha\theta_0\dot{S}_{kk}\right| + \frac{k\,(t_1 - t_{(0)})}{\theta_0}\left|\dot{\vartheta}_{,i}\right|\left|\frac{\dot{\vartheta}}{t_{(0)}\upsilon}\right|. \tag{4.4.29}$$

In order to estimate from above the right-hand side of the inequality (4.4.29), we employ the inequality

$$\sqrt{ab} \le \frac{1}{2}\left(\hat{\varepsilon}a + \hat{\varepsilon}^{-1}b\right), \tag{4.4.30}$$

where $\hat{\varepsilon}$ is a dimensionless positive parameter, while a and b are arbitrary non-negative physical fields of the same dimension, recall eqn (4.3.31).

Letting in eqn (4.4.30)

$$a = (S_{ij},_j)^2,\; b = \left(\frac{\dot{S}_{ij}}{v}\right)^2,\; \hat{\varepsilon} = 1, \tag{4.4.31}$$

we obtain

$$\left|\frac{\dot{S}_{ij}}{v}\right| |S_{jk},_k| \le \frac{1}{2}\left(S_{ij},_j S_{ik},_k + \frac{1}{v^2}\dot{S}_{ij}\dot{S}_{ij}\right). \tag{4.4.32}$$

Letting in eqn (4.4.30)

$$a = (\vartheta,_{ii})^2, \quad b = \left(\frac{\dot{\vartheta},_i}{v}\right)^2, \quad \hat{\varepsilon} = \frac{C_E}{C_S}\left(\frac{t_{(0)}}{t_1}\right)^2, \tag{4.4.33}$$

we obtain

$$|\vartheta,_{kk}|\left|\frac{\dot{\vartheta},_i}{v}\right| \le \frac{1}{2}\left[\frac{C_E}{C_S}\left(\frac{t_{(0)}}{t_1}\right)^2 (\vartheta,_{kk})^2 + \frac{C_E}{C_S}\left(\frac{t_1}{t_{(0)}}\right)^2\left(\frac{\dot{\vartheta},_i}{v}\right)^2\right]. \tag{4.4.34}$$

To estimate the third term on the right-hand side of inequality (4.4.29), we set in eqn (4.4.30)[20]

$$\begin{gathered} a = (\alpha\theta_0)^2\,\dot{S}_{kk}^2, \quad b = \left(\frac{k\dot{\vartheta},_i}{v}\right)^2, \\ \hat{\varepsilon} = \frac{t_{(0)}}{t_1}\left(1 - \frac{C_E}{C_S}\right)^{-1/2}\left[1 - \left(1 - \frac{C_E}{C_S}\right)\frac{t_1}{t_{(0)}}\right] > 0, \end{gathered} \tag{4.4.35}$$

and obtain

$$\begin{aligned} \left|\frac{k\dot{\vartheta}_{,i}}{v}\right| \left|\alpha\theta_0\dot{S}_{kk}\right| \le \frac{1}{2}\Bigg\{ & \frac{t_{(0)}}{t_1}\left(1 - \frac{C_E}{C_S}\right)^{-1/2}\left[1 - \left(1 - \frac{C_E}{C_S}\right)\frac{t_1}{t_{(0)}}\right]\alpha^2\theta_0^2\dot{S}_{kk}^2 \\ & + \frac{t_1}{t_{(0)}}\left(1 - \frac{C_E}{C_S}\right)^{1/2}\left[1 - \left(1 - \frac{C_E}{C_S}\right)\frac{t_1}{t_{(0)}}\right]^{-1}\frac{k^2}{v^2}\left(\dot{\vartheta},_i\right)^2\Bigg\}. \end{aligned} \tag{4.4.36}$$

Finally, letting in eqn (4.4.30)

$$a = \left(\frac{\dot{\vartheta}}{t_{(0)}v}\right)^2,\; b = \left(\dot{\vartheta},_i\right)^2,\; \hat{\varepsilon} = \left(1 + \frac{t_{(0)}}{t_1}\right)^{-1}, \tag{4.4.37}$$

we obtain

$$\left|\frac{\dot{\vartheta}}{t_{(0)}v}\right|\left|\dot{\vartheta},_i\right| \le \frac{1}{2}\left[\left(1 + \frac{t_{(0)}}{t_1}\right)^{-1}\left(\frac{\dot{\vartheta}}{t_{(0)}v}\right)^2 + \left(1 + \frac{t_{(0)}}{t_1}\right)\left(\dot{\vartheta},_i\right)^2\right]. \tag{4.4.38}$$

[20] The inequality $\hat{\varepsilon} > 0$ in eqn (4.4.35) is implied by the inequalities (1.3.55) and (1.3.56).

Thus, in view of eqns (4.4.29), (4.4.32), (4.4.34), (4.4.36) and (4.4.38), we arrive at

$$\begin{aligned}
\upsilon^{-1}\left|R_i\left(x,p_\lambda\left(x\right)\right)\right| &\le \frac{1}{2}\,\rho^{-1}S_{ik},_k\,S_{ij},_j \\
&+\frac{1}{2}\frac{\rho^{-1}}{\upsilon^2}\left[\left(\dot{S}_{ij}-\frac{1}{3}\dot{S}_{kk}\delta_{ij}\right)\left(\dot{S}_{ij}-\frac{1}{3}\dot{S}_{kk}\delta_{ij}\right)+\frac{1}{3}\dot{S}_{kk}^2\right] \\
&+\frac{1}{2}\frac{t_1}{t_{(0)}}\frac{k^2}{C_S\theta_0}\left[\frac{C_E}{C_S}\left(\frac{t_{(0)}}{t_1}\right)^2(\vartheta,_{kk})^2+\frac{C_S}{C_E}\left(\frac{t_1}{t_{(0)}}\right)^2\left(\frac{\dot{\vartheta},_i}{\upsilon}\right)^2\right] \\
&+\frac{1}{2}\frac{t_1}{t_{(0)}}\frac{1}{C_S\theta_0}\left\{\frac{t_{(0)}}{t_1}\left(1-\frac{C_E}{C_S}\right)^{-\frac{1}{2}}\left[1-\left(1-\frac{C_E}{C_S}\right)\frac{t_1}{t_{(0)}}\right]\alpha^2\theta_0^2\dot{S}_{kk}^2\right. \\
&\left.+\frac{t_1}{t_{(0)}}\left(1-\frac{C_E}{C_S}\right)^{1/2}\left[1-\left(1-\frac{C_E}{C_S}\right)\frac{t_1}{t_{(0)}}\right]^{-1}\frac{k^2}{\upsilon^2}(\dot{\vartheta},_i)^2\right\} \\
&+\frac{k\left(t_1-t_{(0)}\right)}{2\theta_0}\left[\left(1+\frac{t_{(0)}}{t_1}\right)^{-1}\left(\frac{\dot{\vartheta}}{t_{(0)}\upsilon}\right)^2+\left(1+\frac{t_{(0)}}{t_1}\right)(\dot{\vartheta},_i)^2\right].
\end{aligned} \tag{4.4.39}$$

If we now employ the identity

$$\frac{\alpha^2\theta_0}{C_S}=\frac{1}{3\left(3\lambda+2\mu\right)}\left(1-\frac{C_E}{C_S}\right), \tag{4.4.40}$$

we transform the inequality (4.4.39) to the form

$$\begin{aligned}
\upsilon^{-1}\left|R_i\left(x,p_\lambda\left(x\right)\right)\right| &\le \frac{1}{2}\rho^{-1}S_{ik},_k\,S_{ij},_j \\
&+\frac{1}{2}\frac{\rho^{-1}}{\upsilon^2}\left[\left(\dot{S}_{ij}-\frac{1}{3}\dot{S}_{kk}\delta_{ij}\right)\left(\dot{S}_{ij}-\frac{1}{3}\dot{S}_{kk}\delta_{ij}\right)+\frac{1}{3}\dot{S}_{kk}^2\right] \\
&+\frac{1}{2}\frac{t_1}{t_{(0)}}\frac{k^2}{C_S\theta_0}\left[\frac{C_E}{C_S}\left(\frac{t_{(0)}}{t_1}\right)^2(\vartheta,_{kk})^2+\frac{C_S}{C_E}\left(\frac{t_1}{t_{(0)}}\right)^2\left(\frac{\dot{\vartheta},_i}{\upsilon}\right)^2\right] \\
&+\frac{1}{2}\left\{\frac{1}{3\left(3\lambda+2\mu\right)}\left(1-\frac{C_E}{C_S}\right)^{1/2}\left[1-\left(1-\frac{C_E}{C_S}\right)\frac{t_1}{t_{(0)}}\right]\dot{S}_{kk}^2\right. \\
&\left.+\left(\frac{t_1}{t_{(0)}}\right)^2\left(1-\frac{C_E}{C_S}\right)^{1/2}\left[1-\left(1-\frac{C_E}{C_S}\right)\frac{t_1}{t_{(0)}}\right]^{-1}\frac{k^2}{C_S\theta_0\upsilon^2}\left(\dot{\vartheta},_i\right)^2\right\} \\
&+\frac{k\left(t_1-t_{(0)}\right)}{2\theta_0}\left[\left(1+\frac{t_{(0)}}{t_1}\right)^{-1}\left(\frac{\dot{\vartheta}}{t_{(0)}\upsilon}\right)^2+\left(1+\frac{t_{(0)}}{t_1}\right)\left(\dot{\vartheta},_i\right)^2\right].
\end{aligned} \tag{4.4.41}$$

On the other hand, using the definition of $P(x,t)$, and the identity (4.4.40), we obtain

$$\frac{1}{2}P(x,p_\lambda(x)) = \frac{1}{2}\left\{\rho^{-1}S_{ik,k}\,S_{ij,j} + \frac{1}{2\mu}\left(\dot{S}_{ij} - \frac{1}{3}\dot{S}_{kk}\delta_{ij}\right)\left(\dot{S}_{ij} - \frac{1}{3}\dot{S}_{kk}\delta_{ij}\right)\right.$$
$$+\frac{1}{3(3\lambda+2\mu)}\left[1-\left(1-\frac{C_E}{C_S}\right)\frac{t_1}{t_{(0)}}\right]\dot{S}_{kk}^2 + \frac{1}{\theta_0 C_S}\left[\frac{t_1}{t_{(0)}}k^2(\vartheta_{,ii})^2\right.$$
$$\left.\left.+\left(\frac{t_1}{t_{(0)}}-1\right)C_S^2\dot{\vartheta}^2 + t_1 k C_S(\dot{\vartheta}_{,i})^2 - 2\left(\frac{t_1}{t_{(0)}}-1\right)kC_S\vartheta_{,ii}\,\dot{\vartheta}\right]\right\}. \tag{4.4.42}$$

Upon integration of eqns (4.4.41 and 42) over Ω, in light of the inequality (4.4.25), we find

$$\left(\frac{1}{2\mu}-\frac{1}{\rho\upsilon^2}\right)\int_\Omega\left(\dot{S}_{ij}-\frac{1}{3}\dot{S}_{kk}\delta_{ij}\right)\left(\dot{S}_{ij}-\frac{1}{3}\dot{S}_{kk}\delta_{ij}\right)\mathrm{d}v$$
$$+\left\{\frac{1}{3(3\lambda+2\mu)}\left[1-\left(1-\frac{C_E}{C_S}\right)\frac{t_1}{t_{(0)}}\right]\left[1-\left(1-\frac{C_E}{C_S}\right)^{1/2}\right]-\frac{1}{3\rho\upsilon^2}\right\}\int_\Omega \dot{S}_{kk}^2\mathrm{d}v$$
$$+\frac{1}{\theta_0 C_S}\int_\Omega\left[\frac{t_1}{t_{(0)}}k^2(\vartheta_{,ii})^2+\left(\frac{t_1}{t_{(0)}}-1\right)C_S^2\,\dot{\vartheta}^2+t_1 kC_S(\dot{\vartheta}_{,i})^2\right]\mathrm{d}v$$
$$-\frac{t_1}{t_{(0)}}\frac{k^2}{\theta_0 C_S}\int_\Omega\left[\frac{C_E}{C_S}\left(\frac{t_{(0)}}{t_1}\right)^2(\vartheta_{,kk})^2+\frac{C_E}{C_S}\left(\frac{t_1}{t_{(0)}}\right)^2\left(\frac{\dot{\vartheta}_{,i}}{\upsilon}\right)^2\right]\mathrm{d}v$$
$$-\left(\frac{t_1}{t_{(0)}}\right)^2\left(1-\frac{C_E}{C_S}\right)^{1/2}\left[1-\left(1-\frac{C_E}{C_S}\right)\frac{t_1}{t_{(0)}}\right]^{-1}\frac{k^2}{\theta_0 C_S}\int_\Omega\left(\frac{\dot{\vartheta}_{,i}}{\upsilon}\right)^2\mathrm{d}v$$
$$-\frac{k\left(t_1-t_{(0)}\right)}{\theta_0}\int_\Omega\left[\left(1+\frac{t_{(0)}}{t_1}\right)^{-1}\left(\frac{\dot{\vartheta}}{\upsilon t_{(0)}}\right)^2+\left(1+\frac{t_{(0)}}{t_1}\right)\left(\dot{\vartheta}_{,i}\right)^2\right]\mathrm{d}v$$
$$\le\frac{2}{\theta_0 C_S}\left(\frac{t_1}{t_{(0)}}-1\right)\int_\Omega k\vartheta_{,ii}\,C_S\dot{\vartheta}\mathrm{d}v. \tag{4.4.43}$$

Letting

$$a=(k\vartheta_{,ii})^2,\quad b=\left(C_S\dot{\vartheta}\right)^2,\quad \hat{\varepsilon}=1+\frac{t_{(0)}}{t_1} \tag{4.4.44}$$

in the inequality (4.4.30), we obtain

$$2\,|k\vartheta_{,ii}|\left|C_S\dot{\vartheta}\right|\le\left(1+\frac{t_{(0)}}{t_1}\right)k^2(\vartheta_{,ii})^2+\left(1+\frac{t_{(0)}}{t_1}\right)^{-1}C_S^2\,\dot{\vartheta}^2. \tag{4.4.45}$$

Thus, using the inequalities (4.4.43) and (4.4.45), we obtain

$$\left(\frac{1}{2\mu}-\frac{1}{\rho\upsilon^2}\right)\int_\Omega\left(\dot{S}_{ij}-\frac{1}{3}\dot{S}_{kk}\delta_{ij}\right)\left(\dot{S}_{ij}-\frac{1}{3}\dot{S}_{kk}\delta_{ij}\right)\mathrm{d}v$$
$$+\left\{\frac{1}{3(3\lambda+2\mu)}\left[1-\left(1-\frac{C_E}{C_S}\right)\frac{t_1}{t_{(0)}}\right]\left[1-\left(1-\frac{C_E}{C_S}\right)^{1/2}\right]-\frac{1}{3\rho\upsilon^2}\right\}\int_\Omega\dot{S}_{kk}^2\mathrm{d}v$$
$$+\frac{k^2}{\theta_0 C_S}\frac{t_{(0)}}{t_1}\left(1-\frac{C_E}{C_S}\right)\int_\Omega(\vartheta_{,ii})^2\,\mathrm{d}v$$
$$+\frac{k}{\theta_0 t_{(0)}}\left(\frac{t_1}{t_{(0)}}-1\right)\left(\frac{t_{(0)}}{t_1}+1\right)^{-1}\left(\frac{C_S t_{(0)}^2}{k t_1}-\frac{1}{\upsilon^2}\right)\int_\Omega\dot{\vartheta}^2\mathrm{d}v$$
$$+\frac{k}{\theta_0}\frac{t_{(0)}^2}{t_1}\left\{1-\frac{k}{C_S t_{(0)}}\left(\frac{t_1}{t_{(0)}}\right)^3\left[\left(1-\frac{C_E}{C_S}\right)^{\frac{1}{2}}\left[1-\left(1-\frac{C_E}{C_S}\right)\frac{t_1}{t_{(0)}}\right]^{-1}\right.\right.$$
$$\left.\left.+\frac{C_S}{C_E}\frac{t_1}{t_{(0)}}\left(1-\frac{C_E}{C_S}\right)^{\frac{1}{2}}\right]\frac{1}{\upsilon^2}\right\}\int_\Omega(\dot{\vartheta}_{,i})^2\mathrm{d}v\le 0. \tag{4.4.46}$$

It follows from the definition of υ (recall eqns (4.4.7–10)) that the coefficients of the first two integrals, as well as the coefficient of the last integral in eqn (4.4.46), are non-negative. Since

$$1-\frac{C_E}{C_S}>0,\ \frac{t_{(0)}}{t_1}>0, \tag{4.4.47}$$

then also the coefficient of the third integral in eqn (4.4.46) is non-negative. We will now show that the coefficient of the fourth integral in eqn (4.4.46) is non-negative. Indeed, the definition of υ implies that

$$\upsilon^2\ge\upsilon_3^2$$
$$=\frac{k}{C_S t_{(0)}}\left(\frac{t_1}{t_{(0)}}\right)^3\left\{\frac{C_S}{C_E}\frac{t_1}{t_{(0)}}+\left(1-\frac{C_E}{C_S}\right)^{1/2}\left[1-\left(1-\frac{C_E}{C_S}\right)\frac{t_1}{t_{(0)}}\right]^{-1}\right\}. \tag{4.4.48}$$

Therefore, since the second term in $\{\}$ is positive, we obtain

$$\upsilon^2\ge\frac{k}{C_S t_{(0)}}\left(\frac{t_1}{t_{(0)}}\right)^3\left\{\frac{C_S}{C_E}\frac{t_1}{t_{(0)}}\right\}. \tag{4.4.49}$$

The latter inequality together with the inequalities

$$\frac{C_S}{C_E}>1,\ \frac{t_1}{t_{(0)}}\ge 1 \tag{4.4.50}$$

implies that

$$v^2 \geq \frac{kt_1}{C_S t_{(0)}^2}, \tag{4.4.51}$$

so that, the coefficient of the fourth integral in eqn (4.4.46) is non-negative. As a result, the left-hand side of the inequality (4.4.46) is a sum of five non-negative terms. Since, in view of eqn (4.4.46) that sum is non-positive, each of these integrals must vanish. As a result, we find, in particular, that

$$\dot{S}_{ij}(x, p_\lambda(x)) = 0, \ \dot{\vartheta}(x, p_\lambda(x)) = 0 \text{ in } \Omega. \tag{4.4.52}$$

Using the definition of $p_\lambda(x)$ and passing in eqn (4.4.52) to the limit $x \to z$, we obtain

$$\dot{S}_{ij}(z, \lambda) = 0, \ \dot{\vartheta}(z, \lambda) = 0. \tag{4.4.53}$$

Since (z, λ) is an arbitrary point of the Cartesian product $\{\mathrm{B} - \mathcal{D}(t)\} \times (0, t)$, and since the pair (S_{ij}, ϑ) is smooth in $\overline{\mathrm{B}} \times [0, \infty)$, from eqn (4.4.53) we infer that

$$\dot{S}_{ij} = 0, \ \dot{\vartheta} = 0 \text{ on } \ \left\{\overline{\mathrm{B}} - \mathcal{D}(t)\right\} \times [0, t]. \tag{4.4.54}$$

Finally, integrating these relations over the time and using the conditions

$$S_{ij}(\cdot, 0) = 0, \ \vartheta(\cdot, 0) = 0 \text{ on } \overline{\mathrm{B}} - \mathcal{D}(t), \tag{4.4.55}$$

we obtain

$$S_{ij} = 0, \ \vartheta = 0 \text{ on } \left\{\overline{\mathrm{B}} - \mathcal{D}(t)\right\} \times [0, t]. \tag{4.4.56}$$

This completes the proof of Theorem 4.4. □

This theorem implies that an upper bound of speeds of the stress–temperature disturbances described by eqns (4.4.1–3) depends on $v_1^{(0)} = (2\mu/\rho)^{1/2}$, $v_2^{(0)} = [(3\lambda + 2\mu)/\rho]^{1/2}$, $v_3^{(0)} = (k/t_0 C_S)^{1/2}$, and on the ratios $t_1/t_{(0)}$ and C_E/C_S. The upper bound goes to infinity as $t_0 \to 0$ $(t_1 > 0)$ and as $t_1 \to 0$, similarly to the case of PTP of the G-L theory in Section 4.2.

For $t_1 = t_0 > 0$ the speeds v_1 and v_2 given by eqns (4.4.8) and (4.4.9) are identical with the speeds v_1 and v_2 given by eqns (4.3.8) and (4.3.9), while the speed v_3 given by eqn (4.4.10) is greater than v_3 given by eqn (4.3.10). From this, we conclude that, for $t_1 = t_0 > 0$, the domain of influence for the NSTP described by eqns (4.4.1–3) is larger than the domain of influence for the SHFP described by eqns (4.3.1–3), provided the supports of thermomechanical loadings are identical and $v = v_3$ in both problems. The latter two hypotheses are satisfied for example when in both problems the thermomechanical loading is specified only in terms of the initial stresses and stress rates on the same support, and when t_0 is a sufficiently small time.

Let us also note that for $\alpha = 0$ $(C_E = C_S)$ the NSTP described by eqns (4.4.1–3) separates into two problems: (a) the problem of propagation of isothermal, elastic stress disturbances described by eqns $(4.4.1)_1$, $(4.4.2)_2$ and $(4.4.3)_1$, at

$\alpha = 0$; (b) the problem of propagation of temperature in a rigid heat conductor described by eqns $(4.4.1)_2$, $(4.4.2)_2$ and $(4.4.3)_2$, at $\alpha = 0$. In the context of the first problem we have a theorem on the domain of influence with a speed $v \geq \max\left(v_1^{(0)}, v_2^{(0)}\right)$, while for the second problem we have a theorem on the domain of influence with the speed $v \geq v_3^{(0)}$ $(t_{(0)} = t_0)$[21].

4.5 The displacement–temperature problem for an inhomogeneous anisotropic body in the L–S and G–L theories

In this section we describe the domain of influence results for inhomogeneous anisotropic thermoelastic solids of the L–S and G–L theories that have been obtained only recently (Ignaczak *et al.*, 1986; Carbonaro and Ignaczak, 1987). The presentation of the results is confined to a formulation of the domain of influence theorems for a mixed displacement–temperature initial boundary value problem, and no proof of the results is given. Instead, an analysis of the results obtained that should prove useful for a wide range of researchers in the field of thermoelastic waves in the non-homogeneous and anisotropic bodies, is discussed.

4.5.1 *A thermoelastic wave propagating in an inhomogeneous anisotropic L–S model*

A mixed displacement–temperature initial-boundary value problem for an inhomogeneous anisotropic body of the L–S theory is described, in components, by eqns (2.1.3)–(2.1.5). An alternative form of the problem, in which direct notation is used, reads.

Find a pair $(\mathbf{u}, \theta)$ that satisfies the field equations

$$\left.\begin{aligned} div\mathbf{C}(\nabla\mathbf{u}) - \rho\ddot{\mathbf{u}} + div(\theta\mathbf{M}) &= -\mathbf{b} \\ div(\mathbf{K}\nabla\theta) - C_E\hat{\dot{\theta}} + \theta_0\mathbf{M}\cdot\nabla\hat{\dot{\mathbf{u}}} &= -\hat{r} \end{aligned}\right\} \text{ on } \mathrm{B}\times[0,\infty), \tag{4.5.1}$$

the initial conditions

$$\left.\begin{aligned} \mathbf{u}(\cdot,0) = \mathbf{u}_0, \quad \dot{\mathbf{u}}(\cdot,0) = \dot{\mathbf{u}}_0 \\ \theta(\cdot,0) = \vartheta_0, \quad \dot{\theta}(\cdot,0) = \dot{\vartheta}_0 \end{aligned}\right\} \text{ on } \mathrm{B}, \tag{4.5.2}$$

and the boundary conditions

$$\begin{aligned} \mathbf{u} = \mathbf{u}' \quad &\text{on } \partial\mathrm{B}_1\times(0,\infty), \\ [\mathbf{C}(\nabla\mathbf{u}) + \theta\mathbf{M}]\mathbf{n} = \mathbf{s}' \quad &\text{on } \partial\mathrm{B}_2\times(0,\infty), \\ \theta = \theta' \quad &\text{on } \partial\mathrm{B}_3\times(0,\infty), \\ -(\mathbf{K}\nabla\theta)\cdot\mathbf{n} = q' \quad &\text{on } \partial\mathrm{B}_4\times(0,\infty). \end{aligned} \tag{4.5.3}$$

Here, $\mathbf{u}$ and θ denote the displacement and temperature fields, respectively, defined on $\bar{\mathrm{B}}\times[0,\infty)$; B is a region of E^3 occupied by the non-homogeneous anisotropic thermoelastic material, and $[0,\infty)$ is the time interval. In addition,

[21] The domain of influence theorem for the problem of propagation of temperature in a rigid heat conductor was proved by Biały (1983), pp. 52–53.

θ_0 is a fixed uniform reference temperature; $\rho = \rho(x)$ are the mass density, $C_E = C_E(x)$ the specific heat for zero strain, and $r = r(x,t)$ the heat supply (scalar) fields, respectively; $\mathbf{b} = \mathbf{b}(x,t)$ is the body force (vector) field; $\mathbf{K} = \mathbf{K}(x)$ and $\mathbf{M} = \mathbf{M}(x)$ are the conductivity and stress–temperature (second-order) tensor fields, respectively; and $\mathbf{C} = \mathbf{C}(x)$ is the elasticity (fourth-order) tensor field. Moreover, the superimposed dot denotes the partial derivative with respect to time t, and the hut operator is defined by

$$\hat{f} = f + t_0\dot{f} \tag{4.5.4}$$

for any function $f = f(x,t)$ on $\bar{\mathrm{B}} \times [0,\infty)$; the parameter t_0 in eqn (4.5.4) is the relaxation time of the L–S theory; $(\partial \mathrm{B}_1,\ \partial \mathrm{B}_2)$ and $(\partial \mathrm{B}_3,\ \partial \mathrm{B}_4)$ in eqns (4.5.3) are two partitions of the boundary $\partial \mathrm{B}$ of B such that

$$\begin{aligned} \partial \mathrm{B} = \partial \mathrm{B}_1 \cup \partial \mathrm{B}_2 = \partial \mathrm{B}_3 \cup \partial \mathrm{B}_4, \\ \partial \mathrm{B}_1 \cap \partial \mathrm{B}_2 = \partial \mathrm{B}_3 \cap \partial \mathrm{B}_4 = \varnothing, \end{aligned} \tag{4.5.5}$$

and $\mathbf{n} = \mathbf{n}(x)$ is the unit outward vector normal to $\partial \mathrm{B}$ at x. Finally, $(\mathbf{u}_0, \dot{\mathbf{u}}_0, \vartheta_0, \dot{\vartheta}_0)$ in eqns (4.5.2) and $(\mathbf{u}', \mathbf{s}', \theta', q')$ in eqns (4.5.3) are prescribed systems of functions that determine the initial and boundary thermomechanical loads, respectively.

Clearly, a displacement–temperature wave corresponding to the problem (4.5.1)–(4.5.3) is produced by an external thermomechanical load that is represented by the system of functions

$$(\mathbf{b}, r, \mathbf{u}_0, \dot{\mathbf{u}}_0, \vartheta_0, \dot{\vartheta}_0, \mathbf{u}', \mathbf{s}', \theta', q'). \tag{4.5.6}$$

Let $\mathrm{B}(t)$ denote a support of the thermomechanical load for a fixed time t, i.e. the set of points of $\bar{\mathrm{B}}$ on which the load does not vanish over the time interval $[0,t]$. Let $C > 0$ be a constant of the velocity dimension that satisfies the inequality

$$C \geq \max(C_1, C_2), \tag{4.5.7}$$

where

$$\begin{aligned} C_1 &= \sup_{\mathrm{B},|\mathbf{m}|=1} \left\{ \frac{1}{2}\left(\frac{\theta_0}{\rho C_E}\right)^{1/2} |\mathbf{M}| + \left\{ |\mathbf{A}| + \left[\frac{1}{2}\left(\frac{\theta_0}{\rho C_E}\right)^{1/2} |\mathbf{M}|\right]^2 \right\}^{1/2} \right\} \\ C_2 &= \sup_{\mathrm{B}} \left\{ \frac{1}{2}\left(\frac{\theta_0}{\rho C_E}\right)^{1/2} |\mathbf{M}| + \left\{ \frac{|\mathbf{K}|}{t_0 C_E} + \left[\frac{1}{2}\left(\frac{\theta_0}{\rho C_E}\right)^{1/2} |\mathbf{M}|\right]^2 \right\}^{1/2} \right\}. \end{aligned} \tag{4.5.8}$$

Here, $\mathbf{A} = \mathbf{A}(x,\mathbf{m})$ is the (second order) "*acoustic tensor in the propagation direction* $\mathbf{m}$" that is defined for any unit vector $\mathbf{m}$ by the relation

$$\mathbf{A}(x,\mathbf{m})\mathbf{a} = \rho^{-1}(x)\mathbf{C}[\mathbf{a} \otimes \mathbf{m}]\mathbf{m}, \tag{4.5.9}$$

where $\mathbf{a}$ is an arbitrary vector.

Moreover, let $S(x, Ct)$ denote an open ball in E^3 with radius Ct and center at x. We shall call the *domain of influence of the thermomechanical load at the instant t for the mixed problem* (4.5.1)–(4.5.3) the set

$$B^*(t) = \{x \in \bar{B} : B(t) \cap \overline{S(x, Ct)} \neq \varnothing\}, \tag{4.5.10}$$

where C is defined by eqns (4.5.7)–(4.5.9). The following theorem shows that on $[0, t]$ the thermomechanical load of the mixed prolem has no influence on points outside of $B^*(t)$.

Theorem 4.5 *(Domain of influence theorem for MDTP of the L–S theory). Let* $(\mathbf{u}, \theta)$ *be a solution to the problem (4.5.1)–(4.5.3). Then*

$$\mathbf{u} = \mathbf{0}, \quad \theta = 0 \quad on \quad \{\bar{B} - B^*(t)\} \times [0, t]. \tag{4.5.11}$$

Proof. (see (Ignaczak *et al.*, 1986)). □

Theorem 4.5 implies that for a finite time t and for a bounded support of the thermomechanical load $B(t)$, the thermoelastic disturbance generated by the pair $(\mathbf{u}, \theta)$ satisfying eqns (4.5.1)–(4.5.3) vanishes outside of a bounded domain that depends on the load support, the bounds for the thermomechanical constitutive fields, and the relaxation time t_0. This theorem also shows that the thermoelastic disturbance propagates as a wave from the domain $B(t)$ with a finite speed equal to or less than the speed C defined by the relations (4.5.7)–(4.5.8). It follows from the definition of C that $C \to \infty$ as $t_0 \to 0 + 0$. Therefore, if the relaxation time tends to zero, the thermoelastic disturbance described by $(\mathbf{u}, \theta)$ gains an infinite speed, as should be expected since in this case the MDTP of the L–S theory reduces to a MDTP of classical hyperbolic-parabolic thermoelasticity.

The definition of velocity C also implies that, for a particular non-homogeneous and anisotropic thermoelastic solid of the L–S theory in which the acoustic and conductivity tensor fields are relatively small, i.e. when

$$|\mathbf{A}| \ll \frac{1}{4}\left(\frac{\theta_0}{\rho C_E}\right)|\mathbf{M}|^2, \qquad |\mathbf{K}| \ll \frac{1}{4}\left(\frac{t_0\theta_0}{\rho}\right)|\mathbf{M}|^2, \tag{4.5.12}$$

the maximum speed of a thermoelastic wave is given by the formula

$$C_0 = \sup_B \left\{ \left(\frac{\theta_0}{\rho C_E}\right)^{1/2} |\mathbf{M}| \right\}. \tag{4.5.13}$$

This formula shows that for a non-homogeneous anisotropic thermoelastic body in which the acoustic and heat conductivity tensor fields are relatively small, the maximum speed of a thermoelastic wave in the L–S theory is dominated by a suitable scaled stress–temperature tensor field.

Also, note that if $|\mathbf{M}|$ is relatively small, the formula for C_1 reduces to that of a domain of influence theorem of classical isothermal elastodynamics (Gurtin, 1972; Eringen and Şuhubi, 1975), while the formula for C_2 reduces to that of a domain of influence theorem for a non-homogeneous anisotropic rigid heat conductor.

Finally, for a finite value of $|\mathbf{M}|$, the velocities C_1 and C_2 represent upper bounds for the velocities of a quasi-mechanical and of a quasi-thermal wave, respectively, propagating in the non-homogeneous anisotropic L–S model.

4.5.2 *A thermoelastic wave propagating in an inhomogeneous anisotropic G–L model*

A mixed displacement–temperature initial-boundary value problem for a non-homogeneous anisotropic body of the G–L theory is described, in components, by eqns (2.1.11)–(2.1.13); when direct notation is used, the statement of the problem reads:

Find a pair $(\mathbf{u}, \theta)$ that satisfies the field equations

$$\left.\begin{aligned} div\mathbf{C}(\nabla\mathbf{u}) - \rho\ddot{\mathbf{u}} + div[\mathbf{M}(\theta + t_1\dot{\theta})] &= -\mathbf{b} \\ div(\mathbf{K}\nabla\theta) - C_E(\dot{\theta} + t_0\ddot{\theta}) + \theta_0\mathbf{M}\cdot\nabla\dot{\mathbf{u}} &= -r \end{aligned}\right\} \text{ on B} \times [0, \infty), \tag{4.5.14}$$

the initial conditions

$$\left.\begin{aligned} \mathbf{u}(\cdot, 0) = \mathbf{u}_0, \quad \dot{\mathbf{u}}(\cdot, 0) = \dot{\mathbf{u}}_0 \\ \theta(\cdot, 0) = \vartheta_0, \quad \dot{\theta}(\cdot, 0) = \dot{\vartheta}_0 \end{aligned}\right\} \text{ on B}, \tag{4.5.15}$$

and the boundary conditions

$$\begin{gathered} \mathbf{u} = \mathbf{u}' \text{ on } \partial\text{B}_1 \times (0, \infty), \\ [\mathbf{C}(\nabla\mathbf{u}) + \mathbf{M}(\theta + t_1\dot{\theta})]\mathbf{n} = \mathbf{s}' \text{ on } \partial\text{B}_2 \times (0, \infty), \\ \theta = \theta' \text{ on } \partial\text{B}_3 \times (0, \infty), \\ -(\mathbf{K}\nabla\theta)\cdot\mathbf{n} = q' \text{ on } \partial\text{B}_4 \times (0, \infty). \end{gathered} \tag{4.5.16}$$

All the symbols in eqns (4.5.14) and (4.5.16) have similar meanings to those of Section 4.5.1.

The existence of two relaxation times t_0 and t_1 ($t_1 \geq t_0 \geq 0$) in eqns (4.5.14) and (4.5.16) makes a difference between the characterizations (4.5.1)–(4.5.3) and (4.5.14)–(4.5.16). Clearly, the set $\text{B}(t)$ from Section 4.5.1 is also a support of the thermomechanical load at an instant t for the problem (4.5.14)–(4.5.16).

A domain of influence of the thermomechanical load at an instant t for the mixed problem (4.5.14)–(4.5.16) is defined as

$$\text{B}^*(t) = \{x \in \bar{\text{B}} : \text{B}(t) \cap \overline{S(x, Ct)} \neq \varnothing\}, \tag{4.5.17}$$

where C is a constant of the velocity dimension that satisfies the inequality

$$C \geq \max(C_1', C_2'), \tag{4.5.18}$$

in which

$$C_1' = \sup_{B,|\mathbf{m}|=1} \left\{ \frac{1}{2}\left(\frac{\theta_0}{\rho C_E}\right)^{1/2} |\mathbf{M}| + \left\{ |\mathbf{A}| + \left[\frac{1}{2}\left(\frac{\theta_0}{\rho C_E}\right)^{1/2} |\mathbf{M}|\right]^2 \right\}^{1/2} \right\}$$

$$C_2' = \sup_{B} \left\{ \frac{1}{2}\frac{t_1}{t_0}\left(\frac{\theta_0}{\rho C_E}\right)^{1/2} |\mathbf{M}| + \left\{ \frac{|\mathbf{K}|}{t_0 C_E} + \left[\frac{1}{2}\frac{t_1}{t_0}\left(\frac{\theta_0}{\rho C_E}\right)^{1/2} |\mathbf{M}|\right]^2 \right\}^{1/2} \right\}. \tag{4.5.19}$$

With regard to the MDTP characterized by eqns (4.5.14)–(4.5.16), the following theorem holds true.

Theorem 4.6 *(Domain of influence theorem for MDTP of the G–L theory). Let* $(\mathbf{u}, \theta)$ *be a solution to the system (4.5.14)–(4.5.16). Then*

$$\mathbf{u} = \mathbf{0}, \quad \theta = 0 \quad on \ \{\bar{B} - B^*(t)\} \times [0, t], \tag{4.5.20}$$

where $B^*(t)$ *is given by the relations (4.5.17)–(4.5.19).*

Proof. (see (Carbonaro and Ignaczak, 1987)). □

A physical interpretation of Theorem 4.6 is similar to that of Theorem 4.5. Moreover, the definition of C [see eqns (4.5.18) and (4.5.19)] implies that the velocities C_1' and C_2' correspond, respectively, to the maximum speed of a quasi-mechanical and of a quasi-thermal wave propagating in the G–L model; and for $|\mathbf{M}| = 0$ they reduce to the maximum speeds of a pure mechanical and a pure thermal wave, respectively.

Also, for a particular non-homogeneous anisotropic thermoelastic solid in which the acoustic and heat conductivity tensor fields are relatively small, i.e. when

$$|\mathbf{A}| \ll \frac{1}{4}\left(\frac{\theta_0}{\rho C_E}\right)|\mathbf{M}|^2, \qquad |\mathbf{K}| \ll \frac{1}{4}\left(\frac{t_1}{t_0}\right)^2\left(\frac{t_0\theta_0}{\rho}\right)|\mathbf{M}|^2, \tag{4.5.21}$$

the maximum speed of a thermoelastic wave is given by

$$C_0' = \sup_{B} \left\{ \frac{t_1}{t_0}\left(\frac{\theta_0}{\rho C_E}\right)^{1/2} |\mathbf{M}| \right\}. \tag{4.5.22}$$

In addition, if (C_1, C_2) stands for a pair of velocities in the L–S theory [see, eqn (4.5.8)], and if the thermomechanical constitutive fields of the L–S and G–L models, such as $\mathbf{K}$ and $\mathbf{M}$, are identical, we have the results $(t_1 \geq t_0 \geq 0)$

$$C_1' = C_1 \quad \text{and} \quad C_2' \geq C_2 \tag{4.5.23}$$

and

$$C_2' = C_2 \Leftrightarrow t_1 = t_0 > 0. \tag{4.5.24}$$

Therefore, the following observations are in order. If the supports of the thermomechanical load in a MDTP of the L–S and G–L theories are the same, then: (i) the domain of influence of the G–L theory is not smaller than that of the L–S theory, and (ii) the domain of influence of the G–L theory coincides with that of the L–S theory if $t_1 = t_0$.

To wind up this section we refer the reader to a survey article on the domain of influence results in generalized thermoelasticity (Ignaczak, 1991) as well as to other survey articles on the subject (Chandrasekharaiah, 1986, 1998).

Remark 4.2 Theorems on the domain of influence for a homogeneous isotropic body given in Sections 4.1–4.4 specify the domains that are not reached by fronts of a thermoelastic disturbance for a time $t > 0$. However, those theorems do not say anything about the nature of these fronts. It follows from the definition of the domain of influence $\mathcal{D}(t)$ that this domain contains the support of thermomechanical loading $\mathcal{D}_0(t)$ and all (open or closed) fronts of thermoelastic disturbances propagating from $\mathcal{D}_0(t)$. Some of those fronts may coincide with the whole or a part of the boundary $\partial\mathcal{D}(t)$. In such cases the outer normal vector on $\partial\mathcal{D}(t)$ is also the direction of a front of thermoelastic disturbances. Similarly, the domains $\mathrm{B}(t)$ and $\mathrm{B}^*(t)$ of Section 4.5 may be used to interpret propagation of thermoelastic waves in a non-homogeneous and anisotropic body.

5

CONVOLUTIONAL VARIATIONAL PRINCIPLES

5.1 Alternative descriptions of a conventional thermoelastic process in the Green–Lindsay theory

According to the definition introduced in Section 2.1, a conventional thermoelastic process in the G–L theory may be specified as an array of functions $(u_i, E_{ij}, S_{ij}, \vartheta, g_i, q_i)$ on $\overline{\mathrm{B}} \times [0, \infty)$, satisfying the relations:[1]

the kinematic and thermal equations

$$E_{ij} = \frac{1}{2}(u_{i,j} + u_{j,i}), \; g_i = \vartheta_{,i} \quad \text{on} \;\; \mathrm{B} \times (0, \infty), \tag{5.1.1}$$

the equations of motion

$$S_{ij,j} + b_i = \rho \ddot{u}_i \quad \text{on} \;\; \mathrm{B} \times (0, \infty), \tag{5.1.2}$$

the energy balance equation

$$-q_{i,i} + r = C_E(\dot{\vartheta} + t_0 \ddot{\vartheta}) - \theta_0 M_{ij} \dot{E}_{ij} \quad \text{on} \;\; \mathrm{B} \times (0, \infty), \tag{5.1.3}$$

the constitutive equations

$$\left.\begin{aligned} S_{ij} &= C_{ijkl} E_{kl} + M_{ij}(\vartheta + t_1 \dot{\vartheta}) \\ q_i &= -k_{ij} g_j \end{aligned}\right\} \quad \text{on} \;\; \mathrm{B} \times (0, \infty), \tag{5.1.4}$$

the initial conditions

$$\left.\begin{aligned} u_i(\cdot, 0) &= u_{i0}, \quad \dot{u}_i(\cdot, 0) = \dot{u}_{i0} \\ \vartheta(\cdot, 0) &= \vartheta_0, \quad \dot{\vartheta}(\cdot, 0) = \dot{\vartheta}_0 \end{aligned}\right\} \quad \text{on} \;\; \mathrm{B}, \tag{5.1.5}$$

and the mixed boundary conditions

$$\begin{aligned} u_i &= u_i' && \text{on} \;\; \partial \mathrm{B}_1 \times (0, \infty), \\ S_{ij} n_j &= s_i' && \text{on} \;\; \partial \mathrm{B}_2 \times (0, \infty), \\ \vartheta &= \vartheta' && \text{on} \;\; \partial \mathrm{B}_3 \times (0, \infty), \\ q_i n_i &= q' && \text{on} \;\; \partial \mathrm{B}_4 \times (0, \infty). \end{aligned} \tag{5.1.6}$$

[1] In Section 1.3 a thermoelastic process of the G–L theory corresponding to a load (b_i, r) is defined as an array of functions $(u_i, E_{ij}, S_{ij}, \vartheta, \eta, q_i)$ that satisfies the field equations (1.3.42–47). Such a definition is equivalent to that in which $(u_i, E_{ij}, S_{ij}, \vartheta, g_i, q_i)$ satisfies (5.1.1–4).

In view of eqns (1.3.48–53), the field equations (5.1.1–4) may be written in the alternative form

$$E_{ij} = \tfrac{1}{2}\,(u_{i},_{j} + u_{j},_{i}\,)\,,\; g_i = \vartheta,_{i} \quad \text{on} \;\; \mathrm{B} \times (0, \infty), \tag{5.1.7}$$

$$S_{ij},_{j} + b_i = \rho \ddot{u}_i \quad \text{on} \;\; \mathrm{B} \times (0, \infty), \tag{5.1.8}$$

$$-q_{i},_{i} + r = C_S(\dot{\vartheta} + t_{(0)}\ddot{\vartheta}) - \theta_0 A_{pq}\dot{S}_{pq} \quad \text{on} \;\; \mathrm{B} \times (0, \infty), \tag{5.1.9}$$

and

$$\left.\begin{aligned} E_{ij} &= K_{ijkl}S_{kl} + A_{ij}(\vartheta + t_1\dot{\vartheta}) \\ q_i &= -\lambda_{ij}q_j \end{aligned}\right. \quad \text{on} \;\; \mathrm{B} \times (0, \infty). \tag{5.1.10}$$

Thus, a conventional thermoelastic process in the G–L theory is described either by the field equations (5.1.1–4) and conditions (5.1.5 and 6), or by the field equations (5.1.7–10) and conditions (5.1.5–6). Similarly to what was done in the previous chapters, the process is assumed to be sufficiently smooth on $\overline{\mathrm{B}} \times [0, \infty)$. This smoothness may be achieved by imposing appropriate conditions on the material functions, thermomechanical loadings and the domain B appearing in the initial-boundary value problem (5.1.1–6) or in the equivalent problem (5.1.7–10) with conditions (5.1.5 and 6).

In order to obtain a variational formulation of the problem (5.1.1–6) we proceed in a way similar to what is done in classical thermoelasticity (Ieşan, 1966). This involves two stages. First, we reduce the initial-boundary value problem (5.1.1–6) to an equivalent boundary value problem by replacing the field equations (5.1.1–4) and initial conditions (5.1.5) by the integro-differential field equations implicitly containing the initial conditions. Next, we derive a variational principle for this reduced problem. The reduced problem is also called an *alternative problem associated with the conventional thermoelastic process of the G–L theory*, and the relations describing it provide an *alternative description of that process.*

In order to obtain the alternative description, we employ a Laplace transform. Let us recall that, if $f = f\,(t)$ is a function defined over $[0, \infty)$, its Laplace transform is the integral

$$\bar{f}\,(p) = \int_0^\infty \mathrm{e}^{-pt} f\,(t)\,\mathrm{d}t, \tag{5.1.11}$$

where p is the parameter of transformation. Applying that transformation to eqns (5.1.2), (5.1.3) and $(5.1.4)_1$, and using the initial conditions (5.1.5), we obtain

$$\bar{S}_{ij},_{j} + \bar{b}_i = \rho(p^2\bar{u}_i - \dot{u}_{i0} - pu_{i0}), \tag{5.1.12}$$

$$-\bar{q}_{i},_{i} + \bar{r} = C_E[p\bar{\vartheta} - \vartheta_0 + t_0(p^2\bar{\vartheta} - \dot{\vartheta}_0 - p\vartheta_0)] - \theta_0 M_{ij}(p\bar{E}_{ij} - E^0_{ij}), \tag{5.1.13}$$

$$\bar{S}_{ij} = C_{ijkl}\bar{E}_{kl} + M_{ij}[\bar{\vartheta} + t_1(p\bar{\vartheta} - \vartheta_0)], \tag{5.1.14}$$

where

$$E^0_{ij} = E_{ij}\left(\cdot, 0\right) = u_{(i0,j)}\,. \tag{5.1.15}$$

We now divide eqn (5.1.12) by p^2, eqn (5.1.13) by $p\,(t_0 p + 1)$, and eqn (5.1.14) by $t_1 p + 1$ to obtain

$$\bar{g}\ \bar{S}_{ij,j} + \bar{b}_i - \rho \bar{u}_i + \bar{l}_i = 0, \tag{5.1.16}$$

$$-\bar{V}_0\ \bar{q}_{i,i} - C_E\ \bar{\vartheta} + \theta_0 M_{ij}\ p\ \bar{V}_0\ \bar{E}_{ij} + \bar{W} = 0, \tag{5.1.17}$$

$$p\bar{V}_1(\bar{S}_{ij} - C_{ijkl}\bar{E}_{kl}) - M_{ij}\bar{\vartheta} + \bar{m}_{ij} = 0, \tag{5.1.18}$$

where

$$\bar{g}\,(p) = p^{-2}, \tag{5.1.19}$$

$$\bar{V}_0\,(p) = \left[p(t_0 p + 1)\right]^{-1}, \tag{5.1.20}$$

$$\bar{V}_1\,(p) = \left[p(t_1 p + 1)\right]^{-1}, \tag{5.1.21}$$

$$\bar{l}_i\,(p) = \bar{g}(\bar{b}_i + \rho \dot{u}_{i0} + \rho p u_{i0}), \tag{5.1.22}$$

$$\bar{W}\,(p) = \bar{V}_0\{\bar{r} + [C_E t_0 \dot{\vartheta}_0 - \theta_0 M_{ij} E^0_{ij} + C_E\,(1 + t_0 p)\,\vartheta_0]\}, \tag{5.1.23}$$

$$\bar{m}_{ij}\,(p) = t_1 \vartheta_0 p \bar{V}_1 M_{ij}. \tag{5.1.24}$$

Applying now the inverse transform to eqns (5.1.16–18), and using the convolution theorem (Mikusiński, 1957), we obtain

$$g * S_{ij,j} - \rho u_i + l_i = 0, \tag{5.1.25}$$

$$-V_0 * q_{i,i} - C_E \vartheta + \theta_0 M_{ij} \dot{V}_0 * E_{ij} + W = 0, \tag{5.1.26}$$

$$\dot{V}_1 * (S_{ij} - C_{ijkl} E_{kl}) - M_{ij}\vartheta + m_{ij} = 0, \tag{5.1.27}$$

where, in view of eqns (5.1.19–24), we have

$$g\,(t) = t, \tag{5.1.28}$$

$$V_0\,(t) = 1 - \mathrm{e}^{-t/t_0}, \quad \dot{V}_0\,(t) = \frac{1}{t_0}\mathrm{e}^{-t/t_0}, \tag{5.1.29}$$

$$V_1\,(t) = 1 - \mathrm{e}^{-t/t_1}, \quad \dot{V}_1\,(t) = \frac{1}{t_1}\mathrm{e}^{-t/t_1}, \tag{5.1.30}$$

$$l_i = t * b_i + \rho(t\dot{u}_{i0} + u_{i0}), \tag{5.1.31}$$

$$W = V_0 * r + V_0(C_E t_0 \dot{\vartheta}_0 - \theta_0 M_{ij} E^0_{ij}) + C_E \vartheta_0, \tag{5.1.32}$$

$$m_{ij} = t_1 \vartheta_0 \dot{V}_1 M_{ij}, \tag{5.1.33}$$

and where $*$ denotes the convolution over time, that is, for any two functions f_1 and f_2 defined over $[0,\infty)$ we have

$$f_1 * f_2 = \int_0^t f_1(\tau)\, f_2(t-\tau)\, \mathrm{d}\tau. \qquad (5.1.34)$$

In this way we arrive at the following lemma [p. 12 in (Gładysz, 1982)]:

Lemma 5.1 *The array of functions $(u_i, E_{ij}, S_{ij}, \vartheta, g_i, q_i)$ represents a conventional thermoelastic process in the G–L theory iff it satisfies the integro-differential field equations*

$$E_{ij} = \frac{1}{2}(u_{i,j} + u_{j,i}) \quad \text{on} \;\; B \times [0,\infty), \qquad (5.1.35)$$

$$g_i = \vartheta_{,i} \quad \text{on} \;\; B \times [0,\infty), \qquad (5.1.36)$$

$$g * S_{ij,j} - \rho u_i + l_i = 0 \quad \text{on} \;\; B \times [0,\infty), \qquad (5.1.37)$$

$$-V_0 * q_{i,i} - C_E \vartheta + \theta_0 M_{ij} \dot{V}_0 * E_{ij} + W = 0 \quad \text{on} \;\; B \times [0,\infty), \qquad (5.1.38)$$

$$\dot{V}_1 * (S_{ij} - C_{ijkl}E_{kl}) - M_{ij}\vartheta + m_{ij} = 0 \quad \text{on} \;\; B \times [0,\infty), \qquad (5.1.39)$$

$$q_i = -k_{ij} g_j \quad \text{on} \; B \times [0,\infty), \qquad (5.1.40)$$

subject to the boundary conditions (5.1.6).

Lemma 5.1 states that the conventional thermoelastic process in the G–L theory may be treated as a solution of the boundary value problem described by relations (5.1.35–40) and (5.1.6). The variational description of that process reduces therefore to the determination of a functional associated with that boundary value problem.

Another, alternative description of the conventional thermoelastic process may be obtained upon taking eqns (5.1.7–10) in place of eqns (5.1.1–4). In the latter case, we apply the Laplace transform to eqns (5.1.8), (5.1.9) and $(5.1.10)_1$ to obtain

$$\bar{S}_{ij,j} + \bar{b}_i = \rho(p^2 \bar{u}_i - \dot{u}_{i0} - p u_{i0}), \qquad (5.1.41)$$

$$\begin{aligned} -\bar{q}_{i,i} + \bar{r} = C_S[(p\bar{\vartheta} - \vartheta_0) + t_{(0)}(p^2\bar{\vartheta} - \dot{\vartheta}_0 - p\vartheta_0)] \\ + \theta_0 A_{pq}\left(p\bar{S}_{pq} - S^0_{pq}\right), \end{aligned} \qquad (5.1.42)$$

$$\bar{E}_{ij} - K_{ijkl}\bar{S}_{kl} = A_{ij}\left[\bar{\vartheta} + t_1\left(p\bar{\vartheta} - \vartheta_0\right)\right], \qquad (5.1.43)$$

where

$$A_{pq}S^0_{pq} = -M_{pq}E^0_{pq} + A_{pq}M_{pq}(\vartheta_0 + t_1\dot{\vartheta}_0). \qquad (5.1.44)$$

Dividing eqn (5.1.41) through by p^2, eqn (5.1.42) by $p\left(t_{(0)}p + 1\right)$, and eqn (5.1.43) by $(pt_1 + 1)$, we obtain

$$\bar{g}\, \bar{S}_{ij,j} - \rho\bar{u}_i + \bar{l}_i = 0, \qquad (5.1.45)$$

$$-\bar{V}p(p^{-1}\bar{q}_{i,i} + \theta_0 A_{pq}\bar{S}_{pq}) - C_S\bar{\vartheta} + \bar{d} = 0, \tag{5.1.46}$$

$$\bar{V}_1\, p\left(\bar{E}_{ij} - K_{ijkl}\bar{S}_{kl}\right) - A_{ij}\bar{\vartheta} + \bar{n}_{ij} = 0, \tag{5.1.47}$$

where the symbols $\bar{g}\,(p)$, $\overline{V}_1(p)$ and $\bar{l}_i\,(p)$ are defined by the formulas (5.1.19), (5.1.21) and (5.1.22), respectively, while the functions $\overline{V}\,(p)$, $\overline{d}\,(p)$ and $\overline{n}_{ij}\,(p)$ are given by

$$\overline{V}\,(p) = \left[p\left(t_{(0)}p + 1\right)\right]^{-1}, \tag{5.1.48}$$

$$\overline{d}\,(p) = \overline{V}\,\overline{r} + C_S\vartheta_0 p^{-1} + \overline{V}(t_{(0)}C_S\dot{\vartheta}_0 + \vartheta_0 A_{pq}S^0_{pq}), \tag{5.1.49}$$

$$\overline{n}_{ij}(p) = t_1\vartheta_0 p\overline{V}_1 A_{ij}. \tag{5.1.50}$$

Retransforming the relations (5.1.45–47) to the time domain, we find

$$g * S_{ij,j} - \rho u_i + l_i = 0, \tag{5.1.51}$$

$$-\dot{V} * (1 * q_{i,i} + \theta_0 A_{pq}S_{pq}) - C_S\vartheta + d = 0, \tag{5.1.52}$$

$$\dot{V}_1 * (E_{ij} - K_{ijkl}S_{kl}) - A_{ij}\vartheta + n_{ij} = 0, \tag{5.1.53}$$

where the functions $g\,(t)$, $V_1\,(t)$ and $l_i\,(t)$ are given by eqns (5.1.28), (5.1.30) and (5.1.31), respectively, while, in light of eqns (5.1.48–50), the functions $V\,(t)$, $d\,(t)$ and $n_{ij}\,(t)$ become

$$V\,(t) = 1 - \mathrm{e}^{-t/t_{(0)}}, \quad \dot{V}\,(t) = \frac{1}{t_{(0)}}\mathrm{e}^{-t/t_{(0)}}, \tag{5.1.54}$$

$$d\,(t) = V * r + C_S\vartheta_0 + V(C_S t_{(0)}\dot{\vartheta}_0 + \theta_0 A_{pq}S^0_{pq}), \tag{5.1.55}$$

$$n_{ij}(t) = t_1\vartheta_0\dot{V}_1 A_{ij}. \tag{5.1.56}$$

As a result, the following lemma holds true [p. 26 in (Gładysz, 1982)]:

Lemma 5.2 *The array of functions $(u_i, E_{ij}, S_{ij}, \vartheta, g_i, q_i)$ is a conventional thermoelastic process in the G–L theory iff it satisfies the field equations*

$$E_{ij} = \frac{1}{2}(u_{i,j} + u_{i,j}) \quad on \;\; B \times [0,\infty), \tag{5.1.57}$$

$$g_i = \vartheta_{,i} \quad on \;\; B \times [0,\infty), \tag{5.1.58}$$

$$g * S_{ij,j} - \rho u_i + l_i = 0 \quad on \;\; B \times [0,\infty), \tag{5.1.59}$$

$$-\dot{V} * (1 * q_{i,i} + \theta_0 A_{pq}S_{pq}) - C_S\vartheta + d = 0 \quad on \;\; B \times [0,\infty), \tag{5.1.60}$$

$$\dot{V}_1 * (E_{ij} - K_{ijkl}S_{kl}) - A_{ij}\vartheta + n_{ij} = 0 \quad on \;\; B \times [0,\infty), \tag{5.1.61}$$

$$g_i = -\lambda_{ij}q_j \quad on \;\; B \times [0,\infty), \tag{5.1.62}$$

subject to the boundary conditions (5.1.6).

Let us note that the field equations appearing in both lemmas must be satisfied on $B \times [0, \infty)$, that is, in particular for $t = 0$ and $x \in B$. The latter condition and smoothness of the process guarantee that these equations replace the original field equations on $B \times [0, \infty)$ along with the initial conditions on B.

We shall next formulate three more lemmas allowing one to describe a conventional thermoelastic process in terms of three pairs of thermomechanical variables: (u_i, ϑ), (S_{ij}, ϑ) and (S_{ij}, q_i)[2]. In accordance with the definition introduced in Section 2.1, a thermomechanical pair $(\cdot, \cdot)$ made of the variables defining a process is the pair corresponding to that process, provided one may reproduce from the pair the remaining variables determined by that process.

With respect to the pair (u_i, ϑ) we have:

Lemma 5.3 *The pair (u_i, ϑ) corresponds to a conventional thermoelastic process in the G–L theory iff*

$$1 * V_1 * (C_{ijkl} u_{k,l})_{,j} + t * (M_{ij}\vartheta)_{,j} - \dot{V}_1 * \rho u_i + l_i^{(1)} = 0 \quad on \quad B \times [0, \infty), \tag{5.1.63}$$

$$V_0 * (k_{ij}\vartheta_j)_{,i} - C_E \vartheta + \theta_0 M_{ij} \dot{V}_0 * u_{i,j} + W = 0 \quad on \; B \times [0, \infty), \tag{5.1.64}$$

and

$$u_i = u_i' \quad on \quad \partial B_1 \times [0, \infty), \tag{5.1.65}$$

$$[C_{ijkl} u_{k,l} + M_{ij}(\vartheta + t_1 \dot{\vartheta})] n_j = s_i' \quad on \quad \partial B_2 \times [0, \infty), \tag{5.1.66}$$

$$\vartheta = \vartheta' \quad on \quad \partial B_3 \times [0, \infty), \tag{5.1.67}$$

$$-k_{ij}\vartheta_{,j} n_i = q' \quad on \quad \partial B_4 \times [0, \infty), \tag{5.1.68}$$

where

$$l_i^{(1)} = \dot{V}_1 * l_i - t * m_{ij,j}\,. \tag{5.1.69}$$

Proof of Lemma 5.3 The proof is based upon the Lemma 5.1. Eliminating the variables E_{ij}, S_{ij}, g_i and q_i from the system (5.1.35–40), we arrive at the system (5.1.63 and 64). The boundary conditions (5.1.65–68) are obtained by using the conditions (5.1.6) along with eqns (5.1.35,36,39,40). In order to recover the relations (5.1.35–40) and the conditions (5.1.6) from the relations (5.1.63–69) it suffices to note that the pair (u_i, ϑ) satisfying those relations generates the fields E_{ij}, S_{ij}, g_i, q_i through the relations (5.1.35 ,36,39,40). □

[2] A conventional thermoelastic process of the G–L theory may also be described by a pair (u_i, q_i) [see p. 24 in (Gładysz, 1982)].

Lemma 5.4 *The pair (S_{ij}, ϑ) corresponds to a conventional thermoelastic process in the G–L theory iff*

$$\dot{V}_1 * \left[t * \left(\rho^{-1} S_{(ik},_{k}\right),_{j)} - K_{ijkl} S_{kl}\right] - A_{ij}\vartheta + n_{ij}^{(1)} = 0 \quad on \quad B \times [0, \infty), \tag{5.1.70}$$

$$\dot{V} * [1 * (k_{ij}\vartheta_{,j}),_{i} - \theta_0 A_{pq} S_{pq}] - C_S \vartheta + d = 0 \quad on \quad B \times [0, \infty), \tag{5.1.71}$$

and

$$\rho^{-1}(t * S_{ik},_{k} + l_i) = u_i' \quad on \quad \partial B_1 \times [0, \infty), \tag{5.1.72}$$

$$S_{ij} n_j = s_i' \quad on \quad \partial B_2 \times [0, \infty), \tag{5.1.73}$$

$$\vartheta = \vartheta' \quad on \quad \partial B_3 \times [0, \infty), \tag{5.1.74}$$

$$-k_{ij}\vartheta,_{j} n_i = q' \quad on \quad \partial B_4 \times [0, \infty), \tag{5.1.75}$$

where

$$n_{ij}^{(1)} = n_{ij} + \dot{V}_1 * \left(\rho^{-1} l_{(i}\right),_{j)} . \tag{5.1.76}$$

Proof of Lemma 5.4 The proof is based upon Lemma 5.2. Elimination of the variables u_i, E_{ij}, g_i and q_i from the system (5.1.57–62), leads to the field equations (5.1.70 and 71). The boundary conditions (5.1.72–75) are obtained by using the conditions (5.1.6) along with the field equations (5.1.58,59,62) extended to the Cartesian product $\overline{\mathrm{B}} \times [0, \infty)$. This completes the proof of the first part of Lemma 5.4. In order to prove that the relations (5.1.70–75) imply eqns (5.1.57–62) and the boundary conditions (5.1.6), it suffices to note that the pair (S_{ij}, ϑ) satisfying eqns (5.1.70–75) generates the fields u_i, E_{ij}, g_i and q_i from the relations (5.1.59,57,58,62). This completes the proof of Lemma 5.4. □

Lemma 5.5 *The pair (S_{ij}, q_i) corresponds to a conventional thermoelastic process in the G–L theory iff it satisfies the relations*

$$\begin{aligned} &\dot{V}_1 * \left[t * \left(\rho^{-1} S_{(ik},_{k}\right),_{j)} - K_{ijkl} S_{kl}\right] \\ &+ C_S^{-1} A_{ij} \dot{V} * (1 * q_k,_{k} + \theta_0 A_{pq} S_{pq}) + n_{ij}^{(2)} = 0 \quad on \quad B \times [0, \infty), \end{aligned} \tag{5.1.77}$$

$$[C_S^{-1} \dot{V} * (1 * q_{k,k} + \theta_0 A_{pq} S_{pq})],_{i} - \lambda_{ij} q_j - \left(C_S^{-1} d\right),_{i} = 0 \quad on \quad B \times [0, \infty), \tag{5.1.78}$$

and

$$\rho^{-1}(t * S_{ik},_{k} + l_i) = u_i' \quad on \quad \partial B_1 \times [0, \infty), \tag{5.1.79}$$

$$S_{ij} n_j = s_i' \quad on \quad \partial B_2 \times [0, \infty), \tag{5.1.80}$$

$$-C_S^{-1} \dot{V} * (1 * q_k,_{k} + \theta_0 A_{pq} S_{pq}) + C_S^{-1} d = \vartheta' \quad on \quad \partial B_3 \times [0, \infty), \tag{5.1.81}$$

$$q_i n_i = q' \quad on \quad \partial B_4 \times [0, \infty), \tag{5.1.82}$$

where

$$n_{ij}^{(2)} = n_{ij}^{(1)} - C_S^{-1} dA_{ij}. \tag{5.1.83}$$

Proof of Lemma 5.5 The proof is also based upon Lemma 5.2. In order to show that eqns (5.1.57–62) and the conditions (5.1.6) appearing in Lemma 5.2 imply eqns (5.1.77–82) and (5.1.79–82), we eliminate the variables u_i, E_{ij}, ϑ and g_i from the relations (5.1.6,57–62), so as to pass to the relations (5.1.77–82).

In order to prove the converse implication, it suffices to note that the pair (S_{ij}, q_i) satisfying the relations (5.1.77–83) generates the fields u_i, E_{ij}, ϑ and g_i from the relations (5.1.59,57,60,62) in such a way that the array $(u_i, E_{ij}, S_{ij}, \vartheta, g_i, q_i)$ is a conventional thermoelastic process in the G–L theory. □

5.2 Variational principles for a conventional thermoelastic process in the Green–Lindsay theory

The present section is devoted to the formulation of five variational principles associated with five alternative descriptions of the conventional thermoelastic process in the G–L theory that were derived in the preceding section. First, we give several basic definitions (Ignaczak, 1980a).

A functional is a real-valued function whose domain is a subset of a linear space of functions. If L is a linear space, K is a subset of L and $\Lambda(\cdot)$ is a functional defined over K, then[3], for

$$p, \tilde{p} \in L, \quad p + \tilde{\lambda}\tilde{p} \in K, \quad \forall \tilde{\lambda} \in \mathbb{R}, \tag{5.2.1}$$

we formally define the *variation of* $\Lambda(p)$ *at* p by

$$\delta_{\tilde{p}} \Lambda(p) = \frac{\mathrm{d}}{\mathrm{d}\tilde{\lambda}} \Lambda\left(p + \tilde{\lambda}\tilde{p}\right)|_{\tilde{\lambda}=0}. \tag{5.2.2}$$

This variation vanishes for a certain $p \in K$, and we write

$$\delta_{\tilde{p}} \Lambda(p) = 0 \text{ on } K, \tag{5.2.3}$$

iff $\delta_{\tilde{p}} \Lambda(p)$ exists and equals zero for every $\tilde{p}$ consistent with eqn (5.2.1).

An ordered set of functions $(u_i, E_{ij}, S_{ij}, \vartheta, g_i, q_i)$, in which ϑ is a sufficiently smooth scalar field over $\overline{\mathrm{B}} \times [0, \infty)$, while u_i, g_i and q_i are sufficiently smooth vector fields over $\overline{\mathrm{B}} \times [0, \infty)$, and E_{ij} and S_{ij} are sufficiently smooth symmetric tensor fields over $\overline{\mathrm{B}} \times [0, \infty)$, is called an *admissible process.* If $p = (u_i, E_{ij}, S_{ij}, \vartheta, g_i, q_i)$ is chosen as an admissible process, and if the operations of addition and multiplication by a scalar are defined as

$$\begin{aligned} p + \tilde{p} &= \left(u_i + \tilde{u}_i, E_{ij} + \tilde{E}_{ij}, S_{ij} + \tilde{S}_{ij}, \vartheta + \tilde{\vartheta}, g_i + \tilde{g}_i, q_i + \tilde{q}_i\right), \\ \tilde{\lambda} p &= \left(\tilde{\lambda} u_i, \tilde{\lambda} E_{ij}, \tilde{\lambda} S_{ij}, \tilde{\lambda}\vartheta, \tilde{\lambda} g_i, \tilde{\lambda} q_i\right), \end{aligned} \tag{5.2.4}$$

[3] The symbol $\mathbb{R}$ in eqn (5.2.1) stands for a set of all real numbers.

then the set of all admissible processes p is a linear space, which may be identified with L. K may then be taken as the set of all p subject to additional restrictions. In the following, we shall denote by p_i and $\tilde{p}_i$ the stress vectors associated with the processes p and $\tilde{p}$, and by q and $\tilde{q}$ the normal components of the heat-flux vectors associated with the processes p and $\tilde{p}$, that is

$$p_i = S_{ij}n_j, \quad \tilde{p}_i = \tilde{S}_{ij}n_j, \tag{5.2.5}$$

and

$$q = q_i n_i, \; \tilde{q} = \tilde{q}_i n_i. \tag{5.2.6}$$

Based on these definitions we shall now prove the following theorem [p. 15 in (Gładysz, 1982)]:

Theorem 5.1 *Let K be the set of all admissible processes p. Let $p = (u_i, E_{ij}, S_{ij}, \vartheta, g_i, q_i)$ and, for every $t \in [0, \infty)$, let us define the functional $\Lambda_t(\cdot)$ over K through*

$$\begin{aligned}
\Lambda_t(p) = \int_B \dot{V}_0 * \Big[t * \left(\frac{1}{2}\dot{V}_1 * C_{ijkl}E_{kl} - m_{ij} \right) * E_{ij} \\
+ \frac{1}{2}\rho \dot{V}_1 * u_i * u_i - t * \dot{V}_1 * S_{ij} * E_{ij} - \dot{V}_1 * (t * S_{ik},_k + l_i) * u_i \Big] \mathrm{d}v \\
- \frac{1}{\theta_0} \int_B t * \Big[\frac{1}{2} V_0 * g_i * k_{ij} g_j + \frac{1}{2} C_E \vartheta * \vartheta + V_0 * g_i * q_i \\
+ (V_0 * q_k,_k - W) * \vartheta - \theta_0 \dot{V}_0 * M_{ij} E_{ij} * \vartheta \Big] \mathrm{d}v \\
+ \int_{\partial B_1} V_0 * V_1 * p_i * u_i' \mathrm{d}a + \int_{\partial B_2} V_0 * V_1 * (p_i - s_i') * u_i \mathrm{d}a \\
+ \frac{1}{\theta_0} \left[\int_{\partial B_3} t * V_0 * q * \vartheta' \mathrm{d}a + \int_{\partial B_4} t * V_0 * (q - q') * \vartheta \mathrm{d}a \right].
\end{aligned} \tag{5.2.7}$$

Then

$$\delta_{\tilde{p}} \Lambda_t(p) = 0 \quad on\ K, \quad t \in [0, \infty), \tag{5.2.8}$$

iff $p = (u_i, E_{ij}, S_{ij}, \vartheta, g_i, q_i)$ is a conventional thermoelastic process in the G–L theory.

Proof. Let $p, \tilde{p} \in K$ and let $p + \tilde{\lambda}\tilde{p} \in K$, $\forall \tilde{\lambda} \in \mathbb{R}$. Then, according to the definition (5.2.2), we obtain

$$\begin{aligned}
\delta_{\tilde{p}}\Lambda_t(p) &= \dot{V}_0 * \int_{\mathrm{B}} \left[\dot{V}_1 * (\rho u_i - t * S_{ij},_j - l_i) * \tilde{u}_i - \dot{V}_1 * t * u_i * \tilde{S}_{ik},_k\right] \mathrm{d}\upsilon \\
&+ \dot{V}_0 * t * \int_{\mathrm{B}} \left[\dot{V}_1 * (C_{ijkl}E_{kl} - S_{ij}) + M_{ij}\vartheta - m_{ij}\right] * \tilde{E}_{ij}\mathrm{d}\upsilon \\
&- \dot{V}_0 * t * \dot{V}_1 * \int_{\mathrm{B}} E_{ij} * \tilde{S}_{ij}\mathrm{d}\upsilon \\
&+ \frac{t}{\theta_0} * \int_{\mathrm{B}} \left[-V_0 * q_k,_k + \theta_0\dot{V}_0 * M_{ij}E_{ij} + W - C_E\vartheta\right] * \tilde{\vartheta}\mathrm{d}\upsilon \\
&- \frac{t}{\theta_0} * V_0 * \left[\int_{\mathrm{B}} (k_{ij}g_j + q_i) * \tilde{g}_i\mathrm{d}\upsilon + \int_{\mathrm{B}} g_i * \tilde{q}_i\mathrm{d}\upsilon + \int_{\mathrm{B}} \tilde{q}_k,_k * \vartheta\mathrm{d}\upsilon\right] \\
&+ V_0 * V_1 * \left\{\int_{\partial\mathrm{B}_1} u_i' * \tilde{p}_i\mathrm{d}a + \int_{\partial\mathrm{B}_2} [(p_i - s_i') * \tilde{u}_i + u_i * \tilde{p}_i]\,\mathrm{d}a\right\} \\
&+ \frac{t * V_0}{\theta_0} * \left\{\int_{\partial\mathrm{B}_3} \vartheta' * \tilde{q}\mathrm{d}a + \int_{\partial\mathrm{B}_4} [(q - \acute{q}) * \tilde{\vartheta} + \vartheta * \tilde{q}]\mathrm{d}a\right\}.
\end{aligned} \tag{5.2.9}$$

If now we note that

$$\int_{\mathrm{B}} u_i * \tilde{S}_{ik},_k \,\mathrm{d}\upsilon = \int_{\partial\mathrm{B}_1} \tilde{p}_i * u_i\mathrm{d}a + \int_{\partial\mathrm{B}_2} \tilde{p}_i * u_i\mathrm{d}a - \int_{\mathrm{B}} u_{(i,j)} * \tilde{S}_{ij}\mathrm{d}\upsilon, \tag{5.2.10}$$

$$\int_{\mathrm{B}} \vartheta * \tilde{q}_k,_k \,\mathrm{d}\upsilon = \int_{\partial\mathrm{B}_3} \vartheta * \tilde{q}\mathrm{d}a + \int_{\partial\mathrm{B}_4} \vartheta * \tilde{q}\mathrm{d}a - \int_{\mathrm{B}} \vartheta,_i * \tilde{q}_i\mathrm{d}\upsilon, \tag{5.2.11}$$

and

$$\dot{V}_0 * t * \dot{V}_1 = V_0 * V_1, \tag{5.2.12}$$

then the formula (5.2.9) becomes

$$\begin{aligned}
\delta_{\tilde{p}}\Lambda_t(p) = -\dot{V}_0 * \dot{V}_1 * \int_B (t * S_{ij},_j - \rho u_i + l_i) * \tilde{u}_i \mathrm{d}v \\
+t * \dot{V}_0 * \int_B \left[\dot{V}_1 * (C_{ijkl}E_{kl} - S_{ij}) + M_{ij}\vartheta - m_{ij}\right] * \tilde{E}_{ij}\mathrm{d}v \\
-V_0 * V_1 * \int_B (E_{ij} - u_{(i,j)}) * \tilde{S}_{ij}\mathrm{d}v \\
+\frac{t}{\theta_0} * \int_B \left(-V_0 * q_{i},_i + \theta_0 \dot{V}_0 * M_{ij}E_{ij} + W - C_E\vartheta\right) * \tilde{\vartheta}\mathrm{d}v \\
-\frac{t}{\theta_0} * V_0 * \left[\int_B (k_{ij}g_j + q_i) * \tilde{g}_i \mathrm{d}v + \int_B (g_i - \vartheta,_i) * \tilde{q}_i\right] \mathrm{d}v \\
+V_0 * V_1 * \left[\int_{\partial B_1} (u_i' - u_i) * \tilde{p}_i \mathrm{d}a + \int_{\partial B_2} (p_i - s_i') * \tilde{u}_i \mathrm{d}a\right] \\
+\frac{t}{\theta_0} * V_0 * \left[\int_{\partial B_3} (\vartheta' - \vartheta) * \tilde{q}\mathrm{d}a + \int_{\partial B_4} (q - q') * \tilde{\vartheta}\mathrm{d}a\right] \quad \forall \tilde{p} \in K.
\end{aligned} \tag{5.2.13}$$

Let us first assume that p is a conventional thermoelastic process in the G–L theory. Then, in view of the Lemma 5.1, the relations (5.1.35–40) along with the boundary conditions (5.1.6), as well as the formula (5.2.13) yield

$$\delta_{\tilde{p}}\Lambda_t(p) = 0 \quad \forall \tilde{p} \in K \quad \forall t \in [0, \infty), \tag{5.2.14}$$

that is, eqn (5.2.8) is satisfied.

Let us now assume that eqn (5.2.8) is satisfied, i.e. the right-hand side of eqn (5.2.13) vanishes. Since this vanishing occurs for every $\tilde{p} \in K$, then, assuming $\tilde{p} = (\tilde{u}_i, 0, 0, 0, 0, 0)$, we obtain the field equation (5.1.37) and the boundary condition $(5.1.6)_2$. If we assume $\tilde{p} = (0, \tilde{E}_{ij}, 0, 0, 0, 0)$, then the vanishing of the right-hand side of eqn (5.2.13) yields the field equation (5.1.39). In a similar manner we recover the remaining field equations and boundary conditions of Lemma 5.1. This completes the proof of Theorem 5.1.

Since, for a function $f(t)$,

$$\lim_{t_i \to 0} \dot{V}_i * f(t) = f(t), \tag{5.2.15}$$

and

$$\lim_{t_i \to 0} V_i * f(t) = 1 * f(t) \qquad (i = 0, 1), \tag{5.2.16}$$

therefore, passing in the functional (5.2.7) to the limit when $(t_0, t_1) \to (0, 0)$, we obtain a variational principle of classical thermoelasticity [p. 339 in (Carlson, 1972)]. □

Finally, note that Theorem 5.1 is a variational principle in which the admissible processes do not have to satisfy any field equations, boundary and initial conditions. It is thus the most general variational formulation of a conventional thermoelastic process of the G–L theory. This description corresponds to the Hu–Washizu variational principle of linear isothermal elastostatics [p. 122 in (Gurtin, 1972)].

The particular variational principles are obtained by placing various restrictions upon the admissible processes. One such "narrower" principle corresponds to that due to Hellinger–Reissner [p. 27 in (Gładysz, 1982)].

Theorem 5.2 *Let $t_{(0)} = const$ (recall eqn (1.3.54)[4]) and let $\mathcal{R}$ be the set of all admissible processes satisfying the strain–displacement relations (5.1.57) and the gradient–temperature relations (5.1.58). Let $z = (u_i, E_{ij}, S_{ij}, \vartheta, g_i, q_i) \in \mathcal{R}$ and, for every $t \in [0, \infty)$, let us define the functional $\Gamma_t(z)$ over $\mathcal{R}$ as follows*

$$
\begin{aligned}
\Gamma_t(z) = \int_B \dot{V} * \Big[t * \left(\frac{1}{2}\dot{V}_1 * K_{ijkl} S_{kl} - n_{ij} \right) * S_{ij} \\
- \frac{1}{2}\rho \dot{V}_1 * u_i * u_i - 1 * V_1 * E_{ij} * S_{ij} + \dot{V}_1 * l_i * u_i \Big] \mathrm{d}v \\
- \frac{1}{\theta_0} \int_B t * \Big(\frac{1}{2} V * \lambda_{ij} * q_j * q_i - \frac{1}{2} C_S \vartheta * \vartheta \\
- \theta_0 \dot{V}_0 * A_{ij} S_{ij} * \vartheta + V * q_i * g_i + \mathrm{d} * \vartheta \Big) \mathrm{d}v \\
+ \int_{\partial B_1} V * V_1 * (u_i - u_i') * p_i \mathrm{d}a + \int_{\partial B_2} V * V_1 * s_i' * u_i \mathrm{d}a \\
+ \frac{1}{\theta_0} \left[\int_{\partial B_3} t * V * (\vartheta - \vartheta') * q \mathrm{d}a + \int_{\partial B_4} t * V * q' * \vartheta \mathrm{d}a \right].
\end{aligned}
\tag{5.2.17}
$$

Then

$$
\delta_{\bar{z}} \Gamma_t(z) = 0 \quad on \ \mathcal{R}, \quad t \in [0, \infty), \tag{5.2.18}
$$

iff z is a conventional thermoelastic process in the G–L theory.

[4] The hypothesis $t_{(0)} = const$ implies that the function V [see eqn (5.1.54)], treated as an operator, commutes with the spatial differentiation.

Proof. The proof is analogous to that of Theorem 5.1, and so, we obtain

$$\begin{aligned}
\delta_{\tilde{z}}\Gamma_t(z) &= \dot{V} * \int_B \dot{V}_1 * (t * S_{ij},_j + l_i - \rho u_i) * \tilde{u}_i \mathrm{d}v \\
&- \dot{V} * \int_B t * \left[\dot{V}_1 * (E_{ij} - K_{ijkl}S_{kl}) - A_{ij}\vartheta - n_{ij}\right] * \tilde{S}_{ij}\mathrm{d}v \\
&- \frac{t}{\theta_0} * \int_B \{V * (\lambda_{ij}q_j + g_i) * \tilde{q}_i \\
&+ \left[-\dot{V} * (1 * q_k,_k + \theta_0 A_{ij}S_{ij}) - C_S\vartheta + d\right] * \tilde{\vartheta}\Big\}\mathrm{d}v \\
&+ V * V_1 * \left[\int_{\partial B_1} (u_i - u_i') * \tilde{p}_i \mathrm{d}a + \int_{\partial B_2} (s_i' - p_i) * \tilde{u}_i \mathrm{d}a\right] \\
&+ \frac{t * V_0}{\theta_0} * \left[\int_{\partial B_3} (\vartheta - \vartheta') * \tilde{q}\mathrm{d}a + \int_{\partial B_4} (q' - q) * \tilde{\vartheta}\mathrm{d}a\right] \quad \forall \tilde{z} \in \mathcal{R}.
\end{aligned} \tag{5.2.19}$$

If z is a conventional thermoelastic process of the G–L theory, then, in view of Lemma 5.2, the right-hand side of eqn (5.2.19) vanishes, that is, the condition (5.2.18) is satisfied. Conversely, since $\tilde{z}$ is an arbitrary element of $\mathcal{R}$, then, taking $\tilde{z} = (\tilde{u}_i, 0, 0, 0, 0, 0)$, from eqn (5.2.18) we infer that eqns (5.1.59) and $(5.1.6)_2$ of Lemma 5.2 are satisfied. Proceeding in a similar way, we reproduce the remaining field equations and boundary conditions appearing in Lemma 5.2. We thus conclude that, if z satisfies eqn (5.2.18), then z is a conventional thermoelastic process. □

Theorems 5.1 and 5.2 provide variational characterizations of the conventional thermoelastic process of the G–L theory, corresponding to the Lemmas 5.1 and 5.2. We shall now give three more variational descriptions of that process, corresponding to Lemmas 5.3, 5.4 and 5.5. First note that Lemma 5.3 offers a description of the conventional thermoelastic process of the G–L theory in terms of the pair (u_i, ϑ). Hence, the variational principle associated with it is called a displacement–temperature variational description of that process. In order to formulate this principle we introduce the concept of a *kinematically and thermally admissible process* $p = (u_i, E_{ij}, S_{ij}, \vartheta, g_i, q_i)$, satisfying the strain–displacement relations (5.1.35), the gradient–temperature relations (5.1.36), the constitutive relations (5.1.39) and the Fourier law (5.1.40), as well as the displacement and temperature boundary conditions $(5.1.6)_{1,3}$. Since the process is described by the fields u_i and ϑ only, in the following it is represented by a pair (u_i, ϑ). For such a process the following theorem holds true.

Theorem 5.3 *(p. 19 in (Gładysz, 1982)) Let $\mathcal{K}'$ be the set of all kinematically and thermally admissible processes p'. Let $p' = (u_i, \vartheta) \in \mathcal{K}'$ and for every*

$t \in [0, \infty)$ *let us define the functional* $\phi_t(\cdot)$ *over* $\mathcal{K}'$ *as follows*

$$\phi_t(p') = \int_B \dot{V}_0 * \left[t * \left(\frac{1}{2}\dot{V}_1 * C_{ijkl}u_{k,l} - m_{ij} \right) * u_{(i,j)} + \frac{1}{2}\rho\dot{V}_1 * u_i * u_i - \dot{V}_1 * l_i * u_i \right] dv$$
$$- \frac{1}{\theta_0}\int_B t * \left(\frac{1}{2}V_0 * k_{ij}\ \vartheta_{,j} * \vartheta_{,i} - \theta_0 \dot{V}_0 * M_{ij}\ u_{i,j} * \vartheta + \frac{1}{2}C_E\ \vartheta * \vartheta - W * \vartheta \right) dv$$
$$+ \int_{\partial B_2} V_0 * V_1 * s'_i * u_i da - \frac{1}{\theta_0}\int_{\partial B_4} t * V_0 * q' * \vartheta da. \tag{5.2.20}$$

Then

$$\delta_{\tilde{p}'}\phi_t(\tilde{p}') = 0 \quad on\ \mathcal{K}',\ \ t \in [0, \infty), \tag{5.2.21}$$

iff $p' = (u_i, \vartheta)$ *is a pair corresponding to a conventional thermoelastic process in the G–L theory.*

Proof. According to the definition of $\mathcal{K}'$, and in view of eqn (5.2.2), we obtain

$$\delta_{\tilde{p}'}\phi_t(p') =$$
$$- \dot{V}_0 * \int_B \left[1 * V_1 * \left(C_{ijkl}u_{(k,l)} \right)_{,j} + t * \left(M_{ij}\vartheta \right)_{,j} - \rho\dot{V}_1 * u_i + l_i^{(1)} \right] * \tilde{u}_i dv$$
$$+ \frac{t}{\theta_0} * \int_B \left[V_0 * \left(k_{ij}\vartheta_{,j} \right)_{,i} - C_E\vartheta + \theta_0\dot{V}_0 * M_{ij}u_{i,j} + W \right] * \tilde{\vartheta} dv$$
$$+ V_0 * V_1 * \int_{\partial B_2} (p_i - s'_i) * \tilde{u}_i da + \frac{t * V_0}{\theta_0} * \int_{\partial B_4} (q - q') * \tilde{\vartheta} da. \tag{5.2.22}$$

Let us now assume that $p' = (u_i, \vartheta)$ is a pair corresponding to a conventional thermoelastic process of the G–L theory. Then, on the basis of Lemma 5.3, the relations (5.1.63–68) are satisfied and the right-hand side of eqn (5.2.22) vanishes, that is, eqn (5.2.21) holds. This completes the proof of the first part of Theorem 5.3.

Let us further assume that the relation (5.2.21) is satisfied, that is, the right-hand side of eqn (5.2.22) vanishes for every admissible $\tilde{p}'$ and $p' \in \mathcal{K}'$. Then, letting $\tilde{p}' = (\tilde{u}_i, 0)$ in eqn (5.2.21), we conclude that the field equation (5.1.63) and the boundary condition (5.1.66) of Lemma 5.3 hold true. On the other hand, letting $\tilde{p}' = (0, \tilde{\vartheta})$ in eqn (5.2.21), we conclude that the field equation (5.1.64) and the condition (5.1.68) of that lemma hold as well. Hence, given that $p' \in \mathcal{K}'$, we infer that p' is a pair corresponding to the conventional thermoelastic process of the G–L theory. □

In order to give a variational description of the pair (S_{ij}, ϑ) corresponding to the conventional thermoelastic process of the G–L theory, we introduce the concept of a *stress and temperature admissible process.* This is an admissible process $p = (u_i, E_{ij}, S_{ij}, \vartheta, g_i, q_i)$ such that the strain–displacement relations (5.1.57), the gradient–temperature relation (5.1.58), the constitutive relations (5.1.61) and the Fourier law (5.1.62), as well as the stress and temperature boundary conditions $(5.1.6)_{2,3}$ are satisfied. Since that process is determined by the fields S_{ij} and ϑ, it shall be denoted by a pair (S_{ij}, ϑ). We now formulate the following variational principle for that process corresponding to the Lemma 5.4:

Theorem 5.4 *Let $t_{(0)} = const$ and let $\mathcal{R}'$ be the set of all stress and temperature admissible processes. Let $z' = (S_{ij}, \vartheta) \in \mathcal{R}'$ and for every $t \in [0, \infty)$ let us define the functional $\psi_t(z)$ over $\mathcal{R}'$ as follows*

$$\begin{aligned}\psi_t(z') = \int_B \dot{V} * \Bigg[\left(\frac{1}{2} \dot{V}_1 * K_{ijkl} S_{kl} - n_{ij} \right) * S_{ij} \\ + \rho^{-1} \dot{V}_1 * \left(\frac{1}{2} t * S_{ij,j} + l_i \right) * S_{ik,k} \Bigg] \mathrm{d}v \\ + \frac{1}{\theta_0} \int_B \left(\frac{1}{2} V * k_{ij} \vartheta_{,j} * \vartheta_{,i} + \frac{1}{2} C_S \vartheta * \vartheta + \theta_0 \dot{V} * A_{ij} S_{ij} * \vartheta - d * \vartheta \right) \mathrm{d}v \\ - \int_{\partial B_1} \dot{V} * \dot{V}_1 * u_i' * p_i da + \frac{1}{\theta_0} \int_{\partial B_4} V * q' * \vartheta \mathrm{d}a.\end{aligned} \tag{5.2.23}$$

Then

$$\delta_{\tilde{z}'} \psi_t(z') = 0 \quad on \ \mathcal{R}', \quad t \in [0, \infty), \tag{5.2.24}$$

iff $z' = (S_{ij}, \vartheta)$ is a pair corresponding to the conventional thermoelastic process in the G–L theory.

Proof. Using the definition of $\mathcal{R}'$, and computing the variation of the functional $\psi_t(\cdot)$ according to eqn (5.2.2), we find

$$\begin{aligned}\delta_{\tilde{z}'} \psi_t(z') = -\dot{V} * \int_{\mathrm{B}} \Big\{ \dot{V}_1 * \left[t * \left(\rho^{-1} S_{(ik,k} \right)_{,j)} - K_{ijkl} S_{kl} \right] \\ - A_{ij} \vartheta + n_{ij}^{(1)} \Big\} * \tilde{S}_{ij} \mathrm{d}v \\ - \frac{1}{\theta_0} * \int_{\mathrm{B}} \left\{ \dot{V} * [1 * (k_{ij} \vartheta_{,j})_{,i} - \theta_0 A_{pq} S_{pq}] - C_S \vartheta + \mathrm{d} \right\} * \tilde{\vartheta} \mathrm{d}v \\ + \dot{V} * \dot{V}_1 * \int_{\partial \mathrm{B}_1} (u_i - u_i') * \tilde{p}_i \mathrm{d}a + \frac{1}{\theta_0} V * \int_{\partial \mathrm{B}_4} (q' - q) * \tilde{\vartheta} \mathrm{d}a \quad \forall \tilde{z}' \in \mathcal{R}'.\end{aligned} \tag{5.2.25}$$

If z' is a pair corresponding to the conventional thermoelastic process in the G–L theory, then, in view of Lemma 5.4, the right-hand side of eqn (5.2.25) vanishes, that is, the relation (5.2.24) holds true. On the other hand, if $z' \in \mathcal{R}'$ and the relation (5.2.24) is satisfied, then an arbitrary choice of $\tilde{z}'$ implies that z' is a pair corresponding to the conventional thermoelastic process of the G–L theory. □

Finally, in order to give a variational description of the pair (S_{ij}, q_i) corresponding to the conventional thermoelastic process of the G–L theory, we introduce the concept of a *stress and heat-flux admissible process*. This is a process that satisfies the strain–displacement relations (5.1.57), the gradient–temperature relations (5.1.58), the energy balance (5.1.60), the constitutive relations (5.1.61), as well as the stress and heat-flux boundary conditions $(5.1.6)_{2,4}$. That process shall be denoted by a pair (S_{ij}, q_i). We now formulate the following variational principle for that process, corresponding to Lemma 5.5:

Theorem 5.5 *Let $t_{(0)} = const$ and let $\mathcal{Q}$ be the set of all stress and heat-flux admissible processes. Let $s = (S_{ij}, q_i) \in \mathcal{Q}$ and for every $t \in [0, \infty)$ let us define the functional $\chi_t(\cdot)$ on $\mathcal{Q}$ as follows*

$$\chi_t(s) = \int_B \Big[C_S^{-1} \left(\frac{1}{2}\theta_0 \dot{V} * A_{ij}S_{ij} - d \right) * A_{pq}S_{pq}$$

$$- \left(\frac{1}{2}\dot{V}_1 * K_{ijkl}S_{kl} - n_{ij} \right) * S_{ij} - \rho^{-1}\dot{V}_1 * \left(\frac{1}{2} t * S_{ij},_j + l_i \right) * S_{ij},_j \Big] \mathrm{d}v$$

$$+ \frac{1}{\theta_0} \int_B 1 * \Big[\frac{1}{2}\lambda_{ij}q_j * q_i + \theta_0 C_S^{-1}\dot{V} * A_{pq}S_{pq} * q_{k,k}$$

$$+ C_S^{-1} \left(\frac{1}{2} V * q_k,_k - d \right) * q_i,_i \Big] \mathrm{d}v + \int_{\partial B_1} \dot{V}_1 * u_i' * p_i \mathrm{d}a + \frac{1}{\theta_0} \int_{\partial B_3} 1 * \vartheta' * q \mathrm{d}a. \tag{5.2.26}$$

Then,

$$\delta_{\tilde{s}'} \chi_t(s) = 0 \quad on\ \mathcal{Q},\ \ t \in [0, \infty), \tag{5.2.27}$$

iff $s = (S_{ij}, q_i)$ is a pair corresponding to a conventional thermoelastic process in the G–L theory.

Proof. In view of the definitions of $\mathcal{Q}$ and the first variation of a functional (recall eqn (5.2.2)), we obtain

$$\delta_{\tilde{s}'} \chi_t(s) = \int_{\mathrm{B}} \Big\{ \dot{V}_1 * \left[t * \left(\rho^{-1} S_{(ik},_k \right),_{j)} + K_{ijkl}S_{kl} \right]$$

$$+ C_S^{-1} A_{ij}\dot{V} * (1 * q_k,_k + \theta_0 A_{pq}S_{pq}) + n_{ij}^{(2)} \Big\} * \tilde{S}_{ij} \mathrm{d}v$$

$$-\frac{1}{\theta_0} * \int_{\mathrm{B}} \left\{ \dot{V} * \left[C_S^{-1} \left(1 * q_{k},_k + \theta_0 A_{pq} S_{pq} \right) \right],_i - \lambda_{ij} q_j - \left(C_S^{-1} d \right),_i \right\} * \tilde{q}_i \mathrm{d}v$$

$$+ \int_{\partial \mathrm{B}_1} \dot{V}_1 * (u_i' - u_i) * \tilde{p}_i \mathrm{d}a + \frac{1}{\theta_0} * \int_{\partial \mathrm{B}_3} (\vartheta' - \vartheta) * \tilde{q} \mathrm{d}a. \tag{5.2.28}$$

If $s = (S_{ij}, q_i)$ is a pair corresponding to the conventional thermoelastic process in the G–L theory, then, in view of the Lemma 5.5, the right-hand side of eqn (5.2.28) vanishes, that is, the relation (5.2.27) holds. On the other hand, eqn (5.2.27) along with the relation $s \in \mathcal{Q}$ and the fact that $\tilde{s}$ is arbitrary, imply that s is a pair corresponding to the conventional thermoelastic process in the G–L theory. □

Remark 5.1 If $\partial \mathrm{B}_1 = \partial \mathrm{B}_3 = \varnothing$, the functional (5.2.26) becomes

$$\hat{\chi}_t(s) = -\frac{1}{2} \int_{\mathrm{B}} \Big[\dot{V}_1 * \left(\rho^{-1} t * S_{ij},_j * S_{ik},_k + K_{ijkl} S_{kl} * S_{ij} \right)$$

$$- \dot{V} \theta_0 * C_S^{-1} A_{ij} S_{ij} * A_{pq} S_{pq} - 2 n_{ij}^{(2)} * S_{ij} \Big] \mathrm{d}v$$

$$+ \frac{1}{2\theta_0} * \int_{\mathrm{B}} \Big[\lambda_{ij} q_j * q_i + C_S^{-1} V * q_{k},_k * q_{i},_i \tag{5.2.29}$$

$$+ 2\theta_0 C_S^{-1} \dot{V} * (A_{pq} S_{pq}) * q_{k},_k + 2 \left(C_S^{-1} d \right),_i * q_i \Big] \mathrm{d}v.$$

Passing in the above to the limit $(t_0, t_1) \to (0, 0)$, and denoting the limit on the left by $-\chi_t^{(0)}(s)$, we find

$$\chi_t^{(0)}(s) = \frac{1}{2} \int_{\mathrm{B}} \left(\rho^{-1} t * S_{ij},_j * S_{ik,k} + K_{ijkl}' S_{kl} * S_{ij} - 2 n_{ij}^{(0)} * S_{ij} \right) \mathrm{d}v$$

$$- \frac{1}{2\theta_0} * \int_{\mathrm{B}} \Big[\lambda_{ij} q_j * q_i + C_S^{-1} * q_{k},_k * q_{i},_i \tag{5.2.30}$$

$$+ 2\theta_0 C_S^{-1} A_{pq} S_{pq} * q_{k},_k + 2 \left(C_S^{-1} d^{(0)} \right),_i * q_i \Big] dv,$$

where[5]

$$K_{ijkl}' = K_{ijkl} - C_S^{-1} \theta_0 A_{ij} A_{kl}, \tag{5.2.31}$$

$$n_{ij}^{(0)} = \left(\rho^{-1} l_{(i)},_{j)} \right) - C_S^{-1} A_{ij} \left(1 * r + \theta_0 A_{pq} S_{pq}^{(0)} + C_S \vartheta_0 \right), \tag{5.2.32}$$

$$d^{(0)} = 1 * r + C_S \vartheta_0 + \theta_0 A_{pq} S_{pq}^{(0)}. \tag{5.2.33}$$

[5] Formula (5.2.31) is identical to eqn (1.1.70).

The functional (5.2.29) is associated with a conventional stress and heat flux initial-boundary value problem of the G–L theory, while the functional (5.2.30) corresponds to an analogous problem of the classical thermoelasticity. Setting $A_{ij} = 0$ and $q_i = 0$ in eqn (5.2.30), we obtain a functional appearing in the second variational principle for a stress field in linear adiabatic elastodynamics [p. 230 in (Gurtin, 1972)].

5.3 Variational principle for a non-conventional thermoelastic process in the Lord–Shulman theory

In this section we give a variational description of a pair (S_{ij}, q_i) as a solution to a natural non-conventional initial-boundary value problem of the L–S theory. This problem consists in finding (S_{ij}, q_i) on $\overline{\mathrm{B}} \times [0, \infty)$, satisfying the field equations [recall eqns (2.1.7–9) in which $\partial \mathrm{B}_1 = \partial \mathrm{B}_3 = \varnothing$]

$$\left.\begin{aligned} &\left(\rho^{-1} S_{(ik,k}\right)_{,j)} - K'_{ijkl}\ddot{S}_{kl} + C_S^{-1} A_{ij}\dot{q}_{k,k} = -F_{(ij)} \\ &[C_S^{-1}(q_{k,k} + \theta_0 A_{pq}\dot{S}_{pq})]_{,i} - \lambda_{ij}\left(\dot{q}_j + t_0 \ddot{q}_j\right) = -f_i \end{aligned}\right\} \quad \text{on } \mathrm{B} \times [0, \infty), \tag{5.3.1}$$

the initial conditions

$$\left.\begin{aligned} S_{ij}(\cdot, 0) = S_{ij}^{(0)},\ \dot{S}_{ij}(\cdot, 0) = \dot{S}_{ij}^{(0)} \\ q_i(\cdot, 0) = q_i^{(0)}, \quad \dot{q}_i(\cdot, 0) = \dot{q}_i^{(0)} \end{aligned}\right\} \quad \text{on } \mathrm{B}, \tag{5.3.2}$$

and the boundary conditions

$$\begin{aligned} S_{ij}n_j = s'_i \quad &\text{on} \quad \partial\mathrm{B} \times (0, \infty), \\ q_i n_i = q' \quad &\text{on} \quad \partial\mathrm{B} \times (0, \infty). \end{aligned} \tag{5.3.3}$$

Here, we have

$$F_{(ij)} = \left(\rho^{-1} b_{(i}\right)_{,j)} - C_S^{-1}\dot{r}A_{ij}, \tag{5.3.4}$$

$$f_i = -\left(C_S^{-1} r\right)_{,i}. \tag{5.3.5}$$

To arrive at the variational description of the problem (5.3.1–3) we proceed in a manner analogous to that taken in Sections 5.1 and 5.2. First, we obtain an alternative description of the problem in which the initial conditions are implicitly incorporated in the field equations, and then we formulate a variational principle associated with this alternative description.

In order to obtain the alternative description of the problem (5.3.1–3), we apply the Laplace transform to eqns $(5.3.1)_{1,2}$, employ the initial conditions (5.3.2), and obtain

$$p^{-2}(\rho^{-1}\bar{S}_{(ik,k})_{,j)} - K'_{ijkl}\bar{S}_{kl} + p^{-1}C_S^{-1}A_{ij}\bar{q}_{k,k} + \bar{f}_{(ij)} = 0, \tag{5.3.6}$$

$$p\bar{V}_0\left[C_S^{-1}\left(p^{-1}\bar{q}_{k,k} + \theta_0 A_{pq}\bar{S}_{pq}\right)\right]_{,i} - \lambda_{ij}\bar{q}_j + \bar{h}_i = 0, \tag{5.3.7}$$

where

$$\bar{f}_{(ij)} = p^{-2}\bar{F}_{(ij)} + p^{-1}K'_{ijkl}S^{(0)}_{kl} + p^{-2}(K'_{ijkl}\dot{S}^{(0)}_{kl} - C_S^{-1}A_{ij}q^{(0)}_{k,k}), \tag{5.3.8}$$

$$\bar{h}_i = \bar{V}_0\bar{f}_i + p^{-1}\lambda_{ij}q^{(0)}_j + \bar{V}_0[t_0\lambda_{ij}\dot{q}^{(0)}_j - \theta_0(C_S^{-1}A_{pq}S^{(0)}_{pq})_{,i}], \tag{5.3.9}$$

and (recall eqn (5.1.20))

$$\bar{V}_0\,(p) = [p\,(t_0 p + 1)]^{-1}. \tag{5.3.10}$$

Applying the inverse Laplace transform to eqns (5.3.6–9), and using the convolution theorem, we obtain

$$\begin{aligned} t * (\rho^{-1}S_{(ik,k})_{,j)} - K'_{ijkl}S_{kl} + C_S^{-1}A_{ij} * q_{k,k} + f_{(ij)} &= 0, \\ \dot{V}_0 * [C_S^{-1}(1 * q_{k,k} + \theta_0 A_{pq}S_{pq})]_{,i} - \lambda_{ij}q_j + h_i &= 0, \end{aligned} \tag{5.3.11}$$

where

$$f_{(ij)} = t * F_{(ij)} + t(K'_{ijkl}\dot{S}^{(0)}_{kl} - C_S^{-1}A_{ij}q^{(0)}_{k,k}) + K'_{ijkl}S^{(0)}_{kl}, \tag{5.3.12}$$

$$h_i = V_0 * f_i + \lambda_{ij}q^{(0)}_j + V_0[t_0\lambda_{ij}\dot{q}^{(0)}_j - \theta_0(C_S^{-1}A_{pq}S^{(0)}_{pq})_{,i}], \tag{5.3.13}$$

as well as (recall eqn (5.1.29))

$$V_0\,(t) = 1 - \exp(-t/t_0). \tag{5.3.14}$$

As a result, we arrive at the following lemma:

Lemma 5.6 *The pair (S_{ij}, q_i) is a solution of the initial-boundary value problem (5.3.1–3) iff it satisfies the field equations*

$$\left.\begin{aligned} t * \left(\rho^{-1}S_{(ik,k}\right)_{,j)} - K'_{ijkl}S_{kl} + C_S^{-1}A_{ij} * q_{k,k} + f_{(ij)} &= 0 \\ \dot{V}_0 * \left[C_S^{-1}(1 * q_{k,k} + \theta_0 A_{pq}S_{pq})\right]_{,i} - \lambda_{ij}q_j + h_i &= 0 \end{aligned}\right\} \quad \text{on } B \times [0,\infty), \tag{5.3.15}$$

and the boundary conditions

$$S_{ij}n_j = s'_i,\ q_i n_i = q' \quad \text{on } \partial B \times [0,\infty). \tag{5.3.16}$$

Similarly to Sections 5.1 and 5.2 we deal with a smooth pair (S_{ij}, q_i) on $\overline{B} \times [0,\infty)$, satisfying the relations (5.3.1–3) or (5.3.15–16).

In order to formulate the variational principle associated with the boundary value problem (5.3.15 and 16), we introduce the notion of a *stress and heat-flux admissible pair.* It is a pair (S_{ij}, q_i) in which S_{ij} is a smooth symmetric tensor field on $\overline{B} \times [0,\infty)$ satisfying the boundary condition $(5.3.16)_1$, while q_i is a smooth vector field on $\overline{B} \times [0,\infty)$ satisfying the boundary condition $(5.3.16)_2$. We can now formulate the following[6]

[6] Theorem 5.6 appears here for the first time.

Theorem 5.6 *Let $\mathcal{Q}$ be the set of all stress and heat-flux admissible pairs. Let $s = (S_{ij}, q_i) \in \mathcal{Q}$ and for every $t \in [0, \infty)$ let us define the functional $\pi_t(\cdot)$ over $\mathcal{Q}$ as follows*

$$\begin{aligned}\pi_t(s) = \\ \int_B \dot{V}_0 * \left(\rho^{-1} t * S_{ij},_j * S_{ik},_k + K'_{ijkl} S_{kl} * S_{ij} - 2 f_{(ij)} * S_{ij}\right) \mathrm{d}v \\ -\frac{1}{2\theta_0} * \int_B (\lambda_{ij} q_j * q_i + C_S^{-1} V_0 * q_k,_k * q_{i,i} \\ +2\theta_0 C_S^{-1} \dot{V}_0 * A_{pq} S_{pq} * q_k,_k - 2h_i * q_i) \mathrm{d}v.\end{aligned} \tag{5.3.17}$$

Then,

$$\delta_{\tilde{s}} \pi_t(s) = 0 \quad on \quad \mathcal{Q}, t \in [0, \infty), \tag{5.3.18}$$

iff $s = (S_{ij}, q_i)$ is a solution of the problem (5.3.1–3).

Proof. Let $s \in \mathcal{Q}$ and $s + \tilde{\lambda}\tilde{s} \in \mathcal{Q}$ for every $\tilde{\lambda} \in \mathbb{R}$. Then

$$\tilde{S}_{ij} n_j = 0, \tilde{q}_i n_i = 0 \quad \text{on} \quad \partial\mathrm{B} \times (0, \infty). \tag{5.3.19}$$

In view of the definition of the first variation of a functional (recall eqn (5.2.2)), we obtain

$$\begin{aligned}\delta_{\tilde{s}} \pi_t(s) = \\ \int_\mathrm{B} \dot{V}_0 * (\rho^{-1} t * S_{ij},_j * \tilde{S}_{ij},_j + K'_{ijkl} S_{kl} * \tilde{S}_{ij} - f_{(ij)} * \tilde{S}_{ij}) \mathrm{d}v \\ -\frac{1}{\theta_0} * \int_\mathrm{B} \Big[\lambda_{ij} q_j * \tilde{q}_i + C_S^{-1} V_0 * q_{k,k} * \tilde{q}_{i,i} \\ +\theta_0 C_S^{-1} \dot{V}_0 * A_{pq} (S_{pq} * \tilde{q}_k,_k + \tilde{S}_{pq} * q_k,_k) - h_i * \tilde{q}_i\Big] \mathrm{d}v.\end{aligned} \tag{5.3.20}$$

In view of the divergence theorem and eqn (5.3.19), we obtain

$$\begin{aligned}\delta_{\tilde{s}} \pi_t(s) = \\ -\dot{V}_0 * \int_\mathrm{B} \left[t * \left(\rho^{-1} S_{(ik,k}\right),_{j)} - K'_{ijkl} S_{kl} + C_S^{-1} A_{ij} * q_k,_k + f_{(ij)}\right] * \tilde{S}_{ij} \mathrm{d}v \\ +\frac{1}{\theta_0} * \int_\mathrm{B} \left\{\dot{V}_0 * \left[C_S^{-1}(1 * q_k,_k + \theta_0 A_{pq} S_{pq})\right],_i - \lambda_{ij} q_j + h_i\right\} * \tilde{q}_i \mathrm{d}v.\end{aligned} \tag{5.3.21}$$

Let us first assume that $s = (S_{ij}, q_i)$ is a solution of the problem (5.3.1–3). Then, in view of the Lemma 5.6, the relations (5.3.15 and 16) hold, and, given eqn (5.3.21), the condition (5.3.18) holds too.

Let us next assume that eqn (5.3.18) holds. Then, the right-hand side of eqn (5.3.21) vanishes, and given the arbitrary character of $\tilde{s} \in \mathcal{Q}$, and the fact that $s \in \mathcal{Q}$, we conclude that s is a pair satisfying eqns (5.3.15 and 16), i.e. eqns (5.3.1–3). □

Remark 5.2 If $t_0 \to 0+0$, the functional $\pi_t(s) \to \pi_t^{(0)}(s)$ such that

$$\pi_t^{(0)}(s) =$$
$$\frac{1}{2}\int_B \left(\rho^{-1}t * S_{ij},_j * S_{ik},_k + K'_{ijkl}S_{kl} * S_{ij} - 2f^{(0)}_{(ij)} * S_{ij}\right) dv$$
$$-\frac{1}{2\theta_0} * \int_B \left(\lambda_{ij}q_j * q_i + C_S^{-1} * q_k,_k * q_i,_i + 2\theta_0 C_S^{-1} A_{pq}S_{pq} * q_k,_k - 2h_i^{(0)} * q_i\right) dv, \tag{5.3.22}$$

where

$$f^{(0)}_{(ij)} = t * F_{(ij)} + K'_{ijkl}S^{(0)}_{kl} + t(K'_{ijkl}\dot{S}^{(0)}_{kl} - C_S^{-1}A_{ij}q^{(0)}_{k,k}), \tag{5.3.23}$$

and

$$h_i^{(0)} = 1 * f_i + \lambda_{ij}q_j^{(0)} - \theta_0(C_S^{-1}A_{pq}S^{(0)}_{pq}),_i\,. \tag{5.3.24}$$

The functional $\pi_t^{(0)}(s)$ is associated with a natural non-conventional stress–heat-flux problem of classical dynamical thermoelasticity. It is similar to the functional $\chi_t^{(0)}(s)$ (5.2.30). Restricting suitably the initial data in the problem (5.3.1–3), we obtain

$$f^{(0)}_{ij} = n^{(0)}_{ij},\ h_i^{(0)} = -(C_S^{-1}d^{(0)}),_i\,, \tag{5.3.25}$$

where $n^{(0)}_{ij}$ and $d^{(0)}$ are given by the formulas (5.2.32 and 33). In this case

$$\pi_t^{(0)}(s) = \chi_t^{(0)}(s)\ \forall s \in \mathcal{Q},\ t \in [0,\infty). \tag{5.3.26}$$

5.4 Variational principle for a non-conventional thermoelastic process in the Green–Lindsay theory

In this section we present a variational principle associated with the natural non-conventional stress–temperature initial-boundary value problem of the G–L theory. This problem consists in finding a pair (S_{ij}, ϑ) on $\bar{B} \times [0,\infty)$ satisfying the field equations[7]

$$\left.\begin{aligned} &(\rho^{-1}S_{(ik,k)}\,,_{j)} - K_{ijkl}\ddot{S}_{kl} - A_{ij}(\ddot{\vartheta} + t_1\dddot{\vartheta}) = -\left(\rho^{-1}b_{(i)}\,,_{j)}\right. \\ &C_S^{-1}[(k_{pq}\vartheta,_q),_p - \theta_0 A_{pq}\dot{S}_{pq}] - (\dot{\vartheta} + t_{(0)}\ddot{\vartheta}) = -C_S^{-1}r \end{aligned}\right\} \text{ on } B \times [0,\infty), \tag{5.4.1}$$

the initial conditions

$$\left.\begin{aligned} &S_{ij}(\cdot,0) = S^{(0)}_{ij},\ \dot{S}_{ij}(\cdot,0) = \dot{S}^{(0)}_{ij} \\ &\vartheta(\cdot,0) = \vartheta^{(0)},\quad \dot{\vartheta}(\cdot,0) = \dot{\vartheta}^{(0)} \end{aligned}\right\} \text{ on } B, \tag{5.4.2}$$

[7] The problem described by eqns (5.4.1–3) is equivalent to that characterized by eqns (2.1.15–17) in which we let $\partial B_1 = \partial B_4 = \varnothing$, since eqn (1.3.58) is equivalent to eqn (1.3.63).

and the boundary conditions

$$S_{ij}n_j = s'_i,\ \vartheta = \vartheta' \quad \text{on} \quad \partial\mathrm{B} \times (0, \infty). \tag{5.4.3}$$

Similar to Section 1.3, we assume $t_{(0)} = const$. Performing the Laplace transform on eqns (5.4.1), and using the initial conditions (5.4.2), we obtain

$$\left.\begin{aligned} p\bar{V}_1\left[p^{-2}\left(\rho^{-1}\bar{S}_{(ik},_k\right),_{j)} - K_{ijkl}\bar{S}_{kl}\right] - A_{ij}\bar{\vartheta} + \bar{l}_{ij} = 0 \\ p\bar{V}\left[p^{-1}\left(k_{pq}\bar{\vartheta},_q\right),_p - \theta_0 A_{pq}\bar{S}_{pq}\right] - C_S\bar{\vartheta} + \bar{m} = 0 \end{aligned}\right\} \quad \text{on} \quad \mathrm{B} \times [0, \infty), \tag{5.4.4}$$

where

$$\begin{aligned} \bar{l}_{ij} = p\bar{V}_1[p^{-2}\left(\rho^{-1}b_{(i}\right),_{j)} + p^{-2}K_{ijkl}(\dot{S}^{(0)}_{kl} + pS^{(0)}_{kl})] \\ + A_{ij}[t_1\ddot{\vartheta}^{(0)}p^{-1}\bar{V}_1 + p^{-1}\vartheta^{(0)} + p^{-2}\dot{\vartheta}^{(0)}], \end{aligned} \tag{5.4.5}$$

$$\bar{m} = \bar{V}\bar{r} + (C_S t_{(0)}\dot{\vartheta}^{(0)} + \theta_0 A_{pq}S_{pq})\bar{V} + C_S p^{-1}\vartheta^{(0)}. \tag{5.4.6}$$

Here, $\bar{V}_1$ and $\bar{V}$ have the same meaning as in Section 5.1 (recall eqns (5.1.21,48)), while the field $\ddot{\vartheta}^{(0)}$ appearing in eqn (5.4.5) is determined by the field equation $(5.4.1)_2$ for $t = 0$, i.e. it is[8]

$$\ddot{\vartheta}^{(0)} = t^{-1}_{(0)}\{C_S^{-1}[(k_{pq}\vartheta^{(0)}_{,q}),_p - \theta_0 A_{pq}\dot{S}^{(0)}_{pq}] + C_S^{-1}r(\cdot, 0) - \dot{\vartheta}^{(0)}\}. \tag{5.4.7}$$

An inversion of the relations (5.5.4–6) leads to

$$\begin{aligned} \dot{V}_1 * \left[t * \left(\rho^{-1}S_{(ik},_k\right),_{j)} - K_{ijkl}S_{kl}\right] - A_{ij}\vartheta + l_{ij} = 0, \\ \dot{V} * \left[1 * \left(k_{pq}\vartheta,_q\right),_p - \theta_0 A_{pq}S_{pq}\right] - C_S\vartheta + m = 0, \end{aligned} \tag{5.4.8}$$

where

$$\begin{aligned} l_{ij} = \dot{V}_1 * [t * \left(\rho^{-1}b_{(i}\right),_{j)} + K_{ijkl}(S^{(0)}_{kl} + t\dot{S}^{(0)}_{kl})] \\ + A_{ij}(t_1\ddot{\vartheta}^{(0)} * V_1 + \vartheta^{(0)} + t\dot{\vartheta}^{(0)}), \end{aligned} \tag{5.4.9}$$

$$m = V * r + V(C_S t_{(0)}\dot{\vartheta}^{(0)} + \theta_0 A_{pq}S^{(0)}_{pq}) + C_S\vartheta^{(0)}. \tag{5.4.10}$$

The functions V_1 and V are given by the formulas

$$V_1 = 1 - \mathrm{e}^{-t/t_1},\ V = 1 - \mathrm{e}^{-t/t_{(0)}}. \tag{5.4.11}$$

We thus arrive at the following lemma[9]

[8] The field $\ddot{\vartheta}^{(0)}$ is defined by the non-conventional thermomechanical load of the problem (5.4.1–3), i.e. by $\vartheta^{(0)}$, $\dot{\vartheta}^{(0)}$, $r(\cdot, 0)$, and $\dot{S}^{(0)}_{pq}$.

[9] The lemma is similar to Lemma 5.4 in which we let $\partial\mathrm{B}_1 = \partial\mathrm{B}_4 = \varnothing$. Equations (5.4.12) are identical with eqns (5.1.70 and 71) if we let $(l_{ij}, m) = (n^{(1)}_{ij}, d)$.

Lemma 5.7 *The pair (S_{ij}, ϑ) is a solution of the initial-boundary value problem (5.4.1–3) iff it satisfies the field equations*

$$\begin{aligned} &\dot{V}_1 * \left[t * \left(\rho^{-1} S_{(ik},{}_{k)}\right),_j - K_{ijkl} S_{kl}\right] - A_{ij}\vartheta + l_{ij} = 0 \\ &\dot{V} * \left[1 * \left(k_{pq}\vartheta,_q\right),_p - \theta_0 A_{pq} S_{pq}\right] - C_S \vartheta + m = 0 \end{aligned} \quad on \quad B \times [0,\infty), \quad (5.4.12)$$

and the boundary conditions

$$S_{ij} n_j = s_i', \quad \vartheta = \vartheta' \quad on \quad \partial B \times [0,\infty). \quad (5.4.13)$$

In order to formulate a variational principle for the problem (5.4.12 and 13), we introduce the notion of a *stress and temperature admissible pair.* It is a pair (S_{ij}, ϑ), in which S_{ij} is a smooth symmetric tensor field on $\overline{\mathrm{B}} \times [0,\infty)$ satisfying the boundary condition $(5.4.13)_1$, while ϑ is a smooth scalar field on $\overline{\mathrm{B}} \times [0,\infty)$ satisfying the boundary condition $(5.4.13)_2$. We can now formulate the following

Theorem 5.7 *Let $\mathcal{R}$ be the set of all stress and temperature admissible pairs. Let $s = (S_{ij}, \vartheta) \in \mathcal{R}$ and, for every $t \in [0,\infty)$, let us define the functional $\mathcal{P}_t(\cdot)$ over $\mathcal{R}$ as follows*

$$\begin{aligned} &\mathcal{P}_t(s) = \\ &\frac{1}{2}\dot{V} * \int_B [\dot{V}_1 * (\rho^{-1} t * S_{ij},_j * S_{ik},_k + K_{ijkl} S_{kl} * S_{ij}) - 2 l_{ij} * S_{ij}] \mathrm{d}v \\ &+ \frac{1}{2\theta_0} \int_B (V * k_{pq}\vartheta,_q * \vartheta,_p + C_S \vartheta * \vartheta + 2\theta_0 \dot{V} * A_{pq} S_{pq} * \vartheta - 2m * \vartheta) \mathrm{d}v. \end{aligned} \quad (5.4.14)$$

Then,

$$\delta_{\tilde{s}} \mathcal{P}_t(s) = 0 \ on \ \mathcal{R}, t \in [0,\infty), \quad (5.4.15)$$

iff $s = (S_{ij}, \vartheta)$ is a solution of the problem (5.4.1–3).

Proof. Let $s \in \mathcal{R}$ and $\tilde{s}$ be chosen such that $s + \tilde{\lambda}\tilde{s} \in \mathcal{R}$ for every $\tilde{\lambda} \in \mathbb{R}$. Then

$$\tilde{S}_{ij} n_j = 0, \tilde{\vartheta} = 0 \quad \text{on} \quad \partial\mathrm{B} \times (0,\infty). \quad (5.4.16)$$

In view of the definition of the first variation of a functional (recall eqn (5.2.2)), we obtain

$$\begin{aligned} &\delta_{\tilde{s}} \mathcal{P}_t(s) = \\ &\dot{V} * \int_B [\dot{V}_1 * (\rho^{-1} t * S_{ij},_j * \tilde{S}_{ik},_k + K_{ijkl} S_{kl} * \tilde{S}_{ij}) - l_{ij} * \tilde{S}_{ij}] \mathrm{d}v \\ &+ \frac{1}{\theta_0} \int_B [V * k_{pq}\vartheta,_q * \tilde{\vartheta},_p + C_S \vartheta * \tilde{\vartheta} \\ &+ \theta_0 \dot{V} * A_{pq}(\tilde{S}_{pq} * \vartheta + S_{pq} * \tilde{\vartheta}) - m * \tilde{\vartheta}] \mathrm{d}v. \end{aligned} \quad (5.4.17)$$

In view of the divergence theorem and eqn (5.4.16), we reduce the above to the form

$$\delta_{\tilde{s}}\mathcal{P}_t(s) =$$
$$-\dot{V} * \int_B \left\{\dot{V}_1 * [t * (\rho^{-1}S_{(ik},_k),_{j)} - K_{ijkl}S_{kl}] - A_{ij}\vartheta + l_{ij}\right\} * \tilde{S}_{ij}\mathrm{d}v$$
$$-\frac{1}{\theta_0}\int_B \left\{\dot{V} * [1 * (k_{pq}\vartheta,_q),_p - \theta_0 A_{pq}S_{pq}] - C_S\vartheta + m\right\} * \tilde{\vartheta}\mathrm{d}v. \tag{5.4.18}$$

Let us first assume that $s = (S_{ij}, q_i)$ is a solution of the problem (5.4.1–3). Then, in view of the Lemma 5.7, s satisfies the relations (5.4.12 and 13), and the right-hand side of eqn (5.4.18) vanishes.

Let us next assume that eqn (5.4.15) holds, i.e. the right-hand side of eqn (5.4.18) vanishes. Then, given the arbitrary character of $\tilde{s}$, and since $s \in \mathcal{R}$, s satisfies eqns (5.4.1–3). □

Remark 5.3 If $(t_0, t_1) \to (0, 0)$, the functional $\mathcal{P}_t(\cdot) \to \mathcal{P}_t^{(0)}(s)$ such that

$$\mathcal{P}_t^{(0)}(s) =$$
$$\frac{1}{2}\int_B \left(\rho^{-1}t * S_{ij},_j * S_{ik},_k + K_{ijkl}S_{kl} * S_{ij} - 2l_{(ij)}^{(0)} * S_{ij}\right)\mathrm{d}v$$
$$+\frac{1}{2\theta_0}\int_B \left(1 * k_{pq}\vartheta,_q * \vartheta,_p + C_S\vartheta * \vartheta + 2\theta_0 A_{pq}S_{pq} * \vartheta - 2m^{(0)} * \vartheta\right)\mathrm{d}v, \tag{5.4.19}$$

where

$$l_{(ij)}^{(0)} = t * (\rho^{-1}b_{(i)}),_{j)} + K_{ijkl}(S_{kl}^{(0)} + t\dot{S}_{kl}^{(0)}) + A_{ij}^{(0)}(\vartheta^{(0)} + t\dot{\vartheta}^{(0)}), \tag{5.4.20}$$

and

$$m^{(0)} = 1 * r + C_S\vartheta^{(0)} - \theta_0 A_{pq}S_{pq}^{(0)}. \tag{5.4.21}$$

The functional $\mathcal{P}_t^{(0)}(s)$ is associated with a natural non-conventional stress–temperature problem of classical dynamical thermoelasticity. Narrowing down suitably the initial data in the problem (5.4.1–3), from the $\mathcal{P}_t^{(0)}(s)$ we may easily obtain the functional for a conventional stress–temperature problem of classical dynamical thermoelasticity. Setting $\vartheta = 0$ in eqn (5.4.19) we obtain the functional associated with a non-conventional variational principle for the stress field of linear isothermal elastodynamics (Hetnarski and Ignaczak, 2004).

Remark 5.4 The variational principles for a generalized thermoelasticity with microstructure were obtained in (Ciumasu and Vieru, 1993), while the variational principles in thermoelasticity without energy dissipation were presented in (Chandrasekharaiah, 1998).

A finite element method was combined with a variational principle of the G–L theory to obtain an approximate solution to a 1D initial-boundary value problem of this theory in (Gładysz, 1986). A variational principle of Hamilton's type for a micropolar thermo-piezo-electro-elastic solid with a relaxation time was obtained in (El-Karamany, 2007).

6

CENTRAL EQUATION OF THERMOELASTICITY WITH FINITE WAVE SPEEDS

6.1 Central equation in the Lord–Shulman and Green–Lindsay theories

In Chapter 4 we formulated a PTP of the L–S as well as the G–L theory, and observed that its role in the analysis of a general problem of that theory is similar to that played by an initial-value problem for the classical wave equation in the analysis of a general problem of isothermal elastodynamics. In particular, certain properties of the solution of PTP carry over onto the solution of the general problem, and some methods of treatment of the PTP may be adapted to a general problem of the L–S and G–L theories. Also, the PTP belongs to one of the few problems of TFWS (thermoelasticity with finite wave speeds) that have been solved in a closed form allowing a full analysis of the thermoelastic process taking place in a given body.

In the present chapter we shall undertake an analysis of a solution to the field equations of PTP in either the L–S or G–L theory. Before undertaking that analysis we shall examine the relationship of the field equations of PTP in the L–S theory to those of PTP in the G–L theory. To this end let us first consider the field equations of the L–S theory, corresponding to vanishing body forces and non-zero heat sources (recall eqn (4.1.8) of Section 4.1)

$$\begin{aligned} &\nabla^2\phi - \frac{\partial^2\phi}{\partial t^2} - \vartheta = 0 \\ &\nabla^2\vartheta - \left(1 + t_0\frac{\partial}{\partial t}\right)\frac{\partial\vartheta}{\partial t} - \epsilon\nabla^2\left(1 + t_0\frac{\partial}{\partial t}\right)\frac{\partial\phi}{\partial t} \\ &\quad = -\left(1 + t_0\frac{\partial}{\partial t}\right) r \end{aligned} \qquad \text{on } \mathrm{B} \times [0, \infty). \qquad (6.1.1)$$

In eqns (6.1.1) ϕ and ϑ denote, respectively, the dimensionless potential of the thermoelastic displacement and the temperature; r is the dimensionless function of heat sources, while ϵ and t_0 are, respectively, the dimensionless parameter coupling the mechanical field with the temperature field and the relaxation time.

By eliminating the function ϑ from eqns (6.1.1), we find the *central equation of the L–S theory*

$$\Lambda\phi = -\left(1 + t_0\frac{\partial}{\partial t}\right) r, \tag{6.1.2}$$

where

$$\Lambda = \left(\nabla^2 - \frac{\partial^2}{\partial t^2}\right)\left(\nabla^2 - t_0\frac{\partial^2}{\partial t^2} - \frac{\partial}{\partial t}\right) - \epsilon\nabla^2\frac{\partial}{\partial t}\left(1 + t_0\frac{\partial}{\partial t}\right) \tag{6.1.3}$$

is called the *central operator of the L–S theory.*

On the other hand, an elimination of the function ϕ from eqns (6.1.1) leads to the following equation for the temperature ϑ

$$\Lambda\vartheta = -\left(1 + t_0\frac{\partial}{\partial t}\right)\left(\nabla^2 - \frac{\partial^2}{\partial t^2}\right) r, \tag{6.1.4}$$

which will be called the *temperature equation of the L–S theory.* Therefore, an alternative form of eqns (6.1.1) is this set of equations

$$\left.\begin{aligned} \Lambda\phi &= -\left(1 + t_0\frac{\partial}{\partial t}\right) r \\ \vartheta &= \left(\nabla^2 - \frac{\partial^2}{\partial t^2}\right)\phi \end{aligned}\right\} \quad \text{on } \mathrm{B} \times [0, \infty). \tag{6.1.5}$$

It follows from eqns (6.1.5) that the knowledge of a function ϕ satisfying eqn $(6.1.5)_1$ allows the determination of temperature ϑ from eqn $(6.1.5)_2$.

Let us now consider the field equations of a PTP in the G–L theory. In the case of vanishing body forces and non-zero heat sources, these equations take the form (recall eqns (4.2.6) of Section 4.2)

$$\left.\begin{aligned} \left(\nabla^2 - \frac{\partial^2}{\partial t^2}\right)\phi - \left(1 + t_1\frac{\partial}{\partial t}\right)\vartheta &= 0 \\ \left(\nabla^2 - t_0\frac{\partial^2}{\partial t^2} - \frac{\partial}{\partial t}\right)\vartheta - \epsilon\frac{\partial}{\partial t}\nabla^2\phi &= -r \end{aligned}\right\} \quad \text{on } \mathrm{B} \times [0, \infty), \tag{6.1.6}$$

where the pair (ϕ, ϑ) and the function r have meanings analogous to those of the L–S theory. Moreover, ϵ and (t_0, t_1) denote, respectively, a coupling parameter and two relaxation times in the G–L theory ($t_1 \geq t_0 > 0$).

By eliminating the function ϑ from eqns (6.1.6), we obtain the *central equation of the G–L theory*

$$\Gamma\phi = -\left(1 + t_1\frac{\partial}{\partial t}\right) r, \tag{6.1.7}$$

where

$$\Gamma = \left(\nabla^2 - \frac{\partial^2}{\partial t^2}\right)\left(\nabla^2 - t_0\frac{\partial^2}{\partial t^2} - \frac{\partial}{\partial t}\right) - \epsilon\nabla^2\frac{\partial}{\partial t}\left(1 + t_1\frac{\partial}{\partial t}\right) \tag{6.1.8}$$

is called the *central operator of the G–L theory.*

Clearly, the elimination of ϕ from eqns (6.1.6) leads to the following *temperature equation of the G–L theory*

$$\Gamma\vartheta = -\left(\nabla^2 - \frac{\partial^2}{\partial t^2}\right) r \quad \text{on} \quad \mathrm{B} \times [0, \infty), \tag{6.1.9}$$

and the following being an alternative system of equations

$$\left.\begin{aligned} \Gamma\phi &= -\left(1 + t_1 \frac{\partial}{\partial t}\right) r \\ \left(1 + t_1 \frac{\partial}{\partial t}\right)\vartheta &= \left(\nabla^2 - \frac{\partial^2}{\partial t^2}\right)\phi \end{aligned}\right\} \quad \text{on} \quad \mathrm{B} \times [0, \infty). \tag{6.1.10}$$

It follows from the definition of Λ and Γ that $\Lambda = \Gamma$ whenever $t_0 = t_1$. Thus, we arrive at the following corollary.

Corollary 6.1 *The central operators of the G–L and L–S theories coincide whenever* $t_1 = t_0 > 0$[1].

It is seen that the condition $t_0 = t_1$ does not imply that the thermodynamic processes are identical in both theories. This follows from the fact that the temperature ϑ in both theories is determined from the function ϕ satisfying eqn $(6.1.5)_1$ (or eqn $(6.1.10)_1$ for $t_1 = t_0$) via two different formulas (recall eqn $(6.1.5)_2$ or eqn $(6.1.10)_2$ for $t_1 = t_0$).

The fact that the temperatures in both theories do not coincide for $t_0 = t_1$ also follows from two different equations governing the temperature: eqn (6.1.4) and (6.1.9) for $t_0 = t_1$.

Now, as the units of the displacement vector, the stress tensor, and the heat flux we take

$$\hat{u}_0 = \hat{\phi}_0 / \widehat{x}_0, \tag{6.1.11}$$

$$\hat{S}_0 = \rho\hat{\phi}_0 / \hat{t}_0^2, \tag{6.1.12}$$

$$\hat{q}_0 = \hat{x}_0 \widehat{r}_0, \tag{6.1.13}$$

where $\widehat{x}_0$, $\hat{t}_0$, $\hat{\phi}_0$, and $\widehat{r}_0$ are the units introduced in Section 4.1 with the help of formulas (4.1.6 and 7). Then, for the dimensionless fields of displacement u_i, stress S_{ij}, and heat flux q_i, which correspond to a pair (ϕ, ϑ) satisfying eqns (6.1.5), we obtain

$$u_i = \phi_{,i} \tag{6.1.14}$$

$$S_{ij} = 2\left(\frac{C_2}{C_1}\right)^2 (\phi_{,ij} - \phi_{,kk}\delta_{ij}) + \ddot{\phi}\delta_{ij} \tag{6.1.15}$$

and

$$q_i + t_0 \dot{q}_i = -\vartheta_{,i}\,, \tag{6.1.16}$$

[1] Proof of Corollary 6.1 was given by Agarwal (1979).

where

$$\frac{1}{C_2^2} = \frac{\rho}{\mu}. \tag{6.1.17}$$

Equations (6.1.14–16) are obtained from the fundamental field equations (1.2.1–6) of Section 1.2 by assuming that: (i) the body is homogeneous and isotropic; (ii) the body forces vanish; and (iii) the displacement field is generated by a scalar potential field.

Note at this point that eqns (6.1.5) along with eqns (6.1.14–16) represent a complete set of equations allowing one to determine a potential–temperature thermoelastic process in the L–S theory.

Using the fundamental equations of the G–L theory [recall formulas (1.3.42–47) of Section 1.3] and taking the same assumptions as introduced in the derivation of eqns (6.1.14–16), we infer that a counterpart of these equations in the G–L theory are the following relations

$$u_i = \phi_{,i}\,, \tag{6.1.18}$$

$$S_{ij} = 2\left(\frac{C_2}{C_1}\right)^2 (\phi_{,ij} - \phi_{,kk}\,\delta_{ij}) + \ddot{\phi}\delta_{ij}, \tag{6.1.19}$$

$$q_i = -\vartheta_{,i}\,. \tag{6.1.20}$$

Thus, a potential–temperature thermoelastic process in the G–L theory is described by eqns (6.1.10) and (6.1.18–20). Since for $t_1 = t_0$ the central operators of both theories coincide, then for $t_0 = t_1$ the displacements u_i and stresses S_{ij} are formally determined via the same formulas (6.1.14 and 15) or (6.1.18 and 19). However, for a uniquely posed PTP, with identical initial-boundary conditions in both theories, and assuming $t_1 = t_0$, the pair (u_i, S_{ij}) of the L–S theory does not coincide with the pair (u_i, S_{ij}) of the G–L theory.

When seeking solutions to the central equation of either L–S or G–L theory, the crucial role is played by a theorem on decomposition, the subject of the next part of this chapter. Since the central equation of the L–S theory is a restricted form of the central equation of the G–L theory, we formulate that theorem for the G–L theory only.

6.2 Decomposition theorem for a central equation of Green–Lindsay theory. Wave-like equations with a convolution

In the case of $r = 0$ in eqn $(6.1.10)_1$, there holds the following:

Theorem 6.1 *Let $\phi = \phi(x, t)$ be a solution of the equation*

$$\Gamma\phi = 0 \quad on \quad B \times (0, \infty), \tag{6.2.1}$$

satisfying the homogeneous initial conditions

$$\frac{\partial^k}{\partial t^k}\phi(\cdot,0)=0 \quad on \quad B,\ k=0,1,2,3. \tag{6.2.2}$$

Let $K=K(t)$ *be a function defined by the formula*

$$K(t)=2\frac{\mathrm{d}}{\mathrm{d}t}\left[\mathrm{e}^{at}\frac{J_1(\beta t)}{\beta t}\right] \quad on \quad [0,\infty), \tag{6.2.3}$$

where $J_1=J_1(t)$ *is the Bessel function of order 1 and of the first kind, and*

$$\begin{aligned}\alpha &= -\left[(1+\epsilon)(t_0+\epsilon t_1)-(1-\epsilon)\right]\triangle^{-1},\\ \beta &= 2\sqrt{\epsilon}\left[1+(1+\epsilon)(t_1-t_0)\right]^{\frac{1}{2}}\triangle^{-1},\end{aligned} \tag{6.2.4}$$

where

$$\triangle=(1-t_0+\epsilon t_1)^2+4\epsilon t_0 t_1=(1+t_0+\epsilon t_1)^2-4t_0>0. \tag{6.2.5}$$

Then

$$\phi=\phi_1+\phi_2 \quad on \quad B\times(0,\infty), \tag{6.2.6}$$

where

$$\left(\nabla^2-\frac{1}{v_1^2}\frac{\partial^2}{\partial t^2}-k_1\frac{\partial}{\partial t}-\lambda-\lambda K*\right)\phi_1=0 \quad on \quad B\times(0,\infty), \tag{6.2.7}$$

$$\left(\nabla^2-\frac{1}{v_2^2}\frac{\partial^2}{\partial t^2}-k_2\frac{\partial}{\partial t}+\lambda+\lambda K*\right)\phi_2=0 \quad on \quad B\times(0,\infty). \tag{6.2.8}$$

Here,

$$\begin{aligned}v_{1.2}^{-2}&=\frac{1}{2}(1+t_0+\epsilon t_1\pm\triangle^{\frac{1}{2}}),\\ k_{1.2}&=\frac{1}{2}(1+\epsilon\mp\alpha\triangle^{\frac{1}{2}}),\\ \lambda&=\frac{1}{4}\beta^2\triangle^{\frac{1}{2}},\end{aligned} \tag{6.2.9}$$

and $*$ *denotes the convolution with respect to time* t, *that is*

$$K*\phi_k(x,t)=\int_0^t K(t-\tau)\,\phi_k(x,\tau)\mathrm{d}\tau, \quad k=1,2. \tag{6.2.10}$$

Clearly, since $t_1>t_0>0$ *and* $\epsilon>0$,

$$v_2>v_1>0 \quad and \quad \lambda>0. \tag{6.2.11}$$

Moreover, if $(t_0,t_1)\to(0,0)$ *for* $t_1\geq t_0\geq 0$, *then*

$$\begin{aligned}(v_1,v_2)&\to(1,\infty), \quad (k_1,k_2)\to(\epsilon,1),\\ \lambda\to\epsilon, &\quad \alpha\to 1-\epsilon, \quad \beta\to 2\sqrt{\epsilon},\end{aligned} \tag{6.2.12}$$

and the theorem reduces to a result of classical thermoelasticity (Brun, 1975; Ignaczak, 1976).

A proof of this theorem is based on four lemmas (Ignaczak, 1978) given in the following.

Lemma 6.1 *Let F_1 denote the set of functions*

$$F_1 = \left\{ \phi(x,t) : \frac{\partial^k}{\partial t^k}\phi(x,0) = 0 \quad for \quad k = 0, 1 \right\}, \tag{6.2.13}$$

and let us introduce the operators

$$\begin{aligned} \hat{L}_1 &= \nabla^2 - \frac{1}{v_1^2}\frac{\partial^2}{\partial t^2} - k_1\frac{\partial}{\partial t} - \lambda - \lambda K* \\ \hat{L}_2 &= \nabla^2 - \frac{1}{v_2^2}\frac{\partial^2}{\partial t^2} - k_2\frac{\partial}{\partial t} + \lambda + \lambda K* \end{aligned} \tag{6.2.14}$$

and

$$L_1 = \hat{L}_1 + \lambda K*, \qquad L_2 = \hat{L}_2 - \lambda K*, \tag{6.2.15}$$

where the parameters v_i, k_i and λ as well as the function $K = K(t)$ have the same meaning as in Theorem 6.1. Then

$$\hat{L}_1\hat{L}_2\phi = \hat{L}_2\hat{L}_1\phi \quad \forall\phi \in F_1. \tag{6.2.16}$$

Proof. Using the definitions (6.2.14 and 15), we obtain

$$\begin{gathered} \hat{L}_1\hat{L}_2\phi = (L_1 - \lambda K*)(L_2 + \lambda K*)\phi = \\ L_1L_2\phi + \lambda L_1(K*\phi) - \lambda K*(L_2\phi) - \lambda^2 K*K*\phi, \end{gathered} \tag{6.2.17}$$

$$\begin{gathered} \hat{L}_2\hat{L}_1\phi = (L_2 + \lambda K*)(L_1 - \lambda K*)\phi = \\ L_2L_1\phi - \lambda L_2(K*\phi) + \lambda K*(L_1\phi) - \lambda^2 K*K*\phi. \end{gathered} \tag{6.2.18}$$

Since for an arbitrary function ϕ,

$$L_1L_2\phi = L_2L_1\phi, \tag{6.2.19}$$

and

$$\nabla^2(K*\phi) = K*\nabla^2\phi, \tag{6.2.20}$$

then, subtracting eqn (6.2.17) from eqn (6.2.18), and using the definitions of operators L_1 and L_2, we find

$$\hat{L}_1\hat{L}_2\phi - \hat{L}_2\hat{L}_1\phi =$$

$$\lambda\left\{K * \left[\left(\frac{1}{v_2^2}\frac{\partial^2}{\partial t^2} + k_2\frac{\partial}{\partial t}\right)\phi\right] - \left[\left(\frac{1}{v_2^2}\frac{\partial^2}{\partial t^2} + k_2\frac{\partial}{\partial t}\right)\right] K * \phi\right. \tag{6.2.21}$$

$$\left. + K * \left[\left(\frac{1}{v_1^2}\frac{\partial^2}{\partial t^2} + k_1\frac{\partial}{\partial t}\right)\phi\right] - \left[\left(\frac{1}{v_1^2}\frac{\partial^2}{\partial t^2} + k_1\frac{\partial}{\partial t}\right)\right] K * \phi\right\}.$$

Moreover, for $\phi \in F_1$, we have (overdot denotes $\partial/\partial t$)

$$K * \dot{\phi} = \dot{\overline{K * \phi}}, \tag{6.2.22}$$

$$K * \ddot{\phi} = \ddot{\overline{K * \phi}}. \tag{6.2.23}$$

The relations (6.2.22) and (6.2.23) imply that the right-hand side of relation (6.2.21) vanishes, that is, the relation (6.2.16) holds. □

Lemma 6.2 *The function $K = K(t)$ defined by eqns (6.2.3–5) satisfies the ordinary differential-integral equation*

$$\frac{\mathrm{d}^2}{\mathrm{d}t^2}K(t) - \alpha\frac{\mathrm{d}}{\mathrm{d}t}K(t) + \frac{\beta^2}{2}K(t) + \frac{\beta^2}{4}K * K(t) = 0 \quad \forall t \in [0, \infty). \tag{6.2.24}$$

Proof. A proof of that lemma was given in (Ignaczak, 1978) with the help of recurrence relations for the Bessel functions. Here, we present another proof, based on the Laplace transform. We first note that, from the definition of the function $K = K(t)$ [recall eqn (6.2.3)] it follows that

$$K(t) = \dot{G}(t), \tag{6.2.25}$$

where

$$G(t) = 2\mathrm{e}^{\alpha t}\frac{J_1(\beta t)}{\beta t}, \quad G(0) = 1. \tag{6.2.26}$$

Upon taking the Laplace transform of eqn (6.2.25), we obtain

$$\bar{K} = p\bar{G} - 1, \tag{6.2.27}$$

where (Mikusiński, 1967; Ignaczak, 1981)

$$\bar{G} = \frac{2}{\beta^2}\left[\sqrt{(p-\alpha)^2 + \beta^2} - (p - \alpha)\right]. \tag{6.2.28}$$

The overbar on K and G in eqns (6.2.27 and 28) denotes the Laplace transform with p being the transform parameter. Raising eqn (6.2.28) to the powers

$k = 1, 2, 3, \ldots$ we obtain

$$\bar{G}^k = \left(\frac{2}{\beta^2}\right)^k \left[\sqrt{(p-\alpha)^2 + \beta^2} - (p-\alpha)\right]^k, \tag{6.2.29}$$

from which (Ignaczak, 1981)

$$G_k(t) = L^{-1}\left(\bar{G}^k\right) = k\left(\frac{2}{\beta}\right)^k \frac{e^{\alpha t}}{t} J_k(\beta t), \tag{6.2.30}$$

where $J_k = J_k(x)$ is the Bessel function of the first kind and order k. From the relations (6.2.27 and 28) we now find

$$\bar{K} = \frac{[\cdot]}{\beta^2}\left\{2\alpha - [\cdot]\right\}, \tag{6.2.31}$$

where $[\cdot]$ denotes the expression standing in the square brackets in eqn (6.2.28), that is

$$[\cdot] = \left(\frac{\beta^2}{2}\right)\bar{G}. \tag{6.2.32}$$

From this

$$\bar{K} = \alpha\bar{G} - \frac{1}{4}\beta^2\bar{G}^2. \tag{6.2.33}$$

Using eqn (6.2.30), we invert the relation (6.2.33) to the form

$$K(t) = 2\beta\frac{e^{\hat{\alpha}\tau}}{\tau}\left[\hat{\alpha}J_1(\tau) - J_2(\tau)\right], \tag{6.2.34}$$

where

$$\tau = \beta t, \quad \widehat{\alpha} = \alpha/\beta. \tag{6.2.35}$$

Multiplying eqn (6.2.33) through by p and passing with p to infinity, while noting that

$$\lim_{p\to\infty} p\bar{K}(p) = K(0), \quad \lim_{p\to\infty} p\bar{G}(p) = G(0) = 1, \tag{6.2.36}$$

we find

$$K(0) = \alpha. \tag{6.2.37}$$

Squaring both sides of eqn (6.2.33), we obtain

$$\bar{K}^2 = \alpha^2\bar{G}^2 - \frac{\alpha\beta^2}{2}\bar{G}^3 + \left(\frac{\beta^2}{4}\right)^2\bar{G}^4. \tag{6.2.38}$$

From this, in view of eqn (6.2.30),

$$K * K = 4\beta\frac{e^{\hat{\alpha}\tau}}{\tau}\left[2\hat{\alpha}^2 J_2(\tau) - 3\hat{\alpha}J_3(\tau) + J_4(\tau)\right]. \tag{6.2.39}$$

In order to find $\dot{K}(t)$, we multiply eqn (6.2.33) through by p and, in view of eqns (6.2.27,33), obtain

$$p\bar{K} = \left(\alpha - \frac{\beta^2}{4}\bar{G}\right)\left(1 + \alpha\bar{G} - \frac{\beta^2}{4}\bar{G}^2\right). \tag{6.2.40}$$

From this and eqn (6.2.37), we arrive at

$$p\bar{K} - K(0) = \left(\alpha^2 - \frac{\beta^2}{4}\right)\bar{G} - \frac{\alpha\beta^2}{2}\bar{G}^2 + \left(\frac{\beta^2}{4}\right)^2\bar{G}^3. \tag{6.2.41}$$

The inversion of this with the help of eqn (6.2.30) leads to the relation

$$\dot{K} = -\frac{\beta^2}{2}\frac{\mathrm{e}^{\hat{\alpha}\tau}}{\tau}\left[\left(1 - 4\hat{\alpha}^2\right)J_1(\tau) + 8\hat{\alpha}J_2(\tau) - 3J_3(\tau)\right]. \tag{6.2.42}$$

By multiplying eqn (6.2.41) through by p and passing with p to infinity, we obtain

$$\dot{K}(0) = \alpha^2 - \frac{\beta^2}{4}. \tag{6.2.43}$$

Finally, in order to determine $\ddot{K}$, we multiply eqn (6.2.41) by p and, using eqns (6.2.27,33,43), we obtain

$$p^2\bar{K} - pK(0) - \dot{K}(0) =$$
$$\alpha\left(\alpha^2 - \frac{3}{4}\beta^2\right)\bar{G} + \left[2\left(\frac{\beta^2}{4}\right)^2 - \frac{3}{4}\alpha^2\beta^2\right]\bar{G}^2 + 3\alpha\left(\frac{\beta^2}{4}\right)^2\bar{G}^3 - \left(\frac{\beta^2}{4}\right)^3\bar{G}^4. \tag{6.2.44}$$

Upon the inverse transformation of eqn (6.2.44), on account of eqn (6.2.30), we find

$$\begin{aligned}\ddot{K} = &-\frac{\beta^3}{4}\frac{\mathrm{e}^{\hat{\alpha}\tau}}{\tau}[\left(3\hat{\alpha} - 4\hat{\alpha}^3\right)J_1(\tau)\\ &-2\left(1 - 6\hat{\alpha}^2\right)J_2(\tau) - 9\hat{\alpha}J_3(\tau) + 2J_4(\tau)].\end{aligned} \tag{6.2.45}$$

Now, inserting the functions (6.2.45), (6.2.42), (6.2.34) and (6.2.39) into the left-hand side of eqn (6.2.24), we verify that K satisfies eqn (6.2.24). □

The above lemma states that the convolution of the kernel K with itself is equivalent to an application of a 2nd-order differential operation to K.

Lemma 6.3 *Let Γ be a central operator of the G–L theory, that is (recall eqn (6.1.8))*

$$\Gamma \equiv \left[\left(\nabla^2 - \frac{\partial^2}{\partial t^2}\right)\left(\nabla^2 - t_0\frac{\partial^2}{\partial t^2} - \frac{\partial}{\partial t}\right) - \epsilon\,\nabla^2\frac{\partial}{\partial t}\left(1 + t_1\frac{\partial}{\partial t}\right)\right]. \tag{6.2.46}$$

Then, for every $\phi \in F_1$,

$$\Gamma\phi = \hat{L}_1 \hat{L}_2 \phi, \tag{6.2.47}$$

where $\hat{L}_1$ *and* $\hat{L}_2$ *are defined by formulas (6.2.14).*

Proof. First, note that from eqns (6.2.14 and 15) we obtain

$$L_1 L_2 \phi = \left\{ \nabla^4 - (k_1 + k_2) \nabla^2 \frac{\partial}{\partial t} - \left(\frac{1}{v_1^2} + \frac{1}{v_2^2} \right) \nabla^2 \frac{\partial^2}{\partial t^2} + \frac{1}{v_1^2 v_2^2} \frac{\partial^4}{\partial t^4} \right.$$
$$\left. + \left(\frac{k_1}{v_2^2} + \frac{k_2}{v_1^2} \right) \frac{\partial^3}{\partial t^3} + \left[k_1 k_2 - \lambda \left(\frac{1}{v_1^2} - \frac{1}{v_2^2} \right) \right] \frac{\partial^2}{\partial t^2} - \lambda (k_1 - k_2) \frac{\partial}{\partial t} - \lambda^2 \right\} \phi. \tag{6.2.48}$$

Next, using the identities

$$k_1 + k_2 = 1 + \epsilon, \tag{6.2.49}$$

$$\frac{1}{v_1^2} + \frac{1}{v_2^2} = 1 + t_0 + \epsilon t_1, \tag{6.2.50}$$

$$\frac{1}{v_1^2} \cdot \frac{1}{v_2^2} = t_0, \tag{6.2.51}$$

$$\frac{k_1}{v_2^2} + \frac{k_2}{v_1^2} = 1, \tag{6.2.52}$$

$$\lambda = k_1 k_2 \left(\frac{1}{v_1^2} - \frac{1}{v_2^2} \right)^{-1}, \tag{6.2.53}$$

we reduce eqn (6.2.48) to the form

$$L_1 L_2 \phi = \left[\nabla^4 - (1 + \epsilon) \nabla^2 \frac{\partial}{\partial t} - (1 + t_0 + \epsilon t_1) \nabla^2 \frac{\partial^2}{\partial t^2} \right.$$
$$\left. + t_0 \frac{\partial^4}{\partial t^4} + \frac{\partial^3}{\partial t^3} - \lambda (k_1 - k_2) \frac{\partial}{\partial t} - \lambda^2 \right] \phi. \tag{6.2.54}$$

In view of the definition of operator Γ (recall eqn (6.2.46)), the last of the above equations may be rewritten in the form

$$L_1 L_2 \phi = \Gamma\phi - \lambda \left[(k_1 - k_2) \frac{\partial}{\partial t} + \lambda \right] \phi. \tag{6.2.55}$$

We next evaluate the expression

$$L_1 (K * \phi) - K * (L_2 \phi) =$$
$$\frac{1}{v_2^2} K * \frac{\partial^2 \phi}{\partial t^2} - \frac{1}{v_1^2} \frac{\partial^2}{\partial t^2} (K * \phi) + k_2 K * \frac{\partial \phi}{\partial t} - k_1 \frac{\partial}{\partial t} (K * \phi) - 2\lambda K * \phi. \tag{6.2.56}$$

Since $\phi \in F_1$, then (recall eqns (6.2.22 and 23))

$$\frac{\partial}{\partial t}(K * \phi) = K * \frac{\partial \phi}{\partial t}, \tag{6.2.57}$$

and

$$\frac{\partial^2}{\partial t^2}(K * \phi) = K * \frac{\partial^2 \phi}{\partial t^2}. \tag{6.2.58}$$

Also,

$$K * \frac{\partial \phi}{\partial t} = \phi * \frac{\partial K}{\partial t} + K_0 \phi, \tag{6.2.59}$$

$$K * \frac{\partial^2 \phi}{\partial t^2} = \phi * \frac{\partial^2 K}{\partial t^2} + \dot{K}_0 \phi + K_0 \frac{\partial \phi}{\partial t}, \tag{6.2.60}$$

where

$$K_0 = K(0), \quad \dot{K}_0 = \dot{K}(0). \tag{6.2.61}$$

Thus, eqn (6.2.56) takes the form

$$\begin{aligned} &L_1(K * \phi) - K * (L_2 \phi) = \\ &-\left(\frac{1}{v_1^2} - \frac{1}{v_2^2}\right)\left(\phi * \frac{\partial^2 K}{\partial t^2} + \dot{K}_0 \phi + K_0 \frac{\partial \phi}{\partial t}\right) \\ &-(k_1 - k_2)\left(\phi * \frac{\partial K}{\partial t} + K_0 \phi\right) - 2\lambda K * \phi. \end{aligned} \tag{6.2.62}$$

Now, using the definitions of $\hat{L}_1$, $\hat{L}_2$ and Γ (recall eqns (6.2.14 and 46)), we obtain

$$\begin{aligned} \hat{L}_1 \hat{L}_2 \phi - \Gamma \phi &= (L_1 - \lambda K*)(L_2 + \lambda K*)\phi - \Gamma \phi \\ &= L_1 L_2 \phi + \lambda [L_1(K * \phi) - K * (L_2 \phi)] \\ &\quad -\lambda^2 K * K * \phi - \Gamma \phi. \end{aligned} \tag{6.2.63}$$

In view of eqns (6.2.55 and 62), this equation becomes

$$\begin{aligned} \hat{L}_1 \hat{L}_2 \phi - \Gamma \phi = -\lambda &\left[(k_1 - k_2) + \left(\frac{1}{v_1^2} - \frac{1}{v_2^2}\right) K_0\right] \frac{\partial \phi}{\partial t} \\ -\lambda &\left[\lambda + (k_1 - k_2) K_0 + \left(\frac{1}{v_1^2} - \frac{1}{v_2^2}\right) \dot{K}_0\right] \phi \\ -\lambda \left[\left(\frac{1}{v_1^2} - \frac{1}{v_2^2}\right) \frac{\mathrm{d}^2 K}{\mathrm{d}t^2} \right.&\left. + (k_1 - k_2)\frac{\mathrm{d}K}{\mathrm{d}t} + 2\lambda K + \lambda K * K\right] * \phi. \end{aligned} \tag{6.2.64}$$

Since

$$K_0 = \alpha = -(k_1 - k_2)\left(\frac{1}{v_1^2} - \frac{1}{v_2^2}\right)^{-1}, \tag{6.2.65}$$

$$\dot{K}_0 = \alpha^2 - \frac{1}{4}\beta^2 = \left[(k_1 - k_2)^2 - k_1 k_2\right]\left(\frac{1}{v_1^2} - \frac{1}{v_2^2}\right)^{-2}, \tag{6.2.66}$$

$$\frac{1}{4}\beta^2 = \lambda\left(\frac{1}{v_1^2} - \frac{1}{v_2^2}\right)^{-1}, \tag{6.2.67}$$

the first two terms on the right-hand side of eqn (6.2.64) vanish and

$$\hat{L}_1\hat{L}_2\phi - \Gamma\phi = -\lambda\left(\frac{1}{v_1^2} - \frac{1}{v_2^2}\right)\left[\frac{\mathrm{d}^2 K}{\mathrm{d}t^2} - \alpha\frac{\mathrm{d}K}{\mathrm{d}t} + \frac{\beta^2}{2}K + \frac{\beta^2}{4}K * K\right] * \phi. \tag{6.2.68}$$

In view of Lemma 6.2, the expression in the square bracket vanishes, and eqn (6.2.68) reduces to eqn (6.2.47). □

The Lemmas 6.1 and 6.3 jointly imply that the operator Γ restricted to the domain F_1 is a product of two commuting operators $\widehat{L}_1$ and $\widehat{L}_2$, that is

$$\Gamma\phi = \hat{L}_1\hat{L}_2\phi = \hat{L}_2\hat{L}_1\phi \quad \forall\phi \in F_1. \tag{6.2.69}$$

Lemma 6.4 *Let $g = g(x,t)$ be a function belonging to F_1, and let us define a function $\phi_1 = \phi_1(x,t)$ as follows*

$$\phi_1 = \left(\frac{1}{v_1^2} - \frac{1}{v_2^2}\right)^{-1} * \left[\mathrm{e}^{\widehat{\alpha}\tau} J_0(\tau)\right] * g, \tag{6.2.70}$$

where $J_0(\tau)$ is the Bessel function of the first kind and order zero, with $\widehat{\alpha}$ and τ being specified through the relations (6.2.35). Then

$$\frac{\partial^k \phi_1}{\partial t^k}(x,0) = 0 \quad \textit{for} \quad k = 0,1,2,3, \tag{6.2.71}$$

and ϕ_1 satisfies the equation

$$\left(\hat{L}_2 - \hat{L}_1\right)\phi_1 = g. \tag{6.2.72}$$

Proof. Let us first introduce the notations

$$\hat{g} = \left(\frac{1}{v_1^2} - \frac{1}{v_2^2}\right)^{-1} g, \tag{6.2.73}$$

$$h_k(t) = \mathrm{e}^{\hat{\alpha}\tau} J_k(\tau), \quad k = 0,1, \tag{6.2.74}$$

and rewrite eqns (6.2.70 and 72) as

$$\phi_1 = 1 * h_0 * \hat{g}, \tag{6.2.75}$$

$$\frac{\partial^2 \phi_1}{\partial t^2} - \alpha \frac{\partial \phi_1}{\partial t} + \frac{\beta^2}{2}\phi_1 + \frac{\beta^2}{2} K * \phi_1 = \hat{g}. \tag{6.2.76}$$

In order to show that ϕ_1 specified via eqn (6.2.75) satisfies eqns (6.2.71) and (6.2.76), we again employ the Laplace transform. Thus, from eqn (6.2.74) we obtain (Mikusiński, 1967)

$$\bar{h}_0 = \frac{1}{\sqrt{(p-\alpha)^2 + \beta^2}}, \tag{6.2.77}$$

$$\bar{h}_1 = \frac{1}{\beta} \frac{1}{\sqrt{(p-\alpha)^2 + \beta^2}} \left[\sqrt{(p-\alpha)^2 + \beta^2} - (p - \alpha) \right]. \tag{6.2.78}$$

From this we get the identities

$$p\bar{h}_0 = 1 + \alpha \bar{h}_0 - \beta \bar{h}_1, \tag{6.2.79}$$

and

$$p^2 \bar{h}_0 = p + \alpha \left(1 + \alpha \bar{h}_0 - \beta \bar{h}_1\right) - \beta p \bar{h}_1. \tag{6.2.80}$$

Transforming eqn (6.2.75), we obtain

$$\overline{\phi}_1 = p^{-1} \bar{h}_0 \overline{\hat{g}}, \tag{6.2.81}$$

which, together with eqns (6.2.79 and 80), leads to

$$p\overline{\phi}_1 = \bar{h}_0 \overline{\hat{g}}, \tag{6.2.82}$$

$$p^2 \bar{\phi}_1 = \overline{\hat{g}} + \alpha \bar{h}_0 \overline{\hat{g}} - \beta \bar{h}_1 \overline{\hat{g}}, \tag{6.2.83}$$

$$p^3 \bar{\phi}_1 = p\overline{\hat{g}} + \alpha \overline{\hat{g}} + \alpha^2 \bar{h}_0 \overline{\hat{g}} - \alpha\beta \bar{h}_1 \overline{\hat{g}} - \beta p \bar{h}_1 \overline{\hat{g}}. \tag{6.2.84}$$

It follows from the definition of ϕ_1 that $\phi_1(\cdot, 0) = 0$. From this and eqn (6.2.82) we obtain

$$\frac{\partial \phi_1}{\partial t} = h_0 * \hat{g}, \tag{6.2.85}$$

which implies

$$\frac{\partial \phi_1}{\partial t}(\cdot, 0) = 0. \tag{6.2.86}$$

Thus, inverting eqn (6.2.83), we find

$$\frac{\partial^2 \phi_1}{\partial t^2} = \hat{g} + \alpha h_0 * \hat{g} - \beta h_1 * \hat{g}. \tag{6.2.87}$$

From this and the fact that $\hat{g} \in F_1$ we find

$$\frac{\partial^2 \phi_1}{\partial t^2}(\cdot, 0) = 0. \tag{6.2.88}$$

Finally, since $\hat{g} \in F_1$ and

$$\frac{\partial^k \phi_1}{\partial t^k}(\cdot, 0) = 0, \quad \text{for} \quad k = 0, 1, 2, \tag{6.2.89}$$

upon inversion of eqn (6.2.84) we obtain

$$\frac{\partial^3 \phi_1}{\partial t^3} = \frac{\partial \hat{g}}{\partial t} + \alpha \hat{g} + \alpha^2 h_0 * \hat{g} - \alpha\beta h_1 * \hat{g} - \beta \hat{g} * \frac{\partial h_1}{\partial t}. \tag{6.2.90}$$

The preceding two equations along with the condition $\hat{g} \in F_1$ imply that

$$\frac{\partial^k \phi_1}{\partial t^k}(\cdot, 0) = 0, \quad \text{for} \quad k = 0, 1, 2, 3, \tag{6.2.91}$$

i.e. the condition (6.2.71) holds. This completes the first part of the proof of Lemma 6.4.

In order to show that ϕ_1 satisfies eqn (6.2.76), let us first note that (recall eqns (6.2.27 and 6.2.28))

$$\frac{\bar{K}}{p} = \frac{2}{\beta^2}\left[\sqrt{(p-a)^2 + \beta^2} - (p-a)\right] - \frac{1}{p}. \tag{6.2.92}$$

Multiplying that relation through by $\bar{h}_0$ and using eqns (6.2.77 and 6.2.78), we obtain

$$\frac{\bar{K}\bar{h}_0}{p} = \frac{2}{\beta}\bar{h}_1 - \frac{\bar{h}_0}{p}. \tag{6.2.93}$$

From this and the definition of ϕ_1 we find

$$K * \phi_1 = \frac{2}{\beta} h_1 * \hat{g} - 1 * h_0 * \hat{g}. \tag{6.2.94}$$

Thus, in view of eqns (6.2.75, 6.2.85, 6.2.87, 6.2.94), the function ϕ_1 satisfies eqn (6.2.72). □

Proof. (Proof of Theorem 6.1) Since ϕ is a given function satisfying eqns (6.2.1 and 6.2.2), it suffices to demonstrate the existence of a field ϕ_1 with the properties

$$\frac{\partial^k \phi_1}{\partial t^k}(\cdot, 0) = 0 \quad \text{for} \quad k = 0, 1, 2, 3, \tag{6.2.95}$$

and

$$\hat{L}_1 \phi_1 = 0, \quad \hat{L}_2(\phi - \phi_1) = 0. \tag{6.2.96}$$

Indeed, if there exists the field ϕ_1, then the existence of ϕ_2 follows from

$$\phi_2 = \phi - \phi_1. \tag{6.2.97}$$

We first note that

$$\hat{L}_2\phi_1 = \hat{L}_1\phi_1 + \left(\hat{L}_2 - \hat{L}_1\right)\phi_1. \tag{6.2.98}$$

Thus, if

$$\hat{L}_1\phi_1 = 0, \tag{6.2.99}$$

then

$$\hat{L}_2\phi_1 = \left(\hat{L}_2 - \hat{L}_1\right)\phi_1. \tag{6.2.100}$$

Hence, eqns (6.2.96) are satisfied so long as

$$\hat{L}_1\phi_1 = 0, \quad \left(\hat{L}_2 - \hat{L}_1\right)\phi_1 = g, \tag{6.2.101}$$

where $g = g(x,t)$ is the given function

$$g = \hat{L}_2\phi. \tag{6.2.102}$$

Since ϕ satisfies the homogeneous initial conditions (6.2.2), then $\phi \in F_1$, and, in view of Lemmas 6.1 and 6.3 (also recall eqn (6.2.69)), we find

$$\hat{L}_1\hat{L}_2\phi = \hat{L}_2\hat{L}_1\phi = \Gamma\phi. \tag{6.2.103}$$

Also, since ϕ satisfies eqn (6.2.1), by operating with $\hat{L}_1$ on both sides of eqn (6.2.102), we find

$$\hat{L}_1 g = 0. \tag{6.2.104}$$

Moreover, eqns (6.2.2) and (6.2.102) imply

$$g \in F_1. \tag{6.2.105}$$

Thus, in order to complete the proof of the theorem, it is sufficient to demonstrate a field ϕ_1 that satisfies eqns (6.2.104) and (6.2.105). To this end, we employ Lemma 6.4 and show that the function ϕ_1 appearing in that lemma for g specified via the formula (6.2.102) satisfies all those conditions. First, we verify that, in view of eqn (6.2.105), g satisfies the assumptions of Lemma 6.4. Thus, ϕ_1 satisfies the homogeneous initial conditions (6.2.95) and the second of eqn (6.2.101). Now, let us note that

$$\begin{aligned}\hat{L}_1\phi_1 &= \nabla^2\phi_1 - \frac{1}{v_1^2}\frac{\partial^2\phi_1}{\partial t^2} - k_1\frac{\partial\phi_1}{\partial t} - \lambda\phi_1 - \lambda K * \phi_1 \\ &= 1 * h_0 * \nabla^2\hat{g} - \frac{1}{v_1^2}\frac{\partial^2\phi_1}{\partial t^2} - k_1\frac{\partial\phi_1}{\partial t} - \lambda\phi_1 - \lambda K * \phi_1,\end{aligned} \tag{6.2.106}$$

and, in view of eqn (6.2.104),

$$\nabla^2\hat{g} = \frac{1}{v_1^2}\frac{\partial^2\hat{g}}{\partial t^2} + k_1\frac{\partial\hat{g}}{\partial t} + \lambda\hat{g} + \lambda K * \hat{g}. \tag{6.2.107}$$

Hence,

$$\begin{aligned}\hat{L}_1\phi_1 = 1 * h_0 * \left(\frac{1}{v_1^2}\frac{\partial^2}{\partial t^2}\hat{g} + k_1\frac{\partial \hat{g}}{\partial t} + \lambda\hat{g} + \lambda K * \hat{g}\right) \\ -\frac{1}{v_1^2}\frac{\partial^2}{\partial t^2}(1 * h_0 * \hat{g}) - k_1\frac{\partial}{\partial t}(1 * h_0 * \hat{g}) \\ -\lambda(1 * h_0 * \hat{g}) - \lambda K * (1 * h_0 * \hat{g}) .\end{aligned} \tag{6.2.108}$$

Since $\hat{g} \in F_1$ (recall eqn (6.2.105)), we have

$$1 * h_0 * \frac{\partial^2 \hat{g}}{\partial t^2} = \frac{\partial^2}{\partial t^2}(1 * h_0 * \hat{g}) , \tag{6.2.109}$$

and

$$1 * h_0 * \frac{\partial \hat{g}}{\partial t} = \frac{\partial}{\partial t}(1 * h_0 * \hat{g}) . \tag{6.2.110}$$

Thus, eqn (6.2.108) reduces to the first equation of eqns (6.2.101), which completes the proof. □

Remark 6.1 Inequalities (6.2.11) and the limit relations (6.2.12) are a direct consequence of formulas determining the parameters v_i, k_i, λ, α and β. Moreover, if $(t_0, t_1) \to (0, 0)$, then the theorem on decomposition reduces to the following theorem (Brun, 1975; Ignaczak, 1976):

Theorem 6.2 *(on decomposition for the central equation of the classical dynamical thermoelasticity) An arbitrary solution of the central equation of the classical dynamical thermoelasticity*

$$\left[\left(\nabla^2 - \frac{\partial^2}{\partial t^2}\right)\left(\nabla^2 - \frac{\partial}{\partial t}\right) - \epsilon\nabla^2\frac{\partial}{\partial t}\right] f = 0 \quad \epsilon \geq 0, \tag{6.2.111}$$

satisfying the homogeneous initial conditions

$$\frac{\partial^k f}{\partial t^k}(\cdot, 0) = 0 \quad \textit{for} \quad k = 0, 1, 2, \tag{6.2.112}$$

takes the form

$$f = f_1 + f_2, \tag{6.2.113}$$

where

$$\left(\nabla^2 - \frac{\partial^2}{\partial t^2} - \epsilon\frac{\partial}{\partial t} - \epsilon - \epsilon K_0 *\right) f_1 = 0, \tag{6.2.114}$$

$$\left(\nabla^2 - \frac{\partial}{\partial t} + \epsilon + \epsilon K_0 *\right) f_2 = 0, \tag{6.2.115}$$

$$K_0(t) = 2\frac{\mathrm{d}}{\mathrm{d}t}\left[\mathrm{e}^{(1-\epsilon)t}\frac{J_1\left(2t\sqrt{\epsilon}\right)}{2t\sqrt{\epsilon}}\right]. \tag{6.2.116}$$

Remark 6.2 In the following, we shall call $\widehat{L}_1$ and $\widehat{L}_2$ [recall eqns (6.2.14)] the *wave-like operators with convolution*, while eqns (6.2.7 and 8) the *wave-like equations with convolution.* A thermoelastic disturbance corresponding to the functions ϕ_i $(i = 1, 2)$ will be called the *ith fundamental thermoelastic disturbance.* For $\lambda > 0$ the parameters v_i and k_i $(i = 1, 2)$ represent the *speed and damping of the ith disturbance.* The parameter λ and the function $K = K(t)$ appearing in the definition of the operator $\widehat{L}_i$ (note eqn (6.2.14)) will be called, respectively, the *coefficient of convolution* and the *kernel of* $\widehat{L}_i$.

Clearly, the parameters v_i, k_i, λ, and the function $K = K(t)$ for each fixed t depend on the three constitutive variables (ϵ, t_0, t_1) according to the the formulas (6.2.3 and 9). These dependencies are analyzed in the subsequent sections of this chapter.

6.3 Speed of a fundamental thermoelastic disturbance in the space of constitutive variables

The theorem on decomposition of Section 6.2 implies that a potential–temperature thermoelastic disturbance in the G–L theory is a sum of two fundamental thermoelastic disturbances propagating with speeds v_1 and v_2, where (recall eqn $(6.2.9)_1$)

$$v_{1.2}^{-2} = \frac{1}{2}\left(1 + t_0 + \epsilon t_1 \pm \Delta^{\frac{1}{2}}\right). \tag{6.3.1}$$

In this equation Δ is given by the formula

$$\Delta = \left(1 - t_0 + \epsilon t_1\right)^2 + 4\epsilon t_1 t_0. \tag{6.3.2}$$

It follows from eqns (6.3.1 and 2) that

$$v_{1.2} = v_{1.2}\left(t_0, \zeta\right), \tag{6.3.3}$$

where ζ is *the reduced constitutive variable*, given as

$$\zeta = \epsilon t_1, \tag{6.3.4}$$

that is, the speeds $v_{1.2}$ may be treated as functions of two variables t_0 and ζ.

In the following, we restrict the domain of functions $\upsilon_{1.2}$ to the region[2]

$$t_0 \geq 1, \quad \zeta \geq \epsilon \geq 0. \tag{6.3.5}$$

Some information on the speeds $\upsilon_{1.2}$ in the space of t_0 and ζ is contained in the following:

Theorem 6.3 *Let $Q = \{(t_0, \zeta) : t_0 \geq 1, \zeta \geq 0\}$. Then, $\forall (t_0, \zeta) \in Q$ the estimates are true*

$$\begin{aligned} \upsilon_1 (t_0, \zeta) &\leq \varphi_1 (t_0, \zeta), \\ \upsilon_2 (t_0, \zeta) &\geq \varphi_2 (t_0, \zeta), \end{aligned} \tag{6.3.6}$$

where the functions $\varphi_{1,2}$ are given by

$$\begin{aligned} \varphi_1 (t_0, \zeta) &= (t_0 + \zeta)^{-\frac{1}{2}}, \\ \varphi_2 (t_0, \zeta) &= (t_0 + \zeta)^{\frac{1}{2}} t_0^{-\frac{1}{2}}. \end{aligned} \tag{6.3.7}$$

Proof. Introduce the notations

$$V_{1.2} = \upsilon_{1.2}^{-2}. \tag{6.3.8}$$

Then, in view of eqns (6.3.1 and 2), we find

$$V_{1.2} = \frac{1}{2}\left(1 + t_0 + \zeta \pm \Delta^{\frac{1}{2}}\right), \tag{6.3.9}$$

where

$$\Delta = [t_0 - (1 + \zeta)]^2 + 4t_0\zeta, \tag{6.3.10}$$

$$\Delta = [t_0 + (1 + \zeta)]^2 - 4t_0, \tag{6.3.11}$$

$$\Delta = [t_0 - (1 - \zeta)]^2 + 4\zeta. \tag{6.3.12}$$

Given eqns (6.2.50 and 6.2.51), we obtain

$$V_1 + V_2 = 1 + t_0 + \zeta, \quad V_1 V_2 = t_0, \tag{6.3.13}$$

$$V_1 - V_2 = \Delta^{\frac{1}{2}}. \tag{6.3.14}$$

Since for every real-valued a and x

$$\left(x^2 + a^2\right)^{\frac{1}{2}} \geq |x|, \tag{6.3.15}$$

and

$$|x| \pm x \geq 0, \tag{6.3.16}$$

[2] As is shown later for an aluminum alloy, the dimensionless time $t_0 = 3.728$; hence, it makes sense to let $t_0 \geq 1$. The inequalities $t_0 \geq 1$, $\epsilon > 0$, and $t_1 \geq t_0$ [see eqn (1.3.23)] imply that $\zeta \geq \epsilon > 0$.

then

$$\varphi(x) \equiv \left(x^2 + a^2\right)^{\frac{1}{2}} \pm x \geq 0. \tag{6.3.17}$$

Setting in eqn (6.3.17)

$$x = [t_0 - (1-\zeta)], \quad a = 2\sqrt{\zeta}, \tag{6.3.18}$$

we find (also recall eqn (6.3.12))

$$\Delta^{1/2} \pm [t_0 - (1-\zeta)] \geq 0. \tag{6.3.19}$$

From this and the definition of $V_{1,2}$ (recall eqn (6.3.9)), we arrive at the estimates

$$V_1 \geq 1, \quad V_2 \leq 1. \tag{6.3.20}$$

Adding V_1 to both sides of the inequality $(6.3.20)_2$ and using eqn $(6.3.13)_1$, we obtain

$$V_1 \geq (t_0 + \zeta). \tag{6.3.21}$$

This and the definition of v_1 leads to

$$v_1(t_0, \zeta) \leq (t_0 + \zeta)^{-\frac{1}{2}}, \tag{6.3.22}$$

that is, the estimate $(6.3.6)_1$ is true.

Next, multiplying eqn (6.3.21) through by $V_2 > 0$, and using eqn $(6.3.13)_2$, we find

$$V_2 \leq t_0(t_0 + \zeta)^{-1}. \tag{6.3.23}$$

This, and the definition of v_2, leads to

$$v_2(t_0, \zeta) \geq (t_0 + \zeta)^{\frac{1}{2}} t_0^{-\frac{1}{2}}, \tag{6.3.24}$$

that is, the estimate $(6.3.6)_2$ is true. □

Graphs of functions $\varphi_1(t_0, \zeta)$ and $\varphi_2(t_0, \zeta)$ for a fixed $t_0 \geq 1$ and $\zeta \in [0, \infty)$ are shown in Figs. 6.1 and 6.2. It is seen that $\forall (t_0, \zeta) \in Q$

$$0 \leq v_1(t_0, \zeta) \leq t_0^{-\frac{1}{2}} \leq 1, \tag{6.3.25}$$

$$1 \leq v_2(t_0, \zeta) \leq \infty, \tag{6.3.26}$$

whereby

$$\lim_{\zeta \to \infty} v_1(t_0, \zeta) = 0 \quad \forall t_0 \geq 1, \tag{6.3.27}$$

$$\lim_{\zeta \to \infty} v_2(t_0, \zeta) = \infty \quad \forall t_0 \geq 1. \tag{6.3.28}$$

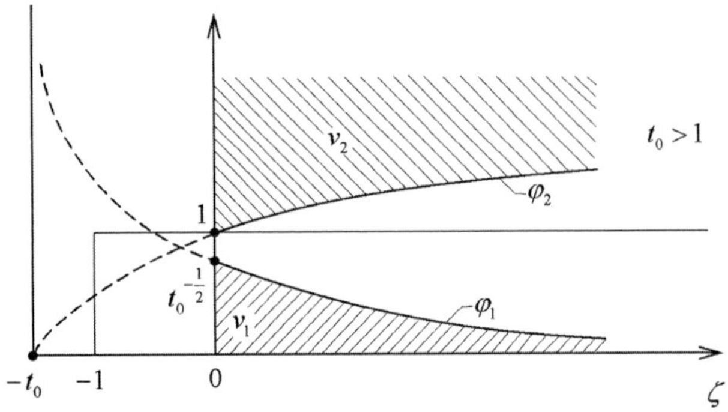

Figure 6.1 Graphs of the functions $\varphi_1 = \varphi_1(t_0, \zeta)$ and $\varphi_2 = \varphi_2(t_0, \zeta)$ for a fixed $t_0 > 1$ and $\zeta \geq 0$.

Moreover, $\forall\,(t_0, \zeta) \in Q$ we have (with $' = \mathrm{d}/\mathrm{d}\zeta$)

$$\varphi_1' \leq 0, \quad \varphi_1'' > 0, \tag{6.3.29}$$

$$\varphi_2' > 0, \quad \varphi_2'' < 0. \tag{6.3.30}$$

Theorem 6.3 states that the speed v_1 is bounded from above by the function φ_1 monotonically decreasing along the axis $\zeta \geq 0$, while v_2 is bounded from below by the function φ_2 monotonically increasing with ζ.

We will now show that, for a fixed $t_0 \geq 1$, the behavior of $v_1(t_0, \zeta)$ (or $v_2(t_0, \zeta)$) is similar to that of $\varphi_1(t_0, \zeta)$ (or $\varphi_2(t_0, \zeta)$) that is, the following theorem is true:

Theorem 6.4 *For every* $(t_0, \zeta) \in Q$

$$v_1' < 0, \quad v_1'' > 0, \tag{6.3.31}$$

$$v_2' > 0, \quad v_2'' < 0. \tag{6.3.32}$$

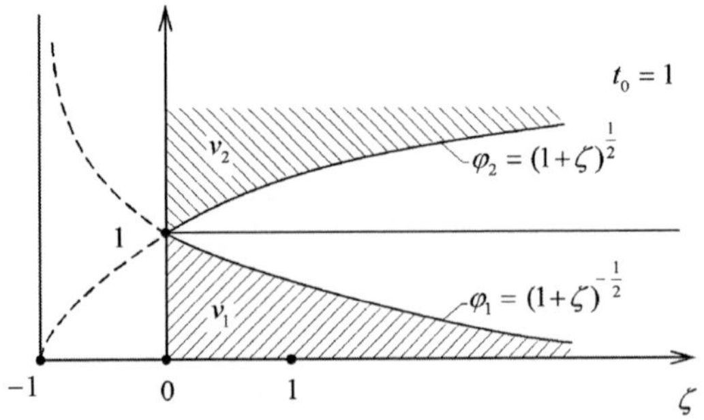

Figure 6.2 Graphs of the functions $\varphi_1 = \varphi_1(t_0, \zeta)$ and $\varphi_2 = \varphi_2(t_0, \zeta)$ for $t_0 = 1$ and $\zeta \geq 0$.

Proof. Differentiating eqns (6.3.13) with respect to ζ, we obtain

$$\begin{aligned} V_1' + V_2' &= 1, \\ (V_1 V_2)' &= 0, \end{aligned} \tag{6.3.33}$$

and

$$\begin{aligned} V_1'' + V_2'' &= 0, \\ (V_1 V_2)'' &= 0. \end{aligned} \tag{6.3.34}$$

From this there follows

$$V_1' = \frac{V_1}{V_1 - V_2}, \quad V_1'' = -\frac{2V_1V_2}{(V_1 - V_2)^3}, \tag{6.3.35}$$

$$V_2' = \frac{V_2}{V_1 - V_2}, \quad V_2'' = \frac{2V_1V_2}{(V_1 - V_2)^3}. \tag{6.3.36}$$

On account of eqns (6.3.8) we find

$$v_{1.2}' = -\frac{1}{2} V_{1.2}^{-\frac{3}{2}} V_{1.2}', \tag{6.3.37}$$

$$v_{1.2}'' = \frac{3}{4} V_{1.2}^{-\frac{5}{2}} (V_{1.2}')^2 - \frac{1}{2} V_{1.2}^{-\frac{3}{2}} V_{1.2}''. \tag{6.3.38}$$

Then, given eqns (6.3.35 and 6.3.36), we find

$$v_1' = -\frac{1}{2} V_1^{-\frac{1}{2}} \frac{1}{V_1 - V_2}, \tag{6.3.39}$$

$$v_1'' = \frac{V_1^{-\frac{1}{2}}}{(V_1 - V_2)^2} \left(\frac{3}{4} + \frac{V_2}{V_1 - V_2} \right), \tag{6.3.40}$$

and

$$v_2' = \frac{1}{2} V_2^{-\frac{1}{2}} \frac{1}{V_1 - V_2}, \tag{6.3.41}$$

$$v_2'' = -\frac{3}{4} \frac{V_2^{-\frac{1}{2}}}{(V_1 - V_2)^3} \left(\frac{1}{3} V_1 + V_2 \right). \tag{6.3.42}$$

Since $V_1 > V_2 > 0$, the formulas (6.3.39–42) imply that the inequalities (6.3.31 and 32) hold. □

Graphs of functions v_1 and v_2 for a fixed $t_0 \geq 1$ and $\zeta \in [0, \infty)$ are shown in Figs. 6.3 and 6.4. Functions φ_1 and φ_2 are graphed there as well. It is seen from Fig. 6.4 that the point $(t_0 = 1, \zeta = 0)$ is a bifurcation point of the four functions v_1, v_2, φ_1 and φ_2. That point should be excluded from an analysis of a thermoelastic body in which $\epsilon > 0$ and $t_1 \geq 1$.

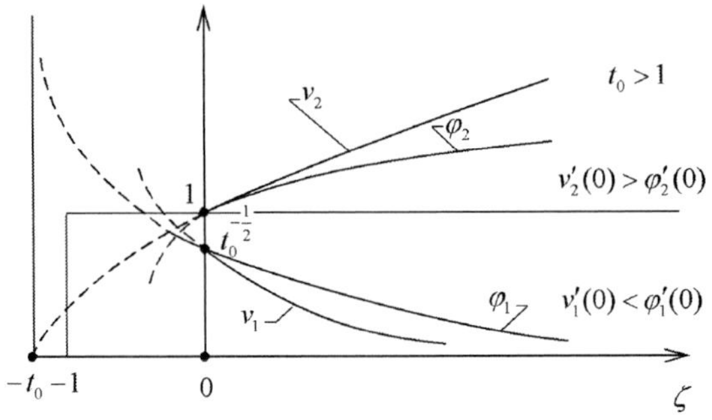

Figure 6.3 Graphs of the functions $v_1 = v_1(t_0, \zeta)$ and $v_2 = v_2(t_0, \zeta)$ for a fixed $t_0 > 1$ and $\zeta \geq 0$.

Employing the formulas (6.3.8, 6.3.9, 6.3.39–42), it may easily be shown that, for $t_0 > 1$ and in a neighborhood of $\zeta = 0$, the speeds v_1 and v_2 possess the expansions

$$v_1 = t_0^{-\frac{1}{2}} \left[1 - \frac{\zeta}{2(t_0 - 1)} + \frac{\zeta^2}{8(t_0 - 1)^3} (3t_0 + 1) + O\left(\zeta^3\right) \right], \tag{6.3.43}$$

$$v_2 = 1 + \frac{\zeta}{2(t_0 - 1)} - \frac{\zeta^2}{8(t_0 - 1)^3} (t_0 + 3) + O\left(\zeta^3\right). \tag{6.3.44}$$

For $t_0 = 1$ and in a neighborhood of $\zeta = \epsilon > 0$ we find

$$\begin{aligned} v_1 = v_1(\epsilon) \Bigg\{ 1 - \frac{(\zeta - \epsilon)}{4\epsilon^{1/2}(1 + \epsilon/4)^{1/2}} \\ + \frac{2 + \epsilon + \epsilon^{1/2}(1 + \epsilon/4)^{1/2}}{32\epsilon^{3/2}(1 + \epsilon/4)^{3/2}} (\zeta - \epsilon)^2 + O\left[(\zeta - \epsilon)^3\right] \Bigg\}, \end{aligned} \tag{6.3.45}$$

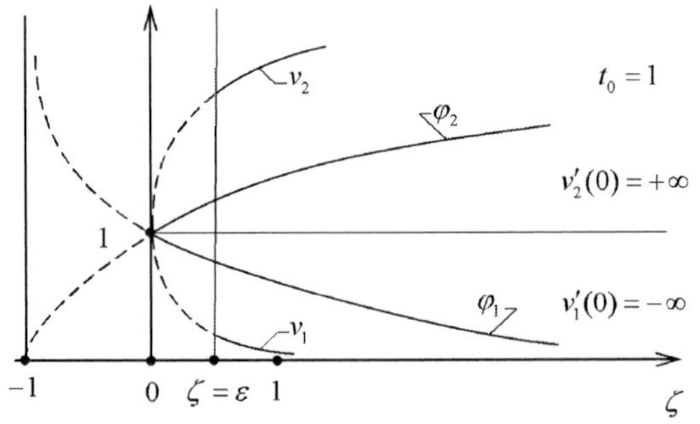

Figure 6.4 Graphs of the functions $v_1 = v_1(t_0, \zeta)$ and $v_2 = v_2(t_0, \zeta)$ for $t_0 = 1$ and $\zeta \geq \epsilon > 0$.

$$
\begin{aligned}
v_2 = v_2(\epsilon)\Bigg\{1 + \frac{(\zeta-\epsilon)}{4\epsilon^{1/2}(1+\epsilon/4)^{1/2}} \\
-\frac{2+\epsilon-\epsilon^{1/2}(1+\epsilon/4)^{1/2}}{32\epsilon^{3/2}(1+\epsilon/4)^{3/2}}(\zeta-\epsilon)^2 + O\left[(\zeta-\epsilon)^3\right]\Bigg\},
\end{aligned}
\tag{6.3.46}
$$

where

$$
v_{1.2}(\epsilon) = \left[1 + \frac{\epsilon}{2} \pm \epsilon^{\frac{1}{2}}\left(1+\frac{\epsilon}{4}\right)^{\frac{1}{2}}\right]^{-\frac{1}{2}}. \tag{6.3.47}
$$

Finally, let us note that for large ζ the following asymptotic formulas apply

$$
v_1(\zeta) = O\left(\zeta^{-\frac{1}{2}}\right) \quad \text{for } \zeta \to \infty, \tag{6.3.48}
$$

$$
v_2(\zeta) = O\left(\zeta^{\frac{1}{2}}\right) \quad \text{for } \zeta \to \infty. \tag{6.3.49}
$$

Thus, for a thermoelastic body in which ϵ is a finite positive number, while $t_1 \to \infty$, the speeds v_1 and v_2 are proportional, respectively, to $(t_1\epsilon)^{-1/2}$ and $(t_1\epsilon)^{1/2}$.

Theorems 6.3 and 6.4 and Figs. 6.3 and 6.4 pertain to the dependence of speeds v_1 and v_2 on ζ for a fixed $t_0 \geq 1$. We now turn to the dependence of v_1 and v_2 on t_0, with a restriction to the domain $Q_0 = \{(t_0, \zeta) : t_0 > 1, \zeta > 0\}$. First, however, we prove the following theorem.

Theorem 6.5 *For every fixed $\zeta > 0$ the functions $V_1(t_0, \zeta)$ and $V_2(t_0, \zeta)$, treated as functions of t_0, satisfy the inequalities*

$$
\begin{aligned}
\dot{V}_1 > 0, \quad \ddot{V}_1 > 0 \\
\dot{V}_2 > 0, \quad \ddot{V}_1 < 0
\end{aligned}
\quad \forall t_0 \geq 1.5. \tag{6.3.50}
$$

Here, the dot denotes a differentiation with respect to t_0.

Proof. Differentiation of eqn (6.3.13) with respect to t_0 yields

$$
\dot{V}_1 + \dot{V}_2 = 1, \qquad (V_1V_2)^{\cdot} = 1, \tag{6.3.51}
$$

$$
\ddot{V}_1 + \ddot{V}_2 = 0, \qquad (V_1V_2)^{\cdot\cdot} = 0. \tag{6.3.52}
$$

Solving eqns (6.3.51) for the pair $(\dot{V}_1, \dot{V}_2)$, and eqns (6.3.52) for $(\ddot{V}_1, \ddot{V}_2)$, we obtain

$$
\dot{V}_1 = \frac{V_1 - 1}{V_1 - V_2}, \quad \dot{V}_2 = \frac{1 - V_2}{V_1 - V_2}, \tag{6.3.53}
$$

$$
\begin{aligned}
\ddot{V}_1 &= \frac{2}{(V_1 - V_2)^3}(V_1 - 1)(1 - V_2), \\
\ddot{V}_2 &= -\frac{2}{(V_1 - V_2)^3}(V_1 - 1)(1 - V_2).
\end{aligned}
\tag{6.3.54}
$$

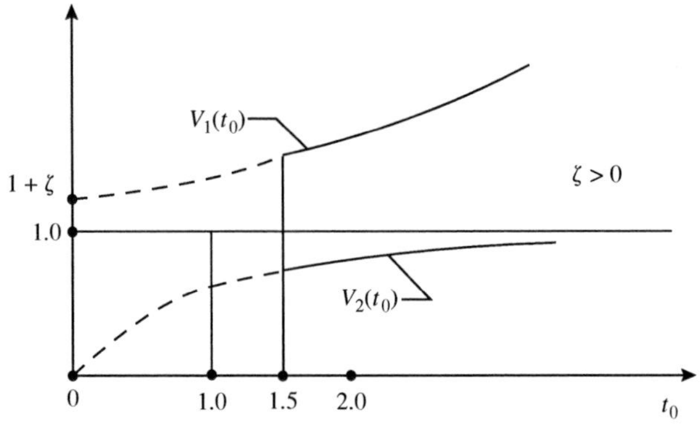

Figure 6.5 Graphs of the functions $V_1 = V_1(t_0, \zeta)$ and $V_2 = V_2(t_0, \zeta)$ for a fixed $\zeta > 0$ and $t_0 > 1.5$.

Since $V_1 > 1$, $1 - V_2 > 0$ and $V_1 - V_2 > 0$, the relations (6.3.53 and 6.3.54) imply the inequalities (6.3.50). □

Using the definition of functions V_1 and V_2 (recall eqns (6.3.9)) and the inequalities (6.3.50), one can readily construct the graphs $V_{1.2}$ for $t_0 > 1.5$, see Fig. 6.5.

In order to examine the dependence of $v_{1.2}$ on t_0, we use the relation

$$v_{1.2} = V_{1.2}^{-1/2}, \tag{6.3.55}$$

$$\dot{v}_{1.2} = -\frac{1}{2} V_{1.2}^{-3/2} \dot{V}_{1.2}, \tag{6.3.56}$$

$$\ddot{v}_{1.2} = \frac{3}{4} V_{1.2}^{-5/2} \left[(\dot{V}_{1.2})^2 - \frac{2}{3} V_{1.2} \ddot{V}_{1.2} \right]. \tag{6.3.57}$$

Replacing the functions $\dot{V}_{1.2}$ and $\ddot{V}_{1.2}$ appearing on the right-hand sides of eqns (6.3.56 and 6.3.57) by the expressions (6.3.53 and 6.3.54), we find

$$\begin{aligned} \dot{v}_1 &= -\frac{1}{2} V_1^{-3/2} \frac{V_1 - 1}{V_1 - V_2}, \\ \ddot{v}_1 &= \frac{3}{4} V_1^{-5/2} \frac{(V_1 - 1)(1 - V_2)}{(V_1 - V_2)^2} \left(\frac{V_1 - 1}{1 - V_2} - \frac{4}{3} \frac{V_1}{V_1 - V_2} \right), \end{aligned} \tag{6.3.58}$$

and

$$\begin{aligned} \dot{v}_2 &= -\frac{1}{2} V_2^{-3/2} \frac{1 - V_2}{V_1 - V_2}, \\ \ddot{v}_2 &= \frac{3}{4} V_2^{-5/2} \frac{(V_1 - 1)(1 - V_2)}{(V_1 - V_2)^2} \left(\frac{1 - V_2}{V_1 - 1} + \frac{4}{3} \frac{V_2}{V_1 - V_2} \right). \end{aligned} \tag{6.3.59}$$

This leads us to the following:

Theorem 6.6 *For every $(t_0, \zeta) \in Q_0$ the speeds υ_1 and υ_2, treated as functions of t_0, satisfy the inequalities*

$$\begin{aligned} \dot{\upsilon}_1 < 0, \quad \ddot{\upsilon}_1 > 0, \\ \dot{\upsilon}_2 < 0, \quad \ddot{\upsilon}_2 > 0. \end{aligned} \tag{6.3.60}$$

Proof. Using the inequalities $V_1 - 1 > 0$, $1 - V_2 > 0$, $V_1 - V_2 > 0$, and the relations (6.3.58 and 6.3.59), it suffices to demonstrate that

$$\frac{V_1 - 1}{1 - V_2} - \frac{4}{3} \frac{V_1}{V_1 - V_2} > 0 \quad \text{on } Q_0. \tag{6.3.61}$$

The inequality (6.3.61) may be written in an equivalent form

$$S \equiv (V_1 - 1)(V_1 - V_2) - \frac{4}{3} V_1 (1 - V_2) > 0. \tag{6.3.62}$$

Since

$$S = V_1^2 - \frac{7}{3} V_1 + \frac{1}{3} V_1 V_2 + V_2, \tag{6.3.63}$$

then, replacing on the right-hand side of eqn (6.3.63) $V_1 V_2$ by t_0 and V_2 by $(1 + \zeta + t_0 - V_1)$ (recall eqns (6.3.13)), we find

$$S = V_1^2 - \frac{10}{3} V_1 + \left(1 + \zeta + \frac{4}{3} t_0\right). \tag{6.3.64}$$

The discriminant of the second-degree polynomial in V_1, given by eqn (6.3.64), is

$$\Delta_0 = 4 \left[\frac{25}{9} - \left(1 + \zeta + \frac{4}{3} t_0\right) \right]. \tag{6.3.65}$$

Since $t_0 > 1.5$ and $\zeta > 0$, then

$$\frac{25}{9} < 3 < 1 + \frac{4}{3} t_0 < 1 + \zeta + \frac{4}{3} t_0$$

and

$$\Delta_0 < 0. \tag{6.3.66}$$

This implies that $S > 0$. □

One can now show that, for a fixed $\zeta > 0$, there exists a point $t_0^* < 1$, where $\ddot{\upsilon}_1$ vanishes.

In view of Theorem 6.6 we conclude that, for a fixed $\zeta > 0$ and $t_0 > 1.5$, the functions υ_1 and υ_2 are decreasing in t_0; their graphs are shown in Fig. 6.6.

Case $t_0 = t_1$

In this case the speeds υ_1 and υ_2 depend on t_0 and ϵ only. Let us introduce the notations

$$\hat{\upsilon}_{1.2}(t_0, \epsilon) = \upsilon_{1.2}(t_0, t_0\epsilon), \qquad \hat{V}_{1.2}(t_0, \epsilon) = V_{1.2}(t_0, t_0\epsilon). \tag{6.3.67}$$

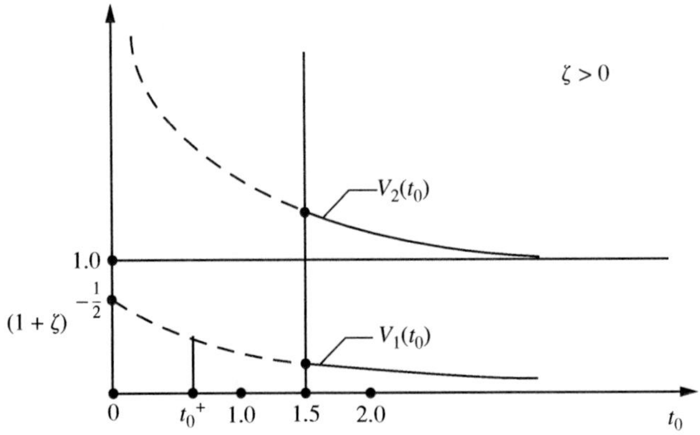

Figure 6.6 Graphs of the functions $\upsilon_1 = \upsilon_1(t_0, \zeta)$ and $\upsilon_2 = \upsilon_2(t_0, \zeta)$ for a fixed $\zeta > 0$ and $t_0 > 1.5$.

As a direct consequence of Theorem 6.3 we now obtain the following bounds on $\hat{\upsilon}_{1.2}$

$$\hat{\upsilon}_1 (t_0, \epsilon) \leq t_0^{-\frac{1}{2}} (1 + \epsilon)^{-\frac{1}{2}}, \qquad \hat{\upsilon}_2 (t_0, \epsilon) \geq (1 + \epsilon)^{\frac{1}{2}}, \tag{6.3.68}$$

while the relations (6.3.13) reduce to the forms

$$\hat{V}_1 + \hat{V}_2 = 1 + (1 + \epsilon) t_0, \qquad \hat{V}_1 \hat{V}_2 = t_0. \tag{6.3.69}$$

Henceforth, denoting the differentiation with respect to ϵ by $'$, we find

$$\hat{V}_1' + \hat{V}_2' = t_0, \qquad (\hat{V}_1 \hat{V}_2)' = 0, \tag{6.3.70}$$

and

$$\hat{V}_1'' + \hat{V}_2'' = 0, \qquad (\hat{V}_1 \hat{V}_2)'' = 0. \tag{6.3.71}$$

Solving the systems (6.3.70 and 71) for the pairs $\left(\hat{V}_1', \hat{V}_2'\right)$ and $\left(\hat{V}_1'', \hat{V}_2''\right)$, we obtain

$$\hat{V}_1' = t_0 \frac{\hat{V}_1}{\hat{V}_1 - \hat{V}_2}, \qquad \hat{V}_2' = -t_0 \frac{\hat{V}_2}{\hat{V}_1 - \hat{V}_2}, \tag{6.3.72}$$

$$\hat{V}_1'' = -2t_0^2 \frac{\hat{V}_1 \hat{V}_2}{(\hat{V}_1 - \hat{V}_2)^3}, \qquad \hat{V}_2'' = 2t_0^2 \frac{\hat{V}_1 \hat{V}_2}{(\hat{V}_1 - \hat{V}_2)^3}. \tag{6.3.73}$$

From this we find the inequalities determining the behavior of functions $\widehat{\upsilon}_{1.2} (\epsilon)$

$$\begin{matrix} \hat{V}_1' > 0, & \hat{V}_1'' < 0 \\ \hat{V}_2' < 0, & \hat{V}_2'' > 0. \end{matrix} \quad \forall t_0 > 1. \tag{6.3.74}$$

For the derivatives of $\hat{v}_{1.2}(\epsilon)$ we then get

$$\hat{v}'_{1.2} = -\frac{1}{2}\hat{V}_{1.2}^{-3/2}\hat{V}'_{1.2}, \tag{6.3.75}$$

$$\begin{aligned} \hat{v}''_1 &= -\frac{1}{4}\hat{V}_1^{-3/2}\hat{V}_2^{-1}\hat{V}'_1\hat{V}'_2\,\frac{3\hat{V}_1+\hat{V}_2}{\hat{V}_1-\hat{V}_2}, \\ \hat{v}''_2 &= \frac{1}{4}\hat{V}_2^{-3/2}\hat{V}_1^{-1}\hat{V}'_1\hat{V}'_2\,\frac{\hat{V}_1+3\hat{V}_2}{\hat{V}_1-\hat{V}_2}. \end{aligned} \tag{6.3.76}$$

Thus, the inequalities describing the behavior of functions $\widehat{v}_{1.2}(\epsilon)$ become

$$\left.\begin{aligned} &\hat{v}'_1 < 0, \quad \hat{v}''_1 > 0 \\ &\hat{v}'_2 > 0, \quad \hat{v}''_2 < 0 \end{aligned}\right. \qquad \forall t_0 > 1. \tag{6.3.77}$$

The latter inequalities imply that, for a fixed $t_0 > 1$, the functions $\hat{v}_1(\epsilon)$ and $\hat{v}_2(\epsilon)$ have the same character as $v_{1.2}(\zeta)$. In particular, $\hat{v}_{1.2}(0) = v_{1.2}(0)$ and $\hat{v}_1(\epsilon)$ ($\widehat{v}_2(\epsilon)$) is a monotonically decreasing (increasing) function of ϵ. The graphs of $\hat{v}_{1.2}(\epsilon)$ are similar to the graphs of $v_{1.2}(\zeta)$ in Fig. 6.3.

We will now examine the dependence of functions $\hat{v}_{1.2}$ on the parameter t_0, at a fixed $\epsilon > 0$. Differentiating eqns (6.3.69) with respect to t_0 and denoting the derivative with respect to t_0 by a dot, we obtain

$$\dot{\hat{V}}_1 + \dot{\hat{V}}_2 = 1+\epsilon, \quad (\hat{V}_1\hat{V}_2)^{\cdot} = 1, \tag{6.3.78}$$

and

$$\ddot{\hat{V}}_1 + \ddot{\hat{V}}_2 = 0, \quad (\hat{V}_1\hat{V}_2)^{\cdot\cdot} = 0. \tag{6.3.79}$$

From eqns (6.3.78) we find

$$\dot{\hat{V}}_1 = \frac{(1+\epsilon)\hat{V}_1 - 1}{\hat{V}_1-\hat{V}_2}, \quad \dot{\hat{V}}_2 = \frac{1-(1+\epsilon)\hat{V}_2}{\hat{V}_1-\hat{V}_2}, \tag{6.3.80}$$

whereas from eqns (6.3.79) we find

$$\ddot{\hat{V}}_1 = 2\frac{\dot{\hat{V}}_1\dot{\hat{V}}_2}{\hat{V}_1-\hat{V}_2}, \quad \ddot{\hat{V}}_2 = -2\frac{\dot{\hat{V}}_1\dot{\hat{V}}_2}{\hat{V}_1-\hat{V}_2}. \tag{6.3.81}$$

From the inequalities $t_0 > 1.5$, $\epsilon > 0$, and inequalities (6.3.68), there follows

$$\begin{aligned} \hat{V}_1 &> (1+\epsilon) > (1+\epsilon)^{-1}, \\ \hat{V}_2 &< (1+\epsilon)^{-1}. \end{aligned} \tag{6.3.82}$$

Hence,

$$(1+\epsilon)\hat{V}_1 - 1 > 0, \quad 1-(1+\epsilon)\hat{V}_2 > 0. \tag{6.3.83}$$

The inequality $\hat{V}_1 - \hat{V}_2 > 0$ together with the inequalities (6.3.83) and the relations (6.3.80 and 6.3.81) then imply that the functions $\hat{V}_{1.2}(t_0)$ satisfy the inequalities

$$\begin{aligned} \dot{\hat{V}}_1 > 0, \quad \ddot{\hat{V}}_1 > 0, \\ \dot{\hat{V}}_2 > 0, \quad \ddot{\hat{V}}_2 < 0. \end{aligned} \tag{6.3.84}$$

For the derivatives of functions $\hat{v}_{1.2}$ we obtain

$$\dot{\hat{v}}_{1.2} = -\frac{1}{2}\hat{V}_1^{-3/2}\dot{\hat{V}}_{1.2}, \tag{6.3.85}$$

$$\ddot{\hat{v}}_{1.2} = \frac{3}{4}\hat{V}_1^{-5/2}(\dot{\hat{V}}_{1.2})^2 - \frac{1}{2}\hat{V}_1^{-3/2}\ddot{\hat{V}}_{1.2}. \tag{6.3.86}$$

From this, in view of eqns (6.3.80 and 6.3.81), we obtain

$$\begin{aligned} \dot{\hat{v}}_1 &= -\frac{1}{2}\hat{V}_1^{-\frac{3}{2}}\left[\frac{(1+\epsilon)\hat{V}_1 - 1}{\hat{V}_1 - \hat{V}_2}\right], \\ \dot{\hat{v}}_2 &= -\frac{1}{2}\hat{V}_2^{-\frac{3}{2}}\left[\frac{1-(1+\epsilon)\hat{V}_2}{\hat{V}_1 - \hat{V}_2}\right], \end{aligned} \tag{6.3.87}$$

$$\begin{aligned} \ddot{\hat{v}}_1 &= \frac{3}{4}\hat{V}_1^{-\frac{5}{2}}\dot{\hat{V}}_1\dot{\hat{V}}_2\left[\frac{(1+\epsilon)\hat{V}_1 - 1}{1-(1+\epsilon)\hat{V}_2} - \frac{4}{3}\frac{\hat{V}_1}{\hat{V}_1 - \hat{V}_2}\right], \\ \ddot{\hat{v}}_2 &= \frac{3}{4}\hat{V}_2^{-\frac{5}{2}}\dot{\hat{V}}_1\dot{\hat{V}}_2\left[\frac{1-(1+\epsilon)\hat{V}_2}{(1+\epsilon)\hat{V}_1 - 1} + \frac{4}{3}\frac{\hat{V}_2}{\hat{V}_1 - \hat{V}_2}\right]. \end{aligned} \tag{6.3.88}$$

We will now show that, for $t_0 > 1.5$ and $\zeta > 0$, the expression in square bracket on the right-hand side of eqn $(6.3.88)_1$ is positive. To prove this it suffices to show that

$$\left[(1+\epsilon)\hat{V}_1 - 1\right](\hat{V}_1 - \hat{V}_2) - \frac{4}{3}\hat{V}_1\left[1-(1+\epsilon)\hat{V}_2\right] > 0, \tag{6.3.89}$$

or

$$(1+\epsilon)\hat{V}_1^2 - \frac{7}{3}\hat{V}_1 + \frac{1}{3}(1+\epsilon)\hat{V}_1\hat{V}_2 + \hat{V}_2 > 0. \tag{6.3.90}$$

On account of eqn (6.3.69), we reduce eqn (6.3.90) to

$$(1+\epsilon)\hat{V}_1^2 - \frac{10}{3}\hat{V}_1 + 1 + \frac{4}{3}(1+\epsilon)t_0 > 0. \tag{6.3.91}$$

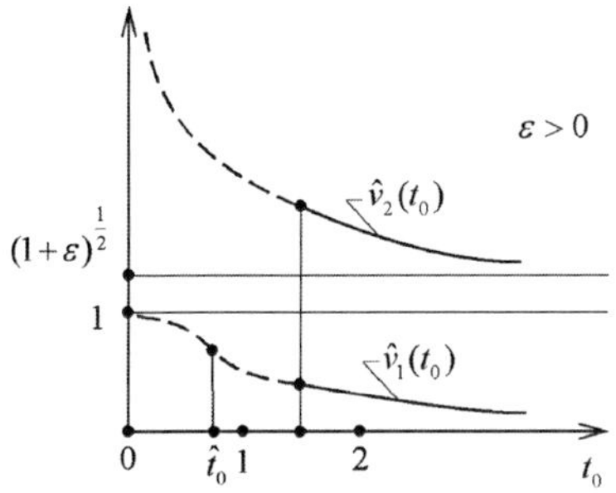

Figure 6.7 Graphs of the functions $\hat{v}_1 = v_1(t_0, t_0\epsilon)$ and $\hat{v}_2 = v_2(t_0, t_0\epsilon)$ for a fixed $\epsilon > 0$ and $t_0 > 1.5$.

The left-hand side of the inequality (6.3.91) is a second-order polynomial in $\hat{V}_1$, whose discriminant is

$$\begin{aligned}\Delta_0 &= \left(\frac{10}{3}\right)^2 - 4(1+\epsilon)\left[1 + \frac{4}{3}(1+\epsilon)\, t_0\right] \\ &= 4\left\{\frac{25}{9} - (1+\epsilon)\left[1 + \frac{4}{3}(1+\epsilon)\, t_0\right]\right\}.\end{aligned} \tag{6.3.92}$$

Since $t_0 > 1.5$ and $\epsilon > 0$, it follows that

$$\frac{25}{9} < 1 + \frac{4}{3}t_0 < (1+\epsilon)\left(1 + \frac{4}{3}t_0\right) < (1+\epsilon)\left[1 + \frac{4}{3}(1+\epsilon)t_0\right]. \tag{6.3.93}$$

Therefore, $\Delta_0 < 0$ and the inequality (6.3.89) holds. From this, in view of the formulas (6.3.87 and 6.3.88), there follow the inequalities

$$\begin{matrix}\dot{\hat{v}}_1 < 0, & \ddot{\hat{v}}_1 > 0 \\ \dot{\hat{v}}_2 < 0, & \ddot{\hat{v}}_2 > 0\end{matrix} \quad \forall t_0 \geq 1.5. \tag{6.3.94}$$

One can show that, for $t_0 < 1$, the function $\hat{v}_1(t_0)$ possesses an inflection point $\hat{t}_0$, where $\ddot{\hat{v}}_1(\hat{t}_0) = 0$.

Schematic graphs of $\hat{v}_1(t_0)$ and $\hat{v}_2(t_0)$ are shown in Fig. 6.7. Comparing the graphs of $\hat{v}_{1.2}(t_0)$ with those of $v_{1.2}(t_0)$, we conclude that $\hat{v}_{1.2}(t_0)$ have the same decreasing character as $v_{1.2}(t_0)$. However, one difference is that the upper bound of $\hat{v}_1$ is greater than that of $v_1(t_0)$, while the lower bound of $\hat{v}_2$ is greater than that of $v_2(t_0)$.

6.4 Attenuation of a fundamental thermoelastic disturbance in the space of constitutive variables

In this section we assume that $t_0 = t_1$, which means taking the attenuation coefficients $\hat{k}_{1.2} = \hat{k}_{1.2}(t_0, \epsilon)$ specified by the formulas (recall eqn $(6.2.9)_2$ for this particular case)

$$\hat{k}_{1.2} = \frac{1}{2}\left(1 + \epsilon \mp \hat{\alpha}\hat{\Delta}^{\frac{1}{2}}\right), \tag{6.4.1}$$

$$\hat{\alpha} = \left[1 - \epsilon - (1+\epsilon)^2 t_0\right]\hat{\Delta}^{-1}, \tag{6.4.2}$$

$$\hat{\Delta} = \left[1 - (1-\epsilon)t_0\right]^2 + 4\epsilon t_0^2. \tag{6.4.3}$$

The coefficients $\hat{k}_1$ and $\hat{k}_2$ are solutions of a system of linear algebraic equations (recall eqns (6.2.49) and (6.2.52))

$$\hat{k}_1 + \hat{k}_2 = 1 + \epsilon, \qquad \hat{k}_1\hat{V}_2 + \hat{k}_2\hat{V}_1 = 1, \tag{6.4.4}$$

where

$$\hat{V}_{1.2} = \frac{1}{2}\left[1 + (1+\epsilon)\,t_0 \pm \hat{\Delta}^{\frac{1}{2}}\right]. \tag{6.4.5}$$

Thus, an analysis of $\hat{k}_1$ and $\hat{k}_2$ may be conducted recalling the properties of $\hat{V}_1$ and $\hat{V}_2$ derived in the preceding section.

6.4.1 *Behavior of functions $\hat{k}_{1.2}$ for a fixed relaxation time t_0*

We shall examine that behavior in a small right neighborhood of $\epsilon = 0$ for $t_0 > 2$[3]. First, we solve eqns (6.4.4) with respect to $(\hat{k}_1, \hat{k}_2)$ and obtain

$$\hat{k}_1 = \frac{(1+\epsilon)\,\hat{V}_1 - 1}{\hat{V}_1 - \hat{V}_2}, \quad \hat{k}_2 = \frac{1 - (1+\epsilon)\,\hat{V}_2}{\hat{V}_1 - \hat{V}_2}. \tag{6.4.6}$$

From this, we obtain ($' = \mathrm{d}/\mathrm{d}\epsilon$)

$$\hat{k}_1'\left(\hat{V}_1 - \hat{V}_2\right) = (1+\epsilon)\,\hat{V}_1' + \hat{V}_1 - \hat{k}_1\left(\hat{V}_1' - \hat{V}_2'\right), \tag{6.4.7}$$

$$\hat{k}_1''\left(\hat{V}_1 - \hat{V}_2\right) = (1+\epsilon)\,\hat{V}_1'' + 2\hat{V}_1' - \hat{k}_1\left(\hat{V}_1'' - \hat{V}_2''\right) - 2\hat{k}_1'\left(\hat{V}_1' - \hat{V}_2'\right), \tag{6.4.8}$$

$$\hat{k}_2' = 1 - \hat{k}_1', \quad \hat{k}_2'' = -\hat{k}_2''. \tag{6.4.9}$$

Introducing, for convenience, the notations

$$\hat{k}_{1.2} = \hat{k}_{1.2}(\epsilon), \quad \hat{V}_{1.2} = \hat{V}_{1.2}(\epsilon), \tag{6.4.10}$$

and employing eqns (6.4.5) as well as eqns (6.3.72 and 6.3.73), we find

$$\begin{aligned}
&\hat{V}_1(0) = t_0, && \hat{V}_2(0) = 1, \\
&\hat{V}_1'(0) = \frac{t_0^2}{t_0 - 1}, && \hat{V}_2'(0) = -\frac{t_0}{t_0 - 1}, \\
&\hat{V}_1''(0) = -2\frac{t_0^3}{(t_0 - 1)^3}, && \hat{V}_2''(0) = 2\frac{t_0^3}{(t_0 - 1)^3}.
\end{aligned} \tag{6.4.11}$$

[3] To justify the hypothesis $t_0 > 2$ see Footnote #2.

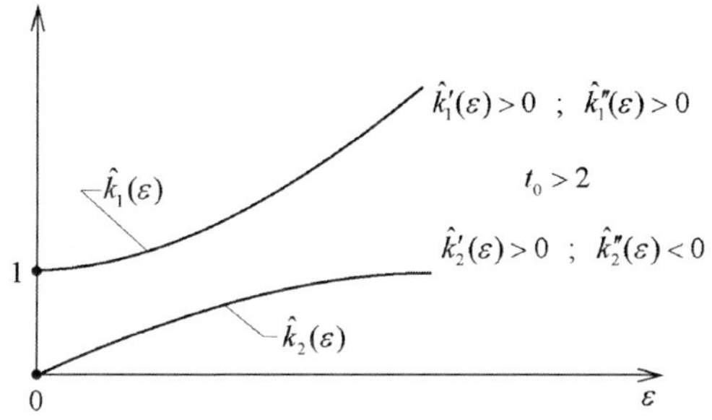

Figure 6.8 Graphs of the functions $\hat{k}_1 = k_1(t_0, t_0\epsilon)$ and $\hat{k}_2 = k_2(t_0, t_0\epsilon)$ for a fixed $t_0 > 2$ and $\epsilon > 0$.

From this, on account of eqns (6.4.6–6.4.9), we find

$$\hat{k}_1(0) = 1, \quad \hat{k}_1'(0) = \frac{t_0(t_0-1)}{(t_0-1)^2}, \quad \hat{k}_1''(0) = \frac{6t_0^2}{(t_0-1)^4}, \tag{6.4.12}$$

$$\hat{k}_2(0) = 0, \quad \hat{k}_2'(0) = \frac{1}{(t_0-1)^2}, \quad \hat{k}_2''(0) = -\frac{6t_0^2}{(t_0-1)^4}. \tag{6.4.13}$$

Therefore, in a small neighborhood of ϵ, for $\hat{k}_{1.2}(\epsilon)$, we may write the expansions

$$\begin{aligned} \hat{k}_1(\epsilon) &= 1 + \frac{t_0(t_0-2)}{(t_0-1)^2}\,\epsilon + \frac{3t_0^2}{(t_0-1)^4}\,\epsilon^2 + O\left(\epsilon^3\right), \\ \hat{k}_2(\epsilon) &= \frac{1}{(t_0-1)^2}\,\epsilon - \frac{3t_0^2}{(t_0-1)^4}\,\epsilon^2 + O\left(\epsilon^3\right). \end{aligned} \tag{6.4.14}$$

Observe that eqns (6.4.14) imply that, for $t_0 > 2$, the functions $\hat{k}_{1.2}(\epsilon)$ are monotonically increasing in a sufficiently small right neighborhood of $\epsilon = 0$. Schematic graphs of $\hat{k}_{1.2}(\epsilon)$ are shown in Fig. 6.8.

6.4.2 *Behavior of functions $\hat{k}_{1.2}$ for a fixed ϵ*

A comparison of formulas (6.4.6) with eqn (6.3.80) indicates that

$$\hat{k}_1 = \dot{\hat{V}}_1, \quad \hat{k}_2 = \dot{\hat{V}}_2 \quad \left(\cdot = \frac{\mathrm{d}}{\mathrm{d}t_0}\right). \tag{6.4.15}$$

Hence [see eqns (6.3.81)],

$$\dot{\hat{k}}_1 = \ddot{\hat{V}}_1 = 2\frac{\hat{k}_1\hat{k}_2}{\hat{V}_1 - \hat{V}_2}, \quad \dot{\hat{k}}_2 = \ddot{\hat{V}}_2 = -2\frac{\hat{k}_1\hat{k}_2}{\hat{V}_1 - \hat{V}_2}, \tag{6.4.16}$$

and

$$\ddot{\hat{k}}_1\frac{\hat{V}_1 - \hat{V}_2}{2} + \dot{\hat{k}}_1\frac{\dot{\hat{V}}_1 - \dot{\hat{V}}_2}{2} = (\hat{k}_1\hat{k}_2)^{\cdot}, \tag{6.4.17}$$

or

$$\ddot{\hat{k}}_1 \frac{\hat{V}_1 - \hat{V}_2}{2} + \dot{\hat{k}}_1 \frac{\hat{k}_1 - \hat{k}_2}{2} = \dot{\hat{k}}_1 \hat{k}_2 + \hat{k}_1 \dot{\hat{k}}_2. \tag{6.4.18}$$

Since

$$\dot{\hat{k}}_1 = -\dot{\hat{k}}_2, \quad \ddot{\hat{k}}_1 = -\ddot{\hat{k}}_2, \tag{6.4.19}$$

on account of eqn (6.4.18) we obtain

$$\ddot{\hat{k}}_1 \frac{(\hat{V}_1 - \hat{V}_2)}{2} = -\frac{3}{2}(\hat{k}_1 - \hat{k}_2)\dot{\hat{k}}_1, \tag{6.4.20}$$

$$\ddot{\hat{k}}_2 \frac{(\hat{V}_1 - \hat{V}_2)}{2} = \frac{3}{2}(\hat{k}_1 - \hat{k}_2)\dot{\hat{k}}_1. \tag{6.4.21}$$

This leads to the following theorem:

Theorem 6.7 *If $\epsilon > 0$ and $t_0 > 1.5$, then*

$$\hat{k}_1 > \hat{k}_2 > 0. \tag{6.4.22}$$

Proof. The formulas (6.4.1–6.4.5) imply

$$\hat{k}_1 - \hat{k}_2 = -\hat{\alpha}(\hat{V}_1 - \hat{V}_2), \tag{6.4.23}$$

where

$$\hat{\alpha} = \left[1 - \epsilon - (1+\epsilon)^2 t_0\right] \hat{\Delta}^{-1}. \tag{6.4.24}$$

Therefore, eqn (6.4.22) is equivalent to the condition $\hat{\alpha} < 0$, that is

$$\frac{1-\epsilon}{(1+\epsilon)^2} < t_0. \tag{6.4.25}$$

Since $\epsilon > 0$ and $t_0 > 1.5$, then

$$\frac{1-\epsilon}{(1+\epsilon)^2} < 1 < 1.5 < t_0, \tag{6.4.26}$$

which implies that eqn (6.4.25) holds. □

The formulas (6.4.15, 6.4.16, 6.4.20 and 6.4.21) together with Theorem 6.7 imply that $\forall (t_0, \epsilon) \in (1.5, \infty)$ the functions $\hat{k}_{1.2}(t_0)$ treated as functions of one variable t_0, satisfy the inequalities

$$\begin{aligned} \dot{\hat{k}}_1 > 0, \quad \ddot{\hat{k}}_1 < 0, \\ \dot{\hat{k}}_2 < 0, \quad \ddot{\hat{k}}_2 > 0. \end{aligned} \tag{6.4.27}$$

Graphs of $\hat{k}_1(t_0)$ and $\hat{k}_2(t_0)$ are displayed in Fig. 6.9, where also the point t_0^* at which $\hat{k}_1 = \hat{k}_2$ is indicated.

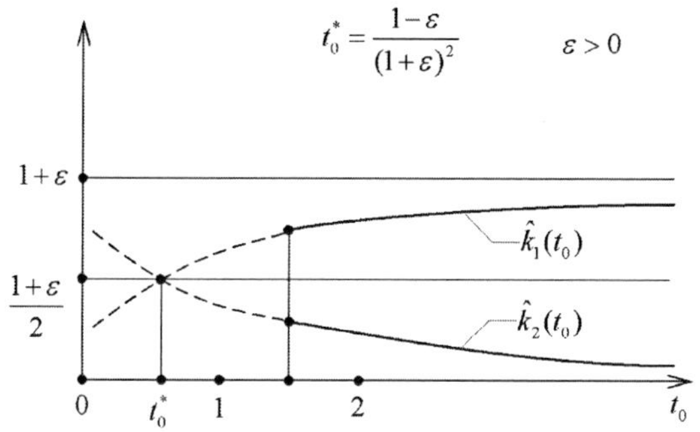

Figure 6.9 Graphs of the functions $\hat{k}_1 = k_1(t_0, t_0\epsilon)$ and $\hat{k}_2 = k_2(t_0, t_0\epsilon)$ for a fixed $\epsilon > 0$ and $t_0 > 1.5$.

6.5 Analysis of the convolution coefficient and kernel

Similar to Section 6.4, the analysis of the coefficient λ and the function $K = K(t)$ will be restricted to the case $t_1 = t_0$. Denoting by $\hat{\lambda}$ and $\hat{K}(t)$, respectively, the value of λ and the function $K(t)$ for $t_0 = t_1$, on account of eqn (6.2.53), we obtain

$$\hat{\lambda} = \hat{k}_1 \hat{k}_2 (\hat{V}_1 - \hat{V}_2)^{-1}, \tag{6.5.1}$$

whereas, on account of, eqn (6.2.34), we have

$$\hat{K}(t) = 2\frac{e^{\hat{\alpha}t}}{\hat{\beta}t}\left[\hat{\alpha}J_1(\hat{\beta}t) - \hat{\beta}J_2(\hat{\beta}t)\right], \tag{6.5.2}$$

where (recall eqns (6.2.65) and (6.2.67))[4]

$$\hat{\alpha} = -(\hat{k}_1 - \hat{k}_2)(\hat{V}_1 - \hat{V}_2)^{-1}, \tag{6.5.3}$$

$$\hat{\beta} = 2\hat{\lambda}^{\frac{1}{2}}(\hat{V}_1 - \hat{V}_2)^{-\frac{1}{2}}. \tag{6.5.4}$$

Since $\hat{k}_{1.2} = \hat{k}_{1.2}(t_0, \epsilon)$ and $\hat{V}_{1.2} = \hat{V}_{1.2}(t_0, \epsilon)$, the coefficient $\hat{\lambda}$ also depends on t_0 and ϵ, that is

$$\hat{\lambda} = \hat{\lambda}(t_0, \epsilon). \tag{6.5.5}$$

6.5.1 *Analysis of $\hat{\lambda}$ at fixed t_0*

Using the formulas (6.4.6) and (6.3.69) we obtain

$$\hat{k}_1 \hat{k}_2 = \epsilon(\hat{V}_1 - \hat{V}_2)^{-2}, \tag{6.5.6}$$

[4] The symbol $\hat{\alpha}$ defined by eqn (6.2.35) must not be confused with $\hat{\alpha}$ in eqn (6.5.3).

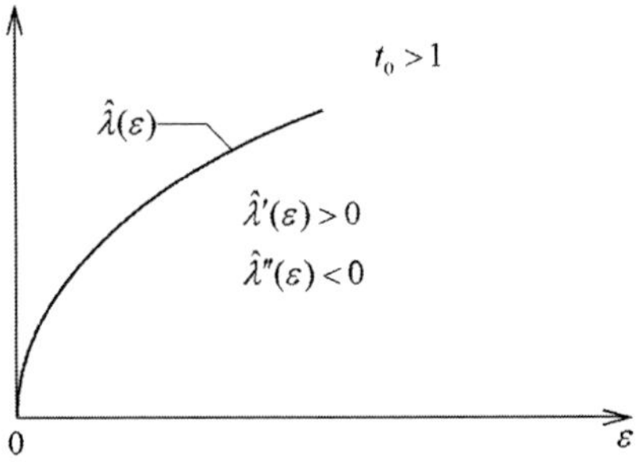

Figure 6.10 Graph of the function $\hat{\lambda} = \lambda(t_0, t_0\epsilon)$ for a fixed $t_0 > 1$ and $\epsilon > 0$.

so that eqn (6.5.1) may be written as

$$\hat{\lambda} = \epsilon(\hat{V}_1 - \hat{V}_2)^{-3}. \tag{6.5.7}$$

Denoting the function (6.5.7) by $\hat{\lambda}(\epsilon)$ for a fixed t_0 and computing the first and second derivatives with respect to ϵ ($' = \mathrm{d}/\mathrm{d}\epsilon$), we find

$$\hat{\lambda}'(\epsilon) = (\hat{V}_1 - \hat{V}_2)^{-4}\left[(\hat{V}_1 - \hat{V}_2) - 3\epsilon(\hat{V}_1' - \hat{V}_2')\right], \tag{6.5.8}$$

$$\begin{aligned}\hat{\lambda}_2''(\epsilon) &= -3(\hat{V}_1 - \hat{V}_2)^{-5}\Big\{2(\hat{V}_1 - \hat{V}_2)(\hat{V}_1' - \hat{V}_2') \\ &+\epsilon\left[(\hat{V}_1 - \hat{V}_2)(\hat{V}_1'' - \hat{V}_2'') - 4(\hat{V}_1' - \hat{V}_2')^2\right]\Big\}.\end{aligned} \tag{6.5.9}$$

From this, on account of formulas (6.4.11), we obtain for $t_0 > 1$

$$\hat{\lambda}'(0) = (t_0 - 1)^{-3}, \tag{6.5.10}$$

$$\hat{\lambda}_2''(0) = -6t_0(t_0 + 1)(t_0 - 1)^{-5}. \tag{6.5.11}$$

Therefore, for $t_0 > 1$ in a right neighborhood of $\epsilon = 0$, the function $\hat{\lambda}(\epsilon)$ has the expansion

$$\hat{\lambda}(\epsilon) = \frac{\epsilon}{(t_0 - 1)^3} - \frac{3t_0(t_0 + 1)}{(t_0 - 1)^5}\epsilon^2 + O(\epsilon^3). \tag{6.5.12}$$

The respective graph of $\hat{\lambda}(\epsilon)$ is given in Fig. 6.10.

6.5.2 *Analysis of $\hat{\lambda}$ at fixed ϵ*

Denoting the function (6.5.7) by $\hat{\lambda}(t_0)$ for a fixed ϵ, we have this theorem:

Theorem 6.8 *For every $t_0 > 2$ and fixed ϵ such that $0 < \epsilon < 1$, the function $\hat{\lambda}(t_0)$ satisfies the inequalities*

$$\hat{\lambda}(t_0) > 0, \quad \dot{\hat{\lambda}}(t_0) < 0, \quad \ddot{\hat{\lambda}}(t_0) > 0. \tag{6.5.13}$$

Proof. The first of the inequalities (6.5.13) is a direct consequence of (6.5.7). In order to prove eqn $(6.5.13)_{2,3}$ we first differentiate eqn (6.5.7) with respect to t_0 and find

$$\epsilon^{-1}\dot{\hat{\lambda}} = -3(\hat{V}_1 - \hat{V}_2)^{-4}(\dot{\hat{V}}_1 - \dot{\hat{V}}_2), \tag{6.5.14}$$

$$\epsilon^{-1}\ddot{\hat{\lambda}} = 12(\hat{V}_1 - \hat{V}_2)^{-5}(\dot{\hat{V}}_1 - \dot{\hat{V}}_2)^2 - 3(\hat{V}_1 - \hat{V}_2)^{-4}(\ddot{\hat{V}}_1 - \ddot{\hat{V}}_2). \tag{6.5.15}$$

On account of $\dot{\hat{V}}_1 - \dot{\hat{V}}_2 = \hat{k}_1 - \hat{k}_2$ (recall eqns (6.4.15)) and $\hat{k}_1 - \hat{k}_2 > 0$, by Theorem 6.7, eqn (6.5.14) implies that the inequality $(6.5.13)_2$ holds. Next, to prove eqn $(6.5.13)_3$ we observe that eqn $(6.4.16)_3$ implies

$$\ddot{\hat{V}}_1 - \ddot{\hat{V}}_2 = \frac{4\hat{k}_1\hat{k}_2}{(\hat{V}_1 - \hat{V}_2)}. \tag{6.5.16}$$

Hence, from eqn (6.5.15) we obtain

$$\epsilon^{-1}\ddot{\hat{\lambda}} = 12(\hat{V}_1 - \hat{V}_2)^{-5}\left[(\hat{k}_1 - \hat{k}_2)^2 - \hat{k}_1\hat{k}_2\right]. \tag{6.5.17}$$

Next, it follows from the formulas (6.5.3) and (6.5.6) that the expression in square brackets on the right of eqn (6.5.17) may be written as

$$[\cdot] = (\hat{V}_1 - \hat{V}_2)^{-2}[(\hat{V}_1 - \hat{V}_2)^4\hat{\alpha}^2 - \epsilon]. \tag{6.5.18}$$

Since, on account of eqn (6.4.23)

$$(\hat{V}_1 - \hat{V}_2)^2\hat{\alpha} = \left[1 - \epsilon - (1+\epsilon)^2 t_0\right], \tag{6.5.19}$$

in order to prove eqn $(6.5.13)_3$ it will suffice to show that

$$\left[(1+\epsilon)^2 t_0 - (1-\epsilon)\right]^2 - \epsilon > 0 \tag{6.5.20}$$

holds. To this end, let

$$t_0^* = \frac{1-\epsilon}{(1+\epsilon)^2}. \tag{6.5.21}$$

Then,

$$(1+\epsilon)^2 t_0 - (1-\epsilon) = (1+\epsilon)^2 (t_0 - t_0^*), \tag{6.5.22}$$

and the inequality (6.5.20) is equivalent to the relation

$$\left[(1+\epsilon)^2 (t_0 - t_0^*) - \sqrt{\epsilon}\right]\left[(1+\epsilon)^2 (t_0 - t_0^*) + \sqrt{\epsilon}\right] > 0. \tag{6.5.23}$$

Since $t_0^* < 1$, while $t_0 > 2$ and $\epsilon > 0$, the condition (6.5.23) is equivalent to

$$t_0 > t_0^* + \frac{\sqrt{\epsilon}}{(1+\epsilon)^2}. \tag{6.5.24}$$

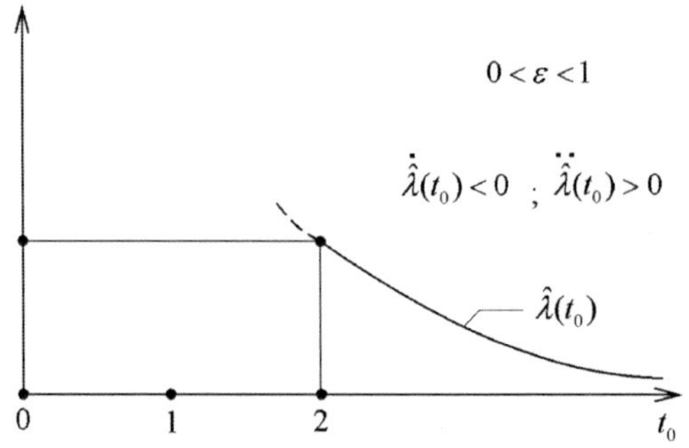

Figure 6.11 Graph of the function $\hat{\lambda} = \lambda(t_0, t_0\epsilon)$ for a fixed $\epsilon \in (0,1)$ and $t_0 > 2$.

Furthermore, since

$$t_0^* + \frac{\sqrt{\epsilon}}{(1+\epsilon)^2} = \frac{1}{(1+\epsilon)^2} + \frac{\sqrt{\epsilon}-\epsilon}{(1+\epsilon)^2} = \frac{1}{(1+\epsilon)^2} + \frac{\epsilon\,(1+\epsilon)}{(\sqrt{\epsilon}+\epsilon)\,(1+\epsilon)}, \tag{6.5.25}$$

and

$$0 < \epsilon < 1, \quad \frac{1}{\sqrt{\epsilon}+\epsilon} < \frac{1}{\epsilon}, \quad \frac{1}{(1+\epsilon)^2} < 1, \tag{6.5.26}$$

from eqn (6.5.25) we find

$$t_0^* + \frac{\sqrt{\epsilon}}{(1+\epsilon)^2} < 2 - \epsilon. \tag{6.5.27}$$

With this and the assumption $t_0 > 2$ we conclude that the inequality (6.5.24) holds. □

A graph of $\hat{\lambda}\,(t_0)$ is shown in Fig. 6.11.

6.5.3 *Analysis of the convolution kernel*

On account of eqn (6.5.2) we have the following theorem:

Theorem 6.9 *The kernel* $\hat{K} = \hat{K}\,(t)$ *is an analytic function of* $t \in [0,\infty)$ $\forall\,(t_0,\epsilon) \in (1,\infty) \times (0,\infty)$. *In addition, the function* $\hat{K}\,(t)$ *satisfies the inequality*

$$|\hat{K}\,(t)\,| \le \hat{k}_0\,(t) \quad \forall t \in [0,\infty), \tag{6.5.28}$$

where

$$\hat{k}_0\,(t) = \mathrm{e}^{\hat{\alpha} t}\left[|\hat{\alpha}| + \frac{1}{4}\hat{\beta}^2 t\right]. \tag{6.5.29}$$

In particular,

$$\hat{K}(t) \to 0 \quad \textit{for} \quad t \to \infty. \tag{6.5.30}$$

Proof. The analyticity of $\hat{K}(t)$ follows from the analytic properties of the functions $J_1(x)/x$ and $J_2(x)/x$ $\forall x \geq 0$ and from the fact that, for $t_0 > 1$ and $\epsilon > 0$, we have $\hat{\alpha} < 0$ (see Theorem 6.7).

The estimate (6.5.28) follows from the definition of the kernel $\hat{K}(t)$ (recall eqn (6.5.2)) and from the inequality

$$|J_n(x)| \leq \frac{x^n}{2^n n!} \qquad \forall x \geq 0, \qquad n \geq 0. \tag{6.5.31}$$

Passing in the inequality (6.5.28) with t to ∞, we obtain eqn (6.5.30).[5] □

For the function $\hat{k}_0(t)$ appearing in Theorem 6.9 there holds the following theorem.

Theorem 6.10 *For every $t_0 > 3$ and $\epsilon > 0$, the function $\hat{k}_0(t)$ satisfies the inequalities*

$$\hat{k}_0(t) > 0, \quad \dot{\hat{k}}_0(t) < 0, \quad \ddot{\hat{k}}_0(t) > 0 \quad \forall t \geq 0. \tag{6.5.32}$$

Proof. The first of the inequalities (6.5.32) is a direct consequence of eqn (6.5.29). In order to prove the remaining two we first differentiate eqn (6.5.29) with respect to t and obtain

$$\dot{\hat{k}}_0 = \mathrm{e}^{\hat{\alpha}t}\left[\hat{\alpha}\left(|\hat{\alpha}| + \frac{1}{4}\hat{\beta}^2 t\right) + \frac{1}{4}\hat{\beta}^2\right], \tag{6.5.33}$$

$$\ddot{\hat{k}}_0 = \mathrm{e}^{\hat{\alpha}t}\left[\hat{\alpha}^2\left(|\hat{\alpha}| + \frac{1}{4}\hat{\beta}^2 t\right) + 2\hat{\alpha}\frac{1}{4}\hat{\beta}^2\right]. \tag{6.5.34}$$

Next, using eqns (6.5.1), (6.5.3) and (6.5.4), we get

$$\frac{1}{4}\hat{\beta}^2 + \hat{\alpha}\,|\hat{\alpha}| = \hat{k}_1\hat{k}_2\left(\hat{V}_1 - \hat{V}_2\right)^{-2} - \left(\hat{k}_1 - \hat{k}_2\right)^2\left(\hat{V}_1 - \hat{V}_2\right)^{-2}, \tag{6.5.35}$$

so that, the inequality $(6.5.13)_3$ of Theorem 6.8 and eqn (6.5.17) imply

$$\frac{1}{4}\hat{\beta}^2 + \hat{\alpha}\,|\hat{\alpha}| < 0. \tag{6.5.36}$$

From this and the formula (6.5.33) it follows that $\dot{\hat{k}}_0(t) < 0$ $\forall t \geq 0$, i.e. that eqn $(6.5.32)_2$ holds.

[5] Theorem 6.9 can be generalized to the case $t_1 \geq t_0 > 1$, $\epsilon > 0$, if one observes that the parameter α, defining the kernel $K = K(t)$ [see eqn (6.2.3)] is negative. In the generalized theorem there occur symbols without huts.

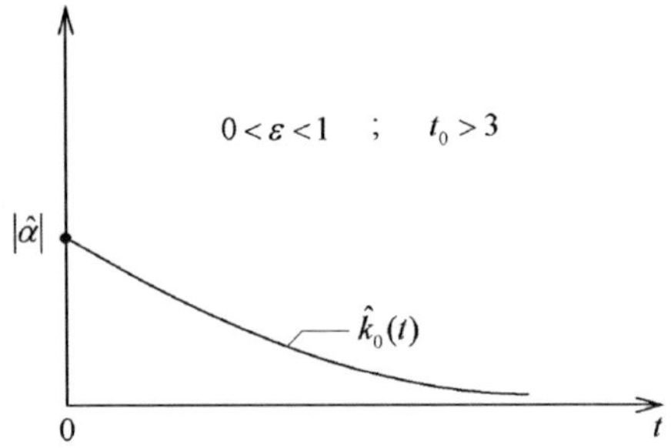

Figure 6.12 Graph of the function $\hat{k}_0 = \hat{k}_0(t)$ for $t > 0$ and fixed values of t_0 and $\epsilon :\ \epsilon \in (0,1)$ and $t_0 > 3$.

In order to prove eqn $(6.5.32)_3$, it will suffice to show that

$$\hat{\alpha}^2 |\hat{\alpha}| + \frac{1}{2}\hat{\alpha}\hat{\beta}^2 > 0. \tag{6.5.37}$$

On account of eqns (6.5.1), (6.5.3) and (6.5.4), we rewrite eqn (6.5.37) as

$$(\hat{V}_1 - \hat{V}_2)^{-3}\left[(\hat{k}_1 - \hat{k}_2)^3 - 2(\hat{k}_1 - \hat{k}_2)\hat{k}_1\hat{k}_2\right] > 0. \tag{6.5.38}$$

In view of Theorem 6.7, this inequality is equivalent to

$$(\hat{k}_1 - \hat{k}_2)^2 - 2\hat{k}_1\hat{k}_2 > 0, \tag{6.5.39}$$

or, given eqns (6.4.23) and (6.4.24) and (6.5.6), it is equivalent to (recall eqn (6.5.23))

$$\left[(1+\epsilon)^2(t_0 - t_0^*) - \sqrt{2\epsilon}\right]\left[(1+\epsilon)^2(t_0 - t_0^*) + \sqrt{2\epsilon}\right] > 0, \tag{6.5.40}$$

where t_0^* is given by eqn (6.5.21).

Since for $\epsilon > 0$

$$t_0^* + \frac{\sqrt{2\epsilon}}{(1+\epsilon)^2} < 3 - \epsilon, \tag{6.5.41}$$

and $t_0 > 3$ by assumption, eqn (6.5.40) is true. □

A graph of $\hat{k}_0(t)$ for a fixed $\epsilon > 0$ and $t_0 > 3$ is shown in Fig. 6.12.

The oscillatory character of the Bessel functions appearing in the kernel $\hat{K}(t)$ along with the Theorems 6.9 and 6.10 lead to the following corollary.

Corollary 6.2 *For every $t_0 > 3$ and $\epsilon > 0$ the kernel $\hat{K}(t)$ is an oscillating function whose amplitudes belong to the interval $\left[-\hat{k}_0(t), \hat{k}_0(t)\right]$ $\forall t \in [0,\infty)$. These oscillations tend to zero as $t \to \infty$.*

In contrast to Corollary 6.2, the following result, based on Theorem 6.2, holds true.

Corollary 6.3 *For classical dynamical thermoelasticity the kernel* $K_0 = K_0(t)$ *[see eqn (6.2.116)] in which the coupling parameter* ϵ *ranges over the interval (0,1) is an oscillating function that goes to infinity as* $t \to \infty$.

Remark 6.3 An admissible domain of the constitutive parameters t_0, t_1 and ϵ is determined by the inequalities

$$t_1 \geq t_0 \geq 0, \quad \epsilon \geq 0. \tag{6.5.42}$$

In the case $t_1 = t_0 = 0$, that is for a classical thermoelastic body, the parameter ϵ is a number in the interval $(0, 1)$ [note Table 1 in (Chadwick, 1960)]: $\epsilon = 3.56 \times 10^{-2}$ for aluminum; $\epsilon = 1.68 \times 10^{-2}$ for copper; $\epsilon = 2.97 \times 10^{-4}$ for iron; $\epsilon = 7.73 \times 10^{-2}$ for lead.

In the case $t_1 = t_0 > 0$ and $\epsilon = 0$, that is for a rigid conductor, the thermal relaxation time $\tilde{t}_0$, measured in seconds, has these values [note Table in (Francis, 1972)[6]]: $\tilde{t}_0 = 8 \times 10^{-12}$ for aluminum; $\tilde{t}_0 = 1.6 \times 10^{-12}$ for carbon steel; $\tilde{t}_0 = 1.5 \times 10^{-12}$ for uranium silicate (U_3Si); $\tilde{t}_0 = 2 \times 10^{-9}$ for liquid helium II.

In order to illustrate an order of magnitude of the dimensionless parameters $\hat{v}_i$, $\hat{k}_i$, $\hat{\lambda}$ and $\hat{\beta}$ characterizing a specific thermoelastic body, let us consider aluminum, for which:

$$\epsilon = 3.56 \times 10^{-2}, \quad \tilde{t}_0 = 8 \times 10^{-12}\,\text{s}. \tag{6.5.43}$$

As units of length and time we choose (recall eqn (4.1.6))

$$\hat{x}_0 = \frac{k}{C_E C_1} \quad \text{and} \quad \hat{t}_0 = \frac{k}{C_E C_1^2}. \tag{6.5.44}$$

In the above, k, C_E and C_1 denote, respectively, the thermal conductivity, the specific heat at zero deformation and the speed of longitudinal waves in aluminum. As a result, we find [recall Table 1 in (Chadwick, 1960)]

$$\hat{x}_0 = 1.36 \times 10^{-6}\,\text{cm}, \quad \hat{t}_0^{-1} = 4.66 \times 10^{11}\,\text{s}^{-1}. \tag{6.5.45}$$

Thus, the chosen units are relatively small: $\hat{x}_0$ is of the order of 100 Å (1 Å $= 10^{-10}$ m), while $\hat{t}_0$ is of the order of 2 ps (1 ps $= 10^{-12}$s). Since $\tilde{t}_0$ is of the order of 8 ps (see (6.5.43)), the dimensionless pair (t_0, ϵ) takes the form

$$(t_0, \epsilon) = (3.728, 0.0356). \tag{6.5.46}$$

Clearly, the pair does belong to the domain of constitutive parameters considered in this chapter. Inserting it into the formulas defining $\hat{v}_i$, $\hat{k}_i$, $\hat{\alpha}$, $\hat{\beta}$ and $\hat{\lambda}$, we find

$$\begin{gathered}
\hat{v}_1 = 0.5059, \quad \hat{v}_2 = 1.0236, \\
\hat{k}_1 = 1.0315, \quad \hat{k}_2 = 0.0040, \\
\hat{\alpha} = -0.3481, \quad \hat{\beta} = 0.0433, \\
\hat{\lambda} = 0.0014.
\end{gathered}$$

[6] The symbol $\tilde{t}_0$ must not be confused with the dimensionless time t_0 of Chapter 4.

Hence, for the quantities $\hat{h}_i$ and $\hat{\lambda}_i$, defined by the formulas

$$\hat{h}_i = \hat{v}_i^2 \hat{k}_i/2, \quad \hat{\lambda}_i = \hat{\lambda}\hat{v}_i^2, \tag{6.5.47}$$

we find

$$\hat{h}_1 = 0.1320, \qquad \hat{h}_2 = 0.0020,$$
$$\hat{\lambda}_1 = 3.583 \times 10^{-4}, \quad \hat{\lambda}_2 = 0.0042.$$

The parameters $\hat{h}_i$ and $\hat{\lambda}_i$, respectively, are called the *reduced attenuation* and the *reduced convolution coefficients* of the ith fundamental thermoelastic disturbance. These two parameters appear in a natural way in the investigation of Neumann-type series solutions of the equation $\hat{L}_i\phi_i = 0$; see Section 7.1.

It follows from the numerical values given above that for aluminum the first fundamental thermoelastic disturbance propagates more slowly and is attenuated more strongly than the second one. This result is consistent with the general considerations of the present chapter (recall Figs. 6.7–6.9). Moreover, the reduced attenuation coefficients are smaller than the conventional ones, while the reduced convolution coefficients are relatively small.

Remark 6.4 By letting $t_0 > 1$ [see eqn (6.3.5)] and introducing the notations

$$c_1 = \lim_{\varepsilon \to 0} v_2, \qquad c_2 = \lim_{\varepsilon \to 0} v_1, \tag{6.5.48}$$

from eqns (6.3.1) we obtain

$$c_1 = 1, \qquad c_2 = t_0^{-1/2}. \tag{6.5.49}$$

Hence, c_1 and c_2 represent the "first" and "second" sound speeds, respectively. Clearly, c_1 is a dimensionless speed of a longitudinal isothermal elastic wave, while c_2 is a dimensionless speed of a heat wave in a rigid heat conductor of the Maxwell–Cattaneo type. As a result, the first and second sound speeds represent the limiting values of speeds of the fundamental thermoelastic disturbances of the G–L theory as $\varepsilon \to 0$. For aluminum $t_0 = 3.728$ [see eqn (6.5.46)]; therefore substituting this value into eqn $(6.5.49)_2$ we obtain

$$c_2 = 0.51791 \simeq \frac{1}{\sqrt{3}} c_1. \tag{6.5.50}$$

The formula (6.5.50) is an approximation of the result obtained by Landau

$$c_2 = \frac{1}{\sqrt{3}} c_1, \tag{6.5.51}$$

where c_1 and c_2 denote the first and second sound velocities in a superfluid helium II in a low temperature (Landau, 1941). Therefore, the velocity formulas of Theorem 6.1 applied for aluminum and taken at $\varepsilon = 0$ lead to an approximate form of the Landau's result for a superfluid helium II at a low temperature, showing an agreement between the Landau and G–L theories as far as tracking the second sound in solids at low temperatures is concerned.

Joseph and Preziosi (1989) expressed some doubts about experiments showing heat waves at low temperatures in liquid helium and in certain dielectric crystals. They wrote there (page 51): "It appears that the response of dielectric crystals to oscillations in temperature is not clear enough to use ultrasound and acoustic methods. The experiments that appear to be successful use pulse inputs whose harmonic content is not perfectly known".

In our opinion, the velocity and attenuation formulas of Theorem 6.1 with the approximate material parameters, referred either to an isothermal elastic solid or to a rigid heat conductor, may be useful in identifying the physical properties of a genuine thermoelastic material tested in a laboratory experiment in a range of low temperatures as well as beyond such a range.

7

EXACT APERIODIC-IN-TIME SOLUTIONS OF GREEN–LINDSAY THEORY

7.1 Fundamental solutions for a 3D bounded domain[1]

A solution of the potential–temperature problem of the G–L theory discussed in Chapter 4 may be obtained with the help of so-called fundamental solutions. In this chapter we introduce two such solutions of the G–L theory corresponding to the two fundamental thermoelastic disturbances of that theory already defined in the previous chapter.

According to the definition introduced in Section 6.2, the *i*th fundamental thermoelastic disturbance in the G–L theory is a potential–temperature disturbance corresponding to the potential ϕ_i that satisfies the wave-like equation

$$\hat{L}_i \phi_i = 0 \; (i = 1, 2), \tag{7.1.1}$$

where (recall eqns (6.2.7), (6.2.8) and (6.2.14))

$$\hat{L}_1 = \nabla^2 - \frac{1}{v_1^2} \frac{\partial^2}{\partial t^2} - k_1 \frac{\partial}{\partial t} - \lambda - \lambda K *, \tag{7.1.2}$$

$$\hat{L}_2 = \nabla^2 - \frac{1}{v_2^2} \frac{\partial^2}{\partial t^2} - k_2 \frac{\partial}{\partial t} + \lambda + \lambda K *. \tag{7.1.3}$$

By virtue of Theorem 6.1 on decomposition a potential–temperature disturbance satisfying homogeneous initial conditions is a sum of two such disturbances. Let B denote a bounded region in a 3D Euclidean space, and let $[0, \infty)$ be a time interval. The function ϕ_i satisfying the relations

$$\hat{L}_i \phi_i = 0 \quad \text{on} \quad \mathrm{B} \times [0, \infty), \tag{7.1.4}$$

$$\phi_i(\cdot, 0) = \dot{\phi}_i(\cdot, 0) = 0 \quad \text{on} \quad \mathrm{B}, \tag{7.1.5}$$

$$\phi_i = f_i \quad \text{on} \quad \partial \mathrm{B} \times [0, \infty), \tag{7.1.6}$$

is called the *ith fundamental thermoelastic disturbance in the G–L theory.* In eqn (7.1.6) f_i $(i = 1, 2)$ is an arbitrary function defined on $\partial \mathrm{B} \times [0, \infty)$. This definition implies that a knowledge of the functional dependence of the first fundamental solution on v_1, k_1, λ and f_1 allows one to determine the second

[1] See (Ignaczak, 1981).

fundamental solution as a function of v_2, k_2, λ and f_2. One can show that for a smooth boundary ∂B and sufficiently smooth functions f_1 and f_2 there exists exactly one fundamental solution.[2]

Our definition of fundamental solutions is motivated by the fact that they may be employed to construct a solution of PTP in the G–L theory, in which the initial conditions are homogeneous and one of the boundary conditions for ϕ is of Dirichlet type.[3] Using the homogeneous initial conditions (7.1.5), one can show that the relations (7.1.4)–(7.1.6) defining the ith fundamental solution are equivalent to the equations

$$\hat{L}_i^* \phi_i = 0 \quad \text{on} \quad \mathrm{B} \times [0, \infty), \tag{7.1.7}$$

$$\phi_i(\cdot, 0) = \dot{\phi}_i(\cdot, 0) = 0 \quad \text{on} \quad \mathrm{B}, \tag{7.1.8}$$

$$\phi_i = f_i \quad \text{on} \quad \partial\mathrm{B} \times [0, \infty), \tag{7.1.9}$$

where $\hat{L}_i^*$ $(i = 1, 2)$ are the wave-like operators

$$\hat{L}_1^* = \nabla^2 - \frac{1}{v_1^2} \frac{\partial^2}{\partial t^2} - k_1 \frac{\partial}{\partial t} - \lambda G * \frac{\partial}{\partial t}, \tag{7.1.10}$$

$$\hat{L}_2^* = \nabla^2 - \frac{1}{v_2^2} \frac{\partial^2}{\partial t^2} - k_2 \frac{\partial}{\partial t} + \lambda G * \frac{\partial}{\partial t}, \tag{7.1.11}$$

in which the kernel $G = G(t)$ is (recall eqn (6.2.26))

$$G(t) = 2e^{\alpha t} \frac{J_1(\beta t)}{\beta t}. \tag{7.1.12}$$

In the following, while looking for the fundamental solutions, we employ the system (7.1.7)–(7.1.9) rather than eqns (7.1.4)–(7.1.6). Thus, we look for the function ϕ_1, satisfying eqns (7.1.7)–(7.1.9) for $i = 1$, in the form of a series

$$\phi_1(x, t) = \sum_j T_{1j}(t) \varphi_j(x), \tag{7.1.13}$$

where $\varphi_j(x)$ $(j = 1, 2, 3, \ldots)$ are the eigenfunctions of the operator $-\nabla^2$ on B satisfying the homogeneous boundary conditions on ∂B, that is [see p. 213 in (Stakgold, 1968)]

$$\nabla^2 \varphi_j + \lambda_j \varphi_j = 0 \quad \text{on} \quad \mathrm{B}, \tag{7.1.14}$$

$$\varphi_j = 0 \quad \text{on} \quad \partial\mathrm{B}. \tag{7.1.15}$$

[2] A uniqueness theorem for the problem (7.1.4)–(7.1.6) is to be proved in the section on expansion of a fundamental solution into a series of eigenfunctions for the operator $-\nabla^2$ (see eqn (7.1.13)).

[3] If a PTP of G–L theory is formulated in such a way that ϕ satisfies the homogeneous initial conditions, and one of the two boundary conditions is of the Neumann type, a fundamental solution, that satisfies eqns (7.1.4) and (7.1.5) subject to a boundary condition of the Neumann type, may also be defined.

The numbers λ_j $(j = 1, 2, 3, \ldots)$ are the eigenvalues of the problem (7.1.14) and (7.1.15) satisfying the inequalities

$$0 < \lambda_1 < \lambda_2 < \lambda_3 < \ldots < \lambda_n < \ldots \tag{7.1.16}$$

The functions $T_{1j}(t)$ depend on time only; they should be chosen in such a way that ϕ_1 given by eqn (7.1.13) satisfies eqns (7.1.7)–(7.1.9) for $i = 1$ and for an arbitrary function f_1 on $\partial\mathrm{B} \times [0, \infty)$. Assuming that the set $\{\varphi_j(x)\}$ forms a complete orthonormal space, from eqn (7.1.13) we obtain

$$T_{1j}(t) = \int_{\mathrm{B}} \phi_1(x,t)\, \phi_j(x)\, \mathrm{d}v. \tag{7.1.17}$$

Given that $\phi_1(x,t)$ satisfies the inhomogeneous boundary condition (7.1.9) for $i = 1$, while $\varphi_j(x)$ vanish at $\partial\mathrm{B}$, the series (7.1.13) cannot be uniformly convergent on $\bar{\mathrm{B}} \times [0, \infty)$. That is, it cannot be differentiated term by term or substituted directly into eqn (7.1.7) for $i = 1$. In order to obtain an equation governing $T_{1j}(t)$, we multiply by $\varphi_j(x)$ both sides of eqn (7.1.7) for $i = 1$, integrate over B so as to find

$$\begin{aligned} &\int_{\mathrm{B}} \varphi_j(x)\, \nabla^2\phi_1 \mathrm{d}v - \frac{1}{v_1^2}\frac{\mathrm{d}^2}{\mathrm{d}t^2}\int_{\mathrm{B}} \varphi_j(x)\, \phi_1 \mathrm{d}v \\ &-k_1 \frac{\mathrm{d}}{\mathrm{d}t}\int_{\mathrm{B}} \varphi_j(x)\, \phi_1 \mathrm{d}v - \lambda G * \frac{\mathrm{d}}{\mathrm{d}t}\int_{\mathrm{B}} \varphi_j(x)\, \phi_1 \mathrm{d}v = 0. \end{aligned} \tag{7.1.18}$$

On account of eqn (7.1.17), from eqn (7.1.18) we find

$$\int_{\mathrm{B}} \varphi_j \nabla^2 \phi_1 \mathrm{d}v - \frac{1}{v_1^2}\frac{\mathrm{d}^2}{\mathrm{d}t^2} T_{1j}(t) - k_1 \frac{\mathrm{d}}{\mathrm{d}t} T_{1j}(t) - \lambda G * \frac{\mathrm{d}}{\mathrm{d}t} T_{1j}(t) = 0. \tag{7.1.19}$$

Since

$$\begin{aligned} \int_{\mathrm{B}} \varphi_j \nabla^2 \phi_1 \mathrm{d}v &= \int_{\mathrm{B}} \phi_1 \nabla^2 \varphi_j \mathrm{d}v + \int_{\partial\mathrm{B}} \left(\varphi_j \frac{\partial \phi_1}{\partial n} - \phi_1 \frac{\partial \varphi_j}{\partial n} \right) \mathrm{d}a, \\ \nabla^2 \varphi_j &= -\lambda_j \varphi_j \quad \text{on B}, \\ \varphi_j &= 0 \quad \text{on } \partial\mathrm{B}, \end{aligned} \tag{7.1.20}$$

and

$$\phi_1 = f_1 \quad \text{on } \partial\mathrm{B} \times [0, \infty), \tag{7.1.21}$$

from eqn (7.1.19) we obtain

$$\frac{1}{v_1^2}\frac{\mathrm{d}^2}{\mathrm{d}t^2} T_{1j}(t) + k_1 \frac{\mathrm{d}}{\mathrm{d}t} T_{1j}(t) + \lambda_j T_{1j}(t) + \lambda G * \frac{\mathrm{d}}{\mathrm{d}t} T_{1j}(t) = -\int_{\partial\mathrm{B}} f_1(x,t) \frac{\partial \varphi_j}{\partial n} \mathrm{d}a. \tag{7.1.22}$$

The homogeneous initial conditions (7.1.8) for $i = 1$ together with the formula (7.1.17) imply the homogeneous conditions for $T_{1j}(t)$

$$T_{1j}(0) = \frac{\mathrm{d}}{\mathrm{d}t} T_{1j}(0) = 0. \tag{7.1.23}$$

We shall now prove that, if there exists a solution $T_{1j} = T_{1j}(t)$ to eqn (7.1.22) under conditions (7.1.23), it is unique (see footnote #2). It will suffice to demonstrate that the equation

$$\frac{1}{v_1^2} \ddot{T}_j(\sigma) + k_1 \dot{T}_j(\sigma) + \lambda_j T_j(\sigma) + \lambda \int_0^\sigma G(\sigma - s) \dot{T}_j(s) \, \mathrm{d}s = 0 \quad \text{for} \quad \sigma \in [0, \infty), \tag{7.1.24}$$

with the conditions ($\cdot = \mathrm{d}/\mathrm{d}\sigma$)

$$T_j(0) = \dot{T}_j(0) = 0, \tag{7.1.25}$$

implies the equation

$$T_j(\sigma) = 0 \text{ for } \sigma \in [0, \infty). \tag{7.1.26}$$

Upon multiplying eqn (7.1.24) through by $\dot{T}_j(\sigma)$ and integrating over the interval $\sigma \in [0, t]$, on account of eqn (7.1.25), we obtain

$$\begin{aligned} &\frac{1}{2v_1^2} \left[\dot{T}_j(t)\right]^2 + k_1 \int_0^t \left[\dot{T}_j(\sigma)\right]^2 \mathrm{d}\sigma + \frac{\lambda_j}{2} \left[T_j(t)\right]^2 \\ &+ \lambda \int_0^t \mathrm{d}\sigma \int_0^\sigma \mathrm{d}s \, G(\sigma - s) \dot{T}_j(s) \dot{T}_j(\sigma) = 0. \end{aligned} \tag{7.1.27}$$

It follows from the definition of the function G on $[0, \infty)$ that[4]

$$|G(\sigma - s)| \leq 1 \text{ for } 0 \leq s \leq \sigma \leq t. \tag{7.1.28}$$

Thus, from eqn (7.1.27) we obtain the inequality

$$\begin{aligned} &\frac{1}{2v_1^2} \left[\dot{T}_j(t)\right]^2 + k_1 \int_0^t \left[\dot{T}_j(\sigma)\right]^2 \mathrm{d}\sigma + \frac{\lambda_j}{2} \left[T_j(t)\right]^2 \\ &\leq \lambda \int_0^t \mathrm{d}\sigma \int_0^\sigma \mathrm{d}s \left|\dot{T}_j(s)\right| \left|\dot{T}_j(\sigma)\right|. \end{aligned} \tag{7.1.29}$$

In the derivation of eqn (7.1.29) we use the fact that $\lambda > 0$ (recall eqn $(6.2.9)_3$).

Now, since

$$\int_0^t \mathrm{d}\sigma \int_0^t \mathrm{d}s \left|\dot{T}_j(s)\right| \left|\dot{T}_j(\sigma)\right| = \frac{1}{2} \left[\int_0^t \left|\dot{T}_j(s)\right| \mathrm{d}s\right]^2, \tag{7.1.30}$$

[4] The inequality (7.1.28) is implied from the inequalities $|J_1(x)| \leq |x|/2$ and $\alpha < 0$ (see footnote #5 of Chapter 6).

and by the Schwartz inequality

$$\left(\int_0^t \left|\dot{T}_j(s)\right| \mathrm{d}s\right)^2 \le t \int_0^t \left[\dot{T}_j(\sigma)\right]^2 \mathrm{d}\sigma, \tag{7.1.31}$$

and $\lambda_j > 0$ (recall eqn (7.1.16)), from eqn (7.1.29) we obtain

$$\frac{1}{2v_1^2}\left[\dot{T}_j(t)\right]^2 + \left(k_1 - \frac{\lambda t}{2}\right)\int_0^t \left[\dot{T}_j(\sigma)\right]^2 \mathrm{d}\sigma \le 0. \tag{7.1.32}$$

Next, we introduce the functions

$$p_j(t) = \int_0^t \left[\dot{T}_j(\sigma)\right]^2 \mathrm{d}\sigma, \tag{7.1.33}$$

$$q(t) = 2v_1^2\left(k_1 - \frac{\lambda t}{2}\right), \tag{7.1.34}$$

and transform the inequality (7.1.32) to

$$\dot{p}_j(t) + q(t)\,p_j(t) \le 0. \tag{7.1.35}$$

Multiplying eqn (7.1.35) through by $\exp\left(\int_0^t q(\tau)\,\mathrm{d}\tau\right)$ we obtain

$$\frac{\mathrm{d}}{\mathrm{d}t}\left[p_j(t)\exp\left(\int_0^t q(\tau)\,\mathrm{d}\tau\right)\right] \le 0. \tag{7.1.36}$$

Integrating this inequality over $(0,t)$ and using the definition of $p_j(t)$ (recall eqn (7.1.33)), we obtain

$$p_j(t) = 0 \text{ for } \ t \in [0,\infty), \tag{7.1.37}$$

that is

$$\dot{T}_j(t) = 0 \text{ for } \ t \in [0,\infty). \tag{7.1.38}$$

From this and eqn $(7.1.25)_1$ we obtain the desired relation (7.1.26) that ends the proof of uniqueness of the solution of eqn (7.1.22) subject to the conditions (7.1.23).

We now turn to the determination of a function $T_{1j}(t)$, satisfying the integro-differential equation (7.1.22) subject to eqn (7.1.23). We will use the Laplace transform with p being the parameter of transformation

$$\bar{f}(p) = \int_0^\infty \mathrm{e}^{-pt} f(t)\,\mathrm{d}t. \tag{7.1.39}$$

In the terminology of the operator calculus of (Mikusiński, 1967), the parameter p is to be treated as a differentiation operator, with the relation (7.1.39) written as

$$\bar{f}(p) = \{f(t)\}. \tag{7.1.40}$$

Upon the transformation of eqn (7.1.22), and using the homogeneous initial conditions (7.1.23), we find

$$\left[\left(\frac{p}{v_1}\right)^2 + k_1 p + \lambda_j + \lambda p \bar{G}\right] \bar{T}_{1j}(p) = -\int_{\partial B} \bar{f}_1 \frac{\partial \varphi_j}{\partial n} da. \tag{7.1.41}$$

From this there follows

$$\begin{aligned} \bar{T}_{1j}(p) &= (-)\left[\left(\frac{p}{v_1}\right)^2 + k_1 p + \lambda_j\right]^{-1} \\ &\times \left\{1 + \lambda p \bar{G}\left[\left(\frac{p}{v_1}\right)^2 + k_1 p + \lambda_j\right]^{-1}\right\}^{-1} \int_{\partial B} \bar{f}_1 \frac{\partial \varphi_j}{\partial n} da, \end{aligned} \tag{7.1.42}$$

or

$$\bar{T}_{1j}(p) = -\sum_{k=0}^{\infty} (-)^k \lambda^k \bar{G}^k \left[\frac{p}{(p/v_1)^2 + k_1 p + \lambda_j}\right]^{k+1} \int_{\partial B} \frac{\bar{f}_1}{p} \frac{\partial \varphi_j}{\partial n} da. \tag{7.1.43}$$

Since (recall eqns (6.2.28)–(6.2.30))

$$\bar{G}(p) = \frac{2}{\beta^2}\left[\sqrt{(p-\alpha)^2 + \beta^2} - (p-\alpha)\right], \tag{7.1.44}$$

therefore

$$\left[\bar{G}(p)\right]^k = \left(\frac{2}{\beta^2}\right)^k \left[\sqrt{(p-\alpha)^2 + \beta^2} - (p-\alpha)\right]^k, \tag{7.1.45}$$

and from the Laplace transform tables (Mikusiński, 1967) we obtain

$$\left[\bar{G}(p)\right]^k = \{G(t)\}^k = \{G_k(t)\}, \tag{7.1.46}$$

where

$$G_0(t) = \delta(t), \tag{7.1.47}$$

and

$$G_k(t) = k\left(\frac{2}{\beta}\right)^k \frac{e^{\alpha t}}{t} J_k(\beta t) \quad \text{for} \quad k = 1, 2, 3, \ldots \tag{7.1.48}$$

Here, $\delta(t)$ denotes the Dirac delta function, while $J_k(x)$ is the Bessel function of the first kind and kth order. Note that[5]

$$G_k(t) = \underbrace{G * G * \ldots * G}_{k \text{ times}} = G^{*k}(t). \tag{7.1.49}$$

[5] The notation for the kth convolutional power of a function is introduced in (Brun, 1975). In the Mikusiński notation (1967) we have $G^{*k}(t) = \{G_k(t)\} = \{G(t)\}^k$.

Next, we introduce the function[6]

$$\bar{h}(p;\lambda,\mu)=\frac{p}{(p+\mu)^2+\lambda^2}, \tag{7.1.50}$$

where $\mu>0$ and λ is a non-negative number or a purely imaginary one with a non-negative imaginary part. Then,

$$\bar{h}(p;\lambda,u)=\{h(t;\lambda,u)\}, \tag{7.1.51}$$

where

$$h(t;\lambda,\mu)=\mathrm{e}^{-\mu t}\left(\cos\lambda t-\mu\lambda^{-1}\sin\lambda t\right), \tag{7.1.52}$$

and

$$\left[\bar{h}(p;\lambda,\mu)\right]^n=\{h(t;\lambda,\mu)\}^n=\{h_n(t;\lambda,\mu)\}, \tag{7.1.53}$$

where

$$h_n(t;\lambda,\mu)=h^{*n}(t;\lambda,\mu)\quad \text{for}\ \ n=1,2,3,\ldots \tag{7.1.54}$$

It is then easy to prove the equality[7]

$$\bar{h}^{n+1}=\frac{1}{2n}\left(\frac{\mu}{\lambda}\frac{\partial}{\partial\lambda}-\frac{\partial}{\partial\mu}\right)\bar{h}^n,\quad n=1,2,3,\ldots, \tag{7.1.55}$$

which implies

$$h_n=\frac{1}{2^{n-1}}\frac{1}{(n-1)!}\left(\frac{\mu}{\lambda}\frac{\partial}{\partial\lambda}-\frac{\partial}{\partial\mu}\right)^{n-1}h. \tag{7.1.56}$$

Letting h_n in the form

$$h_n=\mathrm{e}^{-\mu t}\left(A_n\cos\lambda t+B_n\sin\lambda t\right), \tag{7.1.57}$$

where

$$A_n=A_n(t;\lambda,\mu),\ B_n=B_n(t;\lambda,\mu), \tag{7.1.58}$$

from eqn (7.1.55) we obtain

$$\begin{aligned} A_{n+1}&=\frac{1}{2n}\left[\frac{\mu}{\lambda}\left(A_{n,\lambda}+tB_n\right)-\left(A_{n,\mu}-tA_n\right)\right],\\ B_{n+1}&=\frac{1}{2n}\left[\frac{\mu}{\lambda}\left(B_{n,\lambda}-tA_n\right)-\left(B_{n,\mu}-tB_n\right)\right], \end{aligned} \tag{7.1.59}$$

where the comma indicates a partial differentiation with respect to λ or μ.

The relations (7.1.59) allow determination of the pair (A_{n+1},B_{n+1}) through the pair (A_n,B_n) for any $n=1,2,3,\ldots$ Introducing the notation

$$\zeta=\mu\lambda^{-1},\ t=y\lambda^{-1},\ (\lambda>0), \tag{7.1.60}$$

[6] The parameters μ and λ in eqn (7.1.50) must not be confused with those introduced earlier.
[7] For simplicity, we omit the variables $(p;\lambda,\mu)$ and $(t;\lambda,\mu)$ of the functions $\bar{h}$ and h, respectively.

and employing eqns (7.1.52) and (7.1.57)–(7.1.59), for $n = 1, 2, 3, 4, 5, 6$ and 7, we find

$$A_1 = 1,\ B_1 = -\zeta, \tag{7.1.61}$$

$$A_2 = \frac{y}{2\lambda}\left(1-\zeta^2\right),\ B_2 = \frac{1}{2\lambda}\left(1+\zeta^2-2\zeta y\right), \tag{7.1.62}$$

$$\begin{gathered} A_3 = \frac{y}{(2\lambda)^2\,2!}\left[3\zeta\left(1+\zeta^2\right)+\left(1-3\zeta^2\right)y\right], \\ B_3 = \frac{1}{(2\lambda)^2\,2!}\left[-3\zeta\left(1+\zeta^2\right)+3\left(1+\zeta^2\right)y-\zeta\left(3-\zeta^2\right)y^2\right], \end{gathered} \tag{7.1.63}$$

$$\begin{gathered} A_4 = \frac{y}{(2\lambda)^3\,3!}\left[-3\left(1+6\zeta^2+5\zeta^4\right)+12\zeta\left(1+\zeta^2\right)y\right. \\ \left.+\left(1-6\zeta^2+\zeta^4\right)y^2\right], \\ B_4 = \frac{1}{(2\lambda)^3\,3!}\left[3\left(1+6\zeta^2+5\zeta^4\right)-12\zeta\left(1+\zeta^2\right)y\right. \\ \left.+6\left(1-\zeta^4\right)y^2-4\zeta\left(1-\zeta^2\right)y^3\right], \end{gathered} \tag{7.1.64}$$

$$\begin{gathered} A_5 = \frac{y}{(2\lambda)^4\,4!}[15\zeta\left(3+10\zeta^2+7\zeta^4\right)-15\left(1+6\zeta^2+5\zeta^4\right)y \\ +10\zeta\left(3+2\zeta^2-\zeta^4\right)y^2+\left(1-10\zeta^2+5\zeta^4\right)y^3], \\ B_5 = \frac{1}{(2\lambda)^4\,4!}\left[-15\zeta\left(3+10\zeta^2+7\zeta^4\right)+15\left(1+6\zeta^2+5\zeta^4\right)y\right. \\ \left.-15\zeta\left(1-2\zeta^2-3\zeta^4\right)y^2+10\left(1-2\zeta^2-3\zeta^4\right)y^3-\zeta\left(5-10\zeta^2+\zeta^4\right)y^4\right], \end{gathered} \tag{7.1.65}$$

$$\begin{gathered} A_6 = \frac{y}{(2\lambda)^5\,5!}[-45\zeta\left(1+15\zeta^2+35\zeta^4+21\zeta^6\right) \\ +90\zeta\left(3+10\zeta^2+7\zeta^4\right)y-15\left(3+15\zeta^2+5\zeta^4-7\zeta^6\right)y^2 \\ +60\zeta\left(1-\zeta^4\right)y^3+\left(1-15\zeta^2+15\zeta^4-\zeta^6\right)y^4], \\ B_6 = \frac{1}{(2\lambda)^5\,5!}[45\left(1+15\zeta^2+35\zeta^4+21\zeta^6\right)-90\zeta\left(3+10\zeta^2+7\zeta^4\right)y \\ +30\left(1-15\zeta^4-14\zeta^6\right)y^2+30\zeta\left(1+10\zeta^2+9\zeta^4\right)y^3 \\ \left.+15(1-5\zeta^2-5\zeta^4+\zeta^6)y^4-2\zeta\left(3-10\zeta^2+3\zeta^4\right)y^5\right], \end{gathered} \tag{7.1.66}$$

$$\begin{gathered} A_7 = \frac{y}{(2\lambda)^6\,6!}[315\zeta\left(5+35\zeta^2+63\zeta^4+33\zeta^6\right) \\ -315\left(1+15\zeta^2+35\zeta^4+21\zeta^6\right)y \\ +420\zeta\left(2+5\zeta^2-3\zeta^6\right)y^2-105\left(1+3\zeta^2-5\zeta^4-7\zeta^6\right)y^3 \end{gathered}$$

$$+21\zeta\left(5-5\zeta^2-9\zeta^4+\zeta^6\right)y^4+\left(1-21\zeta^2+35\zeta^4-7\zeta^6\right)y^5],$$

$$\begin{aligned} B_7 = \frac{1}{(2\lambda)^6\,6!}[&-315\zeta\left(5+35\zeta^2+63\zeta^4+33\zeta^6\right) \\ &+315\left(1+15\zeta^2+35\zeta^4+21\zeta^6\right)y \\ &-315\zeta\left(1-5\zeta^2-21\zeta^4-15\zeta^6\right)y^3-420\zeta^2\left(3+10\zeta^2+7\zeta^4\right)y^3 \\ &+210\zeta\left(1+5\zeta^2+31\zeta^4-\zeta^6\right)y^4 \\ &+21\left(1-9\zeta^2-5\zeta^4+5\zeta^6\right)y^5-\zeta\left(7-35\zeta^2+21\zeta^4-\zeta^6\right)y^6]. \end{aligned} \tag{7.1.67}$$

In general, one can show (see Chapter 9) that A_n and B_n are polynomials in y of degree $n-1$ of the form

$$\begin{aligned} A_n &= \frac{1}{(2\lambda)^{n-1}(n-1)!}\sum_{k=0}^{n-1} a_{kn}(\zeta)\,y^k, \\ B_n &= \frac{1}{(2\lambda)^{n-1}(n-1)!}\sum_{k=0}^{n-1} b_{kn}(\zeta)\,y^k, \end{aligned} \tag{7.1.68}$$

where $a_{kn}(\zeta)$ and $b_{kn}(\zeta)$ $(0 \le k \le n-1)$ are polynomials in ζ of degree $\le n$, determined through the recurrence formulas $(n \ge 2)$

$$\begin{aligned} a_{1.n+1} + b_{0.n+1} &= 0 \quad (a_{0.n} = 0), \\ b_{1.n+1} &= (n+1)\,b_{0.n}, \\ a_{n.n+1} &= \zeta b_{n-1.n} + a_{n-1.n}, \\ b_{n.n+1} &= -\zeta a_{n-1.n} + b_{n-1.n}, \\ (k+1)\,a_{k+1.n+1} + b_{k.n+1} &= (n+k+1)\,a_{k.n} - \zeta a_{k-1.n} + b_{k-1.n} \\ &\quad \text{for } 1 \le k \le n-1, \\ (k+1)\,b_{k+1.n+1} - a_{k.n+1} &= (n+k+1)\,b_{k.n} - \zeta b_{k-1.n} - a_{k-1.n} \\ &\quad \text{for } 1 \le k \le n-1. \end{aligned} \tag{7.1.69}$$

The relations (7.1.69) allow finding the pair (a_{kn}, b_{kn}) for $0 \le k \le n-1$ with the help of the pair $(a_{k.n-1}, b_{k.n-1})$ for $0 \le k \le n-2$. Note here that the relations (7.1.68) imply that the functions $A_n(t;\lambda,\mu)$ and $B_n(t;\lambda,\mu)$ determining the function $h_n(t;\lambda,\mu)$ (recall eqn (7.1.57)) are polynomials in t of degree $n-1$ of the form

$$\begin{aligned} A_n &= \frac{1}{(2\lambda)^{n-1}(n-1)!}\sum_{k=0}^{n-1} a_{kn}\left(\frac{\mu}{\lambda}\right)\lambda^k t^k, \\ B_n &= \frac{1}{(2\lambda)^{n-1}(n-1)!}\sum_{k=0}^{n-1} b_{kn}\left(\frac{\mu}{\lambda}\right)\lambda^k t^k. \end{aligned} \tag{7.1.70}$$

These polynomials appear naturally in a construction of the fundamental solutions of the G–L theory. In the following, (A_n, B_n) will be called a pair of *polynomials of thermoelasticity*, whereas (a_{kn}, b_{kn}) for $0 \leq k \leq n-1$ will be called the *associated* pair of *polynomials of thermoelasticity*.

Due to a fundamental importance of these polynomials in the G–L theory, in Chapter 9 we derive a number of recurrence relations for A_n and B_n involving time derivatives of these polynomials, and present a number of general properties of these polynomials.

We shall now show how one may use the polynomials $A_n(t;\lambda,\mu)$ and $B_n(t;\lambda,\mu)$ to obtain a series representation of $T_{1j}(t)$; see eqn (7.1.13). To this end, the expression in square brackets in the sum in eqn (7.1.43) is represented in the form

$$\left[\frac{p}{(p/\upsilon_1)^2 + k_1 p + \lambda_j}\right] = \upsilon_1^2 \left[\frac{p}{(p+\mu_1)^2 + \lambda_{1j}^2}\right], \tag{7.1.71}$$

where

$$\mu_1 = \upsilon_1 h_1,\ h_1 = (\upsilon_1 k_1)/2,\ \lambda_{1j} = \upsilon_1 \sqrt{\lambda_j - h_1^2}. \tag{7.1.72}$$

From this, on account of the definitions of $A_n(t;\lambda,\mu)$ and $B_n(t;\lambda,\mu)$ (recall eqns (7.1.50)–(7.1.57) and (7.1.70)), we find

$$\left[\frac{p}{(p/\upsilon_1)^2 + k_1 p + \lambda_j}\right]^{k+1} = \upsilon_1^{2(k+1)} \left\{H_{jk}^{(1)}(t)\right\}, \tag{7.1.73}$$

in which the function $H_{jk}^{(1)}(t)$ is given by three different formulas, depending on the range of λ_j:

$$H_{jk}^{(1)}(t) = \mathrm{e}^{-\mu_1 t}[A_{k+1}(t; i\lambda_{1j}^*, \mu_1)\, ch\lambda_{1j}^* t + iB_{k+1}(t; i\lambda_{1j}^*, \mu_1)\, sh\lambda_{1j}^* t] \quad \text{for } 0 < \lambda_j < h_1^2, \quad \lambda_{1j}^* = -i\lambda_{1j}, \tag{7.1.74}$$

$$H_{jk}^{(1)}(t) = \mathrm{e}^{-\mu_1 t} \sum_{r=0}^{k+1} \binom{k+1}{r} (-)^r \mu_1^r \frac{t^{k+r}}{(k+r)!} \quad \text{for } \lambda_j = h_1^2, \tag{7.1.75}$$

$$H_{jk}^{(1)}(t) = \mathrm{e}^{-\mu_1 t}[A_{k+1}(t; \lambda_{1j}, \mu_1) \cos \lambda_{1j} t + B_{k+1}(t; \lambda_{1j}, \mu_1) \sin \lambda_{1j} t] \quad \text{for } \lambda_j > h_1^2. \tag{7.1.76}$$

Thus, using eqns (7.1.43) and (7.1.73), we find

$$\{T_{1j}(t)\} = -\sum_{k=0}^{\infty} (-)^k \hat{\lambda}_1^k \{G_k\} \left\{H_{jk}^{(1)}\right\} \int_{\partial \mathrm{B}} \left\{\hat{f}_1\right\} \left\{\frac{\partial \varphi_j}{\partial n}\right\} \mathrm{d}a, \tag{7.1.77}$$

where

$$\hat{\lambda}_1 = \lambda \upsilon_1^2,\ \hat{f}_1 = f_1\, \upsilon_1^2. \tag{7.1.78}$$

Dropping the curly brackets in eqn (7.1.77) leads to

$$T_{1j}(t) = -\sum_{k=0}^{\infty} (-)^k \hat{\lambda}_1^k G_k * H_{jk}^{(1)} * \int_{\partial B} \hat{f}_1 * \frac{\partial \varphi_j}{\partial n} \mathrm{da}, \tag{7.1.79}$$

which, given eqn (7.1.13), leads to the following form of the first fundamental solution of the G–L theory

$$\phi_1(x,t) = -\sum_{j=0}^{\infty} \left[\sum_{k=0}^{\infty} (-)^k \hat{\lambda}_1^k G_k * H_{jk}^{(1)} * \int_{\partial B} \hat{f}_1 * \frac{\partial \varphi_j}{\partial n} \mathrm{da} \right] \varphi_j(x) \\ \forall (x,t) \in \bar{B} \times [0,\infty). \tag{7.1.80}$$

A consideration of the explicit dependence of ϕ_1 on υ_1, k_1 and λ as well as on the function f_1, and a comparison of the definitions of ϕ_1 and ϕ_2 lead to the following formulas for the second fundamental solution of the G–L theory

$$T_{2j}(t) = -\sum_{k=0}^{\infty} \hat{\lambda}_2^k G_k * H_{jk}^{(2)} * \int_{\partial B} \hat{f}_2 * \frac{\partial \varphi_j}{\partial n} \mathrm{da}, \tag{7.1.81}$$

and

$$\phi_2(x,t) = -\sum_{j=0}^{\infty} \left[\sum_{k=0}^{\infty} \hat{\lambda}_2^k G_k * H_{jk}^{(2)} * \int_{\partial B} \hat{f}_2 * \frac{\partial \varphi_j}{\partial n} \mathrm{da} \right] \varphi_j(x) \\ \forall (x,t) \in \bar{B} \times [0,\infty). \tag{7.1.82}$$

The function $H_{jk}^{(2)}(t)$ is given by three different formulas, depending on the range of λ_j:

$$H_{jk}^{(2)}(t) = \mathrm{e}^{-\mu_2 t} \left[A_{k+1}(t; i\lambda_{2j}^*, \mu_2)\, ch\lambda_{2j}^* t + iB_{k+1}(t; i\lambda_{2j}^*, \mu_2)\, sh\lambda_{2j}^* t \right] \\ \text{for } 0 < \lambda_j < h_2^2, \quad \lambda_{2j}^* = -i\lambda_{2j}, \tag{7.1.83}$$

$$H_{jk}^{(2)}(t) = e^{-\mu_2 t} \sum_{r=0}^{k+1} \binom{k+1}{r} (-)^r \mu_2^r \frac{t^{k+r}}{(k+r)!} \\ \text{for } \lambda_j = h_2^2, \tag{7.1.84}$$

$$H_{jk}^{(2)}(t) = \mathrm{e}^{-\mu_2 t} \left[A_{k+1}(t; \lambda_{2j}, \mu_2) \cos \lambda_{2j} t + B_{k+1}(t; \lambda_{2j}, \mu_2) \sin \lambda_{2j} t \right] \\ \text{for } \lambda_j > h_2^2, \tag{7.1.85}$$

where

$$\mu_2 = \upsilon_2 h_2,\ h_2 = \upsilon_2 k_2/2, \tag{7.1.86}$$

$$\lambda_{2j} = \upsilon_2 \sqrt{\lambda_j - h_2^2}. \tag{7.1.87}$$

Moreover, the parameter $\hat{\lambda}_2$ and function $\hat{f}_2$ appearing in eqn (7.1.82) are given by

$$\hat{\lambda}_2 = \lambda \upsilon_2^2, \qquad \hat{f}_2 = f_2 \upsilon_2^2. \tag{7.1.88}$$

In general, since $\lambda_j \to +\infty$ for $j \to \infty$ and $h_i^2 < \infty$ $(i = 1, 2)$, there exists only a finite number of terms in the series $\sum_{j=1}^{\infty} T_{ij}(t)\,\varphi_j(x)$ for which $\lambda_j \leq h_i^2$ $(i = 1, 2)$. This means that only a finite number of terms in the series (7.1.80) and (7.1.82) is determined by hyperbolic functions damped on the time axis. An infinite number of terms of eqn (7.1.80) for $\lambda_j > h_1^2$ and eqn (7.1.82) for $\lambda_j > h_2^2$ is dominated by trigonometric functions damped on the time axis.

Using the definition of $H_{jk}^{(i)}(t)$ [see eqns (7.1.74)–(7.1.76) and (7.1.83)–(7.1.85)], it can be shown that

$$\begin{gathered} H_{jk}^{(i)}(t) \to 0 \quad \text{for} \quad t \to \infty \\ \forall i = 1, 2 \quad j = 1, 2, 3, \ldots \quad k = 0, 1, 2, \ldots \end{gathered} \tag{7.1.89}$$

Moreover, from eqn (7.1.48) we find

$$\begin{gathered} G_k(t) \to 0 \quad \text{for} \quad t \to \infty \\ \forall k = 1, 2, 3, 4, \ldots \quad t_1 \geq t_0 > 1, \quad \epsilon > 0. \end{gathered} \tag{7.1.90}$$

Finally, assuming $\hat{f}_i$ $(i = 1, 2)$ to be a sufficiently smooth function on $\partial\mathrm{B} \times [0, \infty)$, tending to zero for $t \to \infty$, from eqns (7.1.80) and (7.1.82) we find

$$\phi_i(x, t) \to 0 \text{ for } \; t \to \infty \;\; \forall x \in \bar{\mathrm{B}}. \tag{7.1.91}$$

One can also show that, if $\hat{f}_i$ is a sufficiently smooth function on $\partial\mathrm{B} \times [0, \infty)$ and the boundary $\partial\mathrm{B}$ itself is a smooth surface, then the series (7.1.80) and (7.1.82) are of the class C^2 on $\mathrm{B}_{(0)} \times [0, t]$ for any closed domain $\mathrm{B}_{(0)}$ contained in B ($\mathrm{B}_{(0)} \subset \mathrm{B}$) and for any triple of the constitutive parameters (t_1, t_0, ϵ) satisfying the inequalities

$$t_1 \geq t_0 > 1, \, \epsilon > 0. \tag{7.1.92}$$

Furthermore, the series (7.1.80) and (7.1.82) represent classical solutions of the initial-boundary value problems described, respectively, by eqns (7.1.7)–(7.1.9) for $i = 1$ and by eqns (7.1.7)–(7.1.9) for $i = 2$.

Proofs of these results are based on the fact that the functions (7.1.79) and (7.1.81) are the Neumann-type power series satisfying Volterra-type equations with parameters $\hat{\lambda}_1$ and $\hat{\lambda}_2$ on the time axis. Thus, these series, as well as the series obtained by differentiation with respect to time a finite number of times, are convergent for any finite $\hat{\lambda}_1$ and $\hat{\lambda}_2$, and for any point in $[0, t]$.

Remark 7.1 A classical solution ϕ_i of eqns (7.1.7)–(7.1.9) is of the class $C^2(Q) \cap C^1(\bar{Q})$, where $Q = \mathrm{B} \times [0, \infty)$ and $\bar{Q} = \bar{\mathrm{B}} \times [0, \infty)$. This implies that the necessary condition for the existence of such a solution is

$$\hat{f}_i(\cdot,0) = \dot{\hat{f}}_i(\cdot,0) = 0 \text{ on } \partial\mathrm{B}\ \forall i = 1,2. \tag{7.1.93}$$

Remark 7.2 One can show that if $\hat{f}_i$ $(i = 1,2)$ is a sufficiently smooth function on $\partial\mathrm{B}\times[0,\infty)$ and the boundary $\partial\mathrm{B}$ is a sufficiently smooth surface, then the series (7.1.80) and (7.1.82) are of the class $C^4(Q)\cap C^3(\bar{Q})$. In this case, the wave-like equation (7.1.7) and the equation obtained by a differentiation with respect to time (or to a space co-ordinate) can be extended to $\bar{Q}$.

7.2 Solution of a potential–temperature problem for a 3D bounded domain

Let us consider the following PTP of the G–L theory: find a pair (ϕ,ϑ) on $\bar{\mathrm{B}}\times[0,\infty)$ satisfying the field equations

$$\left.\begin{aligned}\nabla^2\phi - \ddot{\phi} - (\vartheta + t_1\dot{\vartheta}) &= 0\\ \nabla^2\vartheta - (\dot{\vartheta} + t_0\ddot{\vartheta}) - \epsilon\nabla^2\dot{\phi} &= 0\end{aligned}\right\} \text{ on } \mathrm{B}\times[0,\infty), \tag{7.2.1}$$

the initial conditions

$$\left.\begin{aligned}\phi(\cdot,0) = 0, \quad \dot{\phi}(\cdot,0) = 0\\ \vartheta(\cdot,0) = 0, \quad \dot{\vartheta}(\cdot,0) = 0\end{aligned}\right\} \text{ on } \bar{\mathrm{B}}, \tag{7.2.2}$$

and the boundary conditions

$$\phi = f,\ \vartheta = g \text{ on } \partial\mathrm{B}\times[0,\infty), \tag{7.2.3}$$

where B is a 3D bounded domain, $[0,\infty)$ is the time interval, while f and g are prescribed functions on $\partial\mathrm{B}\times[0,\infty)$.

The problem described by eqns (7.2.1)–(7.2.3) differs from that described by eqns (4.2.7)–(4.2.9) in that it involves the homogeneous initial conditions while the boundary condition $(4.2.9)_1$ has been replaced by eqn $(7.2.3)_1$. Note that the condition $(7.2.3)_1$ has no direct physical interpretation whenever B is an arbitrary domain in E^3. In the case B is a layer in E^3, the condition $(7.2.3)_1$ implies that there is a mechanical loading prescribed on $\partial\mathrm{B}\times[0,\infty)$; see also Section 7.3.

We shall next assume that f and g are sufficiently smooth functions on $\partial\mathrm{B}\times[0,\infty)$ and the boundary $\partial\mathrm{B}$ is a sufficiently smooth surface, so that a solution to the problem (7.2.1)–(7.2.3) exists in the class $C^4(Q)\cap C^3(\bar{Q})$, where $Q = \mathrm{B}\times[0,\infty)$ and $\bar{Q} = \bar{\mathrm{B}}\times[0,\infty)$. In that case, eqns (7.2.1) and their time derivatives may be extended onto $\bar{Q}$, and the consistency of the conditions (7.2.2) with thus extended equations at $t = 0$, leads to the homogeneous initial conditions for the pair (ϕ,ϑ):

$$\frac{\partial^k}{\partial t^k}(\phi,\vartheta)(\cdot,0) = 0 \text{ on } \bar{\mathrm{B}},\ k = 0,1,2,3. \tag{7.2.4}$$

This implies that a necessary condition for the existence of a solution to the problem (7.2.1)–(7.2.3) takes the form

$$\frac{\partial^k}{\partial t^k}(f,g)(\cdot,0)=0 \text{ on } \partial\bar{\mathrm{B}},\ k=0,1,2,3. \tag{7.2.5}$$

In the following, we shall assume that the functions f and g satisfy the conditions (7.2.5).

Eliminating ϑ from eqns (7.2.1)–(7.2.3), we find that ϕ is a solution to the following problem: find a function ϕ on $\bar{\mathrm{B}}\times[0,\infty)$ satisfying the equation

$$\Gamma\phi=0 \text{ on } \mathrm{B}\times(0,\infty), \tag{7.2.6}$$

subject to the initial conditions

$$\frac{\partial^k\phi}{\partial t^k}(\cdot,0)=0 \quad \text{on } \bar{\mathrm{B}} \quad \text{for } k=0,1,2,3, \tag{7.2.7}$$

and the boundary conditions

$$\phi=f,\quad \nabla^2\phi=\dot{h} \quad \text{on } \partial\mathrm{B}\times[0,\infty). \tag{7.2.8}$$

Here, Γ is the central operator of the G–L theory, determined by the formula (recall eqn (6.1.8))

$$\Gamma=\left(\nabla^2-\frac{\partial^2}{\partial t^2}\right)\left(\nabla^2-t_0\frac{\partial^2}{\partial t^2}-\frac{\partial}{\partial t}\right)-\epsilon\nabla^2\frac{\partial}{\partial t}\left(1+t_1\frac{\partial}{\partial t}\right), \tag{7.2.9}$$

while the function h is given by the formula

$$h=\dot{f}+1*g+t_1 g \quad \text{on } \partial\mathrm{B}\times[0,\infty). \tag{7.2.10}$$

A knowledge of ϕ satisfying eqns (7.2.6)–(7.2.8) allows finding ϑ from eqn $(7.2.1)_1$ subject to the condition $\vartheta(\cdot,0)=0$. Also, the conditions (7.2.5) imply

$$h(\cdot,0)=0 \quad \text{on } \partial\mathrm{B}. \tag{7.2.11}$$

Thus, finding (ϕ,ϑ) satisfying eqns (7.2.1)–(7.2.3) has been reduced to the solution of problem (7.2.6)–(7.2.8). We shall now show that this solution takes the form

$$\phi=\phi_1+\phi_2 \quad \text{on } \bar{\mathrm{B}}\times[0,\infty), \tag{7.2.12}$$

where ϕ_i $(i=1,2)$ is the ith fundamental solution of the G–L theory corresponding to a particular function $\hat{f}_i$, which is determined by f and h appearing in the boundary conditions (7.2.8); see the relations (7.1.80)–(7.1.82).

We shall first demonstrate that the function given by eqn (7.2.12) satisfies the homogeneous initial conditions (7.2.7). To prove this it will suffice to show that

$$\frac{\partial^k}{\partial t^k}\phi_i(\cdot,0)=0 \qquad \text{on } \bar{\mathrm{B}}$$
$$\forall k=0,1,2,3 \quad \text{and} \quad i=1,2. \tag{7.2.13}$$

Let us first note that, on account of the definition of the ith fundamental solution of the G–L theory, the conditions (7.1.13) are satisfied for $k = 0, 1$ and $i = 1, 2$; recall eqns (7.1.7)–(7.1.9). Thus, taking eqn (7.1.7) at $t = 0$, we find that the conditions (7.2.13) are satisfied for $k = 0, 1, 2$ and $i = 1, 2$. Finally, differentiating eqn (7.1.7) with respect to time, setting $t = 0$, and noting that the conditions (7.2.13) are satisfied for $k = 0, 1, 2$ and $i = 1, 2$, we obtain the condition (7.2.13) for $k = 3$ and $i = 1, 2$.

Furthermore, note that, on account of Theorem 6.1 on decomposition, the function (7.2.12) satisfies eqn (7.2.6) on $\mathrm{B} \times (0, \infty)$. We now show how to choose the functions $\hat{f}_1$ and $\hat{f}_2$ appearing in the formulas (7.1.80)–(7.1.82) to make the function (7.2.12) satisfy the boundary conditions (7.2.8). To this end, we apply the Laplace transform to eqns $(7.1.7)_1$ and $(7.1.7)_2$ and get, respectively,

$$\nabla^2 \bar{\phi}_1 - p\left(\frac{p}{v_1^2} + k_1 + \lambda \bar{G}\right)\bar{\phi}_1 = 0 \quad \text{on} \quad \mathrm{B}, \tag{7.2.14}$$

$$\nabla^2 \bar{\phi}_2 - p\left(\frac{p}{v_2^2} + k_2 - \lambda \bar{G}\right)\bar{\phi}_2 = 0 \quad \text{on} \quad \mathrm{B}. \tag{7.2.15}$$

Extending these equations onto $\bar{\mathrm{B}}$ and adding on $\partial\mathrm{B}$, we obtain

$$p\left(\bar{N}_1\bar{\phi}_1 + \bar{N}_2\bar{\phi}_2\right) = \nabla^2\left(\bar{\phi}_1 + \bar{\phi}_2\right) \text{ on } \partial\mathrm{B}, \tag{7.2.16}$$

where

$$\begin{aligned} \bar{N}_1 &= \frac{p}{v_1^2} + k_1 + \lambda\bar{G}, \\ \bar{N}_2 &= \frac{p}{v_2^2} + k_2 - \lambda\bar{G}. \end{aligned} \tag{7.2.17}$$

Thus, the function ϕ determined by eqn (7.2.12) satisfies the boundary conditions (7.2.8) if the functions $\bar{f}_1$ and $\bar{f}_2$ satisfy the system of equations

$$\left.\begin{aligned} \bar{f}_1 + \bar{f}_2 &= \bar{f} \\ \bar{N}_1\bar{f}_1 + \bar{N}_2\bar{f}_2 &= \bar{h} \end{aligned}\right. \quad \text{on} \quad \partial\mathrm{B}, \tag{7.2.18}$$

from which we find

$$\begin{aligned} \bar{f}_1 &= -\frac{\bar{N}_2}{\bar{N}_1 - \bar{N}_2}\bar{f} + \frac{1}{\bar{N}_1 - \bar{N}_2}\bar{h}, \\ \bar{f}_2 &= \frac{\bar{N}_1}{\bar{N}_1 - \bar{N}_2}\bar{f} - \frac{1}{\bar{N}_1 - \bar{N}_2}\bar{h}. \end{aligned} \tag{7.2.19}$$

In view of eqn (7.2.17),

$$\bar{N}_1 - \bar{N}_2 = \left(\frac{1}{v_1^2} - \frac{1}{v_2^2}\right)\left[p + (k_1 - k_2)\left(\frac{1}{v_1^2} - \frac{1}{v_2^2}\right)^{-1} + 2\lambda\left(\frac{1}{v_1^2} - \frac{1}{v_2^2}\right)^{-1}\bar{G}\right]. \tag{7.2.20}$$

From this and from the formulas (6.2.65)–(6.2.67) we obtain

$$\bar{N}_1 - \bar{N}_2 = \left(\frac{1}{\upsilon_1^2} - \frac{1}{\upsilon_2^2}\right)\left(p - \alpha + \frac{\beta^2}{2}\bar{G}\right), \tag{7.2.21}$$

so that, on account of eqn (6.2.28),

$$\bar{N}_1 - \bar{N}_2 = \left(\frac{1}{\upsilon_1^2} - \frac{1}{\upsilon_2^2}\right)\sqrt{(p-\alpha)^2 + \beta^2}. \tag{7.2.22}$$

Furthermore, adding eqns (7.2.17) leads to

$$\bar{N}_1 + \bar{N}_2 = \left(\frac{1}{\upsilon_1^2} + \frac{1}{\upsilon_2^2}\right) p + k_1 + k_2, \tag{7.2.23}$$

and solving eqns (7.2.22) and (7.2.23) for the pair $(\bar{N}_1, \bar{N}_2)$, yields

$$\bar{N}_{1,2} = \frac{1}{2}\left[\left(\frac{1}{\upsilon_1^2} + \frac{1}{\upsilon_2^2}\right) p + (k_1 + k_2) \pm \left(\frac{1}{\upsilon_1^2} - \frac{1}{\upsilon_2^2}\right)\sqrt{(p-\alpha)^2 + \beta^2}\right]. \tag{7.2.24}$$

Substituting that pair into eqn (7.2.19), results in

$$\begin{aligned}
\bar{f}_1 &= \left[\frac{1}{2} - \frac{1}{2}\left(\frac{1}{\upsilon_1^2} - \frac{1}{\upsilon_2^2}\right)^{-1} \frac{\left(\frac{1}{\upsilon_1^2} + \frac{1}{\upsilon_2^2}\right) p + (k_1 + k_2)}{\sqrt{(p-\alpha)^2 + \beta^2}}\right]\bar{f} \\
&\quad + \left(\frac{1}{\upsilon_1^2} - \frac{1}{\upsilon_2^2}\right)^{-1} \frac{\bar{h}}{\sqrt{(p-\alpha)^2 + \beta^2}}, \\
\bar{f}_2 &= \left[\frac{1}{2} + \frac{1}{2}\left(\frac{1}{\upsilon_1^2} - \frac{1}{\upsilon_2^2}\right)^{-1} \frac{\left(\frac{1}{\upsilon_1^2} + \frac{1}{\upsilon_2^2}\right) p + (k_1 + k_2)}{\sqrt{(p-\alpha)^2 + \beta^2}}\right]\bar{f} \\
&\quad - \left(\frac{1}{\upsilon_1^2} - \frac{1}{\upsilon_2^2}\right)^{-1} \frac{\bar{h}}{\sqrt{(p-\alpha)^2 + \beta^2}}.
\end{aligned} \tag{7.2.25}$$

Finally, using the formula

$$\frac{1}{\sqrt{(p-\alpha)^2 + \beta^2}} = \left\{e^{\alpha t} J_0(\beta t)\right\} \equiv \{H_0(t)\}, \tag{7.2.26}$$

where $J_0 = J_0(x)$ is the Bessel function of the first kind and order zero, and the condition (note eqn (7.2.5))

$$f(\cdot, 0) = 0 \text{ on } \partial\mathrm{B}, \tag{7.2.27}$$

and inverting the relations (7.2.25) we obtain

$$\hat{f}_1 = \frac{1}{2}\hat{f}_{11} - \frac{1}{2}\left(\frac{1}{\upsilon_1^2} - \frac{1}{\upsilon_2^2}\right)^{-1}\Bigg[(k_1 + k_2)\,H_0 * \hat{f}_{11}$$
$$+\left(\frac{1}{\upsilon_1^2} + \frac{1}{\upsilon_2^2}\right) H_0 * \dot{\hat{f}}_{11}\Bigg] + \left(\frac{1}{\upsilon_1^2} - \frac{1}{\upsilon_2^2}\right)^{-1} H_0 * \hat{h}_{11},$$
$$\hat{f}_2 = \frac{1}{2}\hat{f}_{22} + \frac{1}{2}\left(\frac{1}{\upsilon_1^2} - \frac{1}{\upsilon_2^2}\right)^{-1}\Bigg[(k_1 + k_2)\,H_0 * \hat{f}_{22}$$
$$+\left(\frac{1}{\upsilon_1^2} + \frac{1}{\upsilon_2^2}\right) H_0 * \dot{\hat{f}}_{22}\Bigg] - \left(\frac{1}{\upsilon_1^2} - \frac{1}{\upsilon_2^2}\right)^{-1} H_0 * \hat{h}_{22}, \tag{7.2.28}$$

where

$$\hat{f}_{11} = f\upsilon_1^2, \quad \hat{h}_{11} = h\upsilon_1^2,$$
$$\hat{f}_{22} = f\upsilon_2^2, \quad \hat{h}_{22} = h\upsilon_2^2, \tag{7.2.29}$$

and (see eqns $(7.1.78)_2$ and $(7.1.88)_2$)

$$\hat{f}_1 = f_1\upsilon_1^2, \quad \hat{f}_2 = f_2\upsilon_1^2. \tag{7.2.30}$$

As a result, the function ϕ given by eqn (7.2.12), where ϕ_1 and ϕ_2 are determined, respectively, by eqns (7.1.80) and (7.1.82), in which $\hat{f}_1$ and $\hat{f}_2$ are given by eqns (7.3.28), is a solution of the initial-boundary value problem (7.2.6)–(7.2.8).

The solution may be represented in a somewhat simpler form, if we note that (see the table of operators in Mikusiński, 1967)

$$\bar{H}_0\bar{G}_k = \{H_0\}\{G_k\} \equiv \{Q_k\}, \tag{7.2.31}$$

where

$$Q_k(t) = \left(\frac{2}{\beta}\right)^k e^{\alpha t} J_k(\beta t), \; k = 0, 1, 2, \ldots \tag{7.2.32}$$

Indeed, multiplying eqn $(7.2.28)_1$ through by G_k and eqn $(7.2.28)_2$ by G_k in the convolution sense, and employing eqn (7.2.31), we find for the convolutions $G_k * \hat{f}_1$ and $G_k * \hat{f}_2$ appearing in eqns (7.1.80) and (7.1.82)

$$M_k^{(1)} \equiv G_k * \hat{f}_1 =$$
$$\frac{1}{2}G_k * \hat{f}_{11} - \frac{1}{2}\left(\frac{1}{\upsilon_1^2} - \frac{1}{\upsilon_2^2}\right)^{-1}\Bigg[(k_1 + k_2)\,Q_k * \hat{f}_{11}$$
$$+\left(\frac{1}{\upsilon_1^2} + \frac{1}{\upsilon_2^2}\right) Q_k * \dot{\hat{f}}_{11}\Bigg] + \left(\frac{1}{\upsilon_1^2} - \frac{1}{\upsilon_2^2}\right)^{-1} Q_k * \hat{h}_{11}, \tag{7.2.33}$$

and

$$M_k^{(2)} \equiv G_k * \hat{f}_2 =$$

$$\frac{1}{2} G_k * \hat{f}_{22} + \frac{1}{2}\left(\frac{1}{v_1^2} - \frac{1}{v_2^2}\right)^{-1} \left[(k_1 + k_2) Q_k * \hat{f}_{22}\right. \tag{7.2.34}$$

$$\left. + \left(\frac{1}{v_1^2} + \frac{1}{v_2^2}\right) Q_k * \dot{\hat{f}}_{22}\right] - \left(\frac{1}{v_1^2} - \frac{1}{v_2^2}\right)^{-1} Q_k * \hat{h}_{22}.$$

Therefore, a solution ϕ of the problem (7.2.6)–(7.2.8) has the form (7.2.12), where ϕ_1 and ϕ_2 are given by

$$\phi_1(x,t) = -\sum_{j=1}^{\infty} \left[\sum_{k=0}^{\infty} (-)^k \hat{\lambda}_1^k H_{jk}^{(1)} * \int_{\partial \mathrm{B}} M_k^{(1)} * \frac{\partial \varphi_j}{\partial n} \mathrm{da}\right] \varphi_j(x), \tag{7.2.35}$$

and

$$\phi_2(x,t) = -\sum_{j=1}^{\infty} \left[\sum_{k=0}^{\infty} \hat{\lambda}_2^k H_{jk}^{(2)} * \int_{\partial \mathrm{B}} M_k^{(2)} * \frac{\partial \varphi_j}{\partial n} \mathrm{da}\right] \varphi_j(x). \tag{7.2.36}$$

Here, $H_{jk}^{(1)}$ and $H_{jk}^{(2)}$ are given by the formulas (7.1.74)–(7.1.76) as well as by (7.1.83)–(7.1.85), while $M_k^{(1)}$ and $M_k^{(2)}$ are specified by eqns (7.2.33) and (7.2.34), respectively.

In the case when $f = 0$ on $\partial \mathrm{B} \times (0, \infty)$, the series (7.2.35) and (7.2.36) reduce to the forms

$$\phi_1(x,t) =$$

$$-\left(\frac{1}{v_1^2} - \frac{1}{v_2^2}\right)^{-1} \sum_{j=1}^{\infty} \left[\sum_{k=0}^{\infty} (-)^k \hat{\lambda}_1^k H_{jk}^{(1)} * Q_k * \int_{\partial \mathrm{B}} \hat{h}_{11} * \frac{\partial \varphi_j}{\partial n} \mathrm{da}\right] \varphi_j(x), \tag{7.2.37}$$

and

$$\phi_2(x,t) =$$

$$\left(\frac{1}{v_1^2} - \frac{1}{v_2^2}\right)^{-1} \sum_{j=1}^{\infty} \left[\sum_{k=0}^{\infty} \hat{\lambda}_2^k H_{jk}^{(2)} * Q_k * \int_{\partial \mathrm{B}} \hat{h}_{22} * \frac{\partial \varphi_j}{\partial n} \mathrm{da}\right] \varphi_j(x), \tag{7.2.38}$$

where

$$\hat{h}_{11} = h v_1^2, \; \hat{h}_{22} = h v_2^2, \tag{7.2.39}$$

$$h = 1 * g + t_1 g. \tag{7.2.40}$$

The knowledge of the function ϕ given by eqn (7.2.12) allows finding ϑ from eqn $(7.2.1)_1$ under the condition $\vartheta(\cdot, 0) = 0$, as well as finding the associated displacements, stresses and heat flux according to the formulas (6.1.18)–(6.1.20).

7.3 Solution for a thermoelastic layer

In this section we show how an exact solution to a 1D initial-boundary value problem of the G–L theory can be obtained by employing a 3D solution to a PTP of Section 7.2; see (Ignaczak, 1981). First, let us take B a layer described by the inequalities:

$$0 \leq x_1 \leq l, \; |x_2| < \infty, \; |x_3| < \infty, \tag{7.3.1}$$

where l is a dimensionless thickness of the layer.[8] Furthermore, we assume the layer to have a "quiescent past" in the sense that the displacement $\mathbf{u}$, the particle velocity $\dot{\mathbf{u}}$, the temperature ϑ and its rate $\dot{\vartheta}$ vanish for $t = 0$ at every point $\mathbf{x} \in \mathrm{B}$. Also, we assume the layer to be loaded on its top and bottom surfaces $x_1 = 0$ and l, respectively, by a smooth-in-time normal traction and temperature that are independent of x_2 and x_3 and have a quiescent past as well, that is they vanish along with their first time derivatives at $t = 0$.

Subject to such loading, the thermoelastic process within the layer is described by a potential $\phi(x_1, t)$, which depends on x_1 and t only, i.e. it is a 1D process. The potential ϕ generates a displacement vector u_i, a stress tensor S_{ij} and a temperature ϑ according to the formulas (recall eqns (6.1.10) and (6.1.18) and (6.1.19), specialized to a 1D process[9])

$$u_1 = \frac{\partial \phi}{\partial x}, \; u_2 = u_3 = 0, \tag{7.3.2}$$

$$\begin{gathered} S_{11} = \frac{\partial^2 \phi}{\partial t^2}, \; S_{22} = S_{33} = \frac{\partial^2 \phi}{\partial t^2} - 2\left(\frac{C_2}{C_1}\right)^2 \frac{\partial^2 \phi}{\partial x^2}, \\ S_{12} = S_{23} = S_{31} = 0, \end{gathered} \tag{7.3.3}$$

$$\left(1 + t_1 \frac{\partial}{\partial t}\right) \vartheta = \left(\frac{\partial^2}{\partial x^2} - \frac{\partial^2}{\partial t^2}\right) \phi. \tag{7.3.4}$$

Furthermore, (recall eqn $(6.1.10)_1$ at $r = 0$)

$$\begin{gathered} \left[\left(\frac{\partial^2}{\partial x^2} - \frac{\partial^2}{\partial t^2}\right)\left(\frac{\partial^2}{\partial x^2} - t_0 \frac{\partial^2}{\partial t^2} - \frac{\partial}{\partial t}\right) - \epsilon \frac{\partial^3}{\partial x^2 \partial t}\left(1 + t_1 \frac{\partial}{\partial t}\right)\right] \phi = 0 \\ \text{for } (x, t) \in [0, l] \times [0, \infty). \end{gathered} \tag{7.3.5}$$

The boundary conditions are

$$S_{11}(0, t) = -\sigma_0(t), \quad S_{11}(l, t) = -\sigma_1(t), \tag{7.3.6}$$

$$\vartheta(0, t) = \vartheta_0(t), \quad \vartheta(l, t) = \vartheta_1(t), \tag{7.3.7}$$

where $\sigma_0(t)$, $\sigma_1(t)$, $\vartheta_0(t)$ and $\vartheta_1(t)$ are functions prescribed on $[0, \infty)$.

[8] Since $\hat{x}_0$ [see eqn (4.1.6)] is the length unit, for a layer of thickness L, $l = L/\hat{x}_0$. In general, $l \neq 1$. For a layer made of aluminum we have $l > 1$ [see the data on aluminum in Section 6.5].

[9] In the following, we identify x_1 with x, i.e. $x = x_1$.

The quiescent past of thermomechanical loading of the layer means that

$$\sigma_0(0) = \dot{\sigma}_0(0) = 0, \quad \sigma_1(0) = \dot{\sigma}_1(0) = 0, \tag{7.3.8}$$

$$\vartheta_0(0) = \dot{\vartheta}_0(0) = 0, \quad \vartheta_1(0) = \dot{\vartheta}_1(0) = 0, \tag{7.3.9}$$

These conditions are consistent with the quiescent past conditions for the layer, which in terms of the function ϕ take the form[10]

$$\frac{\partial^k \phi}{\partial t^k}(x,0) = 0 \; k = 0,1,2,3, \; x \in [0,l]. \tag{7.3.10}$$

The boundary conditions (7.3.6) and (7.3.7), written in terms of ϕ, take the form[11]

$$\begin{aligned} \phi(0,t) &= f_0(t), & \phi(l,t) &= f_1(t), \\ \frac{\partial^2\phi}{\partial x^2}(0,t) &= \dot{h}_0(t), & \frac{\partial^2\phi}{\partial x^2}(l,t) &= \dot{h}_1(t), \end{aligned} \tag{7.3.11}$$

in which

$$\begin{aligned} f_0(t) = -t * \sigma_0, \quad & f_1(t) = -t * \sigma_1, \\ h_0(t) &= 1 * (\vartheta_0 - \sigma_0) + t_1\vartheta_0, \\ h_1(t) &= 1 * (\vartheta_1 - \sigma_1) + t_1\vartheta_1. \end{aligned} \tag{7.3.12}$$

Thus, the problem under consideration reduces to the determination of a function $\phi = \phi(x,t)$ satisfying eqn (7.3.5), the initial conditions (7.3.10) and the boundary conditions (7.3.11). Clearly, it is a 1D equivalent of a PTP of Section 7.2.

The knowledge of the function ϕ allows one to determine ϑ from eqn (7.3.4) subject to the condition $\vartheta(\cdot,0) = 0$. The function ϑ is also a solution of eqn (7.3.5) that satisfies the initial conditions[12]

$$\frac{\partial^k}{\partial t^k}\vartheta(x,0) = 0 \text{ for } k = 0,1,2,3, \; x \in [0,l]. \tag{7.3.13}$$

One can show that the necessary conditions for a pair (ϕ,ϑ) to belong to class $C^4(Q) \cap C^3(\bar{Q})$, where $Q = (0,l) \times (0,\infty)$ and $\bar{Q} = [0,l] \times [0,\infty)$, are, besides eqns (7.3.8) and (7.3.9), the following conditions imposed on the surface temperature[13]

$$\ddot{\vartheta}_0(0) = \dddot{\vartheta}_0(0) = 0, \; \ddot{\vartheta}_1(0) = \dddot{\vartheta}_1(0) = 0. \tag{7.3.14}$$

Thus, for a suitably restricted thermomechanical loading of the layer, the thermoelastic process within it possesses the same smoothness properties as the potential–temperature process of Section 7.2. That is, assuming that the conditions (7.3.8), (7.3.9) and (7.3.14) hold, and proceeding in the same way as

10 The conditions (7.3.10) make a one-dimensional counterpart of eqn (7.2.7).

11 The conditions (7.3.11) make a one-dimensional counterpart of eqn (7.2.8).

12 See eqn (7.2.4).

13 The conditions (7.3.8), (7.3.9) and (7.3.14) correspond to eqn (7.2.5).

in Sections 7.1 and 7.2, as a solution of the 1D problem governed by eqns (7.3.5), (7.3.10) and (7.3.11), we obtain

$$\phi(x,t) = \phi_1(x,t) + \phi_2(x,t), \; (x,t) \in \bar{Q}, \tag{7.3.15}$$

where [see eqns (7.2.35) and (7.2.36)]

$$\phi_1(x,t) = -\sum_{j=1}^{\infty}\left\{\sum_{k=0}^{\infty}(-)^k \hat{\lambda}_1^k H_{jk}^{(1)} * \left[M_k^{(1)} * \varphi_j'(\xi)\right]_{\xi=0}^{\xi=l}\right\}\varphi_j(x), \tag{7.3.16}$$

$$\phi_2(x,t) = -\sum_{j=1}^{\infty}\left\{\sum_{k=0}^{\infty}\hat{\lambda}_2^k H_{jk}^{(2)} * \left[M_k^{(2)} * \varphi_j'(\xi)\right]_{\xi=0}^{\xi=l}\right\}\varphi_j(x). \tag{7.3.17}$$

Here, the prime stands for a derivative with respect to ξ

$$\varphi_j'(\xi) = \frac{d\varphi_j(\xi)}{d\xi}, \tag{7.3.18}$$

while the symbol $[q(\xi)]_{\xi=0}^{\xi=l}0$ denotes

$$[q(\xi)]_{\xi=0}^{\xi=l} = q(l) - q(0). \tag{7.3.19}$$

This expression is equivalent to the surface integral appearing in the formulas (7.2.35) and (7.2.36).

The eigenfunctions $\varphi_j(x)$ and their corresponding eigenvalues λ_j, appearing in the formulas (7.3.16) and (7.3.17), have the form [see p. 216 in (Stakgold, 1968)]

$$\varphi_j(x) = \left(\frac{2}{l}\right)^{\frac{1}{2}}\sin\left(\frac{\pi j x}{l}\right), \; \lambda_j = \left(\frac{\pi j}{l}\right)^2. \tag{7.3.20}$$

The expressions $\left[M_k^{(1)} * \varphi_j'(\xi)\right]_{\xi=0}^{\xi=l}$ and $\left[M_k^{(2)} * \varphi_j'(\xi)\right]_{\xi=0}^{\xi=l}$ appearing in eqns (7.3.16) and (7.3.17), are defined by the formulas [see eqns (7.2.33) and (7.2.34)]

$$\begin{aligned}
v_1^{-2}\left[M_k^{(1)} * \varphi_j'(\xi)\right]_{\xi=0}^{\xi=l} &= \frac{1}{2}G_k * [f_1 * \varphi_j'(l) - f_0 * \varphi_j'(0)] \\
&-\frac{1}{2}\left(\frac{1}{v_1^2} - \frac{1}{v_2^2}\right)^{-1} Q_k * \{(k_1 + k_2)[f_1 * \varphi_j'(l) - f_0 * \varphi_j'(0)] \\
&+\left(\frac{1}{v_1^2} + \frac{1}{v_2^2}\right)[f_1 * \varphi_j'(l) - f_0 * \varphi_j'(0)]\} \\
&+\left(\frac{1}{v_1^2} - \frac{1}{v_2^2}\right)^{-1} Q_k * [h_1 * \varphi_j'(l) - h_0 * \varphi_j'(0)]
\end{aligned} \tag{7.3.21}$$

and

$$
\begin{aligned}
&\upsilon_2^{-2}\left[M_k^{(2)} * \varphi_j'(\xi)\right]_{\xi=0}^{\xi=l} = \frac{1}{2}G_k * \left[f_1 * \varphi_j'(l) - f_0 * \varphi_j'(0)\right] \\
&+\frac{1}{2}\left(\frac{1}{\upsilon_1^2} - \frac{1}{\upsilon_2^2}\right)^{-1} Q_k * \left\{(k_1+k_2)\left[f_1 * \varphi_j'(l) - f_0 * \varphi_j'(0)\right]\right. \\
&\qquad\left.+\left(\frac{1}{\upsilon_1^2} + \frac{1}{\upsilon_2^2}\right)\left[f_1 * \varphi_j'(l) - f_0 * \varphi_j'(0)\right]\right\} \\
&\quad-\left(\frac{1}{\upsilon_1^2} - \frac{1}{\upsilon_2^2}\right)^{-1} Q_k * \left[h_1 * \varphi_j'(l) - h_0 * \varphi_j'(0)\right],
\end{aligned}
\tag{7.3.22}
$$

in which the function $Q_k(t)$ is given by eqn (7.2.32), while $f_0(t)$, $f_1(t)$, $h_0(t)$ and $h_1(t)$ are specified by the formulas (7.3.12). The remaining symbols in the formulas (7.3.16) and (7.3.17) have the same meaning as in eqns (7.2.35) and (7.2.36).

The formulas (7.3.15)–(7.3.22) determine the exact solution of the problem described by eqns (7.3.5), (7.3.10) and (7.3.11). Knowing this solution allows the determination of the displacement vector, stress tensor and temperature via the formulas (7.3.2), (7.3.3) and (7.3.4). The solution (7.3.15)–(7.3.22) simplifies considerably if one assumes the thermomechanical loading of the layer to be symmetric or asymmetric with respect to the middle surface $x = l/2$. In the case of a symmetric loading, that is, for

$$
\sigma_0(t) = \sigma_1(t) \equiv \hat{\sigma}(t), \, \vartheta_0(t) = \vartheta_1(t) \equiv \hat{\vartheta}(t), \tag{7.3.23}
$$

on account of eqn (7.3.12), we obtain

$$
f_0(t) = f_1(t) \equiv f(t), \, h_0(t) = h_1(t) \equiv h(t). \tag{7.3.24}
$$

Since

$$
\varphi_j'(l) = (-)^j \varphi_j'(0), \, \varphi_j'(0) = \left(\frac{2\lambda_j}{l}\right)^{\frac{1}{2}}, \tag{7.3.25}
$$

in this case, for the functions ϕ_1 and ϕ_2, we obtain

$$
\phi_1(x,t) = -\sum_{m=0}^{\infty}\left\{\sum_{k=0}^{\infty}(-)^k \hat{\lambda}_1^k H_{2m+1,k}^{(1)} * \left[M_k^{(1)} * \varphi_{2m+1}'(\xi)\right]_{\xi=0}^{\xi=l}\right\}\varphi_{2m+1}(x) \tag{7.3.26}
$$

and

$$
\phi_2(x,t) = -\sum_{m=0}^{\infty}\left\{\sum_{k=0}^{\infty}\hat{\lambda}_2^k H_{2m+1,k}^{(2)} * \left[M_k^{(2)} * \varphi_{2m+1}'(\xi)\right]_{\xi=0}^{\xi=l}\right\}\varphi_{2m+1}(x), \tag{7.3.27}
$$

where

$$v_{1.2}^{-2}\left[M_k^{(1.2)} * \varphi_{2m+1}'(\xi)\right]_{\xi=0}^{\xi=l} = -\varphi_{2m+1}'(0)\Big\{G_k * 1 * f$$
$$\mp\left(\frac{1}{v_1^2}-\frac{1}{v_2^2}\right)^{-1} Q_k * \left[(k_1+k_2)*f+\left(\frac{1}{v_1^2}+\frac{1}{v_2^2}\right)f\right] \tag{7.3.28}$$
$$\pm 2\left(\frac{1}{v_1^2}-\frac{1}{v_2^2}\right)^{-1} Q_k * 1 * h\Big\}.$$

If, moreover, $\hat{\sigma}(t)=0$, that is when the layer is being heated only at $x=0$ and $x=l$ in a symmetric manner, then eqns (7.3.26) and (7.3.27) reduce to the forms

$$\phi_1(x,t) =$$
$$\frac{4}{l}v_1^2\left(\frac{1}{v_1^2}-\frac{1}{v_2^2}\right)^{-1}\sum_{m=0}^{\infty}\left\{\sum_{k=0}^{\infty}(-)^k\,\hat{\lambda}_1^k\,H_{2m+1,k}^{(1)} * Q_k * 1 * h\right\} \tag{7.3.29}$$
$$\times\left[\frac{(2m+1)\pi}{l}\right]\sin\left[\frac{(2m+1)\pi x}{l}\right],$$

$$\phi_2(x,t) =$$
$$-\frac{4}{l}v_2^2\left(\frac{1}{v_1^2}-\frac{1}{v_2^2}\right)^{-1}\sum_{m=0}^{\infty}\left\{\sum_{k=0}^{\infty}\hat{\lambda}_2^k\,H_{2m+1,k}^{(2)} * Q_k * 1 * h\right\} \tag{7.3.30}$$
$$\times\left[\frac{(2m+1)\pi}{l}\right]\sin\left[\frac{(2m+1)\pi x}{l}\right].$$

On account of eqn $(7.3.3)_1$, the stress component $S_{11}(x,t)$ becomes

$$S_{11}(x,t) =$$
$$\frac{4}{l}\left(\frac{1}{v_1^2}-\frac{1}{v_2^2}\right)^{-1}\sum_{m=0}^{\infty}\left\{\sum_{k=0}^{\infty}\left[(-)^k\,\hat{\lambda}_1^k v_1^2 H_{2m+1,k}^{(1)} - \hat{\lambda}_2^k v_2^2 H_{2m+1,k}^{(2)}\right] * Q_k * \dot{h}\right\}$$
$$\times\left[\frac{(2m+1)\pi}{l}\right]\sin\left[\frac{(2m+1)\pi x}{l}\right], \tag{7.3.31}$$

where

$$\dot{h} = \hat{\vartheta} + t_1\dot{\hat{\vartheta}}. \tag{7.3.32}$$

Furthermore, for the displacement $u_1(x,t)$, in view of eqn (7.3.2), we obtain

$$u_1(x,t) =$$
$$\frac{4}{l}\left(\frac{1}{v_1^2}-\frac{1}{v_2^2}\right)^{-1}\sum_{m=0}^{\infty}\left\{\sum_{k=0}^{\infty}\left[(-)^k\hat{\lambda}_1^k v_1^2 H_{2m+1,k}^{(1)}-\hat{\lambda}_2^k v_2^2 H_{2m+1,k}^{(2)}\right]*Q_k*1*\dot{h}\right\}$$
$$\times\left[\frac{(2m+1)\pi}{l}\right]^2\cos\left[\frac{(2m+1)\pi x}{l}\right]. \tag{7.3.33}$$

It is seen that

$$u_1\left(\frac{l}{2},t\right)=0,\quad u_1(l,t)=-u_1(0,t), \tag{7.3.34}$$

where

$$u_1(0,t) =$$
$$\frac{4}{l}\left(\frac{1}{v_1^2}-\frac{1}{v_2^2}\right)^{-1}\sum_{m=0}^{\infty}\left\{\sum_{k=0}^{\infty}\left[(-)^k\hat{\lambda}_1^k v_1^2 H_{2m+1,k}^{(1)}-\hat{\lambda}_2^k v_2^2 H_{2m+1,k}^{(2)}\right]*Q_k*1*\dot{h}\right\}$$
$$\times\left[\frac{(2m+1)\pi}{l}\right]^2. \tag{7.3.35}$$

It follows from the structure of $H_{2m+1,k}^{(i)}$ and Q_k [recall the formulas (7.1.74)–(7.1.76), (7.1.83)–(7.1.85) and (7.2.32)] that the boundary displacement of the layer is a superposition of damped oscillating functions whose amplitudes tend to zero as time goes to infinity.

7.4 Solution of Nowacki type; spherical wave of a negative order

In Sections 7.2 and 7.3 we presented two exact aperiodic solutions of the G–L theory corresponding to a smooth thermomechanical loading. Now, we turn to the case of a singular thermomechanical loading of a body that produces a closed-form disturbance in the form of a superposition of strong discontinuity thermoelastic waves. This case pertains to a PTP of the G–L theory that in the following will also be called a problem of Nowacki type in the G–L theory. That problem consists in finding a pair (ϕ,ϑ) that satisfies the field equations[14]

$$\left.\begin{aligned}\Gamma\phi &= -\left(1+t^0\frac{\partial}{\partial t}\right)\delta(x-y)\,\delta(t)\\ \left(\nabla^2-\frac{\partial^2}{\partial t^2}\right)\phi &= \left(1+t^0\frac{\partial}{\partial t}\right)\vartheta\end{aligned}\right\}\quad\text{on } E^3\times(0,\infty), \tag{7.4.1}$$

[14] Equations (7.4.1) are obtained from eqn (6.1.10) in which we let $t_1 = t^0$, $r=\delta(x-y)\delta(t)$ and $\mathrm{B}=\mathrm{E}^3$. The replacement of t_1 by t^0 is due to a set of notations characteristic for Section 7.4.

the initial conditions

$$\begin{aligned} \phi(\cdot,0) = \dot{\phi}(\cdot,0) = 0 \\ \vartheta(\cdot,0) = \dot{\vartheta}(\cdot,0) = 0 \end{aligned} \quad \text{on } E^3, x \neq y, \tag{7.4.2}$$

and the appropriate decay conditions at infinity.

In eqn (7.4.1) Γ is the central operator of the G–L theory (recall eqn (6.1.8))

$$\Gamma = \left(\nabla^2 - \frac{\partial^2}{\partial t^2}\right)\left(\nabla^2 - t_0\frac{\partial^2}{\partial t^2} - \frac{\partial}{\partial t}\right) - \epsilon\nabla^2\frac{\partial}{\partial t}\left(1 + t^0\frac{\partial}{\partial t}\right), \tag{7.4.3}$$

where t^0, t_0 and ϵ are the constitutive parameters satisfying the inequalities (recall Section 6.5)

$$t^0 \geq t_0 \geq 1, \quad \epsilon > 0. \tag{7.4.4}$$

Moreover, E^3 is a 3D space, y is a point in that space, and $\delta(x)$ is the Dirac delta. Thus, the pair (ϕ, ϑ) generates disturbances in an infinite thermoelastic body due to a concentrated instantaneous heat source at point y.[15] Of course, a solution of the problem (7.4.1) and (7.4.2) depends on ρ and t only, and, in the following, we let $\phi = \phi(\rho, t)$ and $\vartheta = \vartheta(\rho, t)$, where

$$\rho = |x - y|. \tag{7.4.5}$$

The radial displacement u_ρ and the stresses σ_ρ, σ_θ and σ_φ associated with the pair (ϕ, ϑ) and referred to a spherical co-ordinate system (ρ, θ, φ) centered at y are determined by the formulas (recall eqns (6.1.18) and (6.1.19) specialized to spherical co-ordinates)

$$u_\rho(\rho, t) = \frac{\partial \phi}{\partial \rho} \quad \text{on } E^3 \times (0, \infty). \tag{7.4.6}$$

and

$$\begin{aligned} \sigma_\rho(\rho,t) &= -\frac{2(1-2\nu)}{1-\nu}\frac{1}{\rho}u_\rho + \ddot{\phi} \\ \sigma_\theta(\rho,t) &= \frac{1-2\nu}{1-\nu}\left[\frac{1}{\rho}u_\rho - \left(1 + t^0\frac{\partial}{\partial t}\right)\vartheta\right] + \frac{\nu}{1-\nu}\ddot{\phi} \\ \sigma_\varphi(\rho,t) &= \sigma_\theta(\rho,t) \end{aligned} \quad \text{on } E^3 \times (0,\infty), \tag{7.4.7}$$

where ν is the Poisson ratio. Here, $\sigma_\rho = S_{\rho\rho}$, $\sigma_\theta = S_{\theta\theta}$, and $\sigma_\varphi = S_{\varphi\varphi}$.

It follows from the formulation (7.4.1) and (7.4.2) that the potential ϕ is a solution of the following initial-boundary value problem. Find a function ϕ satisfying the equation

$$\Gamma\phi = -\left(1 + t^0\frac{\partial}{\partial t}\right)\delta(x-y)\,\delta(t) \quad \text{on } E^3 \times (0,\infty), \tag{7.4.8}$$

[15] A closed-form solution to the problem (7.4.1) and (7.4.2) in which $t_1 = t^0 = \epsilon = 0$ was obtained for the first time by Nowacki (1957); see also (Nowacki, 1962, Parkus, 1959).

the initial conditions

$$\frac{\partial^k \phi}{\partial t^k}(\cdot, 0) = 0 \text{ on } E^3 \text{ for } x \neq y \ \forall k = 0, 1, 2, 3, \tag{7.4.9}$$

and appropriate decay conditions at infinity.

Similarly, the temperature ϑ is a solution of the following initial-boundary value problem. Find a function ϑ satisfying the equation[16]

$$\Gamma\vartheta = -\left(\nabla^2 - \frac{\partial^2}{\partial t^2}\right)\delta(x-y)\,\delta(t) \quad \text{on } E^3 \times (0, \infty), \tag{7.4.10}$$

the initial conditions

$$\frac{\partial^k \vartheta}{\partial t^k}(\cdot, 0) = 0 \quad \text{on } E^3 \quad \text{for } x \neq y \quad \forall k = 0, 1, 2, 3, \tag{7.4.11}$$

and appropriate decay conditions at infinity.

Thus, the order of singularity of ϕ is lower by one than the order of singularity of the temperature ϑ. Moreover, ϑ may be found from eqn $(7.4.1)_2$, where ϕ is a known function satisfying the relations (7.4.8) and (7.4.9), or by directly solving the problem (7.4.10) and (7.4.11).

In the following, both problems will be solved in parallel, employing a Laplace transform method presented in (Jakubowska, 1982). Similar to the previous sections, the overbar will denote the transformed function

$$\bar{f}(p) = \int_0^\infty \mathrm{e}^{-pt} f(t)\,\mathrm{d}t, \tag{7.4.12}$$

where $f(t)$ is a function on $[0, \infty)$, while p is the parameter of the transformation. Thus, applying the Laplace transform to eqn (7.4.8) and using the homogeneous initial conditions (7.4.9), we obtain

$$\square_1^2 \square_2^2 \bar{\phi} = -\left(1 + t^0 p\right)\delta(x-y) \text{ on } E^3, \tag{7.4.13}$$

where

$$\square_i^2 = \nabla^2 - s_i^2(p) \ (i = 1, 2), \tag{7.4.14}$$

$$s_{1.2}(p) = \left(\frac{p}{2}\right)^{1/2}\left\{(1+\epsilon) + (1 + t_0 + \epsilon t^o)\,p \pm \Delta^{1/2}\left[(p-\alpha)^2 + \beta^2\right]^{1/2}\right\}^{1/2}, \tag{7.4.15}$$

[16] See eqn (6.1.9) restricted to the case $r = \delta(x-y)\delta(t)$.

and α, β and Δ are the symbols appearing in the theorem on decomposition of the G–L theory (recall the formulas (6.2.4) and (6.2.5))

$$\begin{aligned} \alpha &= -\left[(1+\epsilon)\left(t_0+\epsilon t^0\right)-(1-\epsilon)\right]\Delta^{-1}, \\ \beta &= 2\sqrt{\epsilon}\left[1+(1+\epsilon)\left(t^0-t_0\right)\right]^{1/2}\Delta^{-1}, \\ \Delta &= \left(1-t_0+\epsilon t^0\right)^2+4\epsilon t_0 t^0. \end{aligned} \tag{7.4.16}$$

The only solution of eqn (7.4.13) satisfying the decay conditions at infinity is the function

$$\bar{\phi}(\rho,p) = D\frac{\exp\left[-\rho s_1(p)\right]-\exp\left[-\rho s_2(p)\right]}{4\pi\rho\left[s_1^2(p)-s_2^2(p)\right]}, \tag{7.4.17}$$

where

$$D = 1+t^0 p, \tag{7.4.18}$$

and the complex functions $s_i = s_i(p)$ are chosen so that[17]

$$\mathrm{Re}\left[s_i(p)\right] > 0 \; i=1,2. \tag{7.4.19}$$

Proceeding in the same way when solving the problem (7.4.10) and (7.4.11), we obtain

$$\Box_1^2\Box_2^2\bar{\vartheta} = -\left(\nabla^2-p^2\right)\delta(x-y) \text{ on } E^3, \tag{7.4.20}$$

from which

$$\bar{\vartheta}(\rho,p) = \frac{\left[s_1^2(p)-p^2\right]\exp\left[-\rho s_1(p)\right]-\left[s_2^2(p)-p^2\right]\exp\left[-\rho s_2(p)\right]}{4\pi\rho\left[s_1^2(p)-s_2^2(p)\right]}. \tag{7.4.21}$$

On account of the definition of $s_i(p)$, we reduce the formulas (7.4.17) and (7.4.21) to the form

$$\bar{\phi}(\rho,p) = \frac{1}{4\pi\rho}\left(\frac{1}{v_1^2}-\frac{1}{v_2^2}\right)^{-1}\frac{\mathrm{D}}{p}\left[\bar{M}_1(\rho,p)-\bar{M}_2(\rho,p)\right], \tag{7.4.22}$$

$$\bar{\vartheta}(\rho,p) = \frac{1}{8\pi\rho}\left\{\bar{N}_1(\rho,p)+\bar{N}_2(\rho,p)+\left(\hat{\alpha}+\hat{\beta}p\right)\left[\bar{M}_1(\rho,p)-\bar{M}_2(\rho,p)\right]\right\}, \tag{7.4.23}$$

where the functions $\bar{M}_i(\rho,p)$ and $\bar{N}_i(\rho,p)$ are given by

$$\bar{M}_i(\rho,p) = \frac{\exp\left[-\rho s_i(p)\right]}{\left[(p-\alpha)^2+\beta^2\right]^{1/2}}\;(i=1,2), \tag{7.4.24}$$

and

$$\bar{N}_i(\rho,p) = \exp\left[-\rho s_i(p)\right] \qquad (i=1,2). \tag{7.4.25}$$

[17] $\mathrm{Re}[\bar{f}(p)]$ stands for the real part of the complex-valued function $\bar{f}(p)$.

Moreover, recalling eqn $(6.2.9)_1$,

$$v_{1.2}^{-2} = \frac{1}{2}\left(1 + t_0 + \epsilon t^0 \pm \Delta^{\frac{1}{2}}\right), \tag{7.4.26}$$

and[18]

$$\hat{\alpha} = (1+\epsilon)\,\Delta^{1/2},\ \hat{\beta} = \left(t_0 + \epsilon t^0 - 1\right)\Delta^{1/2}. \tag{7.4.27}$$

It follows from the formulas (7.4.22) and (7.4.23) that $\bar{\phi}(\rho, p)$ and $\bar{\vartheta}(\rho, p)$ take the form

$$\bar{\phi}(\rho, p) = \bar{\phi}_1(\rho, p) + \bar{\phi}_2(\rho, p), \tag{7.4.28}$$

$$\bar{\vartheta}(\rho, p) = \bar{\vartheta}_1(\rho, p) + \bar{\vartheta}_2(\rho, p), \tag{7.4.29}$$

where $\bar{\phi}_i$ and $\bar{\vartheta}_i$ satisfy the equations

$$\Box_i^2 \bar{\phi}_i = 0,\ \Box_i^2 \bar{\vartheta}_i = 0 \text{ for } x \neq y. \tag{7.4.30}$$

Since the operator $\Box_i^2$ corresponds to the wave-type operator $\hat{L}_i$ (recall Section 6.2), the inverse transforms of $\bar{\phi}_i$ and $\bar{\vartheta}_i$ are to be sought in terms of a Neumann-type series for the operator $\hat{L}_i$, that is in terms of a power series with respect to the reduced coefficient of the convolution $\hat{\lambda}_i$ (recall eqn $(6.5.47)_2$). To this end let us recall the definition of the operator $\hat{L}_i$ (recall eqns (7.1.2) and (7.1.3))

$$\hat{L}_{1.2} = \nabla^2 - \frac{1}{v_{1.2}^2}\frac{\partial^2}{\partial t^2} - k_{1.2}\frac{\partial}{\partial t} \mp \lambda \mp K * . \tag{7.4.31}$$

Here, the parameters $v_{1.2}$ are defined by the formulas (7.4.26), while $k_{1.2}$ and λ are

$$k_{1.2} = \frac{1}{2}\left(1 + \epsilon \mp \alpha \Delta^{\frac{1}{2}}\right), \tag{7.4.32}$$

$$\lambda = \frac{1}{4}\beta^2 \Delta^{\frac{1}{2}}, \tag{7.4.33}$$

where α, β and Δ are specified by the formulas (7.4.16). The convolution kernel of $\hat{L}_i$ is given by the formula (recall eqn (6.2.3))

$$K(t) = 2\frac{\mathrm{d}}{\mathrm{d}t}\left[\mathrm{e}^{\alpha t}\frac{J_1(\beta t)}{\beta t}\right]. \tag{7.4.34}$$

[18] The symbols $\hat{\alpha}$ and $\hat{\beta}$ in eqn (7.4.27) must not be confused with such symbols of Sections 6.4 and 6.5.

Applying the Laplace transform to this equation we obtain (see eqns (6.2.27), (6.2.28) and (6.2.31))

$$\bar{K}(p) = \frac{2p}{\beta^2}\bar{\varkappa}(p) - 1 = \frac{1}{\beta^2}\bar{\varkappa}(p)\left[2\alpha - \bar{\varkappa}(p)\right], \tag{7.4.35}$$

$$\bar{\varkappa}(p) = \left[(p-\alpha)^2 + \beta^2\right]^{1/2} - (p-\alpha). \tag{7.4.36}$$

Upon the introduction of notations

$$h_1 = \tfrac{1}{2}k_1 v_1^2,\ \hat{\lambda}_1 = \lambda v_1^2, \tag{7.4.37}$$

$$\omega_1 = \hat{\lambda}_1 - h_1^2,\ b_1^2(p) = \omega_1 + \hat{\lambda}_1\bar{K}(p), \tag{7.4.38}$$

on account of eqn $(7.4.15)_1$, we obtain

$$s_1(p) = \frac{1}{v_1}\left[(p+h_1)^2 + b_1^2(p)\right]^{1/2}. \tag{7.4.39}$$

Similarly, letting

$$h_2 = \tfrac{1}{2}k_2 v_2^2,\ \hat{\lambda}_2 = \lambda v_2^2, \tag{7.4.40}$$

$$\omega_2 = -\hat{\lambda}_2 - h_2^2,\ b_2^2(p) = \omega_2 - \hat{\lambda}_2\bar{K}(p), \tag{7.4.41}$$

eqn $(7.4.15)_2$ is reduced to

$$s_2(p) = \frac{1}{v_2}\left[(p+h_2)^2 + b_2^2(p)\right]^{1/2}. \tag{7.4.42}$$

Clearly, the function $s_2(p)$ may be obtained from $s_1(p)$ by replacing in the formula (7.4.39) v_1 with v_2, k_1 with k_2, and λ with $-\lambda$. Furthermore, the expressions (7.4.37) and (7.4.40) determine, respectively, the damping coefficient and the reduced convolution coefficient for the first and second fundamental thermoelastic disturbances of the G–L theory (recall eqn (6.5.47)).

In order to carry out the inverse transforms of $\bar{\phi}_i$ and $\bar{\vartheta}_i$ (note eqns (7.4.28) and (7.4.29)), we first consider $\bar{N}_i(\rho, p)$ defined by eqn (7.4.25) for $i = 1$. On account of eqn (7.4.39), that function is represented in the form

$$\bar{N}_1(\rho, p) = \bar{n}_1(\rho, p + h_1), \tag{7.4.43}$$

in which

$$\bar{n}_1(\rho, p) = \exp\left\{-\rho_1\left[p^2 + \hat{b}_1^2(p)\right]^{1/2}\right\}, \tag{7.4.44}$$

and

$$\rho_1 = \rho / v_1, \tag{7.4.45}$$

$$\hat{b}_1^2(p) = \omega_1 + \hat{\lambda}_1\bar{K}(p - h_1). \tag{7.4.46}$$

Next, we use the fact that, for arbitrary complex numbers b, z and h, and for $x > 0$, the following formulas are true[19]

$$\exp\left(-x\sqrt{p^2+b^2}\right) = -\frac{\partial}{\partial x}\left[\frac{\exp\left(-x\sqrt{p^2+b^2}\right)}{\sqrt{p^2+b^2}}\right], \tag{7.4.47}$$

$$\frac{\exp\left(-x\sqrt{p^2+b^2}\right)}{\sqrt{p^2+b^2}} = L\left\{J_0\left(b\sqrt{t^2-x^2}\right)H\left(t-x\right)\right\}, \tag{7.4.48}$$

and

$$J_0\left(\sqrt{z+h}\right) = \sum_{n=0}^{\infty}\frac{(-)^n h^n}{2^n n!} z^{-n/2} J_n\left(\sqrt{z}\right). \tag{7.4.49}$$

Here, $J_n = J_n(x)$ $(n \geq 0)$ is the Bessel function of the first kind of order n, $H = H(t)$ is the Heaviside function, while L is the Laplace transform operator

$$Lf(t) = \bar{f}(p) = \int_0^{\infty} \mathrm{e}^{-pt} f(t)\,\mathrm{d}t. \tag{7.4.50}$$

Note here that the formula (7.4.48) holds for an arbitrary complex number b that is not a function of the transform parameter p. Using eqn (7.4.49), we shall now demonstrate that one can generalize that formula onto the case where b is a function of p. Letting

$$z = \left(t^2-\rho_1^2\right)\omega_1,\; h = \left(t^2-\rho_1^2\right)\hat{\lambda}_1\bar{K}\left(p-h_1\right) \tag{7.4.51}$$

in eqn (7.4.49), we obtain

$$J_0\left[\hat{b}_1(p)\sqrt{t^2-\rho_1^2}\right] = \sum_{n=0}^{\infty}\frac{(-)^n\hat{\lambda}_1^n}{2^n n!}\left[\bar{K}\left(p-h_1\right)\right]^n \hat{A}_n\left(\rho_1,t\right), \tag{7.4.52}$$

where

$$\hat{A}_n\left(\rho_1,t\right) = \omega_1^{-n/2}\left(t^2-\rho_1^2\right)^{n/2} J_n\left(\sqrt{\omega_1\left(t^2-\rho_1^2\right)}\right). \tag{7.4.53}$$

Multiplying eqn (7.4.52) through by $\mathrm{e}^{-pt}H\left(t-\rho_1\right)$ and integrating over time from zero to infinity, we find

$$\begin{aligned} &L\left\{J_0\left[\hat{b}_1(p)\sqrt{t^2-\rho_1^2}\right]H\left(t-\rho_1\right)\right\} \\ &= \sum_{n=0}^{\infty}\frac{(-)^n\hat{\lambda}_1^n}{2^n n!}\left[\bar{K}\left(p-h_1\right)\right]^n L\left\{A_n\left(\rho_1,t\right)\right\}, \end{aligned} \tag{7.4.54}$$

[19] The formula (7.4.48) can be found on page 340 in (Mikusiński, 1967), while eqn (7.4.49) is on page 496 in (Watson, 1958).

where

$$A_n(\rho_1, t) = \hat{A}_n(\rho_1, t) H(t - \rho_1). \tag{7.4.55}$$

Hence, an extension of eqn (7.4.48) to the case $b = \hat{b}_1(p)$, $x = \rho_1$, takes the form

$$\frac{\exp\left(-\rho_1\sqrt{p^2 + \hat{b}_1^2(p)}\right)}{\sqrt{p^2 + \hat{b}_1^2(p)}} = \sum_{n=0}^{\infty} \frac{(-)^n \hat{\lambda}_1^n}{2^n n!} \left[\bar{K}(p - h_1)\right]^n L\{A_n(\rho_1, t)\}. \tag{7.4.56}$$

Thus, differentiating eqn (7.4.56) with respect to ρ_1 and using eqn (7.4.44), we conclude that the function $\bar{n}_1(\rho, p)$ possesses the series representation

$$\bar{n}_1(\rho, p) = -\frac{\partial}{\partial \rho_1} \sum_{n=0}^{\infty} \frac{(-)^n \hat{\lambda}_1^n}{2^n n!} \left[\bar{K}(p - h_1)\right]^n L\{A_n(\rho_1, t)\}. \tag{7.4.57}$$

Since, on account of eqn $(7.4.35)_2$,

$$\left[\bar{K}(p - h_1)\right]^n = $$
$$2^n \beta^{-2n} \sum_{k=0}^{n} \binom{n}{k} \frac{(-)^k}{2^k} \alpha^{n-k} \left\{\sqrt{[p - (h_1 + \alpha)]^2 + \beta^2} - [p - (h_1 + \alpha)]\right\}^{n+k}, \tag{7.4.58}$$

and for any $m > 0$ and a complex number a (Mikusiński, 1967)

$$L^{-1}\left(\sqrt{p^2 + a^2} - p\right)^m = m a^m \frac{J_m(at)}{t}, \tag{7.4.59}$$

therefore

$$L^{-1}\left\{\left[\bar{K}(p - h_1)\right]^n\right\} = 2^n h_n^{(1)}(t) \text{ for } n \geq 1, \tag{7.4.60}$$

where

$$h_n^{(1)}(t) = \exp[(\alpha + h_1)t] \sum_{k=0}^{n} \binom{n}{k} \frac{(-)^k}{2^k} \left(\frac{\alpha}{\beta}\right)^{n-k} (n + k) \frac{J_{n+k}(\beta t)}{t}. \tag{7.4.61}$$

From this, given eqn (7.4.43), we obtain the Neumann-type series representation of $N_1(\rho, t)$:

$$N_1(\rho, t) = \delta(t - \rho_1) \exp(-h_1 \rho_1) + P(\rho_1, t), \tag{7.4.62}$$

where

$$P(\rho_1, t) = \rho_1 \exp(-h_1 t) \left\{ -\omega_1 \frac{J_1\left(\sqrt{\omega_1 (t^2 - \rho_1^2)}\right)}{\sqrt{\omega_1 (t^2 - \rho_1^2)}} + \sum_{n=1}^{\infty} \frac{(-)^n}{n!} \frac{\hat{\lambda}_1^n}{\omega_1^{n-1}} \int_{\rho_1}^{t} h_n^{(1)}(t-s) \right.$$
$$\left. \times \left(\sqrt{\omega_1 (s^2 - \rho_1^2)}\right)^{n-1} J_{n-1}\left(\sqrt{\omega_1 (s^2 - \rho_1^2)}\right) ds \right\} H(t - \rho_1). \tag{7.4.63}$$

Evidently, this is a power series in $\hat{\lambda}_1$, the reduced convolution coefficient for the wave-like operator $\hat{L}_1$. This series is formally well defined for $\omega_1 \geq 0$, that is for $\lambda \geq \lambda_0$ where

$$\lambda_0 = h_1^2 v_1^{-2}. \tag{7.4.64}$$

One can show that for $0 < \lambda \leq \lambda_0$ $(\omega_1 \leq 0)$

$$N_1(\rho, t) = \delta(t - \rho_1) \exp(-h_1 \rho_1) + \hat{P}(\rho_1, t), \tag{7.4.65}$$

where

$$\hat{P}(\rho_1, t) = \rho_1 \exp(-h_1 t) \left\{ \hat{\omega}_1 \frac{I_1\left(\sqrt{\hat{\omega}_1 (t^2 - \rho_1^2)}\right)}{\sqrt{\hat{\omega}_1 (t^2 - \rho_1^2)}} + \sum_{n=1}^{\infty} \frac{(-)^n}{n!} \frac{\hat{\lambda}_1^n}{\hat{\omega}_1^{n-1}} \int_{\rho_1}^{t} h_n^{(1)}(t-s) \right.$$
$$\left. \times \left(\sqrt{\hat{\omega}_1 (s^2 - \rho_1^2)}\right)^{n-1} I_{n-1}\left(\sqrt{\hat{\omega}_1 (s^2 - \rho_1^2)}\right) \mathrm{d}s \right\} H(t - \rho_1). \tag{7.4.66}$$

Here, $I_n = I_n(x)$ $(n \geq 0)$ is the modified Bessel function of the first kind of order n, and

$$\hat{\omega}_1 = -\omega_1 \geq 0. \tag{7.4.67}$$

In the limiting case $\lambda = \lambda_0$, that is for $\omega_1 = 0$, from the formula (7.4.62) or (7.4.65) we obtain

$$N_1(\rho, t) = \delta(t - \rho_1) \exp(-h_1 \rho_1) + P_0(\rho_1, t), \tag{7.4.68}$$

where

$$P_0(\rho_1, t) = \rho_1 \exp(-h_1 t) \left\{ \sum_{n=1}^{\infty} \frac{(-)^n}{n!} \frac{\hat{\lambda}_0^n}{2^{n-1}(n-1)!} \right.$$
$$\left. \times \int_{\rho_1}^{t} h_n^{(1)}(t-s) \left(s^2 - \rho_1^2\right)^{n-1} \mathrm{d}s \right\} H(t - \rho_1), \tag{7.4.69}$$

and

$$\hat{\lambda}_0 = \lambda_0 v_1^2. \tag{7.4.70}$$

In order to find the inverse transform of $\bar{N}_2(\rho_1, p)$ given by eqn (7.4.25) for $i = 2$, it suffices to replace v_1 by v_2, k_1 by k_2, and λ by $-\lambda$ in eqn (7.4.65). Thus, we obtain

$$N_2(\rho, t) = \delta(t - \rho_2)\exp(-h_2\rho_2) + R(\rho_2, t), \tag{7.4.71}$$

where

$$R(\rho_2, t) = \rho_2 \exp(-h_2 t)\left\{\hat{\omega}_2 \frac{I_1\left(\sqrt{\hat{\omega}_2(t^2 - \rho_2^2)}\right)}{\sqrt{\hat{\omega}_2(t^2 - \rho_2^2)}} + \sum_{n=1}^{\infty}\frac{\hat{\lambda}_2^n}{n!\hat{\omega}_2^{n-1}}\int_{\rho_2}^{t} h_n^{(2)}(t-s) \times \left(\sqrt{\hat{\omega}_2(s^2 - \rho_2^2)}\right)^{n-1} I_{n-1}\left(\sqrt{\hat{\omega}_2(s^2 - \rho_2^2)}\right) ds\right\} H(t - \rho_2). \tag{7.4.72}$$

The symbols ρ_2, $\hat{\omega}_2$ and $h_n^{(2)}(t)$ appearing in eqn (7.4.72) are given by

$$\rho_2 = \rho/v_2, \; \hat{\omega}_2 = -\omega_2, \tag{7.4.73}$$

$$h_n^{(2)}(t) = \exp[(\alpha + h_2)t]\sum_{k=0}^{n}\binom{n}{k}\frac{(-)^k}{2^k}\left(\frac{\alpha}{\beta}\right)^{n-k}(n+k)\frac{J_{n+k}(\beta t)}{t}. \tag{7.4.74}$$

Clearly, the function $N_2(\rho, t)$ is a Neumann-type series for the operator $\hat{L}_2$, that is a power series in $\hat{\lambda}_2$. In contradistinction to the function $N_1(\rho, t)$, which is determined by two different formulas, each valid separately depending on the range of the parameter $\hat{\lambda}_1$, the function $N_2(\rho, t)$ is represented by a single series (7.4.71).

We shall now invert the function $\bar{M}_i(\rho, p)$ using the method analogous to that used for $\bar{N}_i(\rho, p)$. First, we note that

$$\bar{M}_1(\rho, p) = L\{\exp(-h_1 t)\, m_1(\rho, t)\}, \tag{7.4.75}$$

where

$$m_1(\rho, t) = L^{-1}\left\{\frac{\exp\left[-\rho_1\sqrt{p^2 + \hat{b}_1^2(p)}\right]}{\sqrt{[p - (\alpha + h_1)]^2 + \beta^2}}\right\}. \tag{7.4.76}$$

From this, on account of eqn (7.4.56), we find

$$m_1(\rho, t) = -L^{-1}\left\{\sum_{n=0}^{\infty}\frac{(-)^n\hat{\lambda}_1^n}{2^n n!}\left[\bar{B}(p)\right]^n L\left(\frac{\partial A_n}{\partial \rho_1}\right)\right\}, \tag{7.4.77}$$

where

$$\left[\bar{B}\left(p\right)\right]^{n}=\left[\bar{K}\left(p-h_{1}\right)\right]^{n}\left\{\left[p-\left(\alpha+h_{1}\right)\right]^{2}+\beta^{2}\right\}^{-1/2}. \tag{7.4.78}$$

Since for any complex number a, and for $m \geq 0$ (Mikusiński, 1967)

$$L^{-1}\left\{\frac{\left(\sqrt{p^{2}+a^{2}}-p\right)^{m}}{\sqrt{p^{2}+a^{2}}}\right\}=a^{m}J_{m}\left(at\right), \tag{7.4.79}$$

operating with L^{-1} on eqn (7.4.78) and employing eqn (7.4.58), we obtain

$$L^{-1}\left\{\left[\bar{B}\left(p\right)\right]^{n}\right\}=\exp\left[\left(\alpha+h_{1}\right)t\right]2^{n}\sum_{k=0}^{n}\binom{n}{k}\frac{\left(-\right)^{k}}{2^{k}}\left(\frac{\alpha}{\beta}\right)^{n-k}J_{n+k}\left(\beta t\right). \tag{7.4.80}$$

As a result, the operation L^{-1} on eqn (7.4.75) leads to the following Neumann-type series for $M_1(\rho, t)$

$$\begin{aligned}M_{1}\left(\rho,t\right)=\exp\left(-h_{1}t\right)\Bigg\{&J_{0}\left[\beta\left(t-\rho_{1}\right)\right]\exp\left[\left(\alpha+h_{1}\right)\left(t-\rho_{1}\right)\right]\\&+\rho_{1}\sum_{n=0}^{\infty}\frac{\left(-\right)^{n}\hat{\lambda}_{1}^{n}}{n!\omega_{1}^{n-1}}\int_{\rho_{1}}^{t}g_{n}^{(1)}\left(t-s\right)\\\times\left(\sqrt{\omega_{1}\left(s^{2}-\rho_{1}^{2}\right)}\right)^{n-1}&J_{n-1}\left(\sqrt{\omega_{1}\left(s^{2}-\rho_{1}^{2}\right)}\right)\mathrm{d}s\Bigg\}H\left(t-\rho_{1}\right),\end{aligned} \tag{7.4.81}$$

where

$$g_{n}^{(1)}(t)=\exp\left[\left(\alpha+h_{1}\right)t\right]\sum_{k=0}^{n}\binom{n}{k}\frac{\left(-\right)^{k}}{2^{k}}\left(\frac{\alpha}{\beta}\right)^{n-k}J_{n+k}\left(\beta t\right). \tag{7.4.82}$$

Equation (7.4.81) formally applies for $\lambda \geq \lambda_0$ ($\omega_1 \geq 0$). For $0 < \lambda \leq \lambda_0$ we find

$$\begin{aligned}M_{1}\left(\rho,t\right)=\exp\left(-h_{1}t\right)\Bigg\{&J_{0}\left[\beta\left(t-\rho_{1}\right)\right]\exp\left[\left(\alpha+h_{1}\right)\left(t-\rho_{1}\right)\right]\\&+\rho_{1}\sum_{n=0}^{\infty}\frac{\left(-\right)^{n}\hat{\lambda}_{1}^{n}}{n!\hat{\omega}_{1}^{n-1}}\int_{\rho_{1}}^{t}g_{n}^{(1)}\left(t-s\right)\\\times\left(\sqrt{\hat{\omega}_{1}\left(s^{2}-\rho_{1}^{2}\right)}\right)^{n-1}&I_{n-1}\left(\sqrt{\hat{\omega}_{1}\left(s^{2}-\rho_{1}^{2}\right)}\right)\mathrm{d}s\Bigg\}H\left(t-\rho_{1}\right).\end{aligned} \tag{7.4.83}$$

Taking the limit $\lambda \to \lambda_0$ in eqn (7.4.81) or eqn (7.4.83), yields

$$\begin{aligned}M_{1}\left(\rho,t\right)=\exp\left(-h_{1}t\right)\Bigg\{&J_{0}\left[\beta\left(t-\rho_{1}\right)\right]\exp\left[\left(\alpha+h_{1}\right)\left(t-\rho_{1}\right)\right]\\+\rho_{1}\sum_{n=0}^{\infty}\frac{\left(-\right)^{n}\hat{\lambda}_{0}^{n}}{n!\left(n-1\right)!\,2^{n-1}}&\int_{\rho_{1}}^{t}g_{n}^{(1)}\left(t-s\right)\left(s^{2}-\rho_{1}^{2}\right)^{n-1}\mathrm{d}s\Bigg\}H\left(t-\rho_{1}\right).\end{aligned} \tag{7.4.84}$$

Clearly, the function $M_1(\rho,t)$ is a power series in $\hat{\lambda}_1$, a reduced convolution coefficient for the wave-like operator $\hat{L}_1$. In order to find the inverse transform of $\bar{M}_2(\rho_1,p)$, it suffices to replace υ_1 by υ_2, k_1 by k_2, and λ by $-\lambda$ in eqn (7.4.83). Thus, we find

$$M_2(\rho,t) = \exp(-h_2 t)\left\{J_0[\beta(t-\rho_2)]\exp[(\alpha+h_2)(t-\rho_2)] + \rho_2\sum_{n=0}^{\infty}\frac{\hat{\lambda}_2^n}{n!\hat{\omega}_2^{n-1}}\int_{\rho_2}^{t} g_n^{(2)}(t-s) \times\left(\sqrt{\hat{\omega}_2(s^2-\rho_2^2)}\right)^{n-1} I_{n-1}\left(\sqrt{\hat{\omega}_2(s^2-\rho_2^2)}\right)\mathrm{d}s\right\}H(t-\rho_2), \tag{7.4.85}$$

where

$$g_n^{(2)}(t) = \exp[(\alpha+h_2)t]\sum_{k=0}^{n}\binom{n}{k}\frac{(-)^k}{2^k}\left(\frac{\alpha}{\beta}\right)^{n-k} J_{n+k}(\beta t). \tag{7.4.86}$$

Finally, operating with L^{-1} through on eqns (7.4.22) and (7.4.23), yields the solution to the problem (7.4.1) and (7.4.2)

$$\phi(\rho,t) = \frac{1}{4\pi\rho}\left(\frac{1}{\upsilon_1^2}-\frac{1}{\upsilon_2^2}\right)^{-1} * \left(1+t^0\frac{\partial}{\partial t}\right)[M_1(\rho,t)-M_2(\rho,t)], \tag{7.4.87}$$

and

$$\vartheta(\rho,t) = \frac{1}{8\pi\rho}\left\{N_1(\rho,t)+N_2(\rho,t)+\left(\hat{\alpha}+\hat{\beta}\frac{\partial}{\partial t}\right)[M_1(\rho,t)-M_2(\rho,t)]\right\}. \tag{7.4.88}$$

Here, the function $N_1(\rho,t)$ is given by the formula (7.4.65) for $0<\lambda<\lambda_0$, by eqn (7.4.68) for $\lambda=\lambda_0$, and by eqn (7.4.62) for $\lambda>\lambda_0$. On the other hand, the function $M_1(\rho,t)$ is given by the formula (7.4.83) for $0<\lambda<\lambda_0$, by eqn (7.4.84) for $\lambda=\lambda_0$, and by eqn (7.4.81) for $\lambda>\lambda_0$. Finally, $N_2(\rho,t)$ and $M_2(\rho,t)$ are given, respectively, by eqns (7.4.71) and (7.4.85) for every $\lambda>0$.

Analysis of the solution

An analysis of solution to the problem (7.4.1) and (7.4.2) is to be limited to the case $\lambda>\lambda_0$ ($\omega_1>0$), since for $0<\lambda\le\lambda_0$ ($\omega_1\le 0$) it may be conducted in an analogous way. First, note that the formulas (7.4.87) and (7.4.88) may be represented as follows

$$\phi(\rho,t) = \phi_1(\rho_1,t)+\phi_2(\rho_2,t), \tag{7.4.89}$$

$$\vartheta(\rho,t) = \vartheta_1(\rho_1,t)+\vartheta_2(\rho_2,t), \tag{7.4.90}$$

where

$$\phi_1(\rho_1,t) = \frac{1}{4\pi\upsilon_1}\left(\frac{1}{\upsilon_1^2}-\frac{1}{\upsilon_2^2}\right)^{-1} * \left(1+t^0\frac{\partial}{\partial t}\right) \times \{\exp(-h_1 t)\,\mathcal{M}_1(\rho_1,t)\,H(\zeta_1)\}, \tag{7.4.91}$$

$$\phi_2(\rho_2,t) = -\frac{1}{4\pi\upsilon_2}\left(\frac{1}{\upsilon_1^2}-\frac{1}{\upsilon_2^2}\right)^{-1} * \left(1+t^0\frac{\partial}{\partial t}\right) \times \{\exp(-h_2 t)\,\mathcal{M}_2(\rho_2,t)\,H(\zeta_2)\}, \tag{7.4.92}$$

$$\begin{aligned}\vartheta_1(\rho_1,t) = \frac{1}{8\pi\upsilon_1}\Big\{&\rho_1^{-1}\exp(-h_1\rho_1)\,\delta(\zeta_1) + \exp(-h_1 t)\,\mathcal{N}_1(\rho_1,t)\,H(\zeta_1)\\ &+\left(\hat{\alpha}+\hat{\beta}\frac{\partial}{\partial t}\right)[\exp(-h_1 t)\,\mathcal{M}_1(\rho_1,t)\,H(\zeta_1)]\Big\},\end{aligned} \tag{7.4.93}$$

$$\begin{aligned}\vartheta_2(\rho_2,t) = \frac{1}{8\pi\upsilon_2}\Big\{&\rho_2^{-1}\exp(-h_2\rho_2)\,\delta(\zeta_2) + \exp(-h_2 t)\,\mathcal{N}_2(\rho_2,t)\,H(\zeta_2)\\ &-\left(\hat{\alpha}+\hat{\beta}\frac{\partial}{\partial t}\right)[\exp(-h_2 t)\,\mathcal{M}_2(\rho_2,t)\,H(\zeta_2)]\Big\}.\end{aligned} \tag{7.4.94}$$

In the above $\mathcal{M}_i(\rho_i,t)$ and $\mathcal{N}_i(\rho_i,t)$ $(i=1,2)$ are the following power series in $\hat{\lambda}_i$

$$\mathcal{M}_1(\rho_1,t) = \rho_1^{-1}J_0(\beta\zeta_1)\exp[(\alpha+h_1)\zeta_1] + \sum_{n=0}^{\infty}\frac{(-)^n\hat{\lambda}_1^n}{n!\omega_1^{n-1}}\int_{\rho_1}^{t} g_n^{(1)}(t-s)\,z_1^{n-1}J_{n-1}(z_1)\,\mathrm{d}s, \tag{7.4.95}$$

$$\mathcal{M}_2(\rho_2,t) = \rho_2^{-1}J_0(\beta\zeta_2)\exp[(\alpha+h_2)\zeta_2] + \sum_{n=0}^{\infty}\frac{\hat{\lambda}_2^n}{n!\hat{\omega}_2^{n-1}}\int_{\rho_2}^{t} g_n^{(2)}(t-s)\,z_2^{n-1}I_{n-1}(z_2)\,\mathrm{d}s, \tag{7.4.96}$$

$$\mathcal{N}_1(\rho_1,t) = -\omega_1\frac{J_1(\hat{z}_1)}{\hat{z}_1} + \sum_{n=1}^{\infty}\frac{(-)^n\hat{\lambda}_1^n}{n!\omega_1^{n-1}}\int_{\rho_1}^{t} h_n^{(1)}(t-s)\,z_1^{n-1}J_{n-1}(z_1)\,\mathrm{d}s, \tag{7.4.97}$$

$$\mathcal{N}_2(\rho_2,t) = \hat{\omega}_2\frac{I_1(\hat{z}_2)}{\hat{z}_2} + \sum_{n=1}^{\infty}\frac{\hat{\lambda}_2^n}{n!\hat{\omega}_2^{n-1}}\int_{\rho_2}^{t} h_n^{(2)}(t-s)\,z_2^{n-1}I_{n-1}(z_2)\,\mathrm{d}s, \tag{7.4.98}$$

in which

$$\zeta_i = t - \rho_i \ (i = 1, 2), \tag{7.4.99}$$

and

$$\begin{aligned} z_1 &= \left[\omega_1 \left(s^2 - \rho_1^2\right)\right]^{1/2}, \ z_2 = \left[\hat{\omega}_2 \left(s^2 - \rho_2^2\right)\right]^{1/2}, \\ \hat{z}_1 &= \left[\omega_1 \left(t^2 - \rho_1^2\right)\right]^{1/2}, \ \hat{z}_2 = \left[\hat{\omega}_2 \left(t^2 - \rho_2^2\right)\right]^{1/2}. \end{aligned} \tag{7.4.100}$$

It can be shown that the pair (ϕ, ϑ) given by the formulas (7.4.89)–(7.4.100) is a solution of the problem (7.4.1) and (7.4.2). A proof of this theorem is based on the following lemma.

Lemma 7.1 *The series $\mathcal{M}_i(\rho_i, t)$ and $\mathcal{N}_i(\rho_i, t)$ and their partial derivatives of a finite order are convergent for every $t \geq \rho_i > 0$ $(i = 1, 2)$.*

Proof. The proof is limited to the analysis of convergence of the series $\mathcal{M}_1(\rho_1, t)$, since the other cases are analogous. To this end consider the series defining $\mathcal{M}_1(\rho_1, t)$:[20]

$$Q(\rho_1, t) = \sum_{n=1}^{\infty} \frac{(-)^n \hat{\lambda}_1^n}{n! \omega_1^{n-1}} \int_{\rho_1}^{t} g_n^{(1)}(t - s)\, z_1^{n-1} J_{n-1}(z_1)\, \mathrm{d}s. \tag{7.4.101}$$

Since for every $x > 0$ [note the formula (9.1.62) on page 362 in (Abramowitz and Stegun, 1965)]

$$|J_m(x)| \leq \frac{x^m}{2^m m!}, \quad m \geq 0, \tag{7.4.102}$$

therefore

$$|Q(\rho_1, t)| \leq \int_{\rho_1}^{t} \sum_{n=1}^{\infty} \frac{\hat{\lambda}_1^n}{n!} \frac{\left|g_n^{(1)}(t - s)\right|}{|\omega_1|^{n-1}} \frac{z_1^{2(n-1)}}{2^{n-1}(n-1)!} \mathrm{d}s \tag{7.4.103}$$

or, using the definition of z_1 (recall eqn $(7.4.100)_1$)

$$|Q(\rho_1, t)| \leq \int_{\rho_1}^{t} \sum_{n=1}^{\infty} \frac{\hat{\lambda}_1^n \left(s^2 - \rho_1^2\right)^{n-1}}{n!\, 2^{n-1}(n-1)!} \left|g_n^{(1)}(t - s)\right| \mathrm{d}s. \tag{7.4.104}$$

If we introduce the notation

$$e_1(t) = \exp\left[(\alpha + h_1) t\right], \tag{7.4.105}$$

[20] The series $Q(\rho_1, t)$ differs from $\mathcal{M}_1(\rho_1, t)$ by the first two terms; therefore if $Q(\rho_1, t)$ converges then $\mathcal{M}_1(\rho_1, t)$ also converges.

and use the definition of $g_n^{(1)}(t)$ (recall eqn (7.4.82)) and the estimate (7.4.102), we find that

$$\begin{aligned}\left|g_n^{(1)}(t-s)\right| &\leq e_1(t-s)\sum_{k=0}^{n}\binom{n}{k}\frac{1}{2^k}\left|\frac{\alpha}{\beta}\right|^{n-k}\frac{[\beta(t-s)]^{n+k}}{2^{n+k}(n+k)!} \\ &\leq e_1(t-s)\frac{[\beta(t-s)]^n}{2^n}\sum_{k=0}^{n}\binom{n}{k}\left[\frac{\beta(t-s)}{4}\right]^k\left|\frac{\alpha}{\beta}\right|^{n-k} \\ &= e_1(t-s)\left[\frac{\beta(t-s)}{2}\right]^n\left[\left|\frac{\alpha}{\beta}\right|+\frac{\beta(t-s)}{4}\right]^n.\end{aligned} \tag{7.4.106}$$

Thus, on account of eqns (7.4.104) and (7.4.106), we obtain

$$|Q(\rho_1,t)| \leq \int_{\rho_1}^{t}\hat{e}_1(t-s)\sum_{m=0}^{\infty}\frac{[z(\rho_1,t;s)]^m}{m!}ds, \tag{7.4.107}$$

in which

$$\hat{e}_1(t-s) = e_1(t-s)\,\hat{\lambda}_1\left[\frac{\beta(t-s)}{2}\right]\left[\left|\frac{\alpha}{\beta}\right|+\frac{\beta(t-s)}{4}\right], \tag{7.4.108}$$

and

$$z(\rho_1,t;s) = \hat{\lambda}_1\left(\frac{s^2-\rho_1^2}{2}\right)\left[\frac{\beta(t-s)}{2}\right]\left[\left|\frac{\alpha}{\beta}\right|+\frac{\beta(t-s)}{4}\right]. \tag{7.4.109}$$

Since the functions $\hat{e}_1(t-s)$ and $z(\rho_1,t;s)$, treated as functions of the variable s, are continuous in the interval $[\rho_1,t]$, there exist the functions

$$e_M(\rho_1,t) = \max_{s\epsilon[\rho_1,t]}\{\hat{e}_1(t-s)\}, \tag{7.4.110}$$

$$z_M(\rho_1,t) = \max_{s\epsilon[\rho_1,t]}\{z(\rho_1,t;s)\}. \tag{7.4.111}$$

Hence, on account of eqn (7.4.107), we find

$$|Q(\rho_1,t)| \leq (t-\rho_1)\,e_M(\rho_1,t)\exp[z_M(\rho_1,t)]. \tag{7.4.112}$$

This means that the series $Q(\rho_1,t)$ is bounded from above by a power series convergent for every $t \geq \rho_1 > 0$. Thus, $Q(\rho_1,t)$ is uniformly convergent for every $t \geq \rho_1 > 0$. □

Note that the uniform convergence of $Q(\rho_1,t)$ holds for the parameters t^0, t_0 and ϵ satisfying the inequalities[21]

$$t^0 \geq t_0 > 1,\ \epsilon > 0. \tag{7.4.113}$$

Thus, by Lemma 7.1, the thermoelastic disturbances associated with the pair (ϕ,ϑ) can be represented, outside the surfaces $\zeta_i = 0$ $(i=1,2)$, by a uniformly

[21] The inqualities (7.4.113) imply that $\alpha < 0$, that is a thermoelastic disturbance generated by the pair (ϕ,ϑ) vanishes as $t \to +\infty$.

convergent series in the whole space and for a wide range of parameters satisfying the constitutive inequalities (7.4.113).

In order to examine the displacement–temperature disturbances in a neighborhood of the surface $\zeta_i = 0$, we employ the formulas (7.4.89)–(7.4.94) and express the functions $u_\rho(\rho, t)$ and $\vartheta(\rho, t)$ as

$$\begin{aligned} u_\rho(\rho, t) &= u_\rho^{(1)}(\rho_1, t) + u_\rho^{(2)}(\rho_2, t), \\ \vartheta(\rho, t) &= \vartheta_1(\rho_1, t) + \vartheta_2(\rho_2, t), \end{aligned} \tag{7.4.114}$$

where

$$\begin{aligned} u_\rho^{(1)}(\rho_1, t) &= \frac{1}{4\pi v_1^2}\left(\frac{1}{v_1^2} - \frac{1}{v_2^2}\right)^{-1} * \left(1 + t^0\frac{\partial}{\partial t}\right) \\ &\times\Big[-\rho_1^{-1}\exp(-h_1\rho_1)\,\delta(\zeta_1) + \exp(-h_1 t)\,\mathcal{M}_1'(\rho_1, t)\,H(\zeta_1)\Big], \end{aligned} \tag{7.4.115}$$

$$\begin{aligned} u_\rho^{(2)}(\rho_2, t) &= -\frac{1}{4\pi v_2^2}\left(\frac{1}{v_1^2} - \frac{1}{v_2^2}\right)^{-1} * \left(1 + t^0\frac{\partial}{\partial t}\right) \\ &\times\Big[-\rho_2^{-1}\exp(-h_2\rho_2)\,\delta(\zeta_2) + \exp(-h_2 t)\,\mathcal{M}_2'(\rho_2, t)\,H(\zeta_2)\Big], \end{aligned} \tag{7.4.116}$$

and

$$\begin{aligned} \vartheta_1(\rho_1, t) &= \frac{1}{8\pi v_1}\Big\{(1+\hat{\beta})\rho_1^{-1}\exp(-h_1\rho_1)\,\delta(\zeta_1) \\ &+ \exp(-h_1 t)\Big[\mathcal{N}_1(\rho_1, t) + (\hat{\alpha} - h_1\hat{\beta})\mathcal{M}_1(\rho_1, t) + \hat{\beta}\dot{\mathcal{M}}_1(\rho_1, t)\Big]H(\zeta_1)\Big\}, \end{aligned} \tag{7.4.117}$$

$$\begin{aligned} \vartheta_2(\rho_1, t) &= \frac{1}{8\pi v_2}\Big\{(1-\hat{\beta})\rho_2^{-1}\exp(-h_2\rho_2)\,\delta(\zeta_2) \\ &+ \exp(-h_2 t)\Big[\mathcal{N}_2(\rho_2, t) - (\hat{\alpha} - h_2\hat{\beta})\mathcal{M}_2(\rho_1, t) + \hat{\beta}\dot{\mathcal{M}}_2(\rho_2, t)\Big]H(\zeta_2)\Big\}. \end{aligned} \tag{7.4.118}$$

In the above, the prime denotes a partial derivative with respect to ρ_i $(i = 1, 2)$, while the dot a derivative with respect to t.

The formulas (7.4.115)–(7.4.118) together with the Lemma 7.1 imply that the pair $\left(u_\rho^{(i)}, \vartheta_i\right)$ represents a wave of strong discontinuity whose wavefront is to be identified with the moving surface ζ_i and whose jump across the front is infinite. Integrating these formulas over time and computing the jumps of functions $1 * u_\rho^{(i)}$ and $1 * \vartheta_i$ at $\zeta_i = 0$, we obtain

$$\left[\left[1 * u_\rho^{(1)}\right]\right]_1(t) = -\frac{t^0}{4\pi v_1^2}\left(\frac{1}{v_1^2} - \frac{1}{v_2^2}\right)^{-1}\frac{\exp(-h_1 t)}{t}, \tag{7.4.119}$$

$$\left[\left[1 * u_\rho^{(2)}\right]\right]_2(t) = \frac{t^0}{4\pi v_2^2}\left(\frac{1}{v_1^2} - \frac{1}{v_2^2}\right)^{-1}\frac{\exp(-h_2 t)}{t}, \tag{7.4.120}$$

and

$$[[1 * \vartheta_1]]_1 (t) = \frac{1}{8\pi v_1} \left(1 + \hat{\beta}\right) \frac{\exp(-h_1 t)}{t}, \qquad (7.4.121)$$

$$[[1 * \vartheta_2]]_2 (t) = \frac{1}{8\pi v_2} \left(1 - \hat{\beta}\right) \frac{\exp(-h_2 t)}{t}, \qquad (7.4.122)$$

where the jump of a function $f_i(\rho_i, t)$ on the surface $\zeta_i = 0$ is given by the formula

$$[[f_i]]_i (t) = f_i (t - 0, t) - f_i (t + 0, t). \qquad (7.4.123)$$

The relations (7.4.119)–(7.4.122) imply that the surface $\zeta_i = 0$ is a singular surface of the order zero relative to the pair $\left(1 * u_\rho^{(i)},\ 1 * \vartheta_i\right)$, so that it is a singular surface of the order $n = -1$ relative to the pair $\left(u_\rho^{(i)}, \vartheta_i\right)$; see (Ignaczak, 1985). We conclude from this that the following theorem holds.

Theorem 7.1 *A displacement–temperature wave in the problem of Nowacki type in the G–L theory is a superposition of two spherical waves, each of the order $n = -1$.*

Remark 7.3 If in the formulation of the initial-boundary value problem (7.4.1) and (7.4.2) one takes a concentrated heat source whose intensity is a prescribed function of time $t \geq 0$, then, proceeding in a similar way as in the problem of Nowacki type, we arrive at the following conclusions.[22]

Conclusion 7.1 *The inclusion of a concentrated heat source with a time dependence of the Heaviside function type in the infinite thermoelastic body gives rise to a thermoelastic wave in the form of a superposition of two spherical dislocation-type waves, that is, waves of the order zero with respect to the displacement and temperature fields.*

Conclusion 7.2 *The inclusion of a concentrated heat source with a time dependence of the ramp function type in the infinite thermoelastic body gives rise to a thermoelastic wave being a superposition of two spherical shock waves, that is, waves of the order one with respect to the displacement and temperature fields.*

Conclusion 7.3 *Only the inclusion of a concentrated heat source whose intensity and its first derivative vanish at the initial time in the infinite thermoelastic body gives rise to a thermoelastic wave in the form of a superposition of two weak discontinuity spherical waves, that is waves of the order two with respect to the displacement and temperature fields.*

[22] The conclusions are similar to those obtained by Boley and Hetnarski (1968), where the propagating discontinuities in a thermoelastic half-space with zero relaxation times have been classified according to the discontinuities of a thermomechanical load on the half-space.

7.5 Solution of Danilovskaya type; plane wave of a negative order

This section is devoted to the analysis of a 1D initial-boundary value problem that consists in determining a thermoelastic disturbance propagating in the interior of a half-space due to a singular thermomechanical loading on the boundary.[23] It is assumed that the load is only thermal in nature: a temperature is applied in the form of a pulse to the stress-free boundary $x = 0$ of the half-space $x \geq 0$.[24] This problem, called a *problem of Danilovskaya type*, may be reduced to a 1D PTP of the G–L theory for a pair (ϕ, ϑ). The problem consists in finding a pair (ϕ, ϑ) for $(x, t) \in [0, \infty) \times [0, \infty)$, satisfying the field equations

$$\begin{aligned} &\Gamma\phi = 0 \\ &\left(\frac{\partial^2}{\partial x^2} - \frac{\partial^2}{\partial t^2}\right)\phi = \left(1 + t^0 \frac{\partial}{\partial t}\right)\vartheta \end{aligned} \quad \text{for } (x,t) \in [0,\infty) \times [0,\infty), \tag{7.5.1}$$

the initial conditions

$$\begin{aligned} \phi(x,0) = \dot{\phi}(x,0) = 0 \\ \vartheta(x,0) = \dot{\vartheta}(x,0) = 0 \end{aligned} \quad \text{for } x \in (0,\infty), \tag{7.5.2}$$

and the boundary conditions[25]

$$\begin{aligned} \phi(0,t) = 0 \\ \vartheta(0,t) = \delta(t) \end{aligned} \quad \text{for } t \in (0,\infty). \tag{7.5.3}$$

It is further assumed that, for a fixed time and points sufficiently "far" from the boundary, the solution vanishes.

Here, Γ is the central operator of the G–L theory, specialized to the 1D situation (recall eqn (7.3.5))

$$\Gamma = \left(\frac{\partial^2}{\partial x^2} - \frac{\partial^2}{\partial t^2}\right)\left(\frac{\partial^2}{\partial x^2} - t_0\frac{\partial^2}{\partial t^2} - \frac{\partial}{\partial t}\right) - \epsilon\frac{\partial^3}{\partial x^2 \partial t}\left(1 + t^0\frac{\partial}{\partial t}\right), \tag{7.5.4}$$

and $\delta(t)$ is the Dirac delta.

The displacement $u_1 = u_1(x,t)$ and the stress $S_{11} = S_{11}(x,t)$ are calculated with the help of ϕ from the formulas (recall eqns (7.3.2) and (7.3.3))

$$u_1 = \frac{\partial \phi}{\partial x}, \; S_{11} = \frac{\partial^2 \phi}{\partial t^2}. \tag{7.5.5}$$

Proceeding just like in the problem of Nowacki type (recall Section 7.4), we conclude that the problem (7.5.1)–(7.5.3) may be reduced to the following problem for the function ϕ. Find a function ϕ that satisfies the equation

$$\Gamma\phi = 0 \quad \text{for } (x,t) \in (0,\infty) \times (0,\infty), \tag{7.5.6}$$

[23] The problem is a counterpart of the one of the dynamical theory of thermal stresses solved in a closed form by Danilovskaya (1950).

[24] An analysis of other types of the boundary thermomechanical loads on the half-space can be performed in a similar way.

[25] The condition $(7.5.3)_1$ implies that the boundary $x = 0$ is free from the stress.

the initial conditions

$$\frac{\partial^k \phi}{\partial t^k}(x,0) = 0, \qquad \text{for } x \in (0,\infty), \quad k = 0,1,2,3, \tag{7.5.7}$$

the boundary conditions

$$\begin{aligned} &\phi(0,t) = 0 \\ &\frac{\partial^2 \phi}{\partial x^2}(0,t) = \delta(t) + t^0 \dot{\delta}(t) \end{aligned} \qquad \text{for } t \in (0,\infty), \tag{7.5.8}$$

and appropriate decay conditions as $x \to \infty$.

In view of the singular boundary conditions (7.5.8), the above is a singular initial-boundary value problem. Applying the Laplace transform to eqns (7.5.6) and (7.5.8), and using eqns (7.5.7), we obtain

$$\left[\frac{\partial^2}{\partial x^2} - s_1^2(p)\right]\left[\frac{\partial^2}{\partial x^2} - s_2^2(p)\right]\bar{\phi} = 0, \tag{7.5.9}$$

and

$$\begin{aligned} \bar{\phi}(0,p) &= 0, \\ \frac{\partial^2 \bar{\phi}}{\partial x^2}(0,p) &= 1 + t^0 p, \end{aligned} \tag{7.5.10}$$

where the functions $s_1(p)$ and $s_2(p)$ are defined through the formulas (7.4.15), and

$$\bar{\phi}(x,p) = \int_0^\infty e^{-pt}\phi(x,t)\,dt. \tag{7.5.11}$$

From this we find

$$\bar{\phi}(x,p) = D\frac{\exp[-xs_1(p)] - \exp[-xs_2(p)]}{[s_1^2(p) - s_2^2(p)]}, \tag{7.5.12}$$

where (recall eqn (7.4.18))

$$D = 1 + t^0 p. \tag{7.5.13}$$

Since the formula (7.5.12) is identical to eqn (7.4.17) assuming $\rho = x$, with the accuracy of the coefficient itself, then applying the inverse Laplace transform to eqn (7.5.12), and proceeding in a way similar to that of retransforming eqn (7.4.17), we obtain

$$\phi(x,t) = \phi_1(x,t) + \phi_2(x,t), \tag{7.5.14}$$

where

$$\phi_1(x,t) = \left(\frac{1}{v_1^2} - \frac{1}{v_2^2}\right)^{-1} * \left(1 + t^0 \frac{\partial}{\partial t}\right) \left[\exp(-h_1 t) \hat{\mathcal{M}}_1(x_1,t) H(\hat{\zeta}_1)\right], \quad (7.5.15)$$

$$\phi_2(x,t) = -\left(\frac{1}{v_1^2} - \frac{1}{v_2^2}\right)^{-1} * \left(1 + t^0 \frac{\partial}{\partial t}\right) \left[\exp(-h_2 t) \hat{\mathcal{M}}_2(x_2,t) H(\hat{\zeta}_2)\right]. \quad (7.5.16)$$

In the above

$$\hat{\mathcal{M}}_1(x_1,t) = J_0(\beta\hat{\zeta}_1)\exp(\alpha_1\hat{\zeta}_1) + x_1 \sum_{n=0}^{\infty} \frac{(-)^n \hat{\lambda}_1^n}{n!\, \omega_1^{n-1}} \int_{x_1}^{t} g_n^{(1)}(t-s)\, z_1^{n-1} J_{n-1}(z_1)\, \mathrm{d}s, \quad (7.5.17)$$

$$\hat{\mathcal{M}}_2(x_2,t) = J_0(\beta\hat{\zeta}_2)\exp(\alpha_2\hat{\zeta}_2) + x_2 \sum_{n=0}^{\infty} \frac{\hat{\lambda}_2^n}{n!\, \hat{\omega}_2^{n-1}} \int_{x_2}^{t} g_n^{(2)}(t-s)\, z_2^{n-1} I_{n-1}(z_2)\, \mathrm{d}s, \quad (7.5.18)$$

and

$$x_i = x v_i^{-1},\ \hat{\zeta}_i = t - x_i\ (i = 1,2). \quad (7.5.19)$$

The remaining symbols appearing in eqns (7.5.15)–(7.5.19) have the same meaning as in Section 7.4. In particular, we have

$$g_n^{(i)}(t) = \exp(\alpha_i t) \sum_{k=0}^{n} \binom{n}{k} \frac{(-)^k}{2^k} \left(\frac{\alpha}{\beta}\right)^{n-k} J_{n+k}(\beta t), \quad (7.5.20)$$

$$z_1 = \left[\omega_1 \left(s^2 - x_1^2\right)\right]^{1/2},\ z_2 = \left[\hat{\omega}_2 \left(s^2 - x_2^2\right)\right]^{1/2}, \quad (7.5.21)$$

$$\omega_1 = \hat{\lambda}_1 - h_1^2,\ \hat{\omega}_2 = \hat{\lambda}_2 + h_2^2, \quad (7.5.22)$$

$$\hat{\lambda}_i = \lambda v_i^2,\ h_i = k_i v_i^2/2, \quad (7.5.23)$$

$$\alpha_i = \alpha + h_i\ (i = 1,2). \quad (7.5.24)$$

Using eqns $(7.5.1)_2$ and (7.5.12), the transformed temperature is found as

$$\bar{\vartheta}(x,p) = \frac{\left[s_1^2(p) - p^2\right]\exp\left[-x s_1(p)\right] - \left[s_2^2(p) - p^2\right]\exp\left[-x s_2(p)\right]}{\left[s_1^2(p) - s_2^2(p)\right]}. \quad (7.5.25)$$

Since eqn (7.5.25) is identical to eqn (7.4.21) assuming $\rho = x$, with the accuracy of the coefficient itself, then applying the inverse Laplace transform to eqn (7.5.25), and proceeding in a way similar to that of retransforming eqn (7.4.21), we obtain

$$\vartheta(x,t) = \vartheta_1(x,t) + \vartheta_2(x,t), \quad (7.5.26)$$

where

$$\vartheta_1(x,t) = \frac{1}{2}\Big\{(1+\hat{\beta})\exp(-h_1 x_1)\,\delta(\hat{\zeta}_1) + \exp(-h_1 t)\,[\hat{\mathcal{N}}_1(x_1,t) + (\hat{\alpha} - h_1\hat{\beta})\hat{\mathcal{M}}_1(x_1,t) + \hat{\beta}\dot{\hat{\mathcal{M}}}_1(x_1,t)]H(\hat{\zeta}_1)\Big\}, \tag{7.5.27}$$

$$\vartheta_2(x,t) = \frac{1}{2}\Big\{(1-\hat{\beta})\exp(-h_2 x_2)\,\delta(\hat{\zeta}_2) + \exp(-h_2 t)\,[\hat{\mathcal{N}}_2(x_2,t) - (\hat{\alpha} - h_2\hat{\beta})\hat{\mathcal{M}}_2(x_2,t) - \hat{\beta}\dot{\hat{\mathcal{M}}}_2(x_2,t)]H(\hat{\zeta}_2)\Big\}. \tag{7.5.28}$$

Here,

$$\hat{\mathcal{N}}_1(x_1,t) = x_1\left\{-\omega_1\frac{J_1(\hat{z}_1)}{\hat{z}_1} + \sum_{n=1}^{\infty}\frac{(-)^n\,\hat{\lambda}_1^n}{n!\,\omega_1^{n-1}}\int_{x_1}^{t} h_n^{(1)}(t-s)\,z_1^{n-1}J_{n-1}(z_1)\,\mathrm{d}s\right\}, \tag{7.5.29}$$

$$\hat{\mathcal{N}}_2(x_2,t) = x_2\left\{\hat{\omega}_2\frac{I_1(\hat{z}_2)}{\hat{z}_2} + \sum_{n=1}^{\infty}\frac{\hat{\lambda}_2^n}{n!\,\hat{\omega}_2^{n-1}}\int_{x_2}^{t} h_n^{(2)}(t-s)\,z_2^{n-1}I_{n-1}(z_2)\,\mathrm{d}s\right\}, \tag{7.5.30}$$

in which (recall eqns (7.4.61) and (7.4.74))

$$h_n^{(i)}(t) = \exp(\alpha_i t)\sum_{n=1}^{n}\binom{n}{k}\frac{(-)^k}{2^k}\left(\frac{\alpha}{\beta}\right)^{n-k}(n+k)\,\frac{J_{n+k}(\beta t)}{t}, \tag{7.5.31}$$

$$\hat{z}_1 = \left[\omega_1\left(t^2 - x_1^2\right)\right]^{1/2},\ \hat{z}_2 = \left[\hat{\omega}_2\left(t^2 - x_2^2\right)\right]^{1/2}. \tag{7.5.32}$$

The remaining symbols appearing in eqns (7.5.26)–(7.5.32) have the same meaning as in Section 7.4.

Finally, employing eqns (7.5.5) and (7.5.14), we find

$$u_1(x,t) = u_1^{(1)}(x_1,t) + u_1^{(2)}(x_2,t), \tag{7.5.33}$$

where

$$u_1^{(1)}(x_1,t) = -\frac{1}{v_2}\left(\frac{1}{v_1^2} - \frac{1}{v_2^2}\right)^{-1} * \left(1 + t^0\frac{\partial}{\partial t}\right) \times \Big\{\exp(-h_1 x_1)\,\delta(\hat{\zeta}_1) - \exp(-h_1 t)\,\hat{\mathcal{M}}_1'(x_1,t)\,H(\hat{\zeta}_1)\Big\}, \tag{7.5.34}$$

$$u_1^{(2)}(x_2,t) = \frac{1}{v_2}\left(\frac{1}{v_1^2} - \frac{1}{v_2^2}\right)^{-1} * \left(1 + t^0\frac{\partial}{\partial t}\right) \times \Big\{\exp(-h_2 x_2)\,\delta(\hat{\zeta}_2) - \exp(-h_2 t)\,\hat{\mathcal{M}}_2'(x_2,t)\,H(\hat{\zeta}_2)\Big\} \tag{7.5.35}$$

and

$$S_{11}(x,t) = S_{11}^{(1)}(x_1,t) + S_{11}^{(2)}(x_2,t), \tag{7.5.36}$$

in which

$$\begin{aligned} S_{11}^{(1)}(x_1,t) = & \left(\frac{1}{v_1^2} - \frac{1}{v_2^2}\right)^{-1} \left(1 + t^0 \frac{\partial}{\partial t}\right) \Big\{ \exp(-h_1 x_1)\, \delta(\hat{\zeta}_1) \\ & + \exp(-h_1 t) \left[\dot{\hat{\mathcal{M}}}_1(x_1,t) - h_1 \hat{\mathcal{M}}_1(x_1,t)\right] H(\hat{\zeta}_1) \Big\}, \end{aligned} \tag{7.5.37}$$

$$\begin{aligned} S_{11}^{(2)}(x_2,t) = & -\left(\frac{1}{v_1^2} - \frac{1}{v_2^2}\right)^{-1} \left(1 + t^0 \frac{\partial}{\partial t}\right) \Big\{ \exp(-h_2 x_2)\, \delta(\hat{\zeta}_2) \\ & + \exp(-h_2 t) \left[\dot{\hat{\mathcal{M}}}_2(x_2,t) - h_2 \hat{\mathcal{M}}_2(x_2,t)\right] H(\hat{\zeta}_2) \Big\}. \end{aligned} \tag{7.5.38}$$

The prime and the dot in the formulas (7.5.34) and (7.5.35) and (7.5.37) and (7.5.38) denote, respectively, differentiation with respect to x_i and t.

It follows from the formulas (7.5.26), (7.5.33) and (7.5.36) that the thermoelastic disturbance propagating in the half-space is a superposition of two plane waves of strong discontinuity whose fronts coincide with the planes $\hat{\zeta}_i = 0$ $(i = 1,2)$ that are moving into the half-space at speeds v_i $(i = 1,2)$ $(v_2 > v_1)$. Since

$$[[1 * u_1]]_1(t) = -\frac{t^0}{v_1} \left(\frac{1}{v_1^2} - \frac{1}{v_2^2}\right)^{-1} \exp(-h_1 t), \tag{7.5.39}$$

$$[[1 * u_1]]_2(t) = \frac{t^0}{v_2} \left(\frac{1}{v_1^2} - \frac{1}{v_2^2}\right)^{-1} \exp(-h_2 t), \tag{7.5.40}$$

and

$$[[1 * \vartheta]]_1(t) = \frac{1}{2}\left(1 + \hat{\beta}\right) \exp(-h_1 t), \tag{7.5.41}$$

$$[[1 * \vartheta]]_2(t) = \frac{1}{2}\left(1 - \hat{\beta}\right) \exp(-h_2 t), \tag{7.5.42}$$

both waves are of the order $n = -1$ with respect to the displacement and temperature. This leads to the following theorem.

Theorem 7.2 *A displacement–temperature wave in the problem of Danilovskaya type in the G–L theory, in which the boundary of the half-space is heated by a temperature pulse, is a superposition of two plane waves of strong discontinuity of the order $n = -1$ propagating into its interior at speeds v_i $(i = 1,2)$ $(v_2 > v_1)$. This thermoelastic wave is described by the formulas (7.5.26), (7.5.33) and (7.5.36).*

7.6 Thermoelastic response of a half-space to laser irradiation

The thermoelastic body of G–L type in which the stresses and entropy are sensitive not only to the changes of deformation and temperature but also to the time rates of temperature is a good model for the description of waves in a thermoelastic body subject to a short laser irradiation. In the present section we consider a 1D initial-boundary value problem for a half-space of G–L type in which the thermoelastic wave is generated by a heat source in the whole half-space due to the irradiation of its boundary by a short burst of light energy.

Let us assume the heat source to be represented by

$$r(x,t) = X(x)T(t) \qquad x \geq 0, \qquad t \geq 0, \tag{7.6.1}$$

where $X(x)$ is an exponential function decreasing with depth, while $T(t)$ is a function having the shape of a short time pulse (Strikverda and Scott, 1984). Consistent with the homogeneous displacement–temperature initial conditions and homogeneous stress–temperature boundary conditions, the problem under consideration can be reduced to the following PTP of the G–L theory for a pair (ϕ, ϑ).

Find a pair (ϕ, ϑ) defined for $(x,t) \in [0,\infty) \times [0,\infty)$ and satisfying the field equations

$$\left.\begin{aligned} \Gamma\phi &= -\left(1 + t^0 \frac{\partial}{\partial t}\right) r \\ \left(\frac{\partial^2}{\partial x^2} - \frac{\partial^2}{\partial t^2}\right)\phi &= \left(1 + t^0 \frac{\partial}{\partial t}\right)\vartheta \end{aligned}\right\} \quad \text{for } (x,t) \in (0,\infty) \times (0,\infty), \tag{7.6.2}$$

the initial conditions

$$\left.\begin{aligned} \frac{\partial^k}{\partial t^k}\phi(\cdot,0) &= 0 \\ \vartheta(\cdot,0) = \dot{\vartheta}(\cdot,0) &= 0 \end{aligned}\right\} \quad \text{for } x \in (0,\infty), \quad k = 0,1,2,3, \tag{7.6.3}$$

and the boundary conditions

$$\phi(0,t) = 0, \qquad \vartheta(0,t) = 0 \qquad \text{for } t \in (0,\infty). \tag{7.6.4}$$

Furthermore, the pair (ϕ, ϑ) and its partial derivatives of finite order vanish for $x \to \infty$ and for every fixed t.

The operator Γ appearing in eqn $(7.6.2)_1$ has the form (recall eqn (7.5.4))

$$\Gamma = \left(\frac{\partial^2}{\partial x^2} - \frac{\partial^2}{\partial t^2}\right)\left(\frac{\partial^2}{\partial x^2} - t_0\frac{\partial^2}{\partial t^2} - \frac{\partial}{\partial t}\right) - \epsilon\frac{\partial^3}{\partial x^2 \partial t}\left(1 + t^0\frac{\partial}{\partial t}\right). \tag{7.6.5}$$

The pair (ϕ, ϑ) generates the displacement vector u_i, the stress tensor S_{ij}, the heat flux q_i $(i,j = 1,2,3)$ according to the formulas (recall eqns (6.1.10) and (6.1.18)–(6.1.20))

$$u \equiv u_1(x,t) = \frac{\partial \phi}{\partial x}, \qquad u_2(x,t) = u_3(x,t) = 0, \tag{7.6.6}$$

$$\begin{gathered} S \equiv S_{11}(x,t) = \frac{\partial \phi^2}{\partial t^2}, \\ S_{22}(x,t) = S_{33}(x,t) = \frac{\partial \phi^2}{\partial t^2} - 2\left(\frac{C_2}{C_1}\right)^2 \frac{\partial \phi^2}{\partial x^2}, \\ S_{12}(x,t) = S_{23}(x,t) = S_{31}(x,t) = 0, \end{gathered} \tag{7.6.7}$$

$$q \equiv q_1(x,t) = -\frac{\partial \vartheta}{\partial x}, \qquad q_2(x,t) = q_3(x,t) = 0. \tag{7.6.8}$$

We assume that the function $r(x,t)$ appearing in eqns (7.6.1) and (7.6.2) is smooth for every $x \geq 0$ and $t \geq 0$, and that

$$X(0) > 0, \qquad T(0) = 0. \tag{7.6.9}$$

The initial-boundary value problem (7.6.2)–(7.6.4) is now solved by a Laplace transform technique. Applying that Laplace transform to eqn $(7.6.2)_1$ and employing the homogeneous initial conditions (7.6.3) and $(7.6.9)_2$, we obtain

$$\square_1^2 \square_2^2 \bar{\phi} = -\left(1 + t^0 p\right) X(x) \bar{T}(p) \qquad x \geq 0, \tag{7.6.10}$$

where

$$\square_i^2 = \frac{\partial^2}{\partial x^2} - s_i^2(p). \tag{7.6.11}$$

Here, $s_i(p)$ $(i = 1, 2)$ are functions determined by the formulas (7.4.15), while the overbar denotes the Laplace transform.

Next, transforming the boundary conditions (7.6.4) we obtain

$$\bar{\phi}\,(0,p) = 0, \qquad \bar{\vartheta}(0,p) = 0. \tag{7.6.12}$$

It is clear that a solution to the boundary value problem (7.6.10)–(7.6.12) can be represented in the form

$$\bar{\phi}\,(x,p) = \bar{T}(p) \int_0^\infty \bar{\phi}^*(x,p;x_0) X(x_0) \mathrm{d}x_0, \tag{7.6.13}$$

where the function $\bar{\phi}^* = \bar{\phi}^*(x,p;x_0)$ satisfies the conditions

$$\begin{gathered} \square_1^2 \square_2^2 \bar{\phi}^* = -\left(1 + t^0 p\right) \delta(x - x_0) \qquad x_0 > 0, \\ \bar{\phi}^*(0,p;x_0) = \bar{\vartheta}^*(0,p;x_0) = 0. \end{gathered} \tag{7.6.14}$$

Here, $\delta = \delta(x)$ is the Dirac delta, while $\bar{\vartheta}^*(x,p;x_0)$ is the transform of temperature associated with $\bar{\phi}^*(x,p;x_0)$ through eqn $(7.6.2)_2$, that is

$$\left(1 + t^0 p\right) \bar{\vartheta}^* = \left(\frac{\partial^2 \phi}{\partial x^2} - p^2\right) \bar{\phi}^*. \tag{7.6.15}$$

The temperature transform $\bar{\vartheta}(x,p)$ associated with the transform of the potential, $\bar{\phi}(x,p)$, has the form analogous to eqn (7.6.13), that is

$$\bar{\vartheta}\,(x,p) = \bar{T}(p)\int_0^\infty \bar{\vartheta}^*(x,p;x_0)X(x_0)\mathrm{d}x_0. \tag{7.6.16}$$

In order to determine the pair $(\bar{\phi}^*, \bar{\vartheta}^*)$, we now introduce the operator

$$\Box^2 = \frac{\partial^2}{\partial x^2} - s^2, \tag{7.6.17}$$

where s is a real parameter ($s > 0$). Noting that, for $|x| < \infty$,

$$\frac{1}{\Box^2}\delta(x - x_0) = -\frac{1}{2s}\mathrm{e}^{-s|x-x_0|}, \tag{7.6.18}$$

and

$$\frac{1}{\Box_1^2\Box_2^2} = \frac{1}{s_1^2 - s_2^2}\left(\frac{1}{\Box_1^2} - \frac{1}{\Box_2^2}\right), \tag{7.6.19}$$

we infer that a solution to the problem (7.6.14) takes the form

$$\bar{\phi}^*\,(x,p;x_0) = \bar{\phi}_N(x,p;x_0) + \bar{\phi}_C(x,p;x_0), \tag{7.6.20}$$

where $\bar{\phi}_N$ is the Laplace transform of the 1D potential of the Nowacki type in the G–L theory (recall Section 7.4)

$$\bar{\phi}_N(x,p;x_0) = \frac{\bar{m}}{2\left(s_1^2 - s_2^2\right)}\left(\frac{\mathrm{e}^{-s_1|x-x_0|}}{s_1} - \frac{\mathrm{e}^{-s_2|x-x_0|}}{s_2}\right), \tag{7.6.21}$$

while $\bar{\phi}_C$ is the solution of a homogeneous equation associated with eqn $(7.6.14)_1$

$$\bar{\phi}_C(x,p;x_0) = A_1\mathrm{e}^{-s_1x} + A_2\mathrm{e}^{-s_2x}. \tag{7.6.22}$$

Here,

$$\bar{m} = 1 + t^0p, \tag{7.6.23}$$

while the constants A_1 and A_2 are chosen in such a way that the pair $(\bar{\phi}^*, \bar{\vartheta}^*)$ satisfies the boundary conditions $(7.6.14)_2$ and the conditions of vanishing at infinity.

The temperature transforms $\bar{\vartheta}_N$ and $\bar{\vartheta}_C$ associated, respectively, with the functions $\bar{\phi}_N$ and $\bar{\phi}_C$ are

$$\bar{\vartheta}_N(x,p;x_0) = \frac{1}{2\left(s_1^2 - s_2^2\right)}\left[\frac{s_1^2 - p^2}{s_1}\mathrm{e}^{-s_1|x-x_0|} - \frac{s_2^2 - p^2}{s_2}\mathrm{e}^{-s_2|x-x_0|}\right], \tag{7.6.24}$$

$$\bar{\vartheta}_C(x,p;x_0) = \frac{A_1}{\bar{m}}\left(s_1^2 - p^2\right)\mathrm{e}^{-s_1x} + \frac{A_2}{\bar{m}}\left(s_2^2 - p^2\right)\mathrm{e}^{-s_2x}. \tag{7.6.25}$$

It can be easily verified that the boundary conditions $(7.6.14)_2$ uniquely determine the pair (A_1, A_2) through the pair $[\bar{\phi}_N(0,p;x_0), \bar{\vartheta}_N(0,p;x_0)]$, while the function $\bar{\phi}^*(x,p;x_0)$ takes the form

$$\bar{\phi}^*(x,p;x_0) = \bar{\phi}_N(x,p;x_0) + \frac{\left(s_2^2 - p^2\right) e^{-s_1 x} - \left(s_1^2 - p^2\right) e^{-s_2 x}}{s_1^2 - s_2^2} \bar{\phi}_N(0,p;x_0) - \frac{\bar{m}}{s_1^2 - s_2^2}\left(e^{-s_1 x} - e^{-s_2 x}\right) \bar{\vartheta}_N(0,p;x_0). \tag{7.6.26}$$

From this, and using eqn (7.6.15), we obtain

$$\bar{m}\bar{\vartheta}^*(x,p;x_0) = \bar{m}\bar{\vartheta}_N(x,p;x_0) + \frac{\left(s_1^2 - p^2\right)\left(s_2^2 - p^2\right)}{s_1^2 - s_2^2}\left(e^{-s_1 x} - e^{-s_2 x}\right) \bar{\phi}_N(0,p;x_0) - \frac{\bar{m}}{s_1^2 - s_2^2}\left[\left(s_1^2 - p^2\right) e^{-s_1 x} - \left(s_2^2 - p^2\right) e^{-s_2 x}\right] \bar{\vartheta}_N(0,p;x_0). \tag{7.6.27}$$

Given that [see eqn (8.3.35)]

$$\left(s_1^2 - p^2\right)\left(s_2^2 - p^2\right) = -\epsilon \bar{m} p^3, \tag{7.6.28}$$

the formulas (7.6.26) and (7.6.27) can be represented in the form

$$\bar{\phi}^*(x,p;x_0) = \bar{\phi}_N(x,p;x_0) - \left(e^{-s_1 x} + e^{-s_2 x}\right) \bar{\phi}_N(0,p;x_0) + \frac{\left(s_1^2 - p^2\right) e^{-s_1 x} - \left(s_2^2 - p^2\right) e^{-s_2 x}}{s_1^2 - s_2^2} \bar{\phi}_N(0,p;x_0) - \bar{m}\frac{e^{-s_1 x} - e^{-s_2 x}}{s_1^2 - s_2^2} \bar{\vartheta}_N(0,p;x_0), \tag{7.6.29}$$

and

$$\bar{\vartheta}^*(x,p;x_0) = \bar{\vartheta}_N(x,p;x_0) - \epsilon p^3 \frac{e^{-s_1 x} - e^{-s_2 x}}{s_1^2 - s_2^2} \bar{\phi}_N(0,p;x_0) - \frac{\left(s_1^2 - p^2\right) e^{-s_1 x} - \left(s_2^2 - p^2\right) e^{-s_2 x}}{s_1^2 - s_2^2} \bar{\vartheta}_N(0,p;x_0). \tag{7.6.30}$$

By retransforming the relations (7.6.13) and (7.6.16), we obtain a formal solution to the problem (7.6.2)–(7.6.4)

$$\begin{aligned} \phi(x,t) &= \int_0^\infty X(x_0)\left[\int_0^t \phi^*(x,t-\tau;x_0)T(\tau)\mathrm{d}\tau\right] \mathrm{d}x_0, \\ \vartheta(x,t) &= \int_0^\infty X(x_0)\left[\int_0^t \vartheta^*(x,t-\tau;x_0)T(\tau)\mathrm{d}\tau\right] \mathrm{d}x_0, \end{aligned} \tag{7.6.31}$$

where the pair $[\phi^*(x,t;x_0), \vartheta^*(x,t;x_0)]$ is to be found by an inverse transformation of eqns (7.6.29) and (7.6.30). Clearly, the formulas (7.6.31) are true for arbitrary smooth functions $X(x)$ and $T(x)$ satisfying the conditions (7.6.9), and such that the double integrals in eqns (7.6.31) exist and are bounded for every $x \geq 0$ and $t \geq 0$.

Let us now assume that the heat source that is due to the surface irradiation has the form (recall eqn (7.6.1))

$$r(x,t) = r_0 t \exp(-bt) \exp(-ax), \tag{7.6.32}$$

where $r_0 = const$, b^{-1} defines the width of impulse on the time axis, while a^{-1} is the absorption length of the irradiation on the x-axis ($a > 0,\ b > 0$). From this we find (recall eqn (7.6.1))

$$\begin{aligned} \bar{T}(p) &= r_0 (p+b)^{-2}, \\ X(x) &= \exp(-ax). \end{aligned} \tag{7.6.33}$$

Next, introduce the notations

$$\begin{aligned} \bar{\phi}_0(x,p) &= \bar{T}(p) \int_0^\infty \bar{\phi}_N(x,p;x_0) X(x_0) \mathrm{d}x_0, \\ \bar{\vartheta}_0(x,p) &= \bar{T}(p) \int_0^\infty \bar{\vartheta}_N(x,p;x_0) X(x_0) \mathrm{d}x_0. \end{aligned} \tag{7.6.34}$$

Inserting the function $\bar{\phi}_N(x,p;x_0)$ specified by the formula (7.6.21) into eqn $(7.6.34)_1$ and integrating over x_0, we obtain

$$\bar{\phi}_0(x,p) = \frac{\bar{T}(p)\bar{m}}{s_1^2 - s_2^2} \left\{ \left[\frac{1}{s_1^2 - a^2} - \frac{1}{s_2^2 - a^2} \right] \mathrm{e}^{-ax} - \frac{1}{2} \left[\frac{\mathrm{e}^{-s_1 x}}{s_1(s_1 - a)} - \frac{\mathrm{e}^{-s_2 x}}{s_2(s_2 - a)} \right] \right\}. \tag{7.6.35}$$

Similarly, inserting the function $\bar{\vartheta}_N(x,p;x_0)$ given by eqn (7.6.24) into eqn $(7.6.34)_2$ and integrating with respect to x_0, we obtain

$$\begin{aligned} \bar{\vartheta}_0(x,p) = -\frac{\bar{T}(p)}{s_1^2 - s_2^2} \Bigg\{ & (p^2 - a^2) \left(\frac{1}{s_1^2 - a^2} - \frac{1}{s_2^2 - a^2} \right) \mathrm{e}^{-ax} + \\ + \frac{1}{2} \left[1 + \frac{a}{s_1 - a} - \frac{p^2}{s_1(s_1 - a)} \right] \mathrm{e}^{-s_1 x} & - \frac{1}{2} \left[1 + \frac{a}{s_2 - a} - \frac{p^2}{s_2(s_2 - a)} \right] \mathrm{e}^{-s_2 x} \Bigg\}. \end{aligned} \tag{7.6.36}$$

Now, using the identity (recall eqn (7.4.15))

$$s_1^2 - s_2^2 = \Delta^{1/2} p [(p-\alpha)^2 + \beta^2]^{1/2}, \tag{7.6.37}$$

and the formula specifying $\bar{T}(p)$ (recall eqn $(7.6.33)_1$) for the function $\bar{S}_0(x,p)$, which is a transform of the stress associated with the pair $(\bar{\phi}_0, \bar{\vartheta}_0)$, we obtain

$$\begin{aligned} \bar{S}_0(x,p) &= \frac{r_0}{\Delta^{1/2}} \bar{f}(p) \frac{1}{[(p-\alpha)^2 + \beta^2]^{1/2}} \\ &\times \left\{ \left(\frac{1}{s_1^2 - a^2} - \frac{1}{s_2^2 - a^2} \right) \mathrm{e}^{-ax} - \frac{1}{2} \left[\frac{\mathrm{e}^{-s_1 x}}{s_1(s_1 - a)} - \frac{\mathrm{e}^{-s_2 x}}{s_2(s_2 - a)} \right] \right\}, \end{aligned} \tag{7.6.38}$$

in which

$$\bar{f}(p) = p(1 + t^0 p)(p+b)^{-2}. \tag{7.6.39}$$

Similarly, on account of eqn (7.6.36), we obtain

$$\bar{\vartheta}_0(x,p) = -\frac{r_0}{\Delta^{1/2}}\bar{g}(p)\frac{1}{[(p-\alpha)^2+\beta^2]^{1/2}}$$
$$\times\left\{(p^2-a^2)\left(\frac{1}{s_1^2-a^2}-\frac{1}{s_2^2-a^2}\right)\mathrm{e}^{-ax}\right.$$
$$\left.+\frac{1}{2}\left[1+\frac{a}{s_1-a}-\frac{p^2}{s_1(s_1-a)}\right]\mathrm{e}^{-s_1x}-\frac{1}{2}\left[1+\frac{a}{s_2-a}-\frac{p^2}{s_2(s_2-a)}\right]\mathrm{e}^{-s_2x}\right\}, \tag{7.6.40}$$

where

$$\bar{g}(p) = \bar{f}(p)[\bar{m}p^2]^{-1}. \tag{7.6.41}$$

It follows from the relations (7.6.13), (7.6.16), (7.6.29), (7.6.30) and (7.6.34) that the pair $[\bar{S}(x,p),\bar{\vartheta}(x,p)]$ corresponding to the distribution of heat sources (7.6.32) takes the form

$$\bar{S}(x,p) = \bar{S}_0(x,p) - \left(\mathrm{e}^{-s_1x}+\mathrm{e}^{-s_2x}\right)\bar{S}_0(0,p)$$
$$+\frac{\left(s_1^2-p^2\right)\mathrm{e}^{-s_1x}-\left(s_2^2-p^2\right)\mathrm{e}^{-s_2x}}{s_1^2-s_2^2}\bar{S}_0\,(0,p) - \bar{m}p^2\frac{\mathrm{e}^{-s_1x}-\mathrm{e}^{-s_2x}}{s_1^2-s_2^2}\bar{\vartheta}_0(0,p), \tag{7.6.42}$$

$$\bar{\vartheta}(x,p) = \bar{\vartheta}_0(x,p) - \epsilon p\frac{\mathrm{e}^{-s_1x}-\mathrm{e}^{-s_2x}}{s_1^2-s_2^2}\bar{S}_0(0,p)$$
$$-\frac{\left(s_1^2-p^2\right)\mathrm{e}^{-s_1x}-\left(s_2^2-p^2\right)\mathrm{e}^{-s_2x}}{s_1^2-s_2^2}\bar{\vartheta}_0(0,p), \tag{7.6.43}$$

where the pair $[\bar{S}_0(x,p),\bar{\vartheta}_0(x,p)]$ is specified by the formulas (7.6.38)–(7.6.41). Thus, it is clear that the explicit form of the pair $[S(x,t),\vartheta(x,t)]$ is available provided we can invert in a closed form eqns (7.6.38), (7.6.40), (7.6.42) and (7.6.43). To this end, let us now introduce the functions ($x \geq 0$)

$$N_i(x,t) = L^{-1}(\mathrm{e}^{-s_ix}), \qquad \text{recall eqn (7.4.25)}, \tag{7.6.44}$$

$$M_i(x,t) = L^{-1}\left\{\frac{\mathrm{e}^{-s_ix}}{[(p-\alpha)^2+\beta^2]^{1/2}}\right\}, \qquad \text{recall eqn (7.4.24)}, \tag{7.6.45}$$

$$P_i(x,t) = L^{-1}\left\{\frac{\mathrm{e}^{-s_ix}}{[(p-\alpha)^2+\beta^2]^{1/2}\,(s_i-a)}\right\}, \tag{7.6.46}$$

$$Q_i(x,t) = L^{-1}\left\{\frac{\mathrm{e}^{-s_ix}}{[(p-\alpha)^2+\beta^2]^{1/2}s_i\,(s_i-a)}\right\}, \tag{7.6.47}$$

$$R_i(x,t) = L^{-1}\left\{\frac{\mathrm{e}^{-s_ix}}{[(p-\alpha)^2+\beta^2]^{1/2}\,(s_i^2-a^2)}\right\}, \tag{7.6.48}$$

along with the notations

$$R_i(t) = R_i(0,t), \tag{7.6.49}$$

$$\begin{aligned}
\sigma_0(x,t) &\equiv \frac{\Delta^{1/2}}{r_0} S_0(x,t), \quad \theta_0(x,t) \equiv \frac{\Delta^{1/2}}{r_0} \vartheta_0(x,t), \\
\sigma(x,t) &\equiv \frac{\Delta^{1/2}}{r_0} S(x,t), \quad \theta(x,t) \equiv \frac{\Delta^{1/2}}{r_0} \vartheta(x,t).
\end{aligned} \tag{7.6.50}$$

Equations (7.6.38) and (7.6.40) can then be written as

$$\bar{\sigma}_0(x,p) = \bar{f}(p) \left\{ [\bar{R}_1(p) - \bar{R}_2(p)] \mathrm{e}^{-ax} - \frac{1}{2} [\bar{Q}_1(x,p) - \bar{Q}_2(x,p)] \right\} \tag{7.6.51}$$

and

$$\begin{aligned}
\bar{\theta}_0(x,p) = {} & \left[\bar{g}(p) a^2 - \bar{g}_1(p)\right] \left[\bar{R}_1(p) - \bar{R}_2(p)\right] \mathrm{e}^{-ax} - \frac{1}{2} \bar{g}(p) \left[\bar{M}_1(x,p) + a\bar{P}_1(x,p)\right] \\
& + \frac{1}{2} \bar{g}(p) [\bar{M}_2(x,p) + a\bar{P}_2(x,p)] + \frac{1}{2} \bar{g}_1(p) [\bar{Q}_1(x,p) - \bar{Q}_2(x,p)],
\end{aligned} \tag{7.6.52}$$

where

$$\bar{g}_1(p) = p^2 \bar{g}(p). \tag{7.6.53}$$

Furthermore, eqns (7.6.42) and (7.6.43) take the form

$$\begin{aligned}
\bar{\sigma}(x,p) = {} & \bar{\sigma}_0(x,p) - \frac{1}{2} \Big\{ \bar{N}_1(x,p) + \bar{N}_2(x,p) \\
& - (\hat{\alpha} + p\hat{\beta}) \left[\bar{M}_1(x,p) - \bar{M}_2(x,p)\right] \Big\} \bar{\sigma}_0(0,p) \\
& - \frac{1}{2} \hat{\gamma} p (1 + t^0 p) [\bar{M}_1(x,p) - \bar{M}_2(x,p)] \bar{\theta}_0(0,p),
\end{aligned} \tag{7.6.54}$$

and

$$\begin{aligned}
\bar{\theta}(x,p) = {} & \bar{\theta}_0(x,p) - \frac{1}{2} \{ \bar{N}_1(x,p) + \bar{N}_2(x,p) \\
& + (\hat{\alpha} + p\hat{\beta}) [\bar{M}_1(x,p) - \bar{M}_2(x,p)] \} \bar{\theta}_0(0,p) \\
& - \frac{1}{2} \epsilon \hat{\gamma} [\bar{M}_1(x,p) - \bar{M}_2(x,p)] \bar{\sigma}_0(0,p),
\end{aligned} \tag{7.6.55}$$

where $\hat{\alpha}$ and $\hat{\beta}$ have the same meaning as in Section 7.4 (recall eqns (7.4.27)) and

$$\hat{\gamma} = 2\Delta^{-1/2}. \tag{7.6.56}$$

The inverse transforms of functions $\bar{N}_i(x,p)$ and $\bar{M}_i(x,p)$ were already determined in Section 7.4. Indeed, recalling eqns (7.4.83) and (7.4.85), it follows from the structure of $M_i(x,t)$ that

$$M_i(x,t) = \hat{M}_i(x_i,t) H(\zeta_i), \tag{7.6.57}$$

where $x_i = xv_i^{-1}$, $\zeta_i = t - x_i$, and $\hat{M}_i(x_i, t)$ is a smooth function of both arguments. As we shall see in the following, the functions $P_i(x,t)$, $Q_i(x,t)$, and $R_i(x,t)$ specified through eqns (7.6.46)–(7.6.48) also have the structure of the form (7.6.57), that is

$$\begin{aligned} P_i(x,t) &= \hat{P}_i(x_i,t)H(\zeta_i), \\ Q_i(x,t) &= \hat{Q}_i(x_i,t)H(\zeta_i), \\ R_i(x,t) &= \hat{R}_i(x_i,t)H(\zeta_i), \end{aligned} \tag{7.6.58}$$

with $\hat{P}_i$, $\hat{Q}_i$, and $\hat{R}_i$ being smooth functions of their arguments.

Using eqns (7.6.57) and (7.6.58), and inverting the relations (7.6.51) and (7.6.52), we obtain

$$\begin{aligned} \sigma_0(x,t) = & \left\{ t^0[R_1(t) - R_2(t)] + \int_0^t f_1(t-\tau)[R_1(\tau) - R_2(\tau)]d\tau \right\} \mathrm{e}^{-ax} \\ & -\frac{1}{2}H(\zeta_1)[t^0\hat{Q}_1(x_1,t) + \int_{x_1}^t f_1(t-\tau)\hat{Q}_1(x_1,\tau)\mathrm{d}\tau] \\ & +\frac{1}{2}H(\zeta_2)[t^0\hat{Q}_2(x_2,t) + \int_{x_2}^t f_1(t-\tau)\hat{Q}_2(x_2,\tau)\mathrm{d}\tau], \end{aligned} \tag{7.6.59}$$

and

$$\begin{aligned} \theta_0(x,t) = & \left\{ \int_0^t h(t-\tau)[R_1(\tau) - R_2(\tau)]\mathrm{d}\tau \right\} \mathrm{e}^{-ax} \\ & -\frac{1}{2}H(\zeta_1)\int_{x_1}^t \{g(t-\tau)[\hat{M}_1(x_1,\tau) + a\hat{P}_1(x_1,\tau)] - g_1(t-\tau)\hat{Q}_1(x_1,\tau)\}\mathrm{d}\tau \\ & +\frac{1}{2}H(\zeta_2)\int_{x_2}^t \{g(t-\tau)[\hat{M}_2(x_2,\tau) + a\hat{P}_2(x_2,\tau)] - g_1(t-\tau)\hat{Q}_2(x_2,\tau)\}\mathrm{d}\tau. \end{aligned} \tag{7.6.60}$$

The functions $f_1(t)$, $g_1(t)$, $g(t)$, and $h(t)$ appearing in eqns (7.6.59) and (7.6.60) are given by the relations

$$f_1(t) \equiv f(t) - t^0\delta(t) = [(1 - 2bt^0) - (1 - bt^0)bt]\mathrm{e}^{-bt}, \tag{7.6.61}$$

$$g_1(t) \equiv (1 - bt)\mathrm{e}^{-bt}, \tag{7.6.62}$$

$$g(t) \equiv \frac{1}{b^2}[(1 - a^{-bt}) - bt\mathrm{e}^{-bt}], \tag{7.6.63}$$

$$h(t) \equiv a^2g(t) - g_1(t) = \frac{a^2}{b^2} - \left(1 + \frac{a^2}{b^2}\right)\mathrm{e}^{-bt} + \frac{b^2 - a^2}{b}t\mathrm{e}^{-bt}. \tag{7.6.64}$$

Given that the functions $\hat{M}_i(x_i, t)$ were determined in a closed form in Section 7.4 (recall eqns (7.4.83) and (7.4.85)), the relations (7.6.59) and (7.6.60) allow one to find a closed form of the pair $[\sigma_0(x,t), \theta_0(x,t)]$ so long as the functions $R_i(t)$, $\hat{Q}_i(x_i,t)$ and $\hat{P}_i(x_i,t)$ can also be obtained in closed forms.

Now, the inversion of the relations (7.6.54) and (7.6.55) leads to the equations

$$\begin{aligned}\sigma(x,t) &= \sigma_0(x,t) + \sigma_1(x_1,t) + \sigma_2(x_2,t),\\ \theta(x,t) &= \theta_0(x,t) + \theta_1(x_1,t) + \theta_2(x_2,t),\end{aligned} \tag{7.6.65}$$

where

$$\begin{aligned}\sigma_1(x_1,t) &= -\frac{1}{2}N_1(x,t) * \sigma_0(0,t) + \frac{1}{2}M_1(x,t) * \tilde{\sigma}_0(0,t),\\ \sigma_2(x_2,t) &= -\frac{1}{2}N_2(x,t) * \sigma_0(0,t) - \frac{1}{2}M_2(x,t) * \tilde{\sigma}_0(0,t),\end{aligned} \tag{7.6.66}$$

and

$$\begin{aligned}\theta_1(x_1,t) &= -\frac{1}{2}N_1(x,t) * \theta_0(0,t) - \frac{1}{2}M_1(x,t) * \tilde{\theta}_0(0,t),\\ \theta_2(x_2,t) &= -\frac{1}{2}N_2(x,t) * \theta_0(0,t) + \frac{1}{2}M_2(x,t) * \tilde{\theta}_0(0,t).\end{aligned} \tag{7.6.67}$$

Here,

$$\begin{aligned}\tilde{\sigma}_0(0,t) &= \hat{\alpha}\sigma_0(0,t) + \hat{\beta}\dot{\sigma}_0(0,t) - \hat{\gamma}[\dot{\theta}_0(0,t) + t^0\ddot{\theta}_0(0,t)],\\ \tilde{\theta}_0(0,t) &= \hat{\alpha}\theta_0(0,t) + \hat{\beta}\dot{\theta}_0(0,t) + \epsilon\hat{\gamma}\sigma_0(0,t).\end{aligned} \tag{7.6.68}$$

On account of the relation

$$N_i(x,t) = \delta(\zeta_i)\exp(-h_i x_i) + \hat{N}_i^{(0)}(x_i,t)H(\zeta_i), \tag{7.6.69}$$

where $\hat{N}_i^{(0)}(x_i,t)$ is a smooth function of both arguments [recall the formula (7.4.65) in which $\hat{P}(\rho_1,t)$ is identified with $\hat{N}_1^{(0)}(x_1,t)H(\zeta_1)$, and the formula (7.4.71) in which $R(\rho_2,t)$ is identified with $\hat{N}_2^{(0)}(x_2,t)H(\zeta_2)$], and using eqn (7.6.57), we represent the relations (7.6.66) and (7.6.67) in the form

$$\begin{aligned}\sigma_1(x_1,t) &= -\frac{1}{2}H(\zeta_1)\Bigg\{\exp(-h_1x_1)\sigma_0(0,\zeta_1)\\ &+ \int_{x_1}^{t}[\hat{N}_1^{(0)}(x_1,s)\sigma_0(0,t-s) - \hat{M}_1^{(0)}(x_1,s)\tilde{\sigma}_0(0,t-s)]\mathrm{d}s\Bigg\},\\ \sigma_2(x_2,t) &= -\frac{1}{2}H(\zeta_2)\Bigg\{\exp(-h_2x_2)\sigma_0(0,\zeta_2)\\ &+ \int_{x_2}^{t}[\hat{N}_2^{(0)}(x_2,s)\sigma_0(0,t-s) + \hat{M}_2^{(0)}(x_2,s)\tilde{\sigma}_0(0,t-s)]\mathrm{d}s\Bigg\},\end{aligned} \tag{7.6.70}$$

and

$$\theta_1(x_1,t) = -\frac{1}{2}H(\zeta_1)\Big\{ \exp(-h_1x_1)\theta_0(0,\zeta_1) + \int_{x_1}^{t} [\hat{N}_1^{(0)}(x_1,s)\theta_0(0,t-s) + \hat{M}_1^{(0)}(x_1,s)\tilde{\theta}_0(0,t-s)]\mathrm{d}s\Big\},$$
$$\theta_2(x_2,t) = -\frac{1}{2}H(\zeta_2)\Big\{ \exp(-h_2x_2)\theta_0(0,\zeta_2) + \int_{x_2}^{t} [\hat{N}_2^{(0)}(x_2,s)\theta_0(0,t-s) - \hat{M}_2^{(0)}(x_2,s)\tilde{\theta}_0(0,t-s)]\mathrm{d}s\Big\}. \quad (7.6.71)$$

Given that

$$\hat{M}_1(0,t) = \hat{M}_2(0,t), \qquad \hat{N}_1^{(0)}(0,t) = \hat{N}_2^{(0)}(0,t) = 0, \quad (7.6.72)$$

one can easily verify that the pair $[\sigma(x,t),\theta(x,t)]$ specified by the formulas (7.6.65) satisfies all the conditions of thus formulated initial-boundary value problem. In particular, we have

$$\sigma(0,t) = \theta(0,t) = 0 \qquad \forall t \geq 0. \quad (7.6.73)$$

Clearly, the solution $[\sigma(x,t),\theta(x,t)]$ takes a closed form so long as the functions $\hat{P}_i(x_i,t), \hat{Q}_i(x_i,t)$, and $\hat{R}_i(x_i,t)$ defining the pair $[\sigma_0(x,t),\theta_0(x,t)]$ take a closed form. In the following, we determine the closed forms of these functions.

The function $P_i(x_i,t)$ is specified by the relation (recall eqn (7.6.46))

$$P_i(x,t) = L^{-1}\left\{\frac{\exp[-xs_i(p)]}{[(p-\alpha)^2+\beta^2]^{1/2}[s_i(p)-a]}\right\} \qquad (i=1,2). \quad (7.6.74)$$

It can be rewritten as

$$P_i(x,t) = \mathrm{e}^{-h_it}L^{-1}[\bar{P}_i(x,p-h_i)], \quad (7.6.75)$$

where

$$\bar{P}_i(x,p-h_i) = \frac{\exp[-xs_i(p-h_i)]}{[(p-\alpha_i)^2+\beta^2]^{1/2}[s_i(p-h_i)-a]}. \quad (7.6.76)$$

Here, h_i and α_i have the same meaning as in Sections 7.4 and 7.5 (recall eqn (7.5.24)).

Now, introducing the notations

$$\tilde{s}_i(p-h_i) = v_is_i(p-h_i), \qquad \tilde{a}_i = v_ia, \quad (7.6.77)$$

we find

$$\bar{P}_i(x,p-h_i) = \frac{v_i\exp[-x_i\tilde{s}_i(p-h_i)]}{[(p-\alpha_i)^2+\beta^2]^{1/2}[\tilde{s}_i(p-h_i)-\tilde{a}_i]}. \quad (7.6.78)$$

We shall first carry out an inverse transformation of eqn (7.6.78) for $i = 1$ and $\omega_1 \geq 0$ (recall eqn (7.4.38)). Let us begin with (recall eqns (7.4.76)–(7.4.78))

$$\exp[-x\tilde{s}_1(p-h_1)] = -\sum_{n=0}^{\infty} \frac{(-)^n \hat{\lambda}_1^n}{2^n n!} [\bar{K}(p-h_1)]^n L[A_n'(x_1,t)], \tag{7.6.79}$$

in which the symbols $\hat{\lambda}_1$, $\bar{K}$, and A_n have the same meaning as in Section 7.4, while a prime denotes a differentiation with respect to x_1.

Multiplying eqn (7.6.79) through by $\exp(-\tilde{a}_1 x_1)$ and integrating both sides with respect to x_1 in the interval $[x_1, \infty)$, we obtain

$$\frac{\exp\{-x_1[\tilde{s}_1(p-h_1)+\tilde{a}_1]\}}{\tilde{s}_1(p-h_1)+\tilde{a}_1} = \sum_{n=0}^{\infty} \frac{(-)^n \hat{\lambda}_1^n}{2^n n!} [\bar{K}(p-h_1)]^n L\left[\mathrm{e}^{-\tilde{a}_1 x_1} A_n(x_1,t) - \tilde{a}_1 \int_{x_1}^{\infty} \mathrm{e}^{-\tilde{a}_1 u} A_n(u,t) du\right]. \tag{7.6.80}$$

Multiplying the above formula through by $\exp(\tilde{a}_1 x_1)$ and replacing $\tilde{a}_1$ by $-\tilde{a}_1$, we obtain

$$\frac{\exp[-x_1\tilde{s}_1(p-h_1)]}{\tilde{s}_1(p-h_1)-\tilde{a}_1} = \sum_{n=0}^{\infty} \frac{(-)^n \hat{\lambda}_1^n}{2^n n!} [\bar{K}(p-h_1)]^n L\left[A_n(x_1,t) + \tilde{a}_1 \int_{x_1}^{\infty} \mathrm{e}^{\tilde{a}_1(u-x_1)} A_n(u,t) \mathrm{d}u\right]. \tag{7.6.81}$$

From this and the definition of $P_1(x,t)$ (recall eqn (7.6.75)), we get

$$\begin{aligned} P_1(x,t) &= v_1 \exp(-h_1 t) \sum_{n=0}^{\infty} \frac{(-)^n \hat{\lambda}_1^n}{n!} \\ &\times g_n^{(1)}(t) * \left[A_n(x_1,t) + \tilde{a}_1 \int_{x_1}^{\infty} \mathrm{e}^{\tilde{a}_1(u-x_1)} A_n(u,t) du\right], \end{aligned} \tag{7.6.82}$$

or

$$P_1(x,t) = \hat{P}_1^{(0)}(x_1,t) H(\zeta_1), \tag{7.6.83}$$

where

$$\begin{aligned} \hat{P}_1^{(0)}(x_1,t) &= v_1 \exp(-h_1 t) \sum_{n=0}^{\infty} \frac{(-)^n \hat{\lambda}_1^n}{n!} \int_{x_1}^{t} g_n^{(1)}(t-s) \\ &\times \left\{\hat{A}_n(x_1,s) + a v_1 \int_{x_1}^{s} \exp[a v_1 (u-x_1)] \hat{A}_n(u,s) \mathrm{d}u\right\} \mathrm{d}s, \end{aligned} \tag{7.6.84}$$

and

$$\hat{A}_n(x_1,t) = \omega_1^{-n} \left[\sqrt{\omega_1(t^2-x_1^2)}\right]^n J_n\left(\sqrt{\omega_1(t^2-x_1^2)}\right). \tag{7.6.85}$$

The formula (7.6.83) is true for $\omega_1 \geq 0$. For $\omega_1 \leq 0$ we obtain

$$P_1(x,t) = \hat{P}_1(x_1,t)H(\zeta_1), \tag{7.6.86}$$

where

$$\begin{aligned}\hat{P}_1(x_1,t) = v_1 \exp(-h_1 t) \sum_{n=0}^{\infty} \frac{(-)^n \hat{\lambda}_1^n}{n!} \int_{x_1}^{t} g_n^{(1)}(t-s) \\ \times \left\{ \hat{B}_n(x_1,s) + a v_1 \int_{x_1}^{s} \exp[a v_1 (u - x_1)] \hat{B}_n(u,s) \mathrm{d}u \right\} \mathrm{d}s,\end{aligned} \tag{7.6.87}$$

and

$$\hat{B}_n(x_1,t) = \hat{\omega}_1^{-n} \left[\sqrt{\hat{\omega}_1 (t^2 - x_1^2)} \right]^n I_n \left(\sqrt{\hat{\omega}_1 (t^2 - x_1^2)} \right), \tag{7.6.88}$$

whereby

$$\hat{\omega}_1 = -\omega_1 > 0.$$

Similarly, inverting eqn (7.6.78) for $i = 2$ we find

$$P_2(x,t) = \hat{P}_2(x_2,t)H(\zeta_2), \tag{7.6.89}$$

where

$$\begin{aligned}\hat{P}_2(x_2,t) = v_2 \exp(-h_2 t) \sum_{n=0}^{\infty} \frac{\hat{\lambda}_2^n}{n!} \int_{x_2}^{t} g_n^{(2)}(t-s) \\ \times \left\{ \hat{C}_n(x_2,s) + a v_2 \int_{x_2}^{s} \exp[a v_2 (u - x_2)] \hat{C}_n(u,s) \mathrm{d}u \right\} \mathrm{d}s,\end{aligned} \tag{7.6.90}$$

and

$$\hat{C}_n(x_2,t) = \hat{\omega}_2^{-n} \left[\sqrt{\hat{\omega}_2 (t^2 - x_2^2)} \right]^n I_n \left(\sqrt{\hat{\omega}_2 (t^2 - x_2^2)} \right), \tag{7.6.91}$$

whereby

$$\hat{\omega}_2 = -\omega_2 > 0.$$

Now, we consider the function $Q_i(x,t)$ (recall eqn (7.6.47))

$$Q_i(x,t) = L^{-1} \left\{ \frac{\exp[-x s_i(p)]}{[(p-\alpha)^2 + \beta^2]^{1/2} [s_i(p) - a] s_i(p)} \right\}. \tag{7.6.92}$$

Given that

$$\frac{1}{[s_i(p) - a] s_i(p)} = \frac{1}{a} \left(\frac{1}{s_i(p) - a} - \frac{1}{s_i(p)} \right), \tag{7.6.93}$$

by using a method analogous to that for finding the function $P_i(x,t)$, we obtain

- For $\omega_1 \geq 0$

$$Q_1(x,t) = \hat{Q}_1^{(0)}(x_1,t)H(\zeta_1), \tag{7.6.94}$$

where

$$\hat{Q}_1^{(0)}(x_1,t) = v_1^2 \exp(-h_1 t) \sum_{n=0}^{\infty} \frac{(-)^n \hat{\lambda}_1^n}{n!} \int_{x_1}^{t} g_n^{(1)}(t-s) \times \left\{ \int_{x_1}^{s} \exp[a v_1(u-x_1)] \hat{A}_n(u,s) \mathrm{d}u \right\} \mathrm{d}s. \tag{7.6.95}$$

- For $\omega_1 \leq 0$

$$Q_1(x,t) = \hat{Q}_1(x_1,t) H(\zeta_1), \tag{7.6.96}$$

where

$$\hat{Q}_1(x_1,t) = v_1^2 \exp(-h_1 t) \sum_{n=0}^{\infty} \frac{(-)^n \hat{\lambda}_1^n}{n!} \int_{x_1}^{t} g_n^{(1)}(t-s) \times \left\{ \int_{x_1}^{s} \exp[a v_1(u-x_1)] \hat{B}_n(u,s) \mathrm{d}u \right\} \mathrm{d}s. \tag{7.6.97}$$

Furthermore,

$$Q_2(x,t) = \hat{Q}_2(x_2,t) H(\zeta_2), \tag{7.6.98}$$

where

$$\hat{Q}_2(x_2,t) = v_2^2 \exp(-h_2 t) \sum_{n=0}^{\infty} \frac{\hat{\lambda}_2^n}{n!} \int_{x_2}^{t} g_n^{(2)}(t-s) \times \left\{ \int_{x_2}^{s} \exp[a v_2(u-x_2)] \hat{C}_n(u,s) \mathrm{d}u \right\} \mathrm{d}s. \tag{7.6.99}$$

Finally, let us consider the function (recall eqn (7.6.48))

$$R_i(x,t) = L^{-1} \left\{ \frac{\exp[-x s_i(p)]}{[(p-\alpha)^2 + \beta^2]^{1/2} [s_i^2(p) - a^2]} \right\}. \tag{7.6.100}$$

On account of the fact that

$$\frac{1}{s_i^2(p) - a^2} = \frac{1}{2a} \left[\frac{1}{s_i(p) - a} - \frac{1}{s_i(p) + a} \right], \tag{7.6.101}$$

and using the formula for $P_i(x,t)$, we obtain:

- For $\omega_1 \geq 0$

$$R_1(x,t) = \hat{R}_1^{(0)}(x_1,t) H(\zeta_1), \tag{7.6.102}$$

where

$$\hat{R}_1^{(0)}(x_1,t) = v_1^2 \exp(-h_1 t) \sum_{n=0}^{\infty} \frac{(-)^n \hat{\lambda}_1^n}{n!} \int_{x_1}^{t} g_n^{(1)}(t-s) \times \left\{ \int_{x_1}^{s} \cosh[a v_1(u-x_1)] \hat{A}_n(u,s) \mathrm{d}u \right\} \mathrm{d}s. \tag{7.6.103}$$

- For $\omega_1 \leq 0$

$$R_1(x,t) = \hat{R}_1(x_1,t)H(\zeta_1), \tag{7.6.104}$$

where

$$\begin{aligned}\hat{R}_1(x_1,t) = v_1^2 \exp(-h_1 t)\sum_{n=0}^{\infty}\frac{(-)^n\hat{\lambda}_1^n}{n!}\int_{x_1}^{t} g_n^{(1)}(t-s)\\ \times\left\{\int_{x_1}^{s}\cosh[av_1(u-x_1)]\hat{B}_n(u,s)\mathrm{d}u\right\}\mathrm{d}s.\end{aligned} \tag{7.6.105}$$

Moreover,

$$R_2(x,t) = \hat{R}_2(x_2,t)H(\zeta_2), \tag{7.6.106}$$

where

$$\begin{aligned}\hat{R}_2(x_2,t) = v_2^2 \exp(-h_2 t)\sum_{n=0}^{\infty}\frac{\hat{\lambda}_2^n}{n!}\int_{x_2}^{t} g_n^{(2)}(t-s)\\ \times\left\{\int_{x_2}^{s}\cosh[av_2(u-x_2)]\hat{C}_n(u,s)\mathrm{d}u\right\}\mathrm{d}s.\end{aligned} \tag{7.6.107}$$

Thus, the pair $[\sigma(x,t),\theta(x,t)]$ specified by the formula (7.6.65) has been determined in a closed form.

Let us now consider the particular case of the solution (7.6.65) in which the width of the laser impulse is small, i.e. $b\to\infty$. We restrict the considerations to the case $\omega_1 \leq 0$. Passing with b to infinity in eqns (7.6.65), and using the relations (recall eqns (7.6.61)–(7.6.64))

$$\begin{aligned}f_1(t)\to 0,\ g_1(t)\to 0 \text{ for } b\to\infty,\ t>0,\\ g(t)\to 0,\ \ h(t)\to 0\ \text{ for } b\to\infty,\ t>0,\end{aligned} \tag{7.6.108}$$

we obtain

$$\begin{aligned}\sigma(x,t) = \sigma_0(x,t) + \sigma_1(x_1,t) + \sigma_2(x_2,t),\\ \theta(x,t) = \theta_1(x_1,t) + \theta_2(x_2,t),\end{aligned} \tag{7.6.109}$$

where

$$\sigma_0(x,t) = t^0\left\{[R_1(t) - R_2(t)]\mathrm{e}^{-ax} - \frac{1}{2}[H(\zeta_1)\hat{Q}_1(x_1,t) - H(\zeta_2)\hat{Q}_2(x_2,t)]\right\}, \tag{7.6.110}$$

and

$$\sigma_i(x_i,t) = -\frac{1}{2}H(\zeta_i)\left\{\exp(-h_i x_i)\sigma_0(0,\zeta_i)\right.$$
$$\left. + \int_{x_i}^{t} [\hat{N}_i^{(0)}(x_i,s)\sigma_0(0,t-s) \mp \hat{M}_i(x_i,s)\tilde{\sigma}_0(0,t-s)]\mathrm{d}s\right\}, \qquad (7.6.111)$$

$$\theta_i(x_i,t) = \mp\frac{1}{2}H(\zeta_i)\int_{x_i}^{t} \hat{M}_i(x_i,s)\tilde{\theta}_0(0,t-s)\mathrm{d}s. \qquad (7.6.112)$$

The minus (plus) sign in the formulas (7.6.111) and (7.6.112) corresponds to $i = 1$ ($i = 2$).

Furthermore,

$$\begin{aligned}\tilde{\sigma}_0(0,t) &= \hat{\alpha}\sigma_0(0,t) + \tilde{\beta}\dot{\sigma}_0(0,t),\\ \tilde{\theta}_0(0,t) &= \epsilon\hat{\gamma}\sigma_0(0,t),\end{aligned} \qquad (7.6.113)$$

where

$$\begin{aligned}\sigma_0(0,t) &= t^0[Z_1(t) - Z_2(t)],\\ \dot{\sigma}_0(0,t) &= t^0[\dot{Z}_1(t) - \dot{Z}_2(t)].\end{aligned} \qquad (7.6.114)$$

In the above

$$Z_i(t) = \frac{1}{2}v_i^2\exp(-h_i t)\sum_{n=0}^{\infty}\frac{(\mp\hat{\lambda}_i)^n}{n!}\int_0^t\left\{g_n^{(i)}(t-s)\left[\int_0^s \mathfrak{A}_n^{(i)}(u,s)\mathrm{d}u\right]\right\}\mathrm{d}s,$$
$$\dot{Z}_i(t) = -h_i Z_i(t) + \frac{1}{2}v_i^2\exp(-h_i t)$$
$$\times\left\{\int_0^t \mathfrak{A}_0^{(i)}(u,t)\mathrm{d}u + \sum_{n=0}^{\infty}\frac{(\mp\hat{\lambda}_i)^n}{n!}\int_0^t\left\{\dot{g}_n^{(i)}(t-s)\left[\int_0^s \mathfrak{A}_n^{(i)}(u,s)\mathrm{d}u\right]\right\}\mathrm{d}s\right\}, \qquad (7.6.115)$$

whereby the minus (plus) sign is taken for $i = 1$ ($i = 2$), while the functions $\mathfrak{A}_n^{(i)}(u,s)$ are specified by the relations

$$\mathfrak{A}_n^{(i)}(u,s) = A_n^{(i)}(u,s)\exp(-av_i u), \qquad (7.6.116)$$

in which

$$A_n^{(i)}(u,s) = \hat{\omega}_i^{-n}\left(\sqrt{\hat{\omega}_i(s^2-u^2)}\right)^n I_n\left(\sqrt{\hat{\omega}_i(s^2-u^2)}\right). \qquad (7.6.117)$$

Functions $R_i(t)$ and $\hat{Q}_i(x_i,t)$ appearing in eqn (7.6.110) are given by the formulas (recall eqns (7.6.104)–(7.6.107) for $x_i = 0$, and eqns (7.6.96)–(7.6.99))

$$R_i(t) = v_i^2\exp(-h_i t)\sum_{n=0}^{\infty}\frac{(\mp\hat{\lambda}_i)^n}{n!}\int_0^t\left\{g_n^{(i)}(t-s)\left[\int_0^s B_n^{(i)}(u,s)\mathrm{d}u\right]\right\}\mathrm{d}s, \qquad (7.6.118)$$

and

$$\hat{Q}_i(x_i,t) = v_i^2 \exp(-h_i t) \sum_{n=0}^{\infty} \frac{(\mp\hat{\lambda}_i)^n}{n!} \int_{x_i}^{t} \left\{ g_n^{(i)}(t-s) \left[\int_{x_i}^{s} C_n^{(i)}(x_i;u,s)\mathrm{d}u \right] \right\} \mathrm{d}s, \tag{7.6.119}$$

where

$$\begin{aligned} B_n^{(i)}(u,s) &= \cosh(av_i u) A_n^{(i)}(u,s), \\ C_n^{(i)}(x_i;u,s) &= \exp[av_i(u-x_i)] A_n^{(i)}(u,s). \end{aligned} \tag{7.6.120}$$

Functions $\hat{N}_i^{(0)}(x_i,t)$ and $\hat{M}_i(x_i,t)$ appearing in eqns (7.6.111) and (7.6.112) are given by the formulas (recall eqns (7.4.65) and (7.4.71); and eqns (7.4.83) and (7.4.85))

$$\hat{N}_i^{(0)}(x_i,t) = x_i \exp(-h_i t)[A_{-1}^{(i)}(x_i,t) + \sum_{n=1}^{\infty} \frac{(\mp\hat{\lambda}_i)^n}{n!} \int_{x_i}^{t} h_n^{(i)}(t-s) A_{n-1}^{(i)}(x_i,s)\mathrm{d}s], \tag{7.6.121}$$

and

$$\hat{M}_i(x_i,t) = \exp(-h_i t)[g_0^{(i)}(\zeta_i) + x_i \sum_{n=0}^{\infty} \frac{(\mp\hat{\lambda}_i)^n}{n!} \int_{x_i}^{t} g_n^{(i)}(t-s) A_n^{(i)}(x_i,s)\mathrm{d}s], \tag{7.6.122}$$

where the functions $g_n^{(i)}(t)$ and $h_n^{(i)}(t)$ have the same meaning as in Section 7.4.

In order to examine the behavior of the pair (σ,θ) in a fixed cross-section of the half-space $x = x_0 > 0$, we adopt the following notations

$$\begin{aligned} S(t) &= \sigma(x_0,t)/t^0, \quad \theta(t) = \theta(x_0,t)/(t^0\varepsilon\hat{\gamma}), \\ S_0(t) &= \sigma_0(x_0,t)/t^0, \\ S_i(t) &= \sigma_i(x_i^0,t)/t^0, \ \theta_i(t) = \theta(x_i^0,t)/(t^0\varepsilon\hat{\gamma}), \\ x_i^0 &\equiv t_i^0 = x_0/v_i, \qquad \zeta_i^0 = t - t_i^0, \end{aligned} \tag{7.6.123}$$

and, using eqns (7.6.109)–(7.6.112), we obtain

$$S(t) = S_0(t) + S_1(t) + S_2(t), \qquad \theta(t) = \theta_1(t) + \theta_2(t), \tag{7.6.124}$$

where

$$S_0(t) = [R_1(t) - R_2(t)]\exp(-ax_0) - \frac{1}{2}[H(\zeta_1^0)\hat{Q}_1(t_1^0, t) - H(\zeta_2^0)\hat{Q}_2(t_2^0, t)], \tag{7.6.125}$$

$$S_i(t) = -\frac{1}{2}H(\zeta_i^0)\Big\{ \exp(-h_i t_i^0)\Sigma(\zeta_i^0) + \int_{t_i^0}^{t} [\hat{N}_i^{(0)}(t_i^0, s)\Sigma(t-s) \mp \hat{M}_i^{(0)}(t_i^0, s)\tilde{\Sigma}(t-s)]\mathrm{d}s\Big\}, \tag{7.6.126}$$

$$\theta_i(t) = \mp\frac{1}{2}H(\zeta_i^0)\int_{t_i^0}^{t} \hat{M}_i^{(0)}(t_i^0, s)\Sigma(t-s)\mathrm{d}s. \tag{7.6.127}$$

Functions $\Sigma = \Sigma(t)$ and $\tilde{\Sigma} = \tilde{\Sigma}(t)$ appearing in eqns (7.6.126) and (7.6.127) are given by the formulas

$$\begin{aligned} \Sigma(t) &= Z_1(t) - Z_2(t), \\ \tilde{\Sigma}(t) &= \hat{\alpha}\Sigma(t) + \hat{\beta}\dot{\Sigma}(t). \end{aligned} \tag{7.6.128}$$

It follows from eqns (7.6.124)–(7.6.127) that the discontinuities of the pair $[S(t), \theta(t)]$ may arise at most at the wavefronts $t = t_i^0$ $(i = 1, 2)$.

We shall now demonstrate that also for $t = t_i^0$ this pair is a continuous function. Since $v_2 > v_1 > 0$, then $0 < t_2^0 < t_1^0$ and $\zeta_2^0 > \zeta_1^0$, and the time interval $[0, \infty)$ can be represented in the form

$$[0, \infty) = [0, t_2^0) \cup [t_2^0, t_1^0) \cup [t_1^0, \infty).$$

Case 1: $0 \le t < t_2^0$. In that case $\zeta_i^0 < 0$, and hence

$$\begin{aligned} S_0(t) &= S_0^{(1)}(t) \quad \forall t \in [0, t_2^0), \\ S_i(t) &= \theta_i(t) = 0 \ \forall t \in [0, t_2^0), \end{aligned}$$

where

$$S_0^{(1)}(t) = [R_1(t) - R_2(t)]\exp(-ax_0).$$

Hence

$$S(t) = S_0^{(1)}(t),\ \theta(t) = 0\ \forall t \in [0, t_2^0).$$

Furthermore,

$$\begin{aligned} S_0(t_2^0 - 0) &= S_0^{(1)}(t_2^0), \\ S_i(t_2^0 - 0) &= 0, \qquad \theta_i(t_2^0 - 0) = 0. \end{aligned}$$

from which there follows

$$S(t_2^0 - 0) = S_0^{(1)}(t_2^0),\ \theta(t_2^0 - 0) = 0. \tag{*}$$

Case 2: $t_2^0 < t < t_1^0$. In that case $\zeta_2^0 > 0, \zeta_1^0 < 0$, and hence

$$S_0(t) = S_0^{(2)}(t), \quad S_1(t) = 0, \quad S_2(t) = S_2^{(2)}(t),$$
$$\theta_1(t) = 0, \quad \theta_2(t) = \theta_2^{(2)}(t),$$

where

$$S_0^{(2)}(t) = S_0^{(1)}(t) + \frac{1}{2}\hat{Q}_2(t_2^0, t),$$

$$S_2^{(2)}(t) = -\frac{1}{2}\Big\{ \exp(-h_2 t_2^0)\Sigma(t - t_2^0)$$
$$+ \int_{t_2^0}^{t} [\hat{N}_2^{(0)}(t_2^0, s)\Sigma(t-s) + \hat{M}_2^{(0)}(t_2^0, s)\tilde{\Sigma}(t-s)]\mathrm{d}s \Big\}$$

$$\theta_2^{(2)}(t) = \frac{1}{2}\int_{t_2^0}^{t} \hat{M}_2^{(0)}(t_2^0, s)\Sigma(t-s)\mathrm{d}s.$$

Hence,

$$S(t) = S_0^{(2)}(t) + S_2^{(2)}(t), \; \theta(t) = \theta_2^{(2)}(t).$$

Furthermore,

$$S_0(t_2^0 + 0) = S_0^{(2)}(t_2^0) = S_0^{(1)}(t_2^0),$$
$$S_1(t_2^0 + 0) = 0, \; S_2(t_2^0 + 0) = 0,$$
$$\theta_1(t_2^0 + 0) = 0, \; \theta_2(t_2^0 + 0) = 0,$$

so that

$$S(t_2^0 + 0) = S_0^{(1)}(t_2^0), \; \theta(t_2^0 + 0) = 0. \qquad (**)$$

Next, we compute the limits for $t \to t_1^0 - 0$ and obtain

$$S_0(t_1^0 - 0) = S_0^{(2)}(t_1^0),$$
$$S_1(t_1^0 - 0) = 0, \quad S_2(t_1^0 - 0) = S_2^{(2)}(t_1^0).$$

Thus,

$$S(t_1^0 - 0) = S_0^{(2)}(t_1^0) + S_2^{(2)}(t_1^0),$$
$$\theta(t_1^0 - 0) = \theta_2^{(2)}(t_1^0). \qquad (***)$$

Case 3: $t_1^0 < t < \infty$. In that case $\zeta_i^0 > 0$ and we obtain

$$S_0(t) = S_0^{(3)}(t), \; S_i(t) = S_i^{(3)}(t),$$
$$\theta_i(t) = \theta_i^{(3)}(t),$$

where

$$S_0^{(3)}(t) = S_0^{(2)}(t) - \frac{1}{2}\hat{Q}_1(t_1^0, t),$$

$$S_i^{(3)}(t) = -\frac{1}{2}\Big\{ \exp(-h_i t_i^0)\Sigma(t - t_i^0)$$
$$+ \int_{t_i^0}^{t} [\hat{N}_i^{(0)}(t_i^0, s)\Sigma(t-s) \mp \hat{M}_i(t_i^0, s)\tilde{\Sigma}(t-s)]\mathrm{d}s\Big\},$$

$$\theta_i^{(3)}(t) = \mp\frac{1}{2}\int_{t_i^0}^{t} \hat{M}_i(t_i^0, s)\Sigma(t-s)\mathrm{d}s.$$

Hence

$$S(t) = S_0^{(3)}(t) + S_1^{(3)}(t) + S_2^{(3)}(t), \qquad \theta(t) = \theta_1^{(3)}(t) + \theta_2^{(3)}(t).$$

Furthermore,

$$S_0^{(3)}(t_1^0 + 0) = S_0^{(2)}(t_1^0),$$
$$S_1^{(3)}(t_1^0 + 0) = 0, \qquad S_2^{(3)}(t_1^0 + 0) = S_2^{(3)}(t_1^0),$$
$$\theta_1^{(3)}(t_1^0 + 0) = 0, \qquad \theta_2^{(3)}(t_1^0 + 0) = \theta_2^{(3)}(t_1^0),$$

so that

$$S(t_1^0 + 0) = S_0^{(2)}(t_1^0) + S_2^{(3)}(t_1^0), \qquad \theta(t_1^0 + 0) = \theta_2^{(3)}(t_1^0). \qquad (****)$$

Next, we use the definition of a jump of function $f(t)$ given by the formula (recall eqn (7.4.123))

$$[[f]]\,(t) = f(t-0) - f(t+0) \qquad t \geq 0,$$

from which, on account of (*) and (**), we obtain

$$[[S]]\,(t_2^0) = 0, \qquad [[\theta]]\,(t_2^0) = 0.$$

Also, in view of (***) and (****), we get

$$[[S]]\,(t_1^0) = S_2^{(2)}(t_1^0) - S_2^{(3)}(t_1^0) = 0,$$
$$[[\theta]]\,(t_1^0) = \theta_2^{(2)}(t_1^0) - \theta_2^{(3)}(t_1^0) = 0,$$

which implies that the pair $[S(t), \theta(t)]$ is continuous for $t = t_1^0$ and $t = t_2^0$.

Hence, using eqns (7.6.65) and (7.6.109), we infer that the following theorem holds true:

Theorem 7.3 *A stress–temperature response of the half-space to a laser irradiation corresponding to the heat sources (7.6.32) is a superposition of the three disturbances one of which has a diffusion-type character (for every instant $t > 0$ it occupies the entire half-space $x > 0$), while the other two are plane waves propagating from the boundary $x = 0$ into the half-space at two different speeds v_i,* *attenuations* h_i *and convolution coefficients* λ_i $(i = 1, 2)$. *Each of these disturbances is represented by a power series of the Neumann type involving Bessel functions. For a short laser pulse* $(b \to \infty)$, *a diffusion part of the temperature vanishes, while a diffusion part of the stress field is comparable to its wave part. In that particular case, the stress–temperature response of the half-space is a continuous function for every* $t \geq 0$.

Remark 7.4 A generalization of the solution described in Section 7.6 to include the laser-induced heat of the form

$$r(x,t) = Y(t)\exp(-ax),$$

where

$$Y(t) = Y_0 t^n \exp(-bt^m), \qquad (m > 0, n > 0, b > 0, a > 0, Y_0 > 0),$$

was obtained in (Hetnarski and Ignaczak, 1993, 1994;). Also, an analysis of the laser-induced thermoelastic waves has been presented in (Strikverda and Scott, 1984; Tzou, 1997; Suh and Burger, 1998; Al-Nimr, and Al-Huniti, 2000; Wang and Xu, 2002; Al-Huniti, Al-Nimr, and Megdad, 2003; Tzou, Chen, and Beraun, 2005; El-Karamany and Ezzat, 2005; Al-Qahtani, Datta and Mukdadi, 2005; Yilbas and Ageeli, 2006; Al-Qahtani and Datta, 2008). Hyperbolic heat conduction in a rigid solid due to a mode-locked laser pulse train is discussed in (Hector *et al.*, 1992).

Remark 7.5 A mathematical theory related to the laser-solution obtained by Hetnarski and Ignaczak in 1994 (see Remark 7.4) was presented by the late Prof. Gaetano Fichera in (Fichera, 1997).

8

KIRCHHOFF-TYPE FORMULAS AND INTEGRAL EQUATIONS IN GREEN–LINDSAY THEORY

8.1 Integral representations of fundamental solutions

In Section 7.1 we presented the series forms of fundamental solutions of the GL theory for a 3D bounded region under the assumption that the eigenfunctions and eigenvalues of the operator $-\nabla^2$ are known. We now demonstrate how one can obtain the Kirchhoff-type integral representations for both fundamental solutions. These representations will then be used to show that the fundamental solutions may also be obtained by solving a system of singular integral equations.

According to the definition of Section 7.1, the ith fundamental solution in the G–L theory satisfies the relations (recall eqns (7.1.4)–(7.1.6))

$$\hat{L}_i \phi_i = 0 \quad \text{on} \quad \mathrm{B} \times [0, \infty), \tag{8.1.1}$$

$$\phi_i(\cdot, 0) = \dot{\phi}_i(\cdot, 0) = 0 \quad \text{on} \quad \mathrm{B}, \tag{8.1.2}$$

$$\phi_i = f_i \quad \text{on} \quad \partial\mathrm{B} \times [0, \infty). \tag{8.1.3}$$

Here, $\hat{L}_i$ $(i = 1, 2)$ is the wave-like operator, defined by the formulas (7.1.2) and (7.1.3), while f_i $(i = 1, 2)$ is an arbitrary function defined on $\partial\mathrm{B} \times [0, \infty)$. Let $K_i = K_i(x, y; t)$ be the Green's function for the operator $\hat{L}_i$ defined on $\mathrm{E}^3 \times [0, \infty)$ and satisfying the relations

$$\hat{L}_i K_i(x, y; t) = -\delta(x - y)\,\delta(t) \quad \text{on} \quad \mathrm{E}^3 \times (0, \infty), \tag{8.1.4}$$

$$K_i(x, y; 0) = \dot{K}_i(x, y; 0) = 0 \quad \text{on} \quad \mathrm{E}^3 \text{ for } x \neq y, \tag{8.1.5}$$

$$K_i(x, y; t) \to 0 \quad \text{for } |x| \to \infty \text{ and } \forall t > 0, \tag{8.1.6}$$

where y is a fixed point in E^3.

Applying the Laplace transform to eqns (8.1.1) and (8.1.4), and using the homogeneous initial conditions (8.1.2) and (8.1.5), we obtain

$$\Box_i^2 \bar{\phi}_i = 0 \quad \text{on} \quad \mathrm{B}, \tag{8.1.7}$$

$$\Box_i^2 \bar{K}_i = -\delta(x - y) \quad \text{on} \quad \mathrm{E}^3, \tag{8.1.8}$$

where (recall eqns (7.4.14) and (7.4.15))

$$\Box_i^2 = \nabla^2 - s_i^2(p) \quad i = 1, 2 \tag{8.1.9}$$

and the functions $s_i^2(p)$ are defined by the formula (7.4.15).

Multiplying eqn (8.1.7) through by $\bar{K}_i$, and eqn (8.1.8) by $\bar{\phi}_i$, subtracting one from another, and integrating over B, we obtain

$$\epsilon(y)\,\bar{\phi}_i(y,p) = \int_{\partial \mathrm{B}} \left[\bar{K}_i(x,y;p)\frac{\partial \bar{\phi}_i}{\partial n}(x,p) - \bar{\phi}_i(x,p)\frac{\partial \bar{K}_i}{\partial n}(x,y;p)\right] \mathrm{d}a(x), \tag{8.1.10}$$

where

$$\epsilon(y) = \begin{cases} 1 \text{ for } & y \in \mathrm{B}, \\ 0 \text{ for } & y \in \mathrm{E}^3 - \mathrm{B}. \end{cases} \tag{8.1.11}$$

A solution of eqn (8.1.8) vanishing at infinity takes the form

$$\bar{K}_i = \frac{1}{4\pi\rho}\exp\left[-\rho s_i(p)\right], \tag{8.1.12}$$

where

$$\rho = |x - y|. \tag{8.1.13}$$

Employing the same notation as in Section 7.4 (recall eqns (7.4.25), (7.4.45) and (7.4.73)), that is,

$$\bar{N}_i(\rho,p) = \exp\left[-\rho s_i(p)\right], \tag{8.1.14}$$

$$\rho_i = \rho/\upsilon_i, \tag{8.1.15}$$

we reduce eqn (8.1.12) to the form

$$\bar{K}_i = \frac{1}{4\pi\upsilon_i}\left(\frac{\bar{N}_i}{\rho_i}\right) \quad i = 1,2. \tag{8.1.16}$$

Let us first consider the case $i = 1$. Then, applying the inverse Laplace transform to eqn (8.1.16) at $i = 1$ we obtain [recall eqn (7.4.62)],

$$K_1(\rho_1,t) = \frac{1}{4\pi\upsilon_1}\left\{\rho_1^{-1}\exp(-h_1\rho_1)\,\delta(\zeta_1) + \exp(-h_1 t)\,\mathcal{N}_1(\rho_1,t)\,H(\zeta_1)\right\}, \tag{8.1.17}$$

where (recall eqn (7.4.97))

$$\mathcal{N}_1(\rho_1,t) = -\omega_1\frac{J_1(\hat{z}_1)}{\hat{z}_1} + \sum_{n=1}^{\infty}\frac{(-)^n\,\hat{\lambda}_1^n}{n!\;\omega_1^{n-1}}\int_{\rho_1}^{t} h_n^{(1)}(t-s)\,z_1^{n-1}J_{n-1}(z_1)\,\mathrm{d}s. \tag{8.1.18}$$

For $i = 2$, on account of eqn (8.1.16), we find (recall eqns (7.4.71) and (7.4.98))

$$K_2(\rho_2,t) = \frac{1}{4\pi\upsilon_2}\left\{\rho_2^{-1}\exp(-h_2\rho_2)\,\delta(\zeta_2) + \exp(-h_2 t)\,\mathcal{N}_2(\rho_2,t)\,H(\zeta_2)\right\}, \tag{8.1.19}$$

where

$$\mathcal{N}_2(\rho_2,t)=\hat{\omega}_2\frac{I_1(\hat{z}_2)}{\hat{z}_2}+\sum_{n=1}^{\infty}\frac{\hat{\lambda}_2^n}{n!\,\hat{\omega}_2^{n-1}}\int_{\rho_2}^{t}h_n^{(2)}(t-s)\,z_2^{n-1}I_{n-1}(z_2)\,\mathrm{d}s. \tag{8.1.20}$$

All the symbols appearing in the above formulas have been explained in Section 7.4. In particular

$$\zeta_i=t-\rho_i, \tag{8.1.21}$$

and the surface $\zeta_i=0$ is a carrier of the singularity of kernel $K_i(\rho_i,t)$. Furthermore, (recall formulas (7.4.100))

$$\begin{cases} z_1=\left[\omega_1\left(s^2-\rho_1^2\right)\right]^{1/2},\ z_2=\left[\hat{\omega}_2\left(s^2-\rho_2^2\right)\right]^{1/2},\\ \hat{z}_1=\left[\omega_1\left(t^2-\rho_1^2\right)\right]^{1/2},\ \hat{z}_2=\left[\hat{\omega}_2\left(t^2-\rho_2^2\right)\right]^{1/2}.\end{cases} \tag{8.1.22}$$

Employing the formulas (7.4.17)–(7.4.22) for the partial derivatives $\partial K_i/\partial\rho_i$, we obtain

$$\begin{aligned}\frac{\partial K_1}{\partial\rho_1}&=\exp(-h_1t)\,\mathcal{N}_1'(\rho_1,t)\,H(\zeta_1)\\ &+\frac{1}{4\pi\upsilon_1}\left[-\rho_1^{-1}\dot{\delta}(\zeta_1)+\left(\frac{\omega_1}{2}-\rho_1^{-2}-h_1\rho_1^{-1}\right)\delta(\zeta_1)\right]\exp(-h_1\rho_1),\end{aligned} \tag{8.1.23}$$

$$\begin{aligned}\frac{\partial K_2}{\partial\rho_2}&=\exp(-h_2t)\,\mathcal{N}_2'(\rho_2,t)\,H(\zeta_2)\\ &+\frac{1}{4\pi\upsilon_2}\left[-\rho_2^{-1}\dot{\delta}(\zeta_2)-\left(\frac{\hat{\omega}_2}{2}+\rho_2^{-2}+h_2\rho_2^{-1}\right)\delta(\zeta_2)\right]\exp(-h_2\rho_2).\end{aligned} \tag{8.1.24}$$

Here, the prime indicates a derivative with respect to ρ_i. Thus, applying the inverse Laplace transform to eqn (8.1.10) and employing eqns (8.1.17), (8.1.19), (8.1.23) and (8.1.24) we obtain

$$\begin{aligned}\epsilon(y)\,\phi_1(y,t)&=\frac{1}{4\pi\upsilon_1}\int\limits_{\partial\mathrm{B}\cap S(y,\upsilon_1t)}\exp(-h_1\rho_1)\left\{\frac{1}{\rho_1}\left[\frac{\partial\phi_1}{\partial n}\right]_1-\frac{\partial}{\partial n}\left(\frac{1}{\rho_1}\right)[\phi_1]_1\right.\\ &\left.+\left(\frac{h_1}{\rho_1}-\frac{\omega_1}{2}\right)\frac{\partial\rho_1}{\partial n}[\phi_1]_1+\frac{1}{\rho_1}\frac{\partial\rho_1}{\partial n}\left[\frac{\partial\phi_1}{\partial t}\right]_1\right\}\mathrm{d}a(x)\\ &+\frac{1}{4\pi\upsilon_1}\int\limits_{\partial\mathrm{B}\cap S(y,\upsilon_1t)}\left\{\int_{\rho_1}^{t}\exp(-h_1\tau)\right.\\ &\times\left.\left[\mathcal{N}_1(\rho_1,\tau)\frac{\partial\phi_1}{\partial n}(x,t-\tau)-\phi_1(x,t-\tau)\frac{\partial\mathcal{N}_1}{\partial n}(\rho_1,\tau)\right]\mathrm{d}\tau\right\}\mathrm{d}a(x),\end{aligned} \tag{8.1.25}$$

and

$$\epsilon(y)\,\phi_2(y,t) = \frac{1}{4\pi v_2}\int\limits_{\partial \mathrm{B}\cap S(y,v_2t)} \exp(-h_2\rho_2)\left\{\frac{1}{\rho_2}\left[\frac{\partial\phi_2}{\partial n}\right]_2 - \frac{\partial}{\partial n}\left(\frac{1}{\rho_2}\right)[\phi_2]_2 + \left(\frac{h_2}{\rho_2}+\frac{\hat{\omega}_2}{2}\right)\frac{\partial\rho_2}{\partial n}[\phi_2]_2 + \frac{1}{\rho_2}\frac{\partial\rho_2}{\partial n}\left[\frac{\partial\phi_2}{\partial t}\right]_2\right\}\mathrm{d}a(x)$$
$$+\frac{1}{4\pi v_2}\int\limits_{\partial \mathrm{B}\cap S(y,v_2t)}\left\{\int_{\rho_2}^{t}\exp(-h_2\tau)\times\left[\mathcal{N}_2(\rho_2,\tau)\frac{\partial\phi_2}{\partial n}(x,t-\tau) - \phi_2(x,t-\tau)\frac{\partial\mathcal{N}_2}{\partial n}(\rho_2,\tau)\right]\mathrm{d}\tau\right\}\mathrm{d}a(x). \tag{8.1.26}$$

Here, $[\phi]_i$ denotes the retarded value $\phi(x, t-\rho_i)$ of a function $\phi(x,t)$ relative to the point y. Also, $S(y,d)$ stands for an open ball centered at y having radius d, that is

$$S(y,d) = \left\{x \in E^3 : |x-y| < d\right\}. \tag{8.1.27}$$

Equations (8.1.25) and (8.1.26) are the Kirchhoff-type formulas for the fundamental solutions ϕ_1 and ϕ_2 in the G–L theory. In the case where the wave-like operators $\hat{L}_i$ $(i = 1,2)$ reduce to the classical wave operators with speeds v_i, that is where $h_i = 0$ and $\hat{\lambda}_i = 0$, these formulas turn into classical Kirchhoff formulas.

Let $t > 0$ be a fixed time instant, and let $B_i(t)$ be the set:

$$B_i(t) = \{y \in \mathrm{B} : y \text{ does not belong to a boundary layer of width } v_i t\}. \tag{8.1.28}$$

Then, on account of eqns (8.1.25) and (8.1.26), we obtain

$$\phi_i(y,t) = 0,\ \forall y \in B_i(t),\ \forall t > 0. \tag{8.1.29}$$

Therefore, a theorem on the domain of influence for the fundamental solution ϕ_i holds true, stating that for a fixed instant $t > 0$ a domain of influence due to the boundary disturbance f_i (recall eqn (8.1.3)) is a boundary layer of width $v_i t$, i.e. the set $\mathrm{B} - \mathrm{B}_i(t)$.

Note also that analogous formulas of the Kirchhoff type hold for the "exterior" fundamental solutions $\phi_i^{(e)}$ of the G–L theory. In that case, $\phi_i^{(e)}$ satisfies eqn (8.1.1) on $(\mathrm{E}^3 - \bar{\mathrm{B}}) \times [0,\infty)$, the initial conditions (8.1.2) on $\mathrm{E}^3 - \bar{\mathrm{B}}$, and the boundary condition (8.1.3) on $\partial\mathrm{B} \times [0,\infty)$. The integral representations for the "exterior" fundamental solutions $\phi_1^{(e)}$ and $\phi_2^{(e)}$ are obtained upon replacing $\partial/\partial n$ by $-\partial/\partial n$ in eqns (8.1.25) and (8.1.26), and replacing the function $\epsilon(y)$ by the characteristic function of the set $\mathrm{E}^3 - \bar{\mathrm{B}}$.

For the "exterior" fundamental solution $\phi_i^{(e)}$ there also holds a theorem on the domain of influence, which is analogous to that for the "interior" fundamental solution ϕ_i.

8.2 Integral equations for fundamental solutions

We now show that the "exterior" as well as "interior" fundamental solutions of the G–L theory may be found once we can solve a singular integral equation.

Let $\phi_i^{(i)}$ and $\phi_i^{(e)}$ denote, respectively, the "interior" and "exterior" fundamental solutions. It follows from the formulas (8.1.10) and (8.1.16) that the following integral representations are true for these solutions

$$\epsilon(y)\,\phi_i^{(i)}(y,t) = \int_{\partial \mathrm{B}} \left\{ K_i(x,y;t) * \frac{\partial \phi_i^{(i)}}{\partial n}(x,t) - \phi_i^{(i)}(x,t) * \frac{\partial K_i}{\partial n}(x,y;t) \right\} \mathrm{d}a(x), \tag{8.2.1}$$

and

$$\epsilon_0(y)\,\phi_i^{(e)}(y,t) = -\int_{\partial \mathrm{B}} \left\{ K_i(x,y;t) * \frac{\partial \phi_i^{(e)}}{\partial n}(x,t) - \phi_i^{(e)}(x,t) * \frac{\partial K_i}{\partial n}(x,y;t) \right\} \mathrm{d}a(x), \tag{8.2.2}$$

where $\epsilon(y)$ and $\epsilon_0(y)$ denote, respectively, the characteristic functions of the sets B and $\mathrm{E}^3 - \bar{\mathrm{B}}$, that is

$$\epsilon(y) = \begin{cases} 1 \text{ for } & y \in \mathrm{B}, \\ 0 \text{ for } & y \in \mathrm{E}^3 - \bar{\mathrm{B}}, \end{cases} \tag{8.2.3}$$

$$\epsilon_0(y) = \begin{cases} 0 \text{ for } & y \in \mathrm{B}, \\ 1 \text{ for } & y \in \mathrm{E}^3 - \bar{\mathrm{B}}. \end{cases} \tag{8.2.4}$$

Furthermore, the kernel $K_i = K_i(x,y;t)$ appearing in the formulas (8.2.1) and (8.2.2) is defined by eqns (8.1.17) and (8.1.19).

Since, on account of the definition of the "interior" and "exterior" fundamental solution

$$\phi_i^{(i)} = \phi_i^{(e)} = f_i \text{ on } \partial\mathrm{B} \times [0,\infty), \tag{8.2.5}$$

where f_i is a known function, therefore, adding eqns (8.2.1) and (8.2.2) yields

$$\phi_i(y,t) = \int_{\partial \mathrm{B}} K_i(x,y;t) * \sigma_i(x,t)\,\mathrm{d}a(x) \quad \forall (y,t) \in (\mathrm{E}^3 - \partial\mathrm{B}) \times [0,\infty), \tag{8.2.6}$$

where

$$\phi_i(y,t) = \begin{cases} \phi_i^{(i)}(y,t) \text{ for } & (y,t) \in \mathrm{B} \times [0,\infty), \\ \phi_i^{(e)}(y,t) \text{ for } & (y,t) \in (\mathrm{E}^3 - \bar{\mathrm{B}}) \times [0,\infty), \end{cases} \tag{8.2.7}$$

and

$$\sigma_i(x,t) = \frac{\partial}{\partial n}\left[\phi_i^{(i)}(x,t) - \phi_i^{(e)}(x,t)\right]. \tag{8.2.8}$$

It may be shown that the function $\phi_i(y,t)$ defined by eqn (8.2.6) possesses properties of the potential of a single layer, and, in particular, that is a continuous function for every $(y,t) \in E^3 \times [0,\infty)$. Thus, taking the point y in eqn (8.2.6) to the boundary ∂B and employing the boundary condition (8.2.5), we find the following singular integral equation for the "density" σ_i

$$\int_{\partial B} K_i(x,y;t) * \sigma_i(x,t) \, da(x) = f_i(y,t) \; \forall \, (y,t) \in \partial B \times [0,\infty). \tag{8.2.9}$$

The knowledge of σ_i satisfying eqn (8.2.9) allows the determination of the "interior" and "exterior" fundamental solutions of the G–L theory with the help of formula (8.2.6). Employing eqns (8.1.17) and (8.1.19), and using the notation

$$\varphi_i(y,t) = 4\pi \upsilon_i f_i(y,t), \tag{8.2.10}$$

one can reduce eqn (8.2.9) to the form

$$\varphi_i(y,t) = \int_{\partial B \cap S(y,\upsilon_i t)} \left\{ \frac{\exp(-h_i \rho_i)}{\rho_i} \sigma_i(x, t - \rho_i) \right. \tag{8.2.11}$$
$$\left. + \int_{\rho_i}^{t} \exp(-h_i \tau) \, \mathcal{N}_i(\rho_i, \tau) \, \sigma_i(x, t - \tau) \, d\tau \right\} da(x).$$

Since the function $\mathcal{N}_i(\rho_i, \tau)$ in eqn (8.2.11) is bounded for every $(\rho_i, \tau) \in [0,\infty) \times [\rho_i, t]$, the kernel of this integral equation has a singularity of order ρ_i^{-1} for $\rho_i \to 0$ and for every $t > 0$.

8.3 Integral representation of a solution to a central system of equations

We now derive integral representations of Kirchhoff type for the pair (ϕ, ϑ) satisfying the relations (recall eqns (7.2.1)–(7.2.3))

$$\Gamma \phi = 0$$
$$\left(1 + t^0 \frac{\partial}{\partial t}\right) \vartheta = \left(\nabla^2 - \frac{\partial^2}{\partial t^2}\right) \phi \quad \text{on } B \times [0,\infty), \tag{8.3.1}$$

$$\begin{aligned} \phi(\cdot, 0) &= \dot\phi(\cdot, 0) = 0 \\ \vartheta(\cdot, 0) &= \dot\vartheta(\cdot, 0) = 0 \end{aligned} \quad \text{on } B, \tag{8.3.2}$$

$$\phi = f, \; \vartheta = g \quad \text{on } \partial B \times [0,\infty). \tag{8.3.3}$$

Here, f and g are given functions and Γ is the central operator of the G–L theory (recall eqn (7.4.3))

$$\Gamma = \left(\nabla^2 - \frac{\partial^2}{\partial t^2}\right)\left(\nabla^2 - t_0 \frac{\partial^2}{\partial t^2} - \frac{\partial}{\partial t}\right) - \epsilon \nabla^2 \frac{\partial}{\partial t}\left(1 + t^0 \frac{\partial}{\partial t}\right). \tag{8.3.4}$$

Let (ϕ_N, ϑ_N) be a pair of the Nowacki type – that is, a pair of scalar fields defined on $E^3 \times [0, \infty)$ and satisfying the conditions (recall eqns (7.4.1) and (7.4.2))

$$\left.\begin{aligned} \Gamma\phi_N &= -\left(1 + t^0 \frac{\partial}{\partial t}\right)\delta(x-y)\,\delta(t) \\ \left(1 + t^0\frac{\partial}{\partial t}\right)\vartheta_N &= \left(\nabla^2 - \frac{\partial^2}{\partial t^2}\right)\phi_N \end{aligned}\right\} \text{ on } E^3 \times [0,\infty), \tag{8.3.5}$$

$$\left.\begin{aligned} \phi_N(x, y; 0) &= \dot{\phi}_N(x, y; 0) = 0 \\ \vartheta_N(x, y; 0) &= \dot{\vartheta}_N(x, y; 0) = 0 \end{aligned}\right\} \text{ on } E^3 \text{ and for } x \neq y, \tag{8.3.6}$$

$$(\phi_N, v_N) \to (0, 0) \text{ as } |x| \to \infty \ \forall t > 0. \tag{8.3.7}$$

Here, y is a fixed point in E^3.

It follows from eqns (8.3.1) and (8.3.2) that the potential ϕ satisfies the initial conditions (recall eqn (7.2.4))

$$\frac{\partial^k \phi}{\partial t^k}(\cdot, 0) = 0 \text{ on } B \ \forall k = 0, 1, 2, 3. \tag{8.3.8}$$

Similarly, on account of eqns (8.3.5) and (8.3.6), we obtain (recall eqn (7.4.9))

$$\frac{\partial^k \phi_N}{\partial t^k}(x, y; 0) = 0 \text{ on } E^3 \text{ and for } x \neq y \ \forall k = 0, 1, 2, 3. \tag{8.3.9}$$

Applying the Laplace transform to the system (8.3.1) and employing the homogeneous initial conditions (8.3.2) and (8.3.8), yields

$$\left.\begin{aligned} \Box_1^2 \Box_2^2 \bar{\phi} &= 0 \\ (1 + t^0 p)\,\bar{\vartheta} &= (\nabla^2 - p^2)\,\bar{\phi} \end{aligned}\right\} \text{ on } B, \tag{8.3.10}$$

where

$$\Box_i^2 = \nabla^2 - s_i^2(p), \tag{8.3.11}$$

and the functions $s_i(p)$ are defined by the formulas (7.4.15). Similarly, applying the Laplace transform to the system (8.3.5) and using the conditions (8.3.6) and (8.3.9), leads to

$$\left.\begin{aligned} \Box_1^2 \Box_2^2 \bar{\phi}_N &= -(1 + t^0 p)\,\delta(x-y) \\ (1 + t^0 p)\,\bar{\vartheta}_N &= (\nabla^2 - p^2)\,\bar{\phi}_N \end{aligned}\right\} \text{ on } E^3. \tag{8.3.12}$$

Furthermore, we multiply eqn $(8.3.10)_1$ through by $\bar{\phi}_N$ and eqn $(8.3.12)_1$ by $\bar{\phi}$. Subtracting the resulting equations one from another, upon integration over B, we find

$$(1 + t^0 p)\,\epsilon(y)\,\bar{\phi}(y, p) = \int_B \left(\bar{\phi}_N\, \Box_1^2\, \Box_2^2 \bar{\phi} - \bar{\phi}\, \Box_1^2\, \Box_2^2 \bar{\phi}_N\right) dv(x). \tag{8.3.13}$$

Since

$$\square_1^2\square_2^2 = \nabla^4 - \left(s_1^2 + s_2^2\right)\nabla^2 + s_1^2 s_2^2, \tag{8.3.14}$$

there follows

$$\begin{gathered}\bar{\phi}_N\,\square_1^2\,\square_2^2\bar{\phi} - \bar{\phi}\,\square_1^2\,\square_2^2\bar{\phi}_N = \\ \bar{\phi}_N\nabla^4\bar{\phi} - \bar{\phi}\,\nabla^4\bar{\phi}_N - \left(s_1^2 + s_2^2\right)\left(\bar{\phi}_N\nabla^2\bar{\phi} - \bar{\phi}\nabla^2\bar{\phi}_N\right).\end{gathered} \tag{8.3.15}$$

Also, note that for arbitrary smooth scalar fields φ and ψ defined on B we have the integral identities

$$\begin{gathered}\int\limits_{\mathrm{B}} \left(\varphi\nabla^4\varphi - \psi\nabla^4\varphi\right)\mathrm{d}v\,(x) = \\ \int\limits_{\partial\mathrm{B}} \left[\varphi\frac{\partial}{\partial n}\left(\nabla^2\psi\right) - \psi\frac{\partial}{\partial n}\left(\nabla^2\varphi\right) + \left(\nabla^2\varphi\right)\frac{\partial\psi}{\partial n} - \left(\nabla^2\psi\right)\frac{\partial\varphi}{\partial n}\right]\mathrm{d}a(x),\end{gathered} \tag{8.3.16}$$

and

$$\int\limits_{\mathrm{B}} \left(\varphi\nabla^2\psi - \psi\nabla^2\varphi\right)\mathrm{d}v\,(x) = \int\limits_{\partial\mathrm{B}} \left(\varphi\frac{\partial\psi}{\partial n} - \psi\frac{\partial\varphi}{\partial n}\right)\mathrm{d}a\,(x). \tag{8.3.17}$$

Setting $\varphi = \bar{\phi}_N$ and $\psi = \bar{\phi}$ in these relations, on account of eqns (8.3.13) and (8.3.15), we obtain

$$\begin{aligned}\left(1 + t^0 p\right)\epsilon\,(y)\,\bar{\phi}\,(y,p) = &-\left(s_1^2 + s_2^2\right)\int\limits_{\partial\mathrm{B}} \left(\bar{\phi}_N\frac{\partial\bar{\phi}}{\partial n} - \bar{\phi}\frac{\partial\bar{\phi}_N}{\partial n}\right)\mathrm{d}a\,(x) \\ &+\int\limits_{\partial\mathrm{B}} \left[\bar{\phi}_N\frac{\partial}{\partial n}\left(\nabla^2\bar{\phi}\right) - \bar{\phi}\frac{\partial}{\partial n}\left(\nabla^2\bar{\phi}_N\right) + \left(\nabla^2\bar{\phi}_N\right)\frac{\partial\bar{\phi}}{\partial n} - \left(\nabla^2\bar{\phi}\right)\frac{\partial\bar{\phi}_N}{\partial n}\right]\mathrm{d}a\,(x).\end{aligned} \tag{8.3.18}$$

Since (recall eqns $(7.3.10)_2$ and $(7.3.12)_2$)

$$\nabla^2\bar{\phi} = p^2\bar{\phi} + \left(1 + t^0 p\right)\bar{\vartheta} \text{ on } \partial\mathrm{B}, \tag{8.3.19}$$

and

$$\nabla^2\bar{\phi}_N = p^2\bar{\phi}_N + \left(1 + t^0 p\right)\bar{\vartheta}_N \text{ on } \partial\mathrm{B}, \tag{8.3.20}$$

from eqn (8.3.18) we obtain

$$\begin{gathered}\epsilon\,(y)\,\bar{\phi}\,(y,p) = \\ \int\limits_{\partial\mathrm{B}} \left(\bar{K}_N\frac{\partial\bar{\phi}}{\partial n} - \bar{\phi}\,\frac{\partial\bar{K}_N}{\partial n}\right)\mathrm{d}a\,(x) + \int\limits_{\partial\mathrm{B}} \left(\bar{\phi}_N\frac{\partial\bar{\vartheta}}{\partial n} - \bar{\vartheta}\,\frac{\partial\bar{\phi}_N}{\partial n}\right)\mathrm{d}a\,(x),\end{gathered} \tag{8.3.21}$$

where

$$\bar{K}_N = \frac{2p^2 - \left(s_1^2 + s_2^2\right)}{1 + t^0 p}\,\bar{\phi}_N + \bar{\vartheta}_N. \tag{8.3.22}$$

Equation (8.3.21) is an integral representation of $\bar{\phi}$. Its inverse Laplace transform is the Kirchhoff-type formula for ϕ. Prior to deriving it we first introduce an analogous formula for the function ϑ.

Applying the operation $(\nabla^2 - p^2)$ to both sides of eqn (8.3.21), and employing the formulas $(8.3.10)_2$, (8.3.19), (8.3.20) and (8.3.22), we obtain the following integral representation for $\bar{\vartheta}$

$$\begin{aligned} \epsilon(y)\bar{\vartheta}(y,p) = & \int_{\partial B} \left(\bar{\vartheta}_N \frac{\partial \bar{\vartheta}}{\partial n} - \bar{\vartheta} \frac{\partial \bar{\vartheta}_N}{\partial n} \right) da(x) \\ & + \int_{\partial B} \left(\bar{H}_N \frac{\partial \bar{\phi}}{\partial n} - \bar{\phi} \frac{\partial \bar{H}_N}{\partial n} \right) da(x), \end{aligned} \tag{8.3.23}$$

where

$$\bar{H}_N = \frac{1}{1+t^0 p} \left[\nabla^2 - \left(s_1^2 + s_2^2 - p^2 \right) \right] \bar{\vartheta}_N. \tag{8.3.24}$$

The kernels $\bar{\phi}_N$ and $\bar{\vartheta}_N$ appearing in eqns (8.3.21) and (8.3.23) are given by (recall eqns (7.4.22) and (7.4.23))

$$\bar{\phi}_N(\rho,p) = \frac{1}{4\pi\rho} \left(\frac{1}{\upsilon_1^2} - \frac{1}{\upsilon_2^2} \right)^{-1} \frac{1+t^0 p}{p} \left[\bar{M}_1(\rho,p) - \bar{M}_2(\rho,p) \right], \tag{8.3.25}$$

and

$$\bar{\vartheta}_N(\rho,p) = \frac{1}{8\pi\rho} \left\{ \bar{N}_1(\rho,p) + \bar{N}_2(\rho,p) + \left(\hat{\alpha} + \hat{\beta} p \right) \left[\bar{M}_1(\rho,p) - \bar{M}_2(\rho,p) \right] \right\}, \tag{8.3.26}$$

in which

$$\rho = |x - y|. \tag{8.3.27}$$

The functions $\bar{M}_i$ and $\bar{N}_i$ appearing in eqns (8.3.25) and (8.3.26) are defined by eqns (7.4.24) and (7.4.25), while the symbols υ_i, $\hat{\alpha}$ and $\hat{\beta}$ are given by eqns (7.4.26) and (7.4.27).

Employing the definition of $s_i(p)$ according to eqn (7.4.15), we obtain

$$2p^2 - \left(s_1^2 + s_2^2 \right) = -\left(\frac{1}{\upsilon_1^2} - \frac{1}{\upsilon_2^2} \right) p(\hat{\alpha} + p\hat{\beta}), \tag{8.3.28}$$

from which, on account of eqns (8.3.22), (8.3.25) and (8.3.26), we find the kernel $\bar{K}_N$ appearing in eqn (8.3.21) in the following form

$$\bar{K}_N(\rho,p) = \frac{1}{8\pi\rho} \left\{ \bar{N}_1(\rho,p) + \bar{N}_2(\rho,p) - \left(\hat{\alpha} + \hat{\beta} p \right) \left[\bar{M}_1(\rho,p) - \bar{M}_2(\rho,p) \right] \right\}. \tag{8.3.29}$$

It is seen that the functions $\bar{\phi}_N$, $\bar{\vartheta}_N$ and $\bar{K}_N$ can be retransformed by the method of Section 7.4. In order to reduce the function $\bar{H}_N$ defined by the formula (8.3.24) to a form that can be retransformed by the same method, we first note that

$$\Box_i^2\left[\rho^{-1}\exp\left(-\rho s_i\right)\right]=0 \text{ for } x\neq y, \tag{8.3.30}$$

and, given eqns (8.3.24) and (8.3.26), we find

$$\bar{H}_N=-\frac{1}{4\pi\rho}\frac{\left(s_1^2-p^2\right)\left(s_2^2-p^2\right)}{1+t^0 p}\frac{\left[\exp(-\rho s_1)-\exp(-\rho s_2)\right]}{s_1^2-s_2^2}. \tag{8.3.31}$$

Now, since

$$\left(s_1^2-p^2\right)\left(s_2^2-p^2\right)=s_1^2 s_2^2-\left(s_1^2+s_2^2\right)p^2+p^4, \tag{8.3.32}$$

and, in view of the definition of $s_i\left(p\right)$ (recall eqn (7.4.15))

$$s_1^2+s_2^2=p\left[(1+\epsilon)+\left(1+t_0+\epsilon t^0\right)p\right], \tag{8.3.33}$$

$$s_1^2 s_2^2=p^3\left(1+t_0 p\right), \tag{8.3.34}$$

then

$$\left(s_1^2-p^2\right)\left(s_2^2-p^2\right)=-\epsilon p^3\left(1+t_0 p\right). \tag{8.3.35}$$

Hence, since

$$\frac{\exp(-\rho s_1)-\exp(-\rho s_2)}{s_1^2-s_2^2}=\left(\frac{1}{v_1^2}-\frac{1}{v_2^2}\right)^{-1}\frac{1}{p}\left(\bar{M}_1-\bar{M}_2\right), \tag{8.3.36}$$

for the function $\bar{H}_N$ appearing in eqn (8.3.23) we obtain

$$\bar{H}_N\left(\rho,p\right)=\frac{\epsilon}{4\pi\rho}\left(\frac{1}{v_1^2}-\frac{1}{v_2^2}\right)^{-1}p^2\left[\bar{M}_1\left(\rho,p\right)-\bar{M}_2\left(\rho,p\right)\right]. \tag{8.3.37}$$

Thus, all the kernels appearing in the integral representations (8.3.21) and (8.3.22) may be inverted using the method of Section 7.4. This leads to the formal representation of the Kirchhoff type for a pair (ϕ,ϑ)

$$\begin{aligned}\epsilon\left(y\right)\phi\left(y,t\right)=&\int_{\partial B}\left(K_N*\frac{\partial\phi}{\partial n}-\phi*\frac{\partial K_N}{\partial n}\right)da\left(x\right)\\&+\int_{\partial B}\left(\phi_N*\frac{\partial\vartheta}{\partial n}-\vartheta*\frac{\partial\phi_N}{\partial n}\right)da(x),\end{aligned} \tag{8.3.38}$$

and

$$\begin{aligned}\epsilon\left(y\right)\vartheta\left(y,t\right)=&\int_{\partial B}\left(\vartheta_N*\frac{\partial\vartheta}{\partial n}-\vartheta*\frac{\partial\vartheta_N}{\partial n}\right)da\left(x\right)\\&+\int_{\partial B}\left(H_N*\frac{\partial\phi}{\partial n}-\phi*\frac{\partial H_N}{\partial n}\right)da(x).\end{aligned} \tag{8.3.39}$$

Here, the functions $K_N = K_N(x, y; t)$, $\vartheta_N = \vartheta_N(x, y; t)$, $\phi_N = \phi_N(x, y; t)$ and $H_N = H_N(x, y; t)$ are defined by the formulas

$$K_N = \frac{1}{8\pi\rho}\left[N_1 + N_2 - \left(\hat{\alpha} + \hat{\beta}\frac{\partial}{\partial t}\right)(M_1 - M_2)\right], \tag{8.3.40}$$

$$\vartheta_N = \frac{1}{8\pi\rho}\left[N_1 + N_2 + \left(\hat{\alpha} + \hat{\beta}\frac{\partial}{\partial t}\right)(M_1 - M_2)\right], \tag{8.3.41}$$

$$\phi_N = \frac{1}{4\pi\rho}\left(\frac{1}{v_1^2} - \frac{1}{v_2^2}\right)^{-1} * \left(1 + t^0\frac{\partial}{\partial t}\right)(M_1 - M_2), \tag{8.3.42}$$

$$H_N = \frac{\epsilon}{4\pi\rho}\left(\frac{1}{v_1^2} - \frac{1}{v_2^2}\right)^{-1}\frac{\partial^2}{\partial t^2}(M_1 - M_2), \tag{8.3.43}$$

in which $N_1 = N_1(x, y; t)$, $N_2 = N_2(x, y; t)$, $M_1 = M_1(x, y; t)$ and $M_2 = M_2(x, y; t)$ are defined, respectively, by the formulas (7.4.62), (7.4.71), (7.4.81) and (7.4.85).

Of course, the formulas (8.3.38) and (8.3.39) may also be written in the following form

$$\left.\begin{aligned} \phi &= \phi_1 + \phi_2 \\ \vartheta &= \vartheta_1 + \vartheta_2 \end{aligned}\right\} \quad \text{on } \mathrm{B} \times [0, \infty), \tag{8.3.44}$$

where

$$\phi_i = \int_{\partial \mathrm{B}}\left(K_N^{(i)} * \frac{\partial \phi}{\partial n} - \phi * \frac{\partial K_N^{(i)}}{\partial n}\right) da(x) + \int_{\partial \mathrm{B}}\left(\phi_N^{(i)} * \frac{\partial \vartheta}{\partial n} - \vartheta * \frac{\partial \phi_N^{(i)}}{\partial n}\right) da(x), \tag{8.3.45}$$

$$\vartheta_i = \int_{\partial \mathrm{B}}\left(\vartheta_N^{(i)} * \frac{\partial \vartheta}{\partial n} - \vartheta * \frac{\partial \vartheta_N^{(i)}}{\partial n}\right) da(x) + \int_{\partial \mathrm{B}}\left(H_N^{(i)} * \frac{\partial \phi}{\partial n} - \phi * \frac{\partial H_N^{(i)}}{\partial n}\right) da(x). \tag{8.3.46}$$

Here, the functions $K_N^{(i)}$, $\vartheta_N^{(i)}$, $\phi_N^{(i)}$ and $H_N^{(i)}$ are defined by the formulas

$$K_N^{(i)} = \frac{1}{8\pi v_i}\left[Q_i \mp \left(\hat{\alpha} + \hat{\beta}\frac{\partial}{\partial t}\right)P_i\right], \tag{8.3.47}$$

$$\vartheta_N^{(i)} = \frac{1}{8\pi v_i}\left[Q_i \pm \left(\hat{\alpha} + \hat{\beta}\frac{\partial}{\partial t}\right)P_i\right], \tag{8.3.48}$$

$$\phi_N^{(i)} = \pm \frac{1}{4\pi v_i} \left(\frac{1}{v_1^2} - \frac{1}{v_2^2} \right)^{-1} \left(t^0 + 1* \right) P_i, \tag{8.3.49}$$

$$H_N^{(i)} = \pm \frac{\epsilon}{4\pi v_i} \left(\frac{1}{v_1^2} - \frac{1}{v_2^2} \right)^{-1} \frac{\partial^2}{\partial t^2} P_i, \tag{8.3.50}$$

in which

$$P_i = \exp\left(-h_i t\right) \mathcal{M}_i \left(\rho_i, t\right) H\left(\zeta_i\right), \tag{8.3.51}$$

$$Q_i = \mathcal{M}_i \left(\rho_i, \rho_i\right) \exp\left(-h_i \rho_i\right) \delta\left(\zeta_i\right) + \exp\left(-h_i t\right) \mathcal{N}_i \left(\rho_i, t\right) H\left(\zeta_i\right), \tag{8.3.52}$$

with the functions $\mathcal{M}_i$ and $\mathcal{N}_i$ $(i = 1, 2)$ being defined by eqns (7.4.95)–(7.4.98). The upper (respectively, lower) sign in the formulas (8.3.47)–(8.3.50) stands for $i = 1$ $(i = 2)$. With the notation

$$\mathcal{M}_i^{(0)} = \mathcal{M}_i \left(\rho_i, \rho_i\right), \tag{8.3.53}$$

and in view of the smoothness, i.e. differentiability of $\mathcal{M}_i(\rho_i, t)$ and $\mathcal{N}_i(\rho_i, t)$ $\forall (\rho_i, t) \in (0, \infty) \times (\rho_i, \infty)$, on account of eqns (8.3.47)–(8.3.52), we obtain

$$\begin{aligned} K_N^{(i)} = \frac{1}{8\pi v_i} & \left\{ \left(1 \mp \hat{\beta}\right) \exp\left(-h_i \rho_i\right) \mathcal{M}_i^{(0)} \delta\left(\zeta_i\right) \right. \\ & \left. + \exp\left(-h_i t\right) \left[\mathcal{M}_i \mp \hat{\beta} \dot{\mathcal{M}}_i \mp \left(\hat{\alpha} - h_i \hat{\beta}\right) \mathcal{M}_i \right] H\left(\zeta_i\right) \right\}, \end{aligned} \tag{8.3.54}$$

$$\begin{aligned} \vartheta_N^{(i)} = \frac{1}{8\pi v_i} & \left\{ \left(1 \pm \hat{\beta}\right) \exp\left(-h_i \rho_i\right) \mathcal{M}_i^{(0)} \delta\left(\zeta_i\right) \right. \\ & \left. + \exp\left(-h_i t\right) \left[\mathcal{M}_i \pm \hat{\beta} \dot{\mathcal{M}}_i \pm \left(\hat{\alpha} - h_i \hat{\beta}\right) \mathcal{M}_i \right] H\left(\zeta_i\right) \right\}, \end{aligned} \tag{8.3.55}$$

$$\phi_N^{(i)} = \pm \frac{1}{4\pi v_i} \left(\frac{1}{v_1^2} - \frac{1}{v_2^2} \right)^{-1} \left(t^0 + 1* \right) \left[\exp\left(-h_i t\right) \mathcal{M}_i H\left(\zeta_i\right) \right], \tag{8.3.56}$$

$$\begin{aligned} H_N^{(i)} = \pm \frac{\epsilon}{4\pi v_i} \left(\frac{1}{v_1^2} - \frac{1}{v_2^2} \right)^{-1} & \left[\exp\left(-h_i \rho_i\right) \mathcal{M}_i^{(0)} \dot{\delta}\left(\zeta_i\right) \right. \\ & + \exp\left(-h_i \rho_i\right) \left(\dot{\mathcal{M}}_i^{(0)} - h_i \mathcal{M}_i^{(0)} \right) \delta\left(\zeta_i\right) \\ & \left. + \exp\left(-h_i t\right) \left(\ddot{\mathcal{M}}_i - 2h_i \dot{\mathcal{M}}_i + h_i^2 \mathcal{M}_i \right) H\left(\zeta_i\right) \right], \end{aligned} \tag{8.3.57}$$

where

$$\dot{\mathcal{M}}_i^{(0)} = \dot{\mathcal{M}}_i \left(\rho_i, \rho_i\right). \tag{8.3.58}$$

The normal derivatives of the functions (8.3.54)–(8.3.57) are found as follows

$$\frac{\partial K_N^{(i)}}{\partial n} = \frac{1}{8\pi v_i}\frac{\partial \rho_i}{\partial n}\left\{-\left(1\mp\hat{\beta}\right)\exp\left(-h_i\rho_i\right)\mathcal{M}_i^{(0)}\dot{\delta}\left(\zeta_i\right)\right.$$

$$-\exp\left(-h_i\rho_i\right)\left[\mathcal{N}_i^{(0)}\mp\hat{\beta}\dot{\mathcal{M}}_i^{(0)}\mp\left(\hat{\alpha}-h_i\hat{\beta}\right)\mathcal{M}_i^{(0)}-\left(1\mp\hat{\beta}\right)\left(\frac{\partial}{\partial\rho_i}-h_i\right)\mathcal{M}_i^{(0)}\right]\delta\left(\zeta_i\right)$$

$$\left.+\exp\left(-h_it\right)\left[\mathcal{N}_i'\mp\hat{\beta}\dot{\mathcal{M}}_i'\mp\left(\hat{\alpha}-h_i\hat{\beta}\right)\mathcal{M}_i'\right]H\left(\zeta_i\right)\right\}, \tag{8.3.59}$$

$$\frac{\partial \vartheta_N^{(i)}}{\partial n} = \frac{1}{8\pi v_i}\frac{\partial \rho_i}{\partial n}\left\{-\left(1\pm\hat{\beta}\right)\exp\left(-h_i\rho_i\right)\mathcal{M}_i^{(0)}\dot{\delta}\left(\zeta_i\right)\right.$$

$$-\exp\left(-h_i\rho_i\right)\left[\mathcal{N}_i^{(0)}\pm\hat{\beta}\dot{\mathcal{M}}_i^{(0)}\pm\left(\hat{\alpha}-h_i\hat{\beta}\right)\mathcal{M}_i^{(0)}-\left(1\pm\hat{\beta}\right)\left(\frac{\partial}{\partial\rho_i}-h_i\right)\mathcal{M}_i^{(0)}\right]\delta\left(\zeta_i\right)$$

$$\left.+\exp\left(-h_it\right)\left[\mathcal{N}_i'\pm\hat{\beta}\dot{\mathcal{M}}_i'\pm\left(\hat{\alpha}-h_i\hat{\beta}\right)\mathcal{M}_i'\right]H\left(\zeta_i\right)\right\}, \tag{8.3.60}$$

$$\frac{\partial \phi_N^{(i)}}{\partial n} = \pm\frac{1}{4\pi v_i}\left(\frac{1}{v_1^2}-\frac{1}{v_2^2}\right)^{-1}\left(t^0+1*\right)\frac{\partial\rho_i}{\partial n}$$
$$\times\left[-\exp\left(-h_i\rho_i\right)\mathcal{M}_i^{(0)}\delta\left(\zeta_i\right)+\exp\left(-h_it\right)\mathcal{M}_i'H\left(\zeta_i\right)\right], \tag{8.3.61}$$

$$\frac{\partial H_N^{(i)}}{\partial n} = \pm\frac{\epsilon}{4\pi v_i}\left(\frac{1}{v_1^2}-\frac{1}{v_2^2}\right)^{-1}\frac{\partial\rho_i}{\partial n}\exp\left(-h_i\rho_i\right)$$
$$\times\left\{-\mathcal{M}_i^{(0)}\ddot{\delta}\left(\zeta_1\right)-\left(\dot{\mathcal{M}}_i^{(0)}-\frac{\partial}{\partial\rho_i}\mathcal{M}_i^{(0)}\right)\dot{\delta}\left(\zeta_i\right)\right.$$
$$\left.-\left[\ddot{\mathcal{M}}_i^{(0)}-h_i\dot{\mathcal{M}}_i^{(0)}-\frac{\partial}{\partial\rho_i}\left(\dot{\mathcal{M}}_i^{(0)}-h_i\mathcal{M}_i^{(0)}\right)\right]\delta\left(\zeta_i\right)\right\} \tag{8.3.62}$$
$$\pm\frac{\epsilon}{4\pi v_i}\left(\frac{1}{v_1^2}-\frac{1}{v_2^2}\right)^{-1}\frac{\partial\rho_i}{\partial n}\exp\left(-h_it\right)$$
$$\times\left(\ddot{\mathcal{M}}_i'-2h_i\dot{\mathcal{M}}_i'+h_i^2\mathcal{M}_i'\right)H\left(\zeta_i\right).$$

In the above formulas the prime stands for differentiation with respect to ρ_i, and

$$\mathcal{N}_i^{(0)}=\mathcal{N}_i\left(\rho_i,\rho_i\right),\quad \ddot{\mathcal{M}}_i^{(0)}=\ddot{\mathcal{M}}_i\left(\rho_i,\rho_i\right). \tag{8.3.63}$$

Next, we introduce the notations

$$K_i^{(0)}\left(\rho_i\right)=\frac{1}{8\pi v_i}\left(1\mp\hat{\beta}\right)\exp\left(-h_i\rho_i\right)\mathcal{M}_i^{(0)}, \tag{8.3.64}$$

$$K_i^{(1)}(\rho_i) = \frac{1}{8\pi \upsilon_i} \exp(-h_i\rho_i) \Big[\mathcal{N}_i^{(0)} \mp \hat{\beta}\dot{\mathcal{M}}_i^{(0)} \mp \left(\hat{\alpha} - h_i\hat{\beta}\right)\mathcal{M}_i^{(0)} \\ - \left(1 \mp \hat{\beta}\right)\left(\frac{\partial}{\partial \rho_i} - h_i\right)\mathcal{M}_i^{(0)}\Big], \tag{8.3.65}$$

$$K_i(\rho_i, t) = \frac{1}{8\pi \upsilon_i} \exp(-h_i t) \left[\mathcal{N}_i \mp \hat{\beta}\dot{\mathcal{M}}_i \mp \left(\hat{\alpha} - h_i\hat{\beta}\right)\mathcal{M}_i\right], \tag{8.3.66}$$

$$\theta_i^{(0)}(\rho_i) = \frac{1}{8\pi \upsilon_i}\left(1 \pm \hat{\beta}\right)\exp(-h_i\rho_i)\mathcal{M}_i^{(0)}, \tag{8.3.67}$$

$$\theta_i^{(1)}(\rho_i) = \frac{1}{8\pi \upsilon_i} \exp(-h_i\rho_i) \Big[\mathcal{N}_i^{(0)} \pm \hat{\beta}\dot{\mathcal{M}}_i^{(0)} \pm \left(\hat{\alpha} - h_i\hat{\beta}\right)\mathcal{M}_i^{(0)} \\ - \left(1 \pm \hat{\beta}\right)\left(\frac{\partial}{\partial \rho_i} - h_i\right)\mathcal{M}_i^{(0)}\Big], \tag{8.3.68}$$

$$\theta_i(\rho_i, t) = \frac{1}{8\pi \upsilon_i} \exp(-h_i t) \left[\mathcal{N}_i \pm \hat{\beta}\dot{\mathcal{M}}_i \pm \left(\hat{\alpha} - h_i\hat{\beta}\right)\mathcal{M}_i\right], \tag{8.3.69}$$

$$\varphi_i^{(0)}(\rho_i) = \pm\frac{t^0}{4\pi \upsilon_i}\left(\frac{1}{\upsilon_1^2} - \frac{1}{\upsilon_2^2}\right)^{-1} \exp(-h_i\rho_i)\mathcal{M}_i^{(0)}, \tag{8.3.70}$$

$$\varphi_i(\rho_i, t) = \pm\frac{1}{4\pi \upsilon_i}\left(\frac{1}{\upsilon_1^2} - \frac{1}{\upsilon_2^2}\right)^{-1} \\ \times \left[t^0 \exp(-h_i t)\mathcal{M}_i(\rho_i, t) + \int_{\rho_i}^{t} \exp(-h_i u)\mathcal{M}_i(\rho_i, u)\,\mathrm{d}u\right], \tag{8.3.71}$$

$$H_i^{(0)}(\rho_i) = \pm\frac{\epsilon}{4\pi \upsilon_i}\left(\frac{1}{\upsilon_1^2} - \frac{1}{\upsilon_2^2}\right)^{-1} \exp(-h_i\rho_i)\mathcal{M}_i^{(0)}, \tag{8.3.72}$$

$$H_i^{(1)}(\rho_i) = \pm\frac{\epsilon}{4\pi \upsilon_i}\left(\frac{1}{\upsilon_1^2} - \frac{1}{\upsilon_2^2}\right)^{-1} \exp(-h_i\rho_i)\left(\dot{\mathcal{M}}_i^{(0)} - h_i\mathcal{M}_i^{(0)}\right), \tag{8.3.73}$$

$$H_i^{(2)}(\rho_i) = \pm\frac{\epsilon}{4\pi \upsilon_i}\left(\frac{1}{\upsilon_1^2} - \frac{1}{\upsilon_2^2}\right)^{-1} \exp(-h_i\rho_i)\left(\dot{\mathcal{M}}_i^{(0)} - \frac{\partial}{\partial \rho_i}\mathcal{M}_i^{(0)}\right), \tag{8.3.74}$$

$$H_i^{(3)}(\rho_i) = \pm\frac{\epsilon}{4\pi \upsilon_i}\left(\frac{1}{\upsilon_1^2} - \frac{1}{\upsilon_2^2}\right)^{-1} \exp(-h_i\rho_i) \\ \times \left[\ddot{\mathcal{M}}_i^{(0)} - h_i\mathcal{M}_i^{(0)} - \frac{\partial}{\partial \rho_i}\left(\dot{\mathcal{M}}_i^{(0)} - h_i\mathcal{M}_i^{(0)}\right)\right], \tag{8.3.75}$$

$$H_i(\rho_i, t) = \pm\frac{\epsilon}{4\pi \upsilon_i}\left(\frac{1}{\upsilon_1^2} - \frac{1}{\upsilon_2^2}\right)^{-1} \exp(-h_i t)\left[\ddot{\mathcal{M}}_i - 2h_i\dot{\mathcal{M}}_i + h_i^2\mathcal{M}_i\right]. \tag{8.3.76}$$

Clearly, the kernels $K_i^{(0)}$, $K_i^{(1)}$, $\theta_i^{(0)}$, $\theta_i^{(1)}$, $\varphi_i^{(0)}$, $H_i^{(0)}$, $H_i^{(1)}$, $H_i^{(2)}$ and $H_i^{(3)}$ are smooth functions for $\rho_i > 0$. Moreover, the kernels K_i, θ_i, φ_i and H_i are smooth functions for every $(\rho_i, t) \in (0, \infty) \times [\rho_i, \infty)$. Hence, on account of the formulas (8.3.45) and (8.3.46), we infer that the pair (ϕ_i, ϑ_i) is represented by the following surface integrals

$$
\begin{aligned}
\epsilon(y)\,\phi_i(y,t) = & \int\limits_{\partial B \cap S(y, v_i t)} \left\{ K_i^{(0)}(\rho_i) \left(\left[\frac{\partial \phi}{\partial n} \right]_i + \frac{\partial \rho_i}{\partial n} \left[\frac{\partial \phi}{\partial y} \right]_i \right) \right. \\
& \left. + \frac{\partial \rho_i}{\partial n} \left(K_i^{(1)}(\rho_i) [\phi]_i + \varphi_i^{(0)}(\rho_i) [\vartheta]_i \right) \right\} da(x) \\
& + \int\limits_{\partial B \cap S(y, v_i t)} \left\{ \int\limits_{\rho_i}^{t} \left[K_i(\rho_i, \tau) \frac{\partial \phi}{\partial n}(x, t-\tau) - \frac{\partial K_i}{\partial n}(\rho_i, \tau)\, \phi(x, t-\tau) \right] d\tau \right\} da(x) \\
& + \int\limits_{\partial B \cap S(y, v_i t)} \left\{ \int\limits_{\rho_i}^{t} \left[\varphi_i(\rho_i, \tau) \frac{\partial \vartheta}{\partial n}(x, t-\tau) - \frac{\partial \varphi_i}{\partial n}(\rho_i, \tau)\, \vartheta(x, t-\tau) \right] d\tau \right\} da(x)
\end{aligned}
\tag{8.3.77}
$$

and

$$
\begin{aligned}
\epsilon(y)\,\vartheta_i(y,t) = & \int\limits_{\partial B \cap S(y, v_i t)} \left\{ \theta_i^{(0)}(\rho_i) \left(\left[\frac{\partial \vartheta}{\partial n} \right]_i + \frac{\partial \rho_i}{\partial n} \left[\frac{\partial \vartheta}{\partial y} \right]_i \right) \right. \\
& + \frac{\partial \rho_i}{\partial n} \left(\theta_i^{(1)}(\rho_i) [\vartheta]_i + H_i^{(3)}(\rho_i) [\phi]_i \right) + H_i^{(0)}(\rho_i) \left(\left[\frac{\partial^2 \phi}{\partial n \partial t} \right]_i + \frac{\partial \rho_i}{\partial n} \left[\frac{\partial^2 \phi}{\partial t^2} \right]_i \right) \\
& \left. + H_i^{(1)}(\rho_i) \left[\frac{\partial \phi}{\partial n} \right]_i + H_i^{(2)}(\rho_i) \frac{\partial \rho_i}{\partial n} \left[\frac{\partial \phi}{\partial t} \right]_i \right\} da(x) \\
& + \int\limits_{\partial B \cap S(y, v_i t)} \left\{ \int\limits_{\rho_i}^{t} \left[\theta_i(\rho_i, \tau) \frac{\partial \vartheta}{\partial n}(x, t-\tau) - \frac{\partial \theta_i}{\partial n}(\rho_i, \tau)\, \vartheta(x, t-\tau) \right] d\tau \right\} da(x) \\
& + \int\limits_{\partial B \cap S(y, v_i t)} \left\{ \int\limits_{\rho_i}^{t} \left[H_i(\rho_i, \tau) \frac{\partial \phi}{\partial n}(x, t-\tau) - \frac{\partial H_i}{\partial n}(\rho_i, \tau)\, \phi(x, t-\tau) \right] d\tau \right\} da(x) \\
& \text{for} \quad i = 1, 2.
\end{aligned}
\tag{8.3.78}
$$

Here, similarly to what was done in Section 8.1, the symbol $[f]_i$ denotes the retarded value $f(x, t - \rho_i)$ of a function $f(x,t)$ with respect to a point y. Also, $S(y, v_i t)$ denotes an open ball with its center at y and of radius $v_i t$.

With ϕ_i and ϑ_i defined, respectively, by the integrals (8.3.77) and (8.3.78), the relations (8.3.44) are formulas of the Kirchhoff type for the central system of eqns (8.3.1) with conditions (8.3.2). It follows from these formulas that, for a pair (ϕ, ϑ) satisfying the relations (8.3.1) and (8.3.2), for any time $t > 0$, the domain of influence of a thermomechanical loading applied at the boundary, specified by the pair (f, g) appearing in the condition (8.3.3), is a boundary layer of width $v_2 t$ $(v_2 > v_1)$.

8.4 Integral equations for a potential–temperature problem

Let $(\phi^{(I)}, \vartheta^{(I)})$ be an "interior" solution of PTP in the G–L theory satisfying the relations

$$\begin{gathered} \Gamma\phi^{(I)} = 0 \\ \left(1 + t^0 \frac{\partial}{\partial t}\right) \vartheta^{(I)} = \left(\nabla^2 - \frac{\partial^2}{\partial t^2}\right) \phi^{(I)} \end{gathered} \quad \text{on } \mathrm{B} \times [0, \infty), \tag{8.4.1}$$

$$\begin{gathered} \phi^{(I)}(\cdot, 0) = \dot{\phi}^{(I)}(\cdot, 0) = 0 \\ \vartheta^{(I)}(\cdot, 0) = \dot{\vartheta}^{(I)}(\cdot, 0) = 0, \end{gathered} \quad \text{on } \bar{\mathrm{B}}, \tag{8.4.2}$$

and

$$\phi^{(I)} = f, \ \vartheta^{(I)} = g \quad \text{on } \partial\mathrm{B} \times [0, \infty). \tag{8.4.3}$$

Analogously, let $(\phi^{(E)}, \vartheta^{(E)})$ be an "exterior" solution of PTP in the G–L theory satisfying the relations

$$\begin{gathered} \Gamma\phi^{(E)} = 0 \\ \left(1 + t^0 \frac{\partial}{\partial t}\right) \vartheta^{(E)} = \left(\nabla^2 - \frac{\partial^2}{\partial t^2}\right) \phi^{(E)} \end{gathered} \quad \text{on } (\mathrm{E}^3 - \bar{\mathrm{B}}) \times [0, \infty), \tag{8.4.4}$$

$$\begin{gathered} \phi^{(E)}(\cdot, 0) = \dot{\phi}^{(E)}(\cdot, 0) = 0, \\ \vartheta^{(E)}(\cdot, 0) = \dot{\vartheta}^{(E)}(\cdot, 0) = 0, \end{gathered} \quad \text{on } \mathrm{E}^3 - \bar{\mathrm{B}}, \tag{8.4.5}$$

and

$$\phi^{(E)} = f, \ \ \vartheta^{(E)} = g \quad \text{on } \partial\mathrm{B} \times [0, \infty). \tag{8.4.6}$$

Here, B is a bounded domain in E^3, while f and g are functions prescribed on $\partial\mathrm{B} \times [0, \infty)$.

In accordance with the formulas (8.3.38) and (8.3.39), the pair $(\phi^{(I)}, \vartheta^{(I)})$ may be represented in the form

$$\begin{aligned} \epsilon(y)\,\phi^{(I)}(y, t) = & \int\limits_{\partial\mathrm{B}} \left(K_N * \frac{\partial\phi^{(I)}}{\partial n} - f * \frac{\partial K_N}{\partial n}\right) \mathrm{d}a(x) \\ & + \int\limits_{\partial\mathrm{B}} \left(\phi_N * \frac{\partial\vartheta^{(I)}}{\partial n} - g * \frac{\partial\phi_N}{\partial n}\right) \mathrm{d}a(x), \end{aligned} \tag{8.4.7}$$

$$\epsilon(y)\,\vartheta^{(I)}(y,t) = \int_{\partial B}\left(\vartheta_N * \frac{\partial\vartheta^{(I)}}{\partial n} - g * \frac{\partial\vartheta_N}{\partial n}\right)\mathrm{d}a(x) + \int_{\partial B}\left(H_N * \frac{\partial\phi^{(I)}}{\partial n} - f * \frac{\partial H_N}{\partial n}\right)\mathrm{d}a(x). \tag{8.4.8}$$

Similarly, there hold these integral representations for the pair $(\phi^{(E)}, \vartheta^{(E)})$

$$\epsilon_0(y)\,\phi^{(E)}(y,t) = -\int_{\partial B}\left(K_N * \frac{\partial\phi^{(E)}}{\partial n} - f * \frac{\partial K_N}{\partial n}\right)\mathrm{d}a(x) - \int_{\partial B}\left(\phi_N * \frac{\partial\vartheta^{(E)}}{\partial n} - g * \frac{\partial\phi_N}{\partial n}\right)\mathrm{d}a(x), \tag{8.4.9}$$

$$\epsilon_0(y)\,\vartheta^{(E)}(y,t) = -\int_{\partial B}\left(\vartheta_N * \frac{\partial\vartheta^{(E)}}{\partial n} - g * \frac{\partial\vartheta_N}{\partial n}\right)\mathrm{d}a(x) - \int_{\partial B}\left(H_N * \frac{\partial\phi^{(E)}}{\partial n} - f * \frac{\partial H_N}{\partial n}\right)\mathrm{d}a(x). \tag{8.4.10}$$

$\epsilon(y)$ and $\epsilon_0(y)$ in eqns (8.4.7)–(8.4.10) are the characteristic functions of the sets B and $E^3 - \bar{B}$, respectively, that is

$$\epsilon(y) = \begin{cases} 1 \text{ for } & y \in B, \\ 0 \text{ for } & y \in E^3 - \bar{B}, \end{cases} \tag{8.4.11}$$

$$\epsilon_0(y) = \begin{cases} 0 \text{ for } & y \in B, \\ 1 \text{ for } & y \in E^3 - \bar{B}. \end{cases} \tag{8.4.12}$$

Furthermore, the kernels K_N, ϑ_N, ϕ_N and H_N are specified by the formulas (8.3.40)–(8.3.43).

Now, adding eqn (8.4.7) to eqn (8.4.9) as well as eqn (8.4.8) to eqn (8.4.10), and introducing the notations

$$\left.\begin{aligned} \frac{\partial}{\partial n}\left(\phi^{(I)} - \phi^{(E)}\right) &= \mu_\phi(x,t) \\ \frac{\partial}{\partial n}\left(\vartheta^{(I)} - \vartheta^{(E)}\right) &= \mu_\vartheta(x,t) \end{aligned}\right\} \quad \forall\,(x,t) \in \partial B \times [0,\infty), \tag{8.4.13}$$

we obtain

$$\left.\begin{aligned}\phi(y,t) &= \int\limits_{\partial\mathrm{B}} (K_N * \mu_\phi + \phi_N * \mu_\vartheta)\,\mathrm{d}a(x)\\ \vartheta(y,t) &= \int\limits_{\partial\mathrm{B}} (H_N * \mu_\phi + \vartheta_N * \mu_\vartheta)\,\mathrm{d}a(x)\end{aligned}\right\} \quad \forall (y,t) \in (\mathrm{E}^3 - \partial\mathrm{B}) \times [0,\infty), \tag{8.4.14}$$

where

$$\phi(y,t) = \begin{cases} \phi^{(I)}(y,t) \text{ for } & (y,t) \in \mathrm{B} \times [0,\infty),\\ \phi^{(E)}(y,t) \text{ for } & (y,t) \in (\mathrm{E}^3 - \bar{\mathrm{B}}) \times [0,\infty),\end{cases} \tag{8.4.15}$$

and

$$\vartheta(y,t) = \begin{cases} \vartheta^{(I)}(y,t) \text{ for } & (y,t) \in \mathrm{B} \times [0,\infty),\\ \vartheta^{(E)}(y,t) \text{ for } & (y,t) \in (\mathrm{E}^3 - \bar{\mathrm{B}}) \times [0,\infty).\end{cases} \tag{8.4.16}$$

One can show that the right-hand sides of eqn (8.4.14) are continuous functions $\forall (y,t) \in \mathrm{E}^3 \times [0,\infty)$. Thus, passing with the point y in eqn (8.4.14) to the boundary, and using the conditions (8.4.3) or (8.4.6), we obtain the following integral equations for the unknown functions μ_ϕ and μ_ϑ:

$$\left.\begin{aligned}&\int\limits_{\partial\mathrm{B}} [K_N(x,y;t) * \mu_\phi(x,t) + \phi_N(x,y;t) * \mu_\vartheta(x,t)]\,\mathrm{d}a(x) = f(y,t),\\ &\int\limits_{\partial\mathrm{B}} [H_N(x,y;t) * \mu_\phi(x,t) + \vartheta_N(x,y;t) * \mu_\vartheta(x,t)]\,\mathrm{d}a(x) = g(y,t)\\ &\qquad\qquad \forall (y,t) \in \partial\mathrm{B} \times [0,\infty).\end{aligned}\right. \tag{8.4.17}$$

Equations (8.3.17) represent a system of integral equations of the Fredholm type of the first kind. Knowing its solution, one can determine the unknown pairs $(\phi^{(I)}, \vartheta^{(I)})$ and $(\phi^{(E)}, \vartheta^{(E)})$ from the formula (8.4.14).

In order to examine the character of singularities of the integral equations (8.4.17), observe that the kernels of these equations have the forms (recall eqns (8.3.54)–(8.3.57), (8.3.64), (8.3.66), (8.3.67), (8.3.69), (8.3.71), (8.3.72), (8.3.73) and (8.3.76))

$$K_N = \sum_{i=1}^{2} \left[K_i^{(0)}(\rho_i)\,\delta(\zeta_i) + K_i(\rho_i,t)\,H(\zeta_i)\right], \tag{8.4.18}$$

$$\vartheta_N = \sum_{i=1}^{2} \left[\theta_i^{(0)}(\rho_i)\,\delta(\zeta_i) + \theta_i(\rho_i,t)\,H(\zeta_i)\right], \tag{8.4.19}$$

$$\phi_N = \sum_{i=1}^{2} \varphi_i(\rho_i, t) H(\zeta_i), \tag{8.4.20}$$

$$H_N = \sum_{i=1}^{2} \left[H_i^{(0)}(\rho_i) \dot{\delta}(\zeta_i) + H_i^{(1)}(\rho_i) \delta(\zeta_i) + H_i(\rho_i, t) H(\zeta_i) \right], \tag{8.4.21}$$

where the kernels $K_i^{(0)}$, $\theta_i^{(0)}$, $H_i^{(0)}$ and $H_i^{(1)}$ are smooth functions of ρ_i for every $\rho_i > 0$, while the kernels $K_i(\rho_i, t)$, $\theta_i(\rho_i, t)$, $\varphi_i(\rho_i, t)$, and $H_i(\rho_i, t)$ are smooth functions for every $(\rho_i, t) \in (0, \infty) \times [\rho_i, \infty)$. Hence, using eqns (8.4.18)–(8.4.21), we reduce eqns (8.4.17) to the form

$$\begin{aligned} \sum_{i=1}^{2} \int\limits_{\partial \mathrm{B} \cap S(y, \upsilon_i t)} \Bigg\{ & K_i^{(0)}(\rho_i) [\mu_\phi]_i + \int\limits_{\rho_i}^{t} K_i(\rho_i, \tau) \mu_\phi(x, t-\tau) \, \mathrm{d}\tau \\ & + \int\limits_{\rho_i}^{t} \varphi_i(\rho_i, \tau) \mu_\vartheta(x, t-\tau) \, \mathrm{d}\tau \Bigg\} \mathrm{d}a(x) = f, \end{aligned} \tag{8.4.22}$$

$$\begin{aligned} \sum_{i=1}^{2} \int\limits_{\partial \mathrm{B} \cap S(y, \upsilon_i t)} \Bigg\{ & H_i^{(0)}(\rho_i) [\dot{\mu}_\phi]_i + H_i^{(1)}(\rho_i) [\mu_\phi]_i + \theta_i^{(0)}(\rho_i) [\mu_\vartheta]_i \\ + \int\limits_{\rho_i}^{t} H_i(\rho_i, \tau) \mu_\phi(x, t-\tau) \, \mathrm{d}\tau & + \int\limits_{\rho_i}^{t} \theta_i(\rho_i, \tau) \mu_\vartheta(x, t-\tau) \, \mathrm{d}\tau \Bigg\} \mathrm{d}a(x) = g \\ & \forall (y, t) \in \partial \mathrm{B} \times [0, \infty), \end{aligned} \tag{8.4.23}$$

where $S(y, \upsilon_i t)$ is an open ball centered at y and of radius $\upsilon_i t$, while $[\mu]_i$ denotes the retarded value $\mu(x, t - \rho_i)$ of a function $\mu(x, t)$ with respect to the point y. We now show that the following theorem holds.

Theorem 8.1 *The kernels $K_i^{(0)}(\rho_i)$, $\theta_i^{(0)}(\rho_i)$, $H_i^{(0)}(\rho_i)$ and $H_i^{(1)}(\rho_i)$ possess the singularity of order ρ_i^{-1} for $\rho_i \to 0$, while $K_i(\rho_i, t)$, $\varphi_i(\rho_i, t)$, $H_i(\rho_i, t)$ and $\theta_i(\rho_i, t)$ possess the singularity of order ρ_i^{-1} for $\rho_i \to 0$ for every $t > \rho_i$.*

Proof. The functions $K_i^{(0)}$, $\theta_i^{(0)}$, $H_i^{(0)}$ and $H_i^{(1)}$ are defined by the formulas (recall eqns (8.3.64), (8.3.67), (8.3.72) and (8.3.73))

$$K_i^{(0)}(\rho_i) = \frac{1}{8\pi \upsilon_i} \left(1 \mp \hat{\beta}\right) \exp(-\rho_i h_i) \mathcal{M}_i^{(0)}, \tag{8.4.24}$$

$$\theta_i^{(0)}(\rho_i) = \frac{1}{8\pi \upsilon_i} \left(1 \pm \hat{\beta}\right) \exp(-\rho_i h_i) \mathcal{M}_i^{(0)}, \tag{8.4.25}$$

$$H_i^{(0)}(\rho_i) = \pm\frac{\epsilon}{4\pi\upsilon_i}\left(\frac{1}{\upsilon_1^2}-\frac{1}{\upsilon_2^2}\right)^{-1}\exp(-\rho_i h_i)\,\mathcal{M}_i^{(0)}, \tag{8.4.26}$$

$$H_i^{(1)}(\rho_i) = \pm\frac{\epsilon}{4\pi\upsilon_i}\left(\frac{1}{\upsilon_1^2}-\frac{1}{\upsilon_2^2}\right)^{-1}\exp(-\rho_i h_i)\left(\dot{\mathcal{M}}_i^{(0)} - h_i\mathcal{M}_i^{(0)}\right), \tag{8.4.27}$$

where

$$\mathcal{M}_i^{(0)} = \mathcal{M}_i(\rho_i,\rho_i),\ \dot{\mathcal{M}}_i^{(0)} = \dot{\mathcal{M}}_i(\rho_i,\rho_i) \tag{8.4.28}$$

and (recall eqns (7.4.95) and (7.4.96))

$$\mathcal{M}_1(\rho_1,t) = \rho_1^{-1}J_0(\beta\zeta_1)\exp(\alpha_1\zeta_1) + \sum_{n=0}^{\infty}\frac{(-)^n\hat{\lambda}_1^n}{n!\,\omega_1^{n-1}}\int_{\rho_1}^{t} g_n^{(1)}(t-s)\,z_1^{n-1}J_{n-1}(z_1)\,\mathrm{d}s, \tag{8.4.29}$$

$$\mathcal{M}_2(\rho_2,t) = \rho_2^{-1}J_0(\beta\zeta_2)\exp(\alpha_2\zeta_2) + \sum_{n=0}^{\infty}\frac{\hat{\lambda}_2^n}{n!\,\hat{\omega}_1^{n-1}}\int_{\rho_2}^{t} g_n^{(2)}(t-s)\,z_2^{n-1}I_{n-1}(z_2)\,\mathrm{d}s. \tag{8.4.30}$$

In the following, in order to find the partial derivatives of $\mathcal{M}_i(\rho_i,t)$, we will take advantage of the following recurrence relations for the Bessel functions

$$\begin{aligned}\frac{\mathrm{d}}{\mathrm{d}z}\left[z^nJ_n(z)\right] &= z^nJ_{n-1}(z),\\ \frac{\mathrm{d}}{\mathrm{d}z}\left[z^{-n}J_n(z)\right] &= -z^{-n}J_{n+1}(z),\end{aligned} \tag{8.4.31}$$

and

$$\begin{aligned}\left(\frac{1}{z}\frac{\mathrm{d}}{\mathrm{d}z}\right)^k\left[z^nI_n(z)\right] &= z^{n-k}I_{n-k}(z),\\ \left(\frac{1}{z}\frac{\mathrm{d}}{\mathrm{d}z}\right)^k\left[z^{-n}I_n(z)\right] &= z^{-n-k}I_{n+k}(z),\\ (n=0,1,2,\ldots \quad & k=0,1,2,\ldots).\end{aligned} \tag{8.4.32}$$

Applying the above formulas to compute$\dot{\mathcal{M}}_1$, we obtain

$$\begin{aligned}\dot{\mathcal{M}}_1(\rho_1,t) &= \rho_1^{-1}\left[\alpha_1J_0(\beta\zeta_1)-\beta J_1(\beta\zeta_1)\right]\exp(\alpha_1\zeta_1)\\ &+\sum_{n=0}^{\infty}\frac{(-)^n\hat{\lambda}_1^n}{n!\,\omega_1^{n-1}}g_n^{(1)}(0)\,\hat{z}_1^{n-1}J_{n-1}(\hat{z}_1) + \sum_{n=0}^{\infty}\frac{(-)^n\hat{\lambda}_1^n}{n!\,\omega_1^{n-1}}\int_{\rho_1}^{t}\dot{g}_n^{(1)}(t-s)\,z_1^{n-1}J_{n-1}(z_1)\mathrm{d}s.\end{aligned} \tag{8.4.33}$$

Since (recall eqns (7.4.82) and (7.4.86))

$$\begin{matrix} g_0^{(i)}(0) = 1 \\ g_n^{(i)}(0) = 0 \end{matrix} \quad \text{for } n \geq 1 \ \text{ and } i = 1, 2, \tag{8.4.34}$$

the formula (8.4.33) reduces to the form

$$\begin{aligned} \dot{\mathcal{M}}_1(\rho_1, t) = {} & \rho_1^{-1}\left[\alpha_1 J_0(\beta\zeta_1) - \beta J_1(\beta\zeta_1)\right]\exp(\alpha_1\zeta_1) - \omega_1 \frac{J_1(\hat{z}_1)}{\hat{z}_1} \\ & + \sum_{n=0}^{\infty} \frac{(-)^n \hat{\lambda}_1^n}{n!\ \omega_1^{n-1}} \int_{\rho_1}^{t} \dot{g}_n^{(1)}(t-s)\, z_1^{n-1} J_{n-1}(z_1) \mathrm{d}s. \end{aligned} \tag{8.4.35}$$

In a similar way, upon differentiation of eqn (8.4.30) with respect to t, we find

$$\begin{aligned} \dot{\mathcal{M}}_2(\rho_2, t) = {} & \rho_2^{-1}\left[\alpha_2 J_0(\beta\zeta_2) - \beta J_1(\beta\zeta_2)\right]\exp(\alpha_2\zeta_2) + \hat{\omega}_2 \frac{I_1(\hat{z}_2)}{\hat{z}_2} \\ & + \sum_{n=0}^{\infty} \frac{\hat{\lambda}_2^n}{n!\ \hat{\omega}_2^{n-1}} \int_{\rho_2}^{t} \dot{g}_n^{(2)}(t-s)\, z_2^{n-1} I_{n-1}(z_2) \mathrm{d}s. \end{aligned} \tag{8.4.36}$$

Furthermore, differentiating eqns (8.4.35) and (8.4.36) with respect to t and using the relations

$$\dot{g}_0^{(i)}(0) = \alpha_i,\ \dot{g}_1^{(i)}(0) = \frac{\alpha}{2},\ \dot{g}_n^{(i)}(0) = 0,\ n \geq 2, \tag{8.4.37}$$

we obtain

$$\begin{aligned} & \ddot{\mathcal{M}}_1(\rho_1, t) = \\ & \rho_1^{-1}\left[\left(\alpha_1^2 - \frac{\beta^2}{2}\right) J_0(\beta\zeta_1) - 2\alpha_1\beta J_1(\beta\zeta_1) + \frac{\beta^2}{2} J_2(\beta\zeta_1)\right]\exp(\alpha_1\zeta_1) \\ & + t\ \omega_1^2 \frac{J_2(\hat{z}_1)}{\hat{z}_1^2} - \omega_1\alpha_1 \frac{J_1(\hat{z}_1)}{\hat{z}_1} - \frac{\alpha\hat{\lambda}_1}{2} J_0(\hat{z}_1) \\ & + \sum_{n=0}^{\infty} \frac{(-)^n \hat{\lambda}_1^n}{n!\ \omega_1^{n-1}} \int_{\rho_1}^{t} \ddot{g}_n^{(1)}(t-s)\, z_1^{n-1} J_{n-1}(z_1)\, \mathrm{d}s, \end{aligned} \tag{8.4.38}$$

and

$$\ddot{\mathcal{M}}_2(\rho_2, t) =$$

$$\rho_2^{-1}\left[\left(\alpha_2^2 - \frac{\beta^2}{2}\right) J_0(\beta\zeta_2) - 2\alpha_2\beta J_1(\beta\zeta_2) + \frac{\beta^2}{2} J_2(\beta\zeta_2)\right] \exp(\alpha_2\zeta_2)$$
$$+t\,\hat{\omega}_2^2\frac{I_2(\hat{z}_2)}{\hat{z}_2^2} + \hat{\omega}_2\alpha_2\frac{I_1(\hat{z}_2)}{\hat{z}_2} + \frac{\alpha\hat{\lambda}_2}{2} I_0(\hat{z}_2) \tag{8.4.39}$$
$$+\sum_{n=0}^{\infty}\frac{\hat{\lambda}_2^n}{n!\,\hat{\omega}_2^{n-1}}\int_{\rho_2}^{t}\ddot{g}_n^{(1)}(t-s)\,z_2^{n-1}I_{n-1}(z_2)\,\mathrm{d}s.$$

From the formulas (8.4.29) and (8.4.30) we obtain

$$\mathcal{M}_1^{(0)} = \mathcal{M}_1(\rho_1, \rho_1) = \rho_1^{-1},$$
$$\mathcal{M}_2^{(0)} = \mathcal{M}_2(\rho_2, \rho_2) = \rho_2^{-1}, \tag{8.4.40}$$

while, on account of eqns (8.4.35) and (8.4.36), there follows

$$\dot{\mathcal{M}}_1^{(0)} = \dot{\mathcal{M}}_1(\rho_1, \rho_1) = \alpha_1\rho_1^{-1} - \frac{\omega_1}{2},$$
$$\dot{\mathcal{M}}_2^{(0)} = \dot{\mathcal{M}}_2(\rho_2, \rho_2) = \alpha_2\rho_2^{-1} - \frac{\hat{\omega}_2}{2}. \tag{8.4.41}$$

Hence, given the definitions of kernels $K_i^{(0)}$, $\theta_i^{(0)}$, $H_i^{(0)}$ and $H_i^{(1)}$ (recall eqns (8.4.24)–(8.4.27)), we infer that the first part of Theorem 8.1 is true.

In order to prove the second part, note that (recall eqns (8.3.66), (8.3.71), (8.3.76) and (8.3.69))

$$K_i(\rho_i, t) = \frac{1}{8\pi v_i}\exp(-h_i t)\left[\mathcal{N}_i \mp \hat{\beta}\dot{\mathcal{M}}_i \mp \left(\hat{\alpha} - h_i\hat{\beta}\right)\mathcal{M}_i\right], \tag{8.4.42}$$

$$\varphi_i(\rho_i, t) = \pm\frac{1}{4\pi v_i}\left(\frac{1}{v_1^2} - \frac{1}{v_2^2}\right)^{-1}$$
$$\times\left[t^0\exp(-h_i t)\,\mathcal{M}_i(\rho_i, t) + \int_{\rho_i}^{t}\exp(-h_i u)\,\mathcal{M}_i(\rho_i, u)\,\mathrm{d}u\right], \tag{8.4.43}$$

$$H_i(\rho_i, t) = \pm\frac{\epsilon}{4\pi v_i}\left(\frac{1}{v_1^2} - \frac{1}{v_2^2}\right)^{-1}\exp(-h_i t)\left[\ddot{\mathcal{M}}_i - 2h_i\dot{\mathcal{M}}_i + h_i^2\mathcal{M}_i\right], \tag{8.4.44}$$

and

$$\theta_i(\rho_i, t) = \frac{1}{8\pi v_i}\exp(-h_i t)\left[\mathcal{N}_i \pm \hat{\beta}\dot{\mathcal{M}}_i \pm \left(\hat{\alpha} - h_i\hat{\beta}\right)\mathcal{M}_i\right], \tag{8.4.45}$$

where the functions $\mathcal{M}_i$, $\dot{\mathcal{M}}_i$ and $\ddot{\mathcal{M}}_i$ are specified by the formulas (8.4.29), (8.4.30), (8.4.35), (8.4.36), (8.4.38) and (8.4.39), while the functions $\mathcal{N}_i$ $(i = 1, 2)$ are of the form (recall eqns (7.4.97) and (7.4.98))

$$\mathcal{N}_1(\rho_1, t) = -\omega_1 \frac{J_1(\hat{z}_1)}{\hat{z}_1} + \sum_{n=0}^{\infty} \frac{(-)^n \hat{\lambda}_1^n}{n!\, \omega_1^{n-1}} \int_{\rho_1}^{t} h_n^{(1)}(t-s)\, z_1^{n-1} J_{n-1}(z_1)\, \mathrm{d}s, \quad (8.4.46)$$

$$\mathcal{N}_2(\rho_2, t) = \hat{\omega}_2 \frac{I_1(\hat{z}_2)}{\hat{z}_2} + \sum_{n=0}^{\infty} \frac{\hat{\lambda}_2^n}{n!\, \hat{\omega}_2^{n-1}} \int_{\rho_2}^{t} h_n^{(2)}(t-s)\, z_2^{n-1} I_{n-1}(z_2)\, \mathrm{d}s. \quad (8.4.47)$$

Since $\mathcal{N}_i(\rho_i, t)$ possesses a finite limit for $\rho_i \to 0$ and $t > 0$, while $\mathcal{M}_i(\rho_i, t)$ and $\dot{\mathcal{M}}_i(\rho_i, t)$ have a singularity of order ρ_i^{-1} for $\rho_i \to 0$ and for $t > 0$, then, on account of eqn (8.4.42), the kernel $K_i(\rho_i, t)$ also has a singularity of order ρ_i^{-1} for $\rho_i \to 0$ and for $t > 0$.

We similarly prove that the kernels $\varphi_i(\rho_i, t)$ and $\theta_i(\rho_i, t)$ have singularities of order ρ_i^{-1} for $\rho_i \to 0$ and for $t > 0$.

Finally, in view of eqn (8.4.44) and the fact that $\ddot{\mathcal{M}}_i$, $\dot{\mathcal{M}}_i$ and $\mathcal{M}_i$ all have the same singularity, we conclude that the kernel $H_i(\rho_i, t)$ behaves like ρ_i^{-1} for $\rho_i \to 0$ and for $t > 0$. □

It follows from Theorem 8.1 that an "internal" as well as an "external" potential–temperature problem of the G–L theory may be reduced to that of a system of singular integral equations with kernels of the type of a kernel of a single layer potential for every $t > 0$.

Closing this chapter, we observe that each of these two problems may also be reduced to a solution of the system of singular integral equations with kernels of the type of a kernel of a double-layer potential. In this case, kernels of the integral equations contain first derivatives of the functions $\mathcal{M}_i$ and $\mathcal{N}_i$ with respect to ρ_i, as well as higher-order mixed derivatives of $\mathcal{M}_i$ (recall eqns (8.3.59)–(8.3.62)). These derivatives may be determined using the recurrence formulas (8.4.31) and (8.4.32). In particular, we find

$$\mathcal{M}_1'(\rho_1, t) = p_1(\rho_1, t) - \rho_1 \sum_{n=1}^{\infty} \frac{(-)^n \hat{\lambda}_1^n}{n!\, \omega_1^{n-2}} \int_{\rho_1}^{t} g_n^{(1)}(t-s)\, z_1^{n-2} J_{n-2}(z_1)\, \mathrm{d}s, \quad (8.4.48)$$

and

$$\mathcal{M}_2'(\rho_2, t) = p_2(\rho_2, t) - \rho_2 \sum_{n=1}^{\infty} \frac{\hat{\lambda}_2^n}{n!\, \hat{\omega}_2^{n-2}} \int_{\rho_2}^{t} g_n^{(2)}(t-s)\, z_2^{n-2} I_{n-2}(z_2)\, \mathrm{d}s, \quad (8.4.49)$$

where

$$p_1(\rho_1, t) = -\rho_1^{-2}\left[\left(1 + \alpha_1\rho_1 - \frac{\omega_1}{2}\rho_1^2\right) J_0(\beta\zeta_1)\right.$$
$$\left. -\rho_1\left(\beta + \frac{\alpha}{\beta}\hat{\lambda}_1\rho_1\right) J_1(\beta\zeta_1) + \frac{\hat{\lambda}_1}{2}\rho_1^2 J_2(\beta\zeta_1)\right] \exp(\alpha_1\zeta_1), \tag{8.4.50}$$

and

$$p_2(\rho_2, t) = -\rho_2^{-2}\left[\left(1 + \alpha_2\rho_2 + \frac{\hat{\omega}_2}{2}\rho_2^2\right) J_0(\beta\zeta_2)\right.$$
$$\left. -\rho_2\left(\beta - \frac{\alpha}{\beta}\hat{\lambda}_2\rho_2\right) J_1(\beta\zeta_2) + \frac{\hat{\lambda}_2}{2}\rho_2^2 J_2(\beta\zeta_2)\right] \exp(\alpha_2\zeta_2). \tag{8.4.51}$$

It follows from the formulas (8.4.48)–(8.4.51) that the functions $\mathcal{M}_i'(\rho_i, t)$ $(i = 1, 2)$ have a singularity of order ρ_i^{-2} for $\rho_i \to 0$ and every $t > 0$, which is the singularity of the kernel of a double-layer potential.

9

THERMOELASTIC POLYNOMIALS

9.1 Recurrence relations

It was shown in Chapter 7 that the fundamental solutions of the G–L theory may be determined with the help of polynomial sequences on the time axis, the so-called *polynomials of thermoelasticity.* We now give a number of recurrence relations describing these polynomials (Ignaczak, 1983)). First, let us consider a function $h_n(t; \lambda, \mu)$ specified by the formula (recall (7.1.50)–(7.1.54))

$$h_n(t; \lambda, \mu) = L^{-1}\left\{\frac{p^n}{\left[(p+\mu)^2 + \lambda^2\right]^n}\right\}, \tag{9.1.1}$$

where L is the Laplace transform, p is the parameter of transformation, while μ and λ are positive-valued parameters; $n = 1, 2, 3, \ldots$. The function (9.1.1) may be represented in the form (recall eqns (7.1.57)–(7.1.59))

$$h_n(t; \lambda, \mu) = \mathrm{e}^{-\mu t}\left(A_n \cos \lambda t + B_n \sin \lambda t\right), \tag{9.1.2}$$

where the functions $A_n = A_n(t; \lambda, \mu)$ and $B_n = B_n(t; \lambda, \mu)$ are specified by the recurrence relations

$$\begin{aligned} A_{n+1} &= \frac{1}{2n}\left[\frac{\mu}{\lambda}\left(A_{n,\lambda} + tB_n\right) - \left(A_{n,\mu} - tA_n\right)\right], \\ B_{n+1} &= \frac{1}{2n}\left[\frac{\mu}{\lambda}\left(B_{n,\lambda} - tA_n\right) - \left(B_{n,\mu} - tB_n\right)\right]. \end{aligned} \tag{9.1.3}$$

Here, the comma preceding λ or μ indicates a partial differentiation with respect to λ or μ.

For $n = 1$ we obtain

$$A_1 = 1, \quad B_1 = -\mu\lambda^{-1}. \tag{9.1.4}$$

From this, on account of eqn (9.1.3), for $n = 1$, we find

$$A_2 = \frac{t}{2}\left(1 - \mu^2\lambda^{-2}\right), \quad B_2 = \frac{1}{2\lambda}\left(1 - \mu^2\lambda^{-2} - 2\mu t\right). \tag{9.1.5}$$

In general, $A_n(t;\lambda,\mu)$ and $B_n(t;\lambda,\mu)$ are polynomials in t of degree $(n-1)$, and of the form [recall eqn (7.1.68)]

$$\begin{cases} A_n = \dfrac{1}{(2\lambda)^{n-1}(n-1)!}\displaystyle\sum_{k=0}^{n-1} a_{kn}\left(\frac{\mu}{\lambda}\right)\lambda^k t^k, \\ B_n = \dfrac{1}{(2\lambda)^{n-1}(n-1)!}\displaystyle\sum_{k=0}^{n-1} b_{kn}\left(\frac{\mu}{\lambda}\right)\lambda^k t^k, \end{cases} \tag{9.1.6}$$

where $a_{kn}(\zeta)$ and $b_{kn}(\zeta)$ are polynomials in ζ of degree $\leq n$ $(\zeta=\mu\lambda^{-1})$. The polynomials A_n and B_n are called the *thermoelastic polynomials.* In the following, in order to underline the dependence of h_n, A_n and B_n on t and simplify the notation, we will be omitting λ, μ as arguments, and simply write $A_n(t)$ and $B_n(t)$.

From eqn (9.1.1) we obtain

$$h_n(t) = h_1(t) * h_{n-1}(t) \text{ for } n \geq 2. \tag{9.1.7}$$

Since the functions $h_1(t)$ and $h_{n-1}(t)$ $(n\geq 2)$ are continuous for every $t\geq 0$, then, on account of eqn (9.1.7)

$$h_n(0) = 0 \quad \forall n \geq 2. \tag{9.1.8}$$

Hence, using eqn (9.1.2), we obtain

$$A_n(0) = 0 \quad \forall n \geq 2. \tag{9.1.9}$$

Furthermore, given eqn (9.1.1), for the function $\bar{h}_n(p)$ $(n\geq 1)$ we obtain

$$\bar{h}_n(p) = \left[\bar{h}_1(p)\right]^n, \tag{9.1.10}$$

where

$$\bar{h}_1(p) = \frac{p}{\left[(p+\mu)^2+\lambda^2\right]}. \tag{9.1.11}$$

Next, differentiating eqn (9.1.10) with respect to p and multiplying through by p, we obtain

$$p\frac{\mathrm{d}\bar{h}_n}{\mathrm{d}p} = n\bar{h}_n - 2np\bar{h}_{n+1} - 2n\mu\bar{h}_{n+1}. \tag{9.1.12}$$

Applying the operator L^{-1} to both sides of this equation, and noting that $h_n(0)=0$ for $n\geq 2$, we obtain

$$-\frac{\mathrm{d}}{\mathrm{d}t}(th_n(t)) = nh_n(t) - 2n\dot{h}_{n+1}(t) - 2n\mu h_{n+1}(t) \quad \forall n \geq 1, \tag{9.1.13}$$

from which there follows

$$-h_n(t) - t\dot{h}_n(t) = nh_n(t) - 2n\dot{h}_{n+1}(t) - 2n\mu h_{n+1}(t) \quad \forall n \geq 1, \tag{9.1.14}$$

where the dot indicates differentiation with respect to t. Since $h_1(0) = A_1(0) = 1$ and $h_2(0) = 0$, then, letting $n = 1$ and $t = 0$ in eqn (9.1.14), we obtain

$$\dot{h}_2(0) = 1. \tag{9.1.15}$$

Also, letting $t = 0$ in eqn (9.1.14) and employing eqn (9.1.8), we obtain

$$\dot{h}_n(0) = 0 \text{ for } n \geq 3. \tag{9.1.16}$$

Furthermore, we differentiate eqn (9.1.10) twice with respect to p and multiply through by p^2, so as to get

$$p^2 \frac{\mathrm{d}^2 \bar{h}_n}{\mathrm{d}p^2} = n(n-1)\bar{h}_n - 4n^2\mu\bar{h}_{n+1} + 2np\bar{h}_{n+1} - 4n(n+1)\lambda^2\bar{h}_{n+2}. \tag{9.1.17}$$

The inverse transformation of eqn (9.1.17), and the use of eqn (9.1.8), yields

$$\begin{aligned} \frac{\mathrm{d}^2}{\mathrm{d}t^2}\left[t^2 h_n(t)\right] &= n(n-1)h_n(t) - 4n^2\mu h_{n+1}(t) \\ &+2n\dot{h}_{n+1}(t) - 4n(n+1)\lambda^2 h_{n+2}(t) \qquad \forall n \geq 1. \end{aligned} \tag{9.1.18}$$

Next, we note that

$$\left[(p+\mu)^2 + \lambda^2\right]\bar{h}_n = p\bar{h}_{n-1} \qquad \forall n \geq 2. \tag{9.1.19}$$

Applying L^{-1} to the above, and observing that $h_n(0) = 0$ and $\dot{h}_n(0) = 0$ for $n \geq 3$ [recall eqns (9.1.8) and (9.1.16)], yields

$$\frac{\mathrm{d}^2}{\mathrm{d}t^2}h_n(t) + 2\mu\frac{\mathrm{d}}{\mathrm{d}t}h_n(t) + \left(\mu^2+\lambda^2\right)h_n(t) = \frac{\mathrm{d}}{\mathrm{d}t}h_{n-1}(t) \qquad \forall n \geq 3. \tag{9.1.20}$$

The latter relation may be extended to the range $n \geq 2$, in which the relation (9.1.19) holds. Indeed, for $n = 2$ the relation (9.1.19) takes the form

$$p^2\bar{h}_2 + 2\mu p\bar{h}_2 + \left(\mu^2+\lambda^2\right)\bar{h}_2 = p\bar{h}_1. \tag{9.1.21}$$

Given that $h_2(0) = 0$ and $\dot{h}_2(0) = h_1(0) = 1$, eqn (9.1.21) may be written as

$$p^2\bar{h}_2 - ph_2(0) - \dot{h}_2(0) + 2\mu\left[p\bar{h}_2 - \bar{h}_2(0)\right] + \left(\mu^2+\lambda^2\right)\bar{h}_2 = p\bar{h}_1 - h_1(0). \tag{9.1.22}$$

Applying the operator L^{-1} to eqn (9.1.22) we infer that eqn (9.1.20) holds $\forall n \geq 2$. Thus, in view of eqns (9.1.13), (9.1.18) and (9.1.20)–(9.1.22), we conclude that the following recurrence relations are true $\forall n \geq 2$

$$-\frac{\mathrm{d}}{\mathrm{d}t}(th_n) = nh_n - 2n\dot{h}_{n+1} - 2n\mu h_{n+1}, \tag{9.1.23}$$

$$\ddot{h}_{n+1} + 2\mu\dot{h}_{n+1} + \left(\mu^2+\lambda^2\right)h_{n+1} = \dot{h}_n, \tag{9.1.24}$$

$$\frac{\mathrm{d}^2}{\mathrm{d}t^2}\left(t^2 h_n\right) = n(n-1)h_n - 4n^2\mu h_{n+1} + 2n\dot{h}_{n+1} - 4n(n+1)\lambda^2 h_{n+2}. \tag{9.1.25}$$

Substituting $h_n = h_n(t)$ given by eqn (9.1.2) into the system (9.1.23)–(9.1.25), and setting equal the coefficients of $\cos\lambda t$ and $\sin\lambda t$ on both sides of this system, $\forall n \geq 1$ we obtain the three separated systems containing a pair (A_n, B_n)

$$\begin{aligned} \dot{A}_{n+1} + \lambda B_{n+1} &= \frac{1}{2n}\left[(n+1)A_n + \left(\dot{A}_n - \mu A_n + \lambda B_n\right)t\right], \\ \dot{B}_{n+1} - \lambda A_{n+1} &= \frac{1}{2n}\left[(n+1)B_n + \left(\dot{B}_n - \mu B_n - \lambda A_n\right)t\right], \end{aligned} \tag{9.1.26}$$

$$\begin{aligned} \ddot{A}_{n+1} + 2\lambda\dot{B}_{n+1} &= \dot{A}_n - \mu A_n + \lambda B_n, \\ \ddot{B}_{n+1} - 2\lambda\dot{A}_{n+1} &= \dot{B}_n - \mu B_n + \lambda A_n, \end{aligned} \tag{9.1.27}$$

$$\begin{aligned} 4n(n+1)\lambda^2 A_{n+2} &= 2n\left[\dot{A}_{n+1} - (2n+1)\mu A_{n+1}\right] \\ &+ \left[t^2\left(\lambda^2 - \mu^2\right) + 4\mu t + (n+1)(n-2)\right]A_n - 2t(2-\mu t)\dot{A}_n \\ &- t^2\ddot{A}_n - 2\lambda t^2\dot{B}_n - 2\lambda t(2-\mu t)B_n + 2n\lambda B_{n+1}, \\ 4n(n+1)\lambda^2 B_{n+2} &= 2n\left[\dot{B}_{n+1} - (2n+1)\mu B_{n+1}\right] \\ &+ \left[t^2\left(\lambda^2 - \mu^2\right) + 4\mu t + (n+1)(n-2)\right]B_n - 2t(2-\mu t)\dot{B}_n \\ &- t^2\ddot{B}_n + 2\lambda t^2\dot{A}_n + 2\lambda t(2-\mu t)A_n - 2n\lambda A_{n+1}. \end{aligned} \tag{9.1.28}$$

These are accompanied by the following equivalence relations

$$(9.1.23) \Longleftrightarrow (9.1.26), \quad (9.1.24) \Longleftrightarrow (9.1.27), \quad (9.1.25) \Longleftrightarrow (9.1.28). \tag{9.1.29}$$

Observing that $A_n(0) = 0$ for $\forall n \geq 2$ (recall eqn (9.1.9)) and letting $t = 0$ in eqns $(9.1.26)_2$ and $(9.1.28)_2$, yields

$$2n\dot{B}_{n+1}(0) = (n+1)B_n(0), \tag{9.1.30}$$

and

$$\begin{aligned} 4n(n+1)\lambda^2 B_{n+2}(0) &= 2n\left[\dot{B}_{n+1}(0) - (2n+1)\mu B_{n+1}(0)\right] \\ &+ (n+1)(n-2)B_n(0) \qquad \forall n \geq 1. \end{aligned} \tag{9.1.31}$$

Eliminating $\dot{B}_{n+1}(0)$ from the system (9.1.30) and (9.1.31), we obtain

$$4n(n+1)\lambda^2 B_{n+2}(0) = -2n(2n+1)\mu B_{n+1}(0) + \left(n^2 - 1\right)B_n(0) \qquad \forall n \geq 1. \tag{9.1.32}$$

The relations (9.1.3) and (9.1.26)–(9.1.28) are not the only recurrence relations for the pair of polynomials (A_n, B_n). Prior to deriving further recurrence relations, we shall transform the obtained relations to a more elegant form through the introduction of new variables (recall eqn (7.1.60))

$$\zeta = \mu\lambda^{-1}, \quad y = t\lambda. \tag{9.1.33}$$

Also, we introduce the notations

$$\hat{A}_n = c_n A_n, \quad \hat{B}_n = c_n B_n, \tag{9.1.34}$$

where

$$c_n = (2\lambda)^{n-1}(n-1)! \tag{9.1.35}$$

Employing the formulas (9.1.6), (9.1.34) and (9.1.35) for the pair $(\hat{A}_n, \hat{B}_n)$, we find

$$\hat{A}_n(y,\zeta) = \sum_{k=0}^{n-1} a_{kn}(\zeta)\, y^k, \qquad \hat{B}_n(y,\zeta) = \sum_{k=0}^{n-1} b_{kn}(\zeta)\, y^k. \tag{9.1.36}$$

In the following $(\hat{A}_n, \hat{B}_n)$ will also be called a *pair of thermoelastic polynomials,* and (a_{kn}, b_{kn}) will be called a *pair of associated thermoelastic polynomials.* Clearly, $(\hat{A}_n, \hat{B}_n)$ is a function of two variables (y,ζ) in contradistinction to the pair (A_n, B_n), which is a function of three variables: $(t;\lambda,\mu)$. The functions $\hat{A}_n$ and $\hat{B}_n$ are polynomials of degree $(n-1)$ with respect to y, while A_n and B_n are polynomials of degree $(n-1)$ with respect to t.

In view of the formula (7.1.55), the function h_n specified by eqn (9.1.2) may be given in the form

$$h_n = \frac{1}{2(n-1)}\left(\frac{\mu}{\lambda}\frac{\partial}{\partial\lambda} - \frac{\partial}{\partial\mu}\right) h_{n-1} \qquad \forall n \geq 2. \tag{9.1.37}$$

It follows from the formula (9.1.35) that $c_n = c_n(\lambda)$ and, in view of eqn (9.1.2), an alternative form of h_n is

$$h_n = [c_n(\lambda)]^{-1}\, \hat{h}_n(y,\zeta), \tag{9.1.38}$$

where

$$\hat{h}_n(y,\zeta) = e^{-\zeta y}(\hat{A}_n \cos y + \hat{B}_n \sin y). \tag{9.1.39}$$

The variables ζ and y, specified by the formula (9.1.33) can be treated as functions of the point (λ,μ) by writing

$$\zeta = \zeta(\lambda,\mu) \equiv \mu\lambda^{-1}, \qquad y = y(\lambda,\mu) \equiv \lambda t. \tag{9.1.40}$$

From this we obtain

$$\begin{aligned} \frac{\partial\zeta}{\partial\lambda} &= -\zeta\lambda^{-1}, & \frac{\partial\zeta}{\partial\mu} &= \lambda^{-1}, \\ \frac{\partial y}{\partial\lambda} &= t, & \frac{\partial y}{\partial\mu} &= 0. \end{aligned} \tag{9.1.41}$$

On account of eqn (9.1.38), for the partial derivatives of h_n we find

$$\frac{\partial h_n}{\partial\lambda} = -\left[c_n(\lambda)\right]^{-2} c_n'(\lambda)\hat{h}_n + \left[c_n(\lambda)\right]^{-1}\left[\frac{\partial\hat{h}_n}{\partial y}t + \frac{\partial\hat{h}_n}{\partial\zeta}(-\zeta)\frac{1}{\lambda}\right], \tag{9.1.42}$$

$$\frac{\partial h_n}{\partial\mu} = \left[c_n(\lambda)\right]^{-1}\left(\frac{\partial\hat{h}_n}{\partial y}\cdot 0 + \frac{\partial\hat{h}_n}{\partial\zeta}\frac{1}{\lambda}\right), \tag{9.1.43}$$

from which it follows that

$$\frac{\partial h_n}{\partial \lambda} = c_n^{-1}\lambda^{-1}\left(y\frac{\partial \hat{h}_n}{\partial y} - \zeta\frac{\partial \hat{h}_n}{\partial \zeta} - \frac{c_n'}{c_n}\lambda \hat{h}_n\right), \tag{9.1.44}$$

and

$$\frac{\partial h_n}{\partial \mu} = c_n^{-1}\lambda^{-1}\frac{\partial \hat{h}_n}{\partial \zeta}. \tag{9.1.45}$$

On the other hand, in view of eqn (9.1.35),

$$c_n = 2^{n-1}\,(n-1)!\lambda^{n-1}. \tag{9.1.46}$$

Therefore,

$$\lambda c_n' = 2^{n-1}\,(n-1)!(n-1)\lambda^{n-1}, \tag{9.1.47}$$

and

$$\lambda\frac{c_n'}{c_n} = n-1, \tag{9.1.48}$$

and, given eqns (9.1.44) and (9.1.45), we obtain

$$\left(\frac{\mu}{\lambda}\frac{\partial}{\partial \lambda} - \frac{\partial}{\partial \mu}\right) h_n = c_n^{-1}\lambda^{-1}\left[\zeta y\frac{\partial}{\partial y} - (1+\zeta^2)\frac{\partial}{\partial \zeta} - (n-1)\zeta\right]\hat{h}_n. \tag{9.1.49}$$

From this and eqn (9.1.37) we find

$$\begin{gathered} e^{-\zeta y}c_n^{-1}(\hat{A}_n\cos y + \hat{B}_n\sin y) = \\ [2(n-1)]^{-1}\,c_{n-1}^{-1}\lambda^{-1}\left[\zeta y\frac{\partial}{\partial y} - (1+\zeta^2)\frac{\partial}{\partial \zeta} - (n-2)\zeta\right] \\ \times \mathrm{e}^{-\zeta y}(\hat{A}_{n-1}\cos y + \hat{B}_{n-1}\sin y). \end{gathered} \tag{9.1.50}$$

Since

$$2\,(n-1)\,c_{n-1}\lambda = c_n, \tag{9.1.51}$$

from eqn (9.1.50) we obtain

$$\begin{gathered} \hat{A}_n\cos y + \hat{B}_n\sin y = \\ \mathrm{e}^{\zeta y}\left[\zeta y\frac{\partial}{\partial y} - \left(1+\zeta^2\right)\frac{\partial}{\partial \zeta} - (n-2)\,\zeta\right]\mathrm{e}^{-\zeta y}(\hat{A}_{n-1}\cos y + \hat{B}_{n-1}\sin y). \end{gathered} \tag{9.1.52}$$

Equating the coefficients of $\cos y$ and $\sin y$ on both sides of eqn (9.1.52), we arrive at the recurrence formulas for the pair $(\hat{A}_n, \hat{B}_n)$, an analog of eqn (9.1.3) for the pair (A_n, B_n)

$$\begin{aligned}\hat{A}_{n+1} &= L_n\hat{A}_n + \zeta y\hat{B}_n,\\ \hat{B}_{n+1} &= -\zeta y\hat{A}_n + L_n\hat{B}_n,\end{aligned} \tag{9.1.53}$$

where

$$L_n = \zeta y\frac{\partial}{\partial y} - (1+\zeta^2)\frac{\partial}{\partial\zeta} + y - (n-1)\zeta \qquad \forall n \geq 1, \tag{9.1.54}$$

and

$$\hat{A}_1 = 1, \quad \hat{B}_1 = -\zeta. \tag{9.1.55}$$

Note that the relations (9.1.53)–(9.1.55) may also be adopted as the definition of a pair of the thermoelastic polynomials $(\hat{A}_n, \hat{B}_n)$ $\forall n \geq 1$. An immediate consequence of that definition is the following theorem.

Theorem 9.1 *The three successive polynomials $\hat{A}_{n+2}$, $\hat{A}_{n+1}$ and $\hat{A}_n$ or $\hat{B}_{n+2}$, $\hat{B}_{n+1}$ and $\hat{B}_n$, treated as functions of y and ζ, satisfy the recurrence relations*

$$\hat{A}_{n+2} - 2\left(L_n + \frac{\zeta^{-1}+\zeta}{2}\right)\hat{A}_{n+1} + \left(L_n^2 + \zeta^{-1}L_n + \zeta^2y^2\right)\hat{A}_n = 0, \tag{9.1.56}$$

or

$$\hat{B}_{n+2} - 2\left(L_n + \frac{\zeta^{-1}+\zeta}{2}\right)\hat{B}_{n+1} + \left(L_n^2 + \zeta^{-1}L_n + \zeta^2y^2\right)\hat{B}_n = 0, \tag{9.1.57}$$

where L_n $(n \geq 1)$ is the operator specified by eqn (9.1.54), while L_n^2 is its square.[1]

The proof of eqn (9.1.56) (or eqn (9.1.57)) consists of the elimination of polynomials $\hat{B}_{n+1}$ and $\hat{B}_n$ (respectively, $\hat{A}_{n+1}$ and $\hat{A}_n$) from the recurrence formulas (9.1.53).

Also, employing eqns (9.1.53)–(9.1.55), we obtain three further theorems as counterparts of the recurrence systems (9.1.26), (9.1.27) and (9.1.28).

Theorem 9.2 *The pair $(\hat{A}_n, \hat{B}_n)$, treated as a function of y, for any fixed ζ, satisfies the recurrence relations*

$$\begin{aligned}\hat{A}'_{n+1} + \hat{B}_{n+1} &= (n+1)\hat{A}_n + (\hat{A}'_n - \zeta\hat{A}_n + \hat{B}_n)y,\\ \hat{B}'_{n+1} - \hat{A}_{n+1} &= (n+1)\hat{B}_n + (\hat{B}'_n - \zeta\hat{B}_n - \hat{A}_n)y,\end{aligned} \tag{9.1.58}$$

where the prime denotes the derivative with respect to y.

[1] One can show that $\zeta = 0$ is not a singular point of the recurrence relations (9.1.56) and (9.1.57).

Theorem 9.3 *The pair $(\hat{A}_n, \hat{B}_n)$, treated as a function of y, for any fixed ζ, satisfies the recurrence relations*

$$\begin{aligned}\hat{A}''_{n+1} + 2\hat{B}'_{n+1} &= 2n(\hat{A}'_n - \zeta\hat{A}_n + \hat{B}_n),\\ \hat{B}''_{n+1} - 2\hat{A}'_{n+1} &= 2n(\hat{B}'_n - \zeta\hat{B}_n - \hat{A}_n).\end{aligned} \tag{9.1.59}$$

Theorem 9.4 *The pair $(\hat{A}_n, \hat{B}_n)$, treated as a function of y, for any fixed ζ, satisfies the recurrence relations*

$$\begin{aligned}\hat{A}_{n+2} &= \hat{A}'_{n+1} - (2n+1)\zeta\hat{A}_{n+1} + [(1-\zeta^2)y^2 + 4\zeta y + (n+1)(n-2)]\hat{A}_n\\ &\quad -2y(2-\zeta y)\hat{A}'_n - y^2\hat{A}''_n - 2y^2\hat{B}'_n - 2y(2-\zeta y)\hat{B}_n + \hat{B}_{n+1},\\ \hat{B}_{n+2} &= \hat{B}'_{n+1} - (2n+1)\zeta\hat{B}_{n+1} + [(1-\zeta^2)y^2 + 4\zeta y + (n+1)(n-2)]\hat{B}_n\\ &\quad -2y(2-\zeta y)\hat{B}'_n - y^2\hat{B}''_n + 2y^2\hat{A}'_n + 2y(2-\zeta y)\hat{A}_n - \hat{A}_{n+1}.\end{aligned} \tag{9.1.60}$$

The relations (9.1.58) and (9.1.59) couple together two successive pairs of thermoelastic polynomials in two different ways. The relation (9.1.60) couples three successive pairs of thermoelastic polynomials. We end this section by giving the system of recurrence relations coupling four successive pairs of thermoelastic polynomials.

Theorem 9.5 *The pair $(\hat{A}_n, \hat{B}_n)$, treated as a function of y, for any fixed ζ, satisfies the recurrence relations*

$$\begin{aligned}&\hat{A}_{n+3} + (2n+3)\zeta\hat{A}_{n+2} - [(1-\zeta^2)y^2 + 3\zeta y + n(n+2)]\hat{A}_{n+1}\\ &-(2\zeta y - 2n - 3)y^2\hat{A}'_n + [2\zeta^2y^2 - (4n+5)\zeta y + 3(n+1)]y\hat{A}_n\\ &\qquad -(2\zeta y - 2n - 3)y^2\hat{B}_n = 0,\\ &\hat{B}_{n+3} + (2n+3)\zeta\hat{B}_{n+2} - [(1-\zeta^2)y^2 + 3\zeta y + n(n+2)]\hat{B}_{n+1}\\ &-(2\zeta y - 2n - 3)y^2\hat{B}'_n + [2\zeta^2y^2 - (4n+5)\zeta y + 3(n+1)]y\hat{B}_n\\ &\qquad +(2\zeta y - 2n - 3)y^2\hat{A}_n = 0.\end{aligned} \tag{9.1.61}$$

Proof. To prove the relations (9.1.61) we rewrite the recurrence relations (9.1.60) as

$$\begin{aligned}\hat{A}_{n+2} &= \underline{\hat{A}'_{n+1} + \hat{B}_{n+1}} - (2n+1)\zeta\hat{A}_{n+1}\\ &\quad +[(1-\zeta^2)y^2 + 4\zeta y + (n+1)(n-2)]\hat{A}_n\\ &\quad -2y(2-\zeta y)(\hat{A}'_n + \hat{B}_n) - y^2(\hat{A}''_n + 2\hat{B}'_n),\\ \hat{B}_{n+2} &= \underline{\hat{B}'_{n+1} - \hat{A}_{n+1}} - (2n+1)\zeta\hat{B}_{n+1}\\ &\quad +[(1-\zeta^2)y^2 + 4\zeta y + (n+1)(n-2)]\hat{B}_n\\ &\quad -2y(2-\zeta y)(\hat{B}'_n - \hat{A}_n) - y^2(\hat{B}''_n - 2\hat{A}'_n).\end{aligned} \tag{9.1.62}$$

Now, substituting the right-hand sides of eqns $(9.1.58)_1$ and $(9.1.58)_2$ in place of the underlined terms in eqns $(9.1.62)_1$ and $(9.1.62)_2$, respectively, we obtain

$$\begin{aligned}\hat{A}_{n+2} &= -(2n+1)\zeta\hat{A}_{n+1} + [(1-\zeta^2)y^2 + 3\zeta y + (n^2-1)]\hat{A}_n \\ &\quad -y(3-2\zeta y)(\hat{A}'_n + \hat{B}_n) - y^2(\hat{A}''_n + 2\hat{B}'_n), \\ \hat{B}_{n+2} &= -(2n+1)\zeta\hat{B}_{n+1} + [(1-\zeta^2)y^2 + 3\zeta y + (n^2-1)]\hat{B}_n \\ &\quad -y(3-2\zeta y)(\hat{B}'_n - \hat{A}_n) - y^2(\hat{B}''_n - 2\hat{A}'_n).\end{aligned} \tag{9.1.63}$$

Interchanging n with $n+1$ in eqn (9.1.63), yields

$$\begin{aligned}\hat{A}_{n+3} &= -(2n+3)\zeta\hat{A}_{n+2} + [(1-\zeta^2)y^2 + 3\zeta y + n(n+2)]\hat{A}_{n+1} \\ &\quad -y(3-2\zeta y)\underline{(\hat{A}'_{n+1} + \hat{B}_{n+1})} - y^2\underleftrightarrow{(\hat{A}''_{n+1} + 2\hat{B}'_{n+1})}, \\ \hat{B}_{n+3} &= -(2n+3)\zeta\hat{B}_{n+2} + [(1-\zeta^2)y^2 + 3\zeta y + n(n+2)]\hat{B}_{n+1} \\ &\quad -y(3-2\zeta y)\underline{(\hat{B}'_{n+1} - \hat{A}_{n+1})} - y^2\underleftrightarrow{(\hat{B}''_{n+1} - 2\hat{A}'_{n+1})}.\end{aligned} \tag{9.1.64}$$

Finally, expressing the terms underlined by lines and arrows by the right-hand sides of eqns (9.1.58) and (9.1.59), respectively, leads to the recurrence relations (9.1.61). □

9.2 Differential equation

In this section we show that a pair of thermoelastic polynomials $(\hat{A}_n, \hat{B}_n)$ can be identified with an element of the null space of a linear ordinary differential operator. More specifically, we shall prove the following theorem.

Theorem 9.6 *The pair $(\hat{A}_n, \hat{B}_n)$ is a solution of the following system of ordinary differential equations*

$$\begin{aligned}&y\hat{A}'''_n - (n-3+\zeta y)\hat{A}''_n + 2[(n-1)\zeta - y)]\hat{A}'_n + 2(n-1)\hat{A}_n \\ &\quad +3y\hat{B}''_n - 2(n-3+\zeta y)\hat{B}'_n + 2(n-1)\zeta\hat{B}_n = 0, \\ &y\hat{B}'''_n - (n-3+\zeta y)\hat{B}''_n + 2[(n-1)\zeta - y]\hat{B}'_n + 2(n-1)\hat{B}_n \\ &\quad -3y\hat{A}''_n + 2(n-3+\zeta y)\hat{A}'_n - 2(n-1)\zeta\hat{A}_n = 0.\end{aligned} \tag{9.2.1}$$

Proof. In the proof a key role is played by the relations (9.1.58) and (9.1.59). The first of these have the form

$$\begin{aligned}\hat{A}'_{n+1} + \hat{B}_{n+1} &= M_n\hat{A}_n + y\hat{B}_n, \\ -\hat{A}_{n+1} + \hat{B}'_{n+1} &= -y\hat{A}_n + M_n\hat{B}_n,\end{aligned} \tag{9.2.2}$$

where

$$M_n = y\frac{\partial}{\partial y} - \zeta y + (n+1). \tag{9.2.3}$$

Furthermore, the relations (9.1.59) can be rewritten as

$$\begin{aligned} \hat{A}''_{n+1} + 2\hat{B}'_{n+1} &= 2n(\hat{A}'_n - \zeta\hat{A}_n + \hat{B}_n), \\ -2\hat{A}'_{n+1} + \hat{B}''_{n+1} &= 2n(-\hat{A}_n + \hat{B}'_n - \zeta\hat{B}_n). \end{aligned} \tag{9.2.4}$$

Differentiating eqn $(9.2.2)_1$ with respect to y, and combining the result with eqn $(9.2.4)_1$, yields

$$\begin{aligned} \hat{A}''_{n+1} + \hat{B}'_{n+1} &= (M_n\hat{A}_n + y\hat{B}_n)', \\ \hat{A}''_{n+1} + 2\hat{B}'_{n+1} &= 2n(\hat{A}'_n - \zeta\hat{A}_n + \hat{B}_n). \end{aligned} \tag{9.2.5}$$

An analogous differentiation of eqn $(9.2.2)_2$ with respect to y, in combination with eqn $(9.2.4)_2$, yields

$$\begin{aligned} -\hat{A}'_{n+1} + \hat{B}''_{n+1} &= (-y\hat{A}_n + M_n\hat{B}_n)', \\ -2\hat{A}'_{n+1} + \hat{B}''_{n+1} &= 2n(-\hat{A}_n + \hat{B}'_n - \zeta\hat{B}_n). \end{aligned} \tag{9.2.6}$$

The system (9.2.5) is equivalent to

$$\begin{aligned} \hat{A}''_{n+1} &= -2n(\hat{A}'_n - \zeta\hat{A}_n + \hat{B}_n) + 2(M_n\hat{A}_n + y\hat{B}_n)', \\ \hat{B}'_{n+1} &= 2n(\hat{A}'_n - \zeta\hat{A}_n + \hat{B}_n) - (M_n\hat{A}_n + y\hat{B}_n)', \end{aligned} \tag{9.2.7}$$

while the system (9.2.6) is equivalent to

$$\begin{aligned} \hat{A}'_{n+1} &= -2n(-\hat{A}_n + \hat{B}'_n - \zeta\hat{B}_n) + (-y\hat{A}_n + M_n\hat{B}_n)', \\ \hat{B}''_{n+1} &= -2n(-\hat{A}_n + \hat{B}'_n - \zeta\hat{B}_n) + 2(-y\hat{A}_n + M_n\hat{B}_n)'. \end{aligned} \tag{9.2.8}$$

We now differentiate eqn $(9.2.7)_2$ with respect to y and set the right-hand side of the resulting equation equal to the right-hand side of eqn $(9.2.8)_2$. Similarly, we differentiate eqn $(9.2.8)_1$ with respect to y and set the right-hand side of the resulting equation equal to the right-hand side of eqn $(9.2.7)_1$. As a result, we find the system

$$\begin{aligned} &[2n(\hat{A}'_n - \zeta\hat{A}_n + \hat{B}_n) - (M_n\hat{A}_n + y\hat{B}_n)']' \\ &+2n(-\hat{A}_n + \hat{B}'_n - \zeta\hat{B}_n) - 2(-y\hat{A}_n + M_n\hat{B}_n)' = 0, \\ &[-2n(-\hat{A}_n + \hat{B}'_n - \zeta\hat{B}_n) + (-y\hat{A}_n + M_n\hat{B}_n)']' \\ &+2n(\hat{A}'_n - \zeta\hat{A}_n + \hat{B}_n) - 2(M_n\hat{A}_n + y\hat{B}_n)' = 0. \end{aligned} \tag{9.2.9}$$

Given the definition (9.2.3) of the operator M_n, we conclude that eqn (9.2.9) is equivalent to eqn (9.2.1). □

Observe that the system (9.2.1) can be written in a matrix form

$$\begin{bmatrix} R_n, & S_n \\ -S_n, & R_n \end{bmatrix} \begin{pmatrix} \hat{A}_n \\ \hat{B}_n \end{pmatrix} = \begin{pmatrix} 0 \\ 0 \end{pmatrix}, \tag{9.2.10}$$

where

$$R_n = y\frac{\partial^3}{\partial y^3} - (n-3+\zeta y)\frac{\partial^2}{\partial y^2} + 2[(n-1)\zeta - y]\frac{\partial}{\partial y} + 2(n-1),$$
$$S_n = 3y\frac{\partial^2}{\partial y^2} - 2(n-3+\zeta y)\frac{\partial}{\partial y} + 2(n-1)\zeta, \tag{9.2.11}$$

and the formula (9.2.1)$\Longleftrightarrow$(9.2.10) states that the pair of thermoelastic polynomials $(\hat{A}_n, \hat{B}_n)$ is an element of the null space of a linear matrix operator containing only the derivatives with respect to y. It follows from the relations (9.1.58)–(9.1.60) that such an element satisfies the following initial conditions of the Cauchy type

$$\left.\begin{pmatrix} \hat{A}_n \\ \hat{B}_n \end{pmatrix}\right|_{y=0} = \begin{pmatrix} \hat{A}_n^{(0)} \\ \hat{B}_n^{(0)} \end{pmatrix}, \tag{9.2.12}$$

$$\left.\begin{pmatrix} \hat{A}'_n \\ \hat{B}'_n \end{pmatrix}\right|_{y=0} = \begin{pmatrix} \hat{A}_n^{(1)} \\ \hat{B}_n^{(1)} \end{pmatrix}, \tag{9.2.13}$$

$$\left.\begin{pmatrix} \hat{A}''_n \\ \hat{B}''_n \end{pmatrix}\right|_{y=0} = \begin{pmatrix} \hat{A}_n^{(2)} \\ \hat{B}_n^{(2)} \end{pmatrix}, \tag{9.2.14}$$

where

$$\hat{A}_1^{(0)} = 1, \quad \hat{A}_n^{(0)} = 0 \ \text{ for } n \geq 2, \tag{9.2.15}$$

$$\hat{B}_1^{(0)} = -\zeta, \ \hat{B}_2^{(0)} = 1+\zeta^2,$$
$$\hat{B}_n^{(0)} = -(2n-3)\zeta\hat{B}_{n-1}^{(0)} + (n-3)(n-1)\hat{B}_{n-2}^{(0)} \ \text{ for } n \geq 3, \tag{9.2.16}$$

$$\hat{A}_1^{(1)} = 0, \ \hat{A}_2^{(1)} = 1-\zeta^2,$$
$$\hat{A}_n^{(1)} = (2n-3)\zeta\hat{B}_{n-1}^{(0)} - (n-3)(n-1)\hat{B}_{n-2}^{(0)} \ \text{ for } n \geq 3, \tag{9.2.17}$$

$$\hat{B}_1^{(1)} = 0, \ \hat{B}_2^{(1)} = -2\zeta, \quad \hat{B}_n^{(1)} = n\hat{B}_{n-1}^{(0)} \ \text{ for } n \geq 3, \tag{9.2.18}$$

$$\hat{A}_1^{(2)} = 0, \ \hat{A}_2^{(2)} = 0, \ \hat{A}_3^{(2)} = 2(1-3\zeta^2),$$
$$\hat{A}_n^{(2)} = -2n\hat{B}_{n-1}^{(0)} \ \text{ for } n \geq 4, \tag{9.2.19}$$

$$\hat{B}_1^{(2)} = 0, \ \hat{B}_2^{(2)} = 0,$$
$$\hat{B}_n^{(2)} = 2\left[(n-2)\zeta\hat{B}_{n-1}^{(0)} + 2(n-1)\hat{B}_{n-2}^{(0)}\right] \ \text{ for } n \geq 3. \tag{9.2.20}$$

We are thus led to the following theorem.

Theorem 9.7 *The pair of thermoelastic polynomials $(\hat{A}_n, \hat{B}_n)$ is a solution of the Cauchy problem described by the matrix equation (9.2.10) for $y > 0$ under the conditions (9.2.12)–(9.2.14) for $y = 0$.*

9.3 Integral relation

We now proceed to prove the following theorem.

Theorem 9.8 *The pair of thermoelastic polynomials* $(\hat{A}_n, \hat{B}_n)$ *is self-equilibrated on the semi-axis* $y > 0$ *in the following sense*

$$\int_0^\infty \mathrm{e}^{-\zeta y} \left[\hat{A}_{k+1}(y,\zeta)\cos y + \hat{B}_{k+1}(y,\zeta)\sin y\right] \mathrm{d}y = 0, \quad \forall \zeta > 0 \quad \text{and} \quad \forall k = 0,1,2,3,... \tag{9.3.1}$$

Proof. We proceed by induction. If we introduce the notation

$$f_{k+1}(y,\zeta) = \mathrm{e}^{-\zeta y}\left[\hat{A}_{k+1}(y,\zeta)\cos y + \hat{B}_{k+1}(y,\zeta)\sin y\right], \tag{9.3.2}$$

the relation (9.3.1) takes the form

$$\int_0^\infty f_{k+1}(y,\zeta)\mathrm{d}y = 0 \quad \forall \zeta > 0 \text{ and } \forall k = 0,1,2,3,... \tag{9.3.3}$$

First, we check that eqn (9.3.3) holds for $k = 0$. To this end, we employ the relations

$$\int_0^\infty \mathrm{e}^{-\zeta y}\cos y\mathrm{d}y = \zeta(\zeta^2+1)^{-1}\ (\zeta > 0), \qquad \int_0^\infty \mathrm{e}^{-\zeta y}\sin y\mathrm{d}y = (\zeta^2+1)^{-1}. \tag{9.3.4}$$

For $k = 0$ the left-hand side of eqn (9.3.3) becomes

$$\int_0^\infty f_1(y,\zeta)\mathrm{d}y = \int_0^\infty \mathrm{e}^{-\zeta y}(\cos y - \zeta \sin y)\mathrm{d}y. \tag{9.3.5}$$

From this, on account of eqn (9.3.4), we infer that eqn (9.3.3) holds for $k = 0$:

$$\int_0^\infty f_1(y,\zeta)\mathrm{d}y = 0. \tag{9.3.6}$$

Furthermore, we verify that it also holds for $k = 1$. In that case, we obtain

$$\int_0^\infty f_2(y,\zeta)\mathrm{d}y = \int_0^\infty \mathrm{e}^{-\zeta y}\{(1-\zeta^2)y\cos y + [(1+\zeta^2) - 2\zeta y]\sin y\}\mathrm{d}y. \tag{9.3.7}$$

Differentiating eqns $(9.3.4)_1$ and $(9.3.4)_2$ with respect to ζ, we obtain, respectively

$$\int_0^\infty \mathrm{e}^{-\zeta y} y\cos y\mathrm{d}y = \frac{\zeta^2-1}{(\zeta^2+1)^2}, \qquad \int_0^\infty \mathrm{e}^{-\zeta y} y\sin y\mathrm{d}y = \frac{2\zeta}{(\zeta^2+1)^2}. \tag{9.3.8}$$

Thus, calculating the integral on the RHS of eqn (9.3.7) with the use of eqns (9.3.4) and (9.3.8), we find that eqn (9.3.3) holds for $k = 1$.

Now, let us assume that

$$\int_0^\infty f_k(y, \zeta)\mathrm{d}y = 0 \quad \forall \zeta > 0 \quad \text{and} \quad k \geq 3. \tag{9.3.9}$$

In order to prove Theorem 9.8, it will suffice to show that eqn (9.3.9) implies eqn (9.3.3). To this end, note from eqn (9.1.53) that the pair $(\hat{A}_n, \hat{B}_n)$ is specified through the recurrence relations

$$\begin{aligned} \hat{A}_{k+1} &= L_k \hat{A}_k + \zeta y \hat{B}_k, \\ \hat{B}_{k+1} &= -\zeta y \hat{A}_{k+1} + L_k \hat{B}_k, \end{aligned} \tag{9.3.10}$$

where L_k is a first-order linear partial differential operator

$$L_k = \zeta y \frac{\partial}{\partial y} - (1 + \zeta^2)\frac{\partial}{\partial \zeta} + [y - (k-1)\zeta] \quad \forall k \geq 1. \tag{9.3.11}$$

With the above relations, we represent the function $f_{k+1}(y, \zeta)$ given by eqn (9.3.2) in the form

$$f_{k+1}(y, \zeta) = \mathrm{e}^{-\zeta y}[(L_k \hat{A}_k + \zeta y \hat{B}_k)\cos y + (-\zeta y \hat{A}_{k+1} + L_k \hat{B}_k)\sin y]. \tag{9.3.12}$$

Furthermore, we compute the integrals

$$\begin{aligned} &\int_0^\infty \mathrm{e}^{-\zeta y}(L_k \hat{A}_k)\cos y \mathrm{d}y \\ = &\int_0^\infty \mathrm{e}^{-\zeta y}\left\{\zeta y \frac{\partial}{\partial y}\hat{A}_k + y\hat{A}_k - \left[(1+\zeta^2)\frac{\partial}{\partial \zeta} + (k-1)\zeta\right]\hat{A}_k\right\}\cos y \mathrm{d}y \end{aligned} \tag{9.3.13}$$

and

$$\begin{aligned} &\int_0^\infty \mathrm{e}^{-\zeta y}(L_k \hat{B}_k)\sin y \mathrm{d}y \\ = &\int_0^\infty \mathrm{e}^{-\zeta y}\left\{\zeta y \frac{\partial}{\partial y}\hat{B}_k + y\hat{B}_k - \left[(1+\zeta^2)\frac{\partial}{\partial \zeta} + (k-1)\zeta\right]\hat{B}_k\right\}\sin y \mathrm{d}y. \end{aligned} \tag{9.3.14}$$

Since

$$\begin{aligned} &\int_0^\infty \mathrm{e}^{-\zeta y}\left(\zeta y \frac{\partial}{\partial y}\hat{A}_k\right)\cos y \mathrm{d}y \\ &= \zeta \int_0^\infty \mathrm{e}^{-\zeta y}[(\zeta y - 1)\cos y + y \sin y]\hat{A}_k \mathrm{d}y \end{aligned} \tag{9.3.15}$$

and

$$\int_0^\infty \mathrm{e}^{-\zeta y}\left(\zeta y\frac{\partial}{\partial y}\hat{B}_k\right)\sin y\mathrm{d}y$$
$$= \zeta\int_0^\infty e^{-\zeta y}[-y\cos y + (\zeta y - 1)\sin y]\hat{B}_k\mathrm{d}y, \tag{9.3.16}$$

upon carrying out the integration of eqn (9.3.12) from $y = 0$ to $y = \infty$, and using eqns (9.3.13)–(9.316), we obtain

$$\int_0^\infty f_{k+1}(y,\zeta)\mathrm{d}y$$
$$= -k\zeta\int_0^\infty f_k(y,\zeta)\mathrm{d}y - (1+\zeta^2)\frac{\partial}{\partial\zeta}\int_0^\infty f_k(y,\zeta)\mathrm{d}y. \tag{9.3.17}$$

From this and eqn (9.3.9) it follows that eqn (9.3.3) is true. □

9.4 Associated thermoelastic polynomials

Consistent with the definition adopted in Section 9.1, the pair (a_{kn}, b_{kn}), $0 \leq k \leq n-1$, appearing in eqn (9.1.36) is called the pair of associated thermoelastic polynomials. This pair determines uniquely the pair $(\hat{A}_n, , \hat{B}_n)$ with the help of relations

$$\left.\begin{aligned}\hat{A}_n(y,\zeta) &= \sum_{k=0}^{n-1} a_{kn}(\zeta)\, y^k\\ \hat{B}_n(y,\zeta) &= \sum_{k=0}^{n-1} b_{kn}(\zeta)\, y^k\end{aligned}\right\} \quad n \geq 1. \tag{9.4.1}$$

For $n = 1, 2, 3$ and 4 these relations take the form [recall eqns (7.1.61)–(7.1.64) and eqns (9.1.34) and (9.1.35)]

$$\hat{A}_1 = 1, \quad \hat{B}_1 = -\zeta, \tag{9.4.2}$$

$$\hat{A}_2 = \left(1-\zeta^2\right), \quad \hat{B}_2 = (1+\zeta^2) - 2\zeta y, \tag{9.4.3}$$

$$\begin{aligned}\hat{A}_3 &= 3\zeta\left(1+\zeta^2\right)y + \left(1-3\zeta^2\right)y^2,\\ \hat{B}_3 &= -3\zeta\left(1+\zeta^2\right) + 3\left(1+\zeta^2\right)y - \zeta\left(3-\zeta^2\right)y^2,\end{aligned} \tag{9.4.4}$$

$$\begin{aligned}\hat{A}_4 &= -3\left(1+6\zeta^2+5\zeta^4\right)y + 12\zeta\left(1+\zeta^2\right)y^2 + \left(1-6\zeta^2+\zeta^4\right)y^3,\\ \hat{B}_4 &= 3\left(1+6\zeta^2+5\zeta^4\right) - 12\zeta\left(1+\zeta^2\right)y + 6\left(1-\zeta^4\right)y^2 - 4\zeta\left(1-\zeta^2\right)y^3.\end{aligned} \tag{9.4.5}$$

The above formulas show that the functions $a_{kn}(\zeta)$ and $b_{kn}(\zeta)$ for $0 \leq k \leq n-1$ are the polynomials in ζ of degree not greater than n.

A substitution of eqn (9.4.1) into a homogeneous form of eqns (9.1.53), and setting the coefficients of like powers of y to zero, leads to the following recurrence formulas specifying a pair $(a_{k,n+1}, b_{k,n+1})$, $0 \leq k \leq n$, in terms of (a_{kn}, b_{kn}), $0 \leq k \leq n-1$,

$$\begin{aligned}
&\left.\begin{aligned} a_{0,n+1} &= -(1+\zeta^2)a'_{0n} - (n-1)\zeta a_{0n} \\ b_{0,n+1} &= -(1+\zeta^2)b'_{0n} - (n-1)\zeta b_{0n} \end{aligned}\right\} \quad (n \geq 1)\,, \\
&\left.\begin{aligned} a_{k,n+1} &= -(1+\zeta^2)a'_{kn} \\ &\quad -(n-1-k)\zeta a_{kn} + a_{k-1,n} + \zeta b_{k-1,n} \\ b_{k,n+1} &= -(1+\zeta^2)b'_{kn} \\ &\quad -(n-1-k)\zeta b_{kn} + b_{k-1,n} - \zeta a_{k-1,n} \end{aligned}\right\} \quad (1 \leq k \leq n-1)\,, \\
&\left.\begin{aligned} a_{n,n+1} &= a_{n-1,n} + \zeta b_{n-1,n} \\ b_{n,n+1} &= b_{n-1,n} - \zeta a_{n-1,n} \end{aligned}\right\} \quad (n \geq 1)\,,
\end{aligned} \tag{9.4.6}$$

where the prime indicates the differentiation with respect to ζ.

Similarly, substituting eqn (9.4.1) into a homogeneous form of eqns (9.1.58), and setting the coefficients of like powers of y to zero, yields [see eqn (7.1.69)]

$$\begin{aligned}
&\left.\begin{aligned} a_{1,n+1} + b_{0,n+1} &= 0 \\ b_{1,n+1} &= (n+1)b_{0n} \end{aligned}\right\} \quad (n \geq 2)\,, \\
&\left.\begin{aligned} &(k+1)a_{k+1,n+1} + b_{k,n+1} \\ &= (n+k+1)a_{kn} - \zeta a_{k-1,n} + b_{k-1,n} \\ &(k+1)b_{k+1,n+1} - a_{k,n+1} \\ &= (n+k+1)b_{kn} - \zeta b_{k-1,n} - a_{k-1,n} \end{aligned}\right\} \quad (1 \leq k \leq n-1)\,, \\
&\left.\begin{aligned} a_{n,n+1} &= a_{n-1,n} + \zeta b_{n-1,n} \\ b_{n,n+1} &= b_{n-1,n} - \zeta a_{n-1,n} \end{aligned}\right\} \quad (n \geq 1)\,.
\end{aligned} \tag{9.4.7}$$

Also, substituting eqn (9.4.1) into a homogeneous form of eqns (9.1.59), and setting the coefficients of like powers of y to zero, yields

$$\left.\begin{aligned}
&(k+1)(k+2)a_{k+2,n+1} + 2(k+1)b_{k+1,n+1} \\
&\quad = 2n[(k+1)a_{k+1,n} - \zeta a_{kn} + b_{kn}], \\
&(k+1)(k+2)b_{k+2,n+1} - 2(k+1)a_{k+1,n+1} \\
&\quad = 2n[(k+1)b_{k+1,n} - \zeta b_{kn} - a_{kn}], \\
&a_{n,n+1} = a_{n-1,n} + \zeta b_{n-1,n}, \\
&b_{n,n+1} = b_{n-1,n} - \zeta a_{n-1,n},
\end{aligned}\right\} \quad (0 \leq k \leq n-2)\,. \tag{9.4.8}$$

Finally, substituting eqn (9.4.1) into eqn (9.2.1), we obtain

$$\begin{aligned}
&(k+1)(k+2)(n-3-k)a_{k+2,n} - \zeta(k+1)(2n-2-k)a_{k+1,n}\\
&+(k+1)(2n-6-3k)b_{k+1,n} - 2(n-1-k)(a_{kn}+\zeta b_{kn}) = 0,\\
&(k+1)(k+2)(n-3-k)b_{k+2,n} - \zeta(k+1)(2n-2-k)b_{k+1,n}\\
&-(k+1)(2n-6-3k)a_{k+1,n} + 2(n-1-k)(\zeta a_{kn} - b_{kn}) = 0,\\
&\qquad (0 \leq k \leq n-3)\,,\\
&n(n-1)(\zeta a_{n-1,n} + b_{n-1,n}) + 2(a_{n-2,n} + \zeta b_{n-2,n}) = 0,\\
&n(n-1)(\zeta b_{n-1,n} - a_{n-1,n}) + 2(b_{n-2,n} - \zeta a_{n-2,n}) = 0,\\
&\qquad (n \geq 2)\,.
\end{aligned} \tag{9.4.9}$$

Clearly, the formulas (9.4.6)–(9.4.8) can be used for determination of a pair (a_{kn}, b_{kn}), $0 \leq k \leq n-1$, with the help of $(a_{k,n-1}, b_{k,n-1})$, $0 \leq k \leq n-2$. The relations (9.4.9) involving the pair (a_{kn}, b_{kn}), $0 \leq k \leq n-1$, can be employed in a general analysis of thermoelastic polynomials.

Concluding this chapter we note that, by substituting eqn (9.4.1) into eqn (9.3.1), we obtain a global relation for the pair of associated thermoelastic polynomials

$$\sum_{i=0}^{k} (-)^i \left[a_{i,k+1}(\zeta) \frac{\mathrm{d}^i}{\mathrm{d}\zeta^i} \left(\frac{\zeta}{\zeta^2+1} \right) + b_{i,k+1}(\zeta) \frac{\mathrm{d}^i}{\mathrm{d}\zeta^i} \left(\frac{1}{\zeta^2+1} \right) \right] = 0 \tag{9.4.10}$$

$$\forall \zeta > 0, \quad \forall k = 0, 1, 2, 3, \ldots$$

10

MOVING DISCONTINUITY SURFACES

In Section 7.4 an exact solution of the Nowacki type for the G–L theory, consisting of a superposition of two spherical waves of strong discontinuity of order $n = -1$ was obtained. The speeds and reduced damping coefficients of the "amplitudes" of these waves were identified on the basis of the exact solution. An analogous solution of the Danilovskaya type, consisting of a superposition of two plane waves of strong discontinuity of order $n = -1$, was considered in Section 7.5.

In this chapter we analyze the moving discontinuity surfaces in models of the L–S and G–L type on the basis of a theory of propagation of singular surfaces in continuum mechanics (e.g. Gurtin, 1972), without solving any specific initial-boundary value problems. In the first section of the chapter we shall introduce the concept of a singular surface propagating in a thermoelastic medium as well as the concept of a thermoelastic wave of order n ($\gtrless 0$), both of which are generalizations of such notions from classical linear isothermal elastodynamics (e.g. Ignaczak, 1985).

In the second and third sections of the chapter we shall analyze the discontinuities of a plane stress–temperature wave of order $n = 0$ for the L–S body as well as the discontinuities of a plane displacement–temperature wave of order $n = 2$ in the G–L theory.

10.1 Singular surfaces propagating in a thermoelastic medium; thermoelastic wave of order n ($\gtrless 0$)

Definition 10.1 *The singular surface propagating in a thermoelastic medium*[1] *is a surface on which certain kinematic and thermal quantities possess discontinuities.*

Let us assume that a thermoelastic disturbance $f = f(\mathbf{x}, t)$[2] is defined on $\mathrm{B} \times [0, \infty)$, where B is a domain in E^3, while $[0, \infty)$ is the time interval.

[1] By a thermoelastic medium we mean here an L–S or G–L model.

[2] By a thermoelastic disturbance we mean a solution to an initial-boundary value problem for the L–S or G–L model. For example, the function $f = f(x, t)$ may be identified with a pair $(\mathbf{u}, \vartheta)$, where $\mathbf{u}$ and ϑ stand for the displacement and temperature, respectively. In particular, the pair may represent the first or second fundamental thermoelastic disturbance of the G–L theory (see Section 6.2).

Furthermore, let W denote a singular surface appearing in Definition 10.1. That surface may be described by the equation

$$t = \psi(\mathbf{x}) \quad \text{for} \quad (\mathbf{x}, t) \in \mathrm{B} \times [0, \infty), \tag{10.1.1}$$

where ψ is a scalar field on B, while t denotes time. If $|\nabla\psi| \neq 0$ on B, the surface moves in the direction

$$\mathbf{m}(\mathbf{x}, \mathbf{t}) = \frac{\nabla\psi(\mathbf{x})}{|\nabla\psi(\mathbf{x})|}, \tag{10.1.2}$$

at the speed

$$c(\mathbf{x}, \mathbf{t}) = \frac{1}{|\nabla\psi(\mathbf{x})|}. \tag{10.1.3}$$

Thus, W is a smooth 3D manifold in the space-time $\mathrm{E}^3 \times [0, \infty)$, whose normal is the four-component vector of the form

$$\mathbf{m}^{(4)} = (\mathbf{m}, -c), \tag{10.1.4}$$

where $|\mathbf{m}| = 1$ and $c > 0$.

Definition 10.2 *If f is a continuous function on $B \times [0, \infty) - W$ and f undergoes a jump-like discontinuity on W, then W is a singular surface of order $n = 0$ relative to f.*

The jump in f on W is given by the formula

$$[[f]](\mathbf{x}, t) = \lim_{h \to 0+} \{f(\mathbf{x}, t + h) - f(\mathbf{x}, t - h)\}. \tag{10.1.5}$$

Since the function f is continuous on both sides of W, one can compute $[[f]]$ by taking the limit along $\mathbf{m}^{(4)}$ according to

$$[[f]]\left(\boldsymbol{\xi}^{(4)}\right) = \lim_{h \to 0+} \left\{f\left(\boldsymbol{\xi}^{(4)} - h\mathbf{m}^{(4)}\right) - f\left(\boldsymbol{\xi}^{(4)} + h\mathbf{m}^{(4)}\right)\right\}, \tag{10.1.6}$$

where $\mathbf{m}^{(4)}$ is given by the formula (10.1.4) and $\boldsymbol{\xi}^{(4)} = (\mathbf{x}, t) \in W$. Thus, the jump in f is the difference between its values at two points of $\mathbf{m}^{(4)}$ placed directly in front and behind of W.

In order to prove that (10.1.5)$\Longleftrightarrow$(10.1.6), we use the definition (10.1.4) and obtain

$$f\left(\boldsymbol{\xi}^{(4)} - h\mathbf{m}^{(4)}\right) = f(\mathbf{x} - h\mathbf{m}, t + hc), \tag{10.1.7}$$

$$f\left(\boldsymbol{\xi}^{(4)} + h\mathbf{m}^{(4)}\right) = f(\mathbf{x} + h\mathbf{m}, t - hc). \tag{10.1.8}$$

Since f is continuous outside W, upon passing in eqns (10.1.7 and 10.1.8) to the limit $h \to 0$, we obtain

$$\lim_{h\to 0+} f\left(\boldsymbol{\xi}^{(4)} - h\mathbf{m}^{(4)}\right) = \lim_{h\to 0+} f\left(\mathbf{x}, t + hc\right), \tag{10.1.9}$$

$$\lim_{h\to 0+} f\left(\boldsymbol{\xi}^{(4)} + h\mathbf{m}^{(4)}\right) = \lim_{h\to 0+} f\left(\mathbf{x}, t - hc\right). \tag{10.1.10}$$

Subtracting eqn (10.1.9) from eqn (10.1.10), we infer that (10.1.6)⇒(10.1.5). The proof of the statement (10.1.5)⇒(10.1.6) follows from the fact that, given the continuity of f outside W, the expression $\{\cdot\}$ in eqn (10.1.5) may be replaced by the expression $\{\cdot\}$ in eqn (10.1.6).

Remark 10.1 In the case when ψ depends on one variable only, say $x_1 = x$, and when ψ possesses an inverse function, then the equation

$$x = \psi^{-1}(t) \tag{10.1.11}$$

represents a plane moving in the positive direction of the x-axis with the speed

$$v = \left| \frac{\mathrm{d}}{\mathrm{d}t} \left[\psi^{-1}(t)\right] \right|, \tag{10.1.12}$$

and the jump of a function $f = f(x,t)$ on that plane is a time-dependent function (recall eqn (10.1.5))

$$[[f]]\,(t) = \lim_{h\to 0+} \left\{f\left[\psi^{-1}(t), t + h\right] - f\left[\psi^{-1}(t), t - h\right]\right\}, \tag{10.1.13}$$

or (recall eqn (10.1.6))

$$[[f]]\,(t) = \lim_{h\to 0+} \left\{f\left[\psi^{-1}(t) - h, t\right] - f\left[\psi^{-1}(t) + h, t\right]\right\}. \tag{10.1.14}$$

Here, ψ^{-1} is a function inverse to ψ.

Remark 10.2 If the function f depends on the radial co-ordinate ρ and the time t only, that is

$$f = f(\rho, t), \tag{10.1.15}$$

where $\rho = |\mathbf{x} - \mathbf{y}|$ and $\mathbf{x}, \mathbf{y} \in \mathrm{E}^3$, and if the equation of a singular surface W is of the form

$$\zeta = 0, \tag{10.1.16}$$

where $\zeta = t - \rho$, then the formula (10.1.6) becomes (recall eqn (7.4.123))

$$[[f]]\,(t) = f(t - 0, t) - f(t + 0, t). \tag{10.1.17}$$

In that case, eqn (10.1.16) represents a spherical surface of order zero relative to f, with the center at y, which propagates in the direction $\mathbf{m} = \nabla\rho$ at the unit velocity.[3]

Definition 10.3 *W is a singular surface of order $n \geq 1$ relative to f, provided*

(a) f is of class C^{n-1} on $B \times [0, \infty)$ and of class C^n on $B \times [0, \infty) - W$;
(b) the nth-order derivatives of f undergo jump discontinuities on W.

Definition 10.4 *W is a singular surface of order $n = -1$ relative to f, whenever W is a singular surface of order zero relative to $1 * f$. Here, $*$ denotes a convolution on the time axis, that is*

$$(1 * f)(\mathbf{x}, t) = \int_0^t f(\mathbf{x}, \tau)\mathrm{d}\tau. \tag{10.1.18}$$

The latter definition offers a hint on how to specify a singular surface of order $n = -k$ for $k = 1, 2, 3, \ \ldots$

Since a thermoelastic process taking place in the body B at any $t \geq 0$ can be described in terms of various pairs of thermomechanical variables (recall Chapter 2), one may therefore consider a thermoelastic wave of order n corresponding to a given pair of such variables.

In the following, we shall give the definition of a displacement–temperature thermoelastic wave of order $n(\gtreqless 0)$ for the G–L model, and the definition of a stress–temperature thermoelastic wave of order $n \geq 0$ for the L–S model.

Definition 10.5 *The displacement–temperature thermoelastic wave of order $n(\gtreqless 0)$ for the G–L model is a solution of the field equations of the G–L theory for which there exists a moving singular surface of order n relative to the displacement and temperature fields.*

Definition 10.6 *The stress–temperature thermoelastic wave of order $n(\gtreqless 0)$ for the L–S model is a solution of the field equations of the L–S theory for which there exists a moving singular surface of order n relative to the stress and temperature fields.*

Note that the existence of spherical and plane displacement–temperature waves of order $n = -1$ for the G–L model was proved in Sections 7.4 and 7.5, respectively. The existence of plane displacement–temperature waves of order $n = 2$ for the G–L model and the plane stress–temperature waves of order $n = 0$ for the L–S model will be proved in later sections of this chapter.

Definition 10.7 *The displacement–temperature wave of order $n \leq 1$ in the G–L model is called a thermoelastic wave of strong discontinuity. The thermoelastic displacement–temperature wave of order $n = 1$ is called a shock wave, while*

[3] Note that the independent variables ρ and t of Remark 10.2 are dimensionless.

the displacement–temperature wave of order $n = 0$ is called a thermoelastic dislocation wave. The displacement–temperature wave of order $n \geq 2$ is called a thermoelastic wave of weak discontinuity. The thermoelastic displacement–temperature wave of order $n = 2$ is also called a thermoelastic acceleration wave.

Remark 10.3 Definition 10.7 pertains to the G–L model only.

10.2 Propagation of a plane shock wave in a thermoelastic half-space with one relaxation time

Let us consider a 1D thermoelastic process in the L–S theory, taking place in an isotropic thermoelastic half-space $x_1 = x \geq 0$ under the action of a discontinuous-in-time thermomechanical loading applied at the boundary $x = 0$. Let us also assume that the process begins at time $t = 0$ from a certain state of the half-space corresponding to the homogeneous initial conditions, and that there are neither heat sources nor body forces within the half-space.

Upon the introduction of non-dimensional independent variables, thermomechanical fields and constitutive parameters, we conclude just like in Sections 4.1 (recall eqns (4.1.5–4.1.10)) and 6.1 (recall eqns (6.1.11–6.1.17)) that the process is described by the system of dimensionless functions $(u, e, S, \vartheta, \eta, q)$ defined for every $(x, t) \in [0, \infty) \times [0, \infty)$ and satisfying the following field equations:

the strain–displacement equation

$$e = u_x, \tag{10.2.1}$$

the equation of motion

$$S_x = u_{tt}, \tag{10.2.2}$$

the energy balance

$$\eta_t = -q_x, \tag{10.2.3}$$

the constitutive equations

$$S = e - \vartheta, \tag{10.2.4}$$

$$\eta = \epsilon e + \vartheta, \tag{10.2.5}$$

$$q + t_0 q_t = -\vartheta_x. \tag{10.2.6}$$

Here, u, e and S denote, respectively, the displacement, strain and stress in the x direction, while ϑ, η and q stand for the temperature difference, the entropy and the heat flux, also along x. The parameters ϵ and t_0, appearing in eqns (10.2.5) and (10.2.6) represent, respectively, the thermomechanical coupling parameter and the relaxation time ($\epsilon > 0, t_0 > 0$). Furthermore, the subscripts x and t indicate partial derivatives of a function $f = f(x, t)$

$$f_x = \frac{\partial f}{\partial x}, \qquad f_t = \frac{\partial f}{\partial t}. \tag{10.2.7}$$

If we introduce the particle velocity in the x-direction as

$$v = u_t, \tag{10.2.8}$$

the equation of motion (10.2.2) can be rewritten in the form

$$S_x = v_t. \tag{10.2.9}$$

It can also be shown that, if

$$u = \phi_x, \tag{10.2.10}$$

where ϕ is a scalar field, then the pair (u, ϑ) corresponds to the process $(u, e, S, \vartheta, \eta, q)$, provided the pair (ϕ, ϑ) satisfies the 1D potential–temperature equations of the L–S theory, corresponding to zero body forces and heat sources (recall eqns (4.1.11) specialized to coupled fields dependent on x and t only).

Since the thermomechanical boundary loading depends on time only, a plane thermoelastic wave in the form of a sum of plane waves propagates into the half-space. The fronts of these waves are given by

$$\frac{x}{V} = t, \tag{10.2.11}$$

where $V > 0$ is the speed of a given wavefront that propagates in the positive x-direction (recall eqns (10.1.11) and (10.1.12)).

Let us now assume that the boundary loading gives rise to a superposition of plane thermoelastic waves of order $n = 0$ relative to the pair (S, ϑ) (recall Definition 10.6 of Section 10.1), that is, the waves for which

$$[[S]](t) \neq 0, \qquad [[\vartheta]](t) \neq 0, \tag{10.2.12}$$

at the front given by eqn (10.2.11).

In the relations (10.2.12) the symbol $[[f]](t)$ denotes the jump of a function $f = f(x,t)$ on the surface (10.2.11), which is given by (recall eqn (10.1.14))

$$[[f]](t) = f(Vt - 0, t) - f(Vt + 0, t). \tag{10.2.13}$$

It follows from the definition of a wave of order $n = 0$ relative to the pair (S, ϑ) that

$$[[u]](t) = 0 \tag{10.2.14}$$

and, for an arbitrary surface $x = x_0$ that does not coincide with the wavefront (10.2.11) at time $t > 0$, we have

$$\Big[\Big[\int_{x_0}^{x} v(\xi, t)\mathrm{d}\xi\Big]\Big](t) = 0, \tag{10.2.15}$$

$$\Big[\Big[\int_{x_0}^{x} \eta(\xi, t)\mathrm{d}\xi\Big]\Big](t) = 0, \tag{10.2.16}$$

and

$$\left[\left[\int_0^t q(x,\tau)\mathrm{d}\tau\right]\right](t) = 0. \tag{10.2.17}$$

Furthermore, besides the relations (10.2.12), we have[4]

$$[[e]]\,(t) \neq 0, \qquad [[v]]\,(t) \neq 0, \tag{10.2.18}$$

$$[[\eta]]\,(t) \neq 0, \qquad [[q]]\,(t) \neq 0. \tag{10.2.19}$$

In order to demonstrate the existence of a wavefront (10.2.11) on which there hold the relations (10.2.12) and (10.2.14)–(10.2.19), we employ the fact that, for a function $f = f(x,t)$ such that $[[f]]\,(t) \neq 0$, $[[f_t]]\,(t) \neq 0$ and $[[f_x]]\,(t) \neq 0$, there holds the kinematic compatibility relation (Achenbach, 1968)

$$\frac{\mathrm{d}}{\mathrm{d}t}\left\{[[f]]\,(t)\right\} = [[f_t]]\,(t) + V\,[[f_x]]\,(t). \tag{10.2.20}$$

The relation (10.2.20) is also true whenever any of the three jumps appearing in it vanishes.

In what follows, for the sake of simplicity, we will simply write

$$[[f]] = [[f]]\,(t). \tag{10.2.21}$$

Now, let f be a function defined by

$$f(x,t) = f(x_0,t) + \int_{x_0}^{x} v(\xi,t)\mathrm{d}\xi, \tag{10.2.22}$$

where $x_0 \neq Vt, \forall t \geq 0$.

Because $[[f(x_0,t)]] = 0$ and the condition (10.2.15) is satisfied, therefore

$$[[f]] = 0. \tag{10.2.23}$$

From this and eqn (10.2.20) we obtain

$$[[f_t]] + V\,[[f_x]] = 0. \tag{10.2.24}$$

On account of eqns (10.2.9) and (10.2.22) we find

$$\begin{aligned} f_t(x,t) &= f_t(x_0,t) + \int_{x_0}^{x} v_t(\xi,t)\mathrm{d}\xi = \\ f_t(x_0,t) &+ \int_{x_0}^{x} S_\xi(\xi,t)\mathrm{d}\xi = f_t(x_0,t) + S(x,t) - S(x_0,t). \end{aligned} \tag{10.2.25}$$

[4] The conditions (10.2.14), (10.2.15), and (10.2.18) occur in the definition of a shock wave in linear isothermal elastodynamics. The concept of an isothermal shock wave has been extended in Section 10.2 to a thermoelastic shock wave by including the conditions (10.2.16), (10.2.17), and (10.2.19).

From this, given that $x_0 \neq Vt$, we arrive at

$$[[f_t]] = [[S]]\,. \tag{10.2.26}$$

Moreover, differentiating eqn (10.2.22) with respect to x, we get

$$f_x = v. \tag{10.2.27}$$

Thus, in view of eqns (10.2.24), (10.2.26) and (10.2.27), we find the following relation between the jumps $[[S]]$ and $[[v]]$

$$[[S]] = -V\,[[v]]\,. \tag{10.2.28}$$

This relation is identical with the compatibility condition for a plane isothermal shock wave [see p. 254 in (Gurtin, 1972)].

Let us now consider a function $g = g(x,t)$ defined by

$$g(x,t) = g(x_0,t) - \int_{x_0}^{x} \eta(\xi,t)\mathrm{d}\xi. \tag{10.2.29}$$

On account of eqn (10.2.16) and noting that $x_0 \neq Vt$, we obtain

$$[[g]] = 0. \tag{10.2.30}$$

Differentiating this relation with respect to t and employing eqn (10.2.20), we find

$$[[g_t]] + V\,[[g_x]] = 0. \tag{10.2.31}$$

On account of eqns (10.2.3) and (10.2.29) we find

$$g_t(x,t) = g_t(x_0,t) + \int_{x_0}^{x} q_\xi(\xi,t)\mathrm{d}\xi = g_t(x_0,t) + q(x,t) - q(x_0,t). \tag{10.2.32}$$

From this, given that $x_0 \neq Vt$, we get

$$[[g_t]] = [[q]]\,. \tag{10.2.33}$$

Also, on account of eqn (10.2.29) we obtain

$$[[g_x]] = -\,[[\eta]]\,. \tag{10.2.34}$$

Thus, using eqns (10.2.31), (10.2.33) and (10.2.34) we obtain the following relation between the jumps $[[\eta]]$ and $[[q]]$

$$[[q]] = V\,[[\eta]]\,. \tag{10.2.35}$$

This relation is a thermal equivalent of the kinematic compatibility condition (10.2.28).

Finally, we introduce the function $Q = Q(x,t)$ through

$$Q_t = q + t_0 q_t. \tag{10.2.36}$$

Integrating this relation with respect to time, we obtain

$$Q(x,t) - Q(x,0) = \int_0^t q(x,\tau)\mathrm{d}\tau + t_0[q(x,t) - q(x,0)]. \tag{10.2.37}$$

From this, given eqn (10.2.17), we find

$$[[Q]] = t_0\,[[q]]\,. \tag{10.2.38}$$

Now, let us observe that the relations (10.2.6) and (10.2.36) imply the relation

$$Q_t = -\vartheta_x, \tag{10.2.39}$$

which is analogous to the relation (10.2.3) in the sense that the pair (η, q) corresponds to the pair (Q, ϑ). Thus, proceeding in the same manner as in the derivation of eqn (10.2.35), we obtain

$$[[\vartheta]] = V\,[[Q]]\,. \tag{10.2.40}$$

Hence, on account of eqn (10.2.38) we find the following relation between the jumps $[[\vartheta]]$ and $[[q]]$

$$[[\vartheta]] = t_0 V\,[[q]]\,. \tag{10.2.41}$$

Furthermore, let us consider the constitutive relations (10.2.4) and (10.2.5). Operating with $[[\cdot]]$ on them, we find

$$[[S]] = [[e]] - [[\vartheta]]\,, \tag{10.2.42}$$

$$[[\eta]] = \epsilon\,[[e]] + [[\vartheta]]\,. \tag{10.2.43}$$

We also note the relation

$$[[v]] = -V\,[[e]]\,, \tag{10.2.44}$$

which may be obtained by differentiating eqn (10.2.14) with respect to time and using eqns (10.2.1), (10.2.8) and (10.2.20).

From the relations (10.2.28) and (10.2.44) we now get

$$[[S]] = V^2\,[[e]]\,. \tag{10.2.45}$$

Thus, eliminating the jump $[[S]]$ from eqns (10.2.42) and (10.2.45) we obtain

$$[[\vartheta]] = -(V^2 - 1)\,[[e]]\,. \tag{10.2.46}$$

On the other hand, eliminating the jumps $[[\eta]]$ and $[[q]]$ from eqns (10.2.35), (10.2.41) and (10.2.43), we obtain

$$(V^2 - t_0^{-1})\,[[\vartheta]] = -\epsilon V^2\,[[e]]\,. \tag{10.2.47}$$

From this, on account of eqns (10.2.46) and (10.2.47), the necessary and sufficient condition for the propagation of a plane wave front described by eqn (10.2.11) is that V be a positive root of the algebraic equation

$$(V^2 - t_0^{-1})(V^2 - 1) - V^2\epsilon = 0. \tag{10.2.48}$$

It can be shown that eqn (10.2.48) has two positive roots. Denoting these roots by $\hat{v}_1 = \hat{v}_1(t_0, \epsilon)$ and $\hat{v}_2 = \hat{v}_2(t_0, \epsilon)$, we obtain

$$\hat{v}_1^{-2} = \frac{1}{2}\left\{1 + (1+\epsilon)t_0 + \sqrt{[1 + (1+\epsilon)t_0]^2 - 4t_0}\right\}, \tag{10.2.49}$$

$$\hat{v}_2^{-2} = \frac{1}{2}\left\{1 + (1+\epsilon)t_0 - \sqrt{[1 + (1+\epsilon)t_0]^2 - 4t_0}\right\}. \tag{10.2.50}$$

Comparing the formulas (10.2.49)–(10.2.50) with eqns $(6.3.1)_1$ and $(6.3.1)_2$ for $t_1 = t_0$, we conclude that the speeds $\hat{v}_1$ and $\hat{v}_2$ are identical, respectively, with the speeds of the first and second thermoelastic disturbances in the G–L theory in which $t_0 = t_1$. Thus, we are led to the following theorem.

Theorem 10.1 *A discontinuous-in-time stress–temperature loading applied to the boundary of a half-space $x \geq 0$ gives rise to a disturbance consisting of two plane stress–temperature waves of order $n = 0$ propagating in the positive direction of the x-axis with speeds of the first and second thermoelastic disturbance in the G–L theory in which $t_0 = t_1$.*

Schematic graphs of the functions $\hat{v}_1(t_0)$ and $\hat{v}_2(t_0)$ for a fixed $\epsilon > 0$ and $t_0 > 1.5$ are shown in Fig. 6.7. The graphs of the functions $\hat{v}_1(\epsilon)$ and $\hat{v}_2(\epsilon)$ for a fixed $t_0 > 1$ and $\epsilon > 0$ are shown in Fig. 6.3 in which $\xi = \epsilon t_0$.

Let us now assume the boundary loading to be of the form

$$S(0,t) = S_0 H(t), \qquad \vartheta(0,t) = \vartheta_0 H(t), \tag{10.2.51}$$

where S_0 and ϑ_0 are constant, while $H(t)$ is the Heaviside function. Since a total disturbance of the half-space is the sum of two plane waves of order zero relative to the pair (S, ϑ), some of the externally applied discontinuities propagate into the half-space at the speed $\hat{v}_1$, while other discontinuities propagate at the speed $\hat{v}_2$. Denoting these discontinuities by $(S_1,\ \vartheta_1)$ and $(S_2,\ \vartheta_2)$, for $x = 0$ and $t = 0$, we obtain

$$S_0 = S_1 + S_2, \qquad \vartheta_0 = \vartheta_1 + \vartheta_2. \tag{10.2.52}$$

Eliminating the jump $[[e]]$ from eqns (10.2.45) and (10.2.47), and taking the result at $t = x = 0$ and $V = \hat{v}_i$ $(i = 1, 2)$, we obtain

$$\vartheta_1 = -\frac{\epsilon}{\hat{v}_1^2 - t_0^{-1}} S_1, \qquad \vartheta_2 = -\frac{\epsilon}{\hat{v}_2^2 - t_0^{-1}} S_2. \tag{10.2.53}$$

In the above

$$S_0 = [[S]]\,(0), \qquad \vartheta_0 = [[\vartheta]]\,(0), \tag{10.2.54}$$

$$S_i = [[S]]_i\,(0), \qquad \vartheta_i = [[\vartheta]]_i\,(0), \tag{10.2.55}$$

and $[[\cdot]]_i\,(t)$ is the jump at the ith wavefront $(i = 1, 2)$.

Equations (10.2.52) and (10.2.53) may be treated as a system of four linear equations for four unknowns S_1, S_2, ϑ_1 and ϑ_2, whenever the pair (S_0, ϑ_0) is known. In the case when $S_0 \neq 0$ and $\vartheta_0 = 0$, the solution of that system is

$$S_1 = \frac{\hat{v}_1^2 - t_0^{-1}}{\hat{v}_1^2 - \hat{v}_2^2} S_0, \qquad S_2 = -\frac{\hat{v}_1^2 - t_0^{-1}}{\hat{v}_1^2 - \hat{v}_2^2} S_0, \tag{10.2.56}$$

$$\vartheta_1 = -\frac{\epsilon}{\hat{v}_1^2 - \hat{v}_2^2} S_0, \qquad \vartheta_2 = \frac{\epsilon}{\hat{v}_1^2 - \hat{v}_2^2} S_0. \tag{10.2.57}$$

Because we have (recall eqn (6.3.25))

$$\hat{v}_1^2 - t_0^{-1} < 0, \tag{10.2.58}$$

and, for $t_0 > 1$, (recall Fig. 6.7)

$$\hat{v}_2^2 - t_0^{-1} > 0, \tag{10.2.59}$$

$$\hat{v}_2 > \hat{v}_1, \tag{10.2.60}$$

then, in the case when $S_0 < 0$, on account of eqns (10.2.56) and (10.2.57) we obtain

$$S_1 < 0, \qquad S_2 < 0, \tag{10.2.61}$$

$$\vartheta_1 < 0, \qquad \vartheta_2 > 0. \tag{10.2.62}$$

It follows from the relations (10.2.46) and (10.2.48) that, for $t_0 > 0$,

$$[[\vartheta]]_i = -(\hat{v}_i^2 - 1)\,[[e]]_i\,, \tag{10.2.63}$$

and

$$[[\vartheta]]_i = -\frac{\epsilon \hat{v}_i^2}{\hat{v}_i^2 - t_0^{-1}}\,[[e]]_i \qquad (i = 1, 2). \tag{10.2.64}$$

From this, in view of eqns (10.2.58) and (10.2.59) $\forall t > 0$ and $t_0 > 1$ we obtain

$$[[\vartheta]]_1\,[[e]]_1 > 0, \tag{10.2.65}$$

and

$$[[\vartheta]]_2\,[[e]]_2 < 0. \tag{10.2.66}$$

Thus, a rise in temperature at the wavefront $t = x/\hat{v}_1$ ($[[\vartheta]]_1 < 0$) is accompanied by a compression ($[[e]]_1 < 0$), while a rise in temperature at the wavefront $t = x/\hat{v}_2$ ($[[\vartheta]]_2 < 0$) is accompanied by a tension ($[[e]]_2 > 0$).

We shall now show that the "amplitude" $[[S]]_i$ (or $[[\vartheta]]_i$) decreases exponentially in time and the exponent describing that decay is identical with the reduced damping coefficient for the ith fundamental thermoelastic disturbance in the G–L theory restricted to $t_1 = t_0$ (recall eqn $(6.5.47)_1$).

Theorem 10.2 *In the problem under consideration the amplitude* $[[S]]_i$ *(or* $[[\vartheta]]_i$*) satisfies the linear ordinary first-order differential equation*

$$\left(\frac{\mathrm{d}}{\mathrm{d}t} + \hat{h}_i\right)[[S]]_i = 0 \qquad \forall t \geq 0, \tag{10.2.67}$$

where (recall eqns (6.5.47) and (6.4.1)–(6.4.3))

$$\hat{h}_i = \frac{1}{2}\hat{v}_i^2 \hat{k}_i > 0 \qquad (i = 1, 2), \tag{10.2.68}$$

$$\hat{k}_{1,2} = \frac{1}{2}(1 + \epsilon \mp \hat{\alpha}\ \hat{\Delta}^{1/2}) > 0, \tag{10.2.69}$$

$$\hat{\alpha} = [1 - \epsilon - (1+\epsilon)^2 t_0]\hat{\Delta}^{-1}, \tag{10.2.70}$$

$$\hat{\Delta} = [1 - (1-\epsilon)t_0]^2 + 4\epsilon t_0^2. \tag{10.2.71}$$

Proof. On account of eqn (10.2.20) we have

$$\frac{\mathrm{d}}{\mathrm{d}t}[[q]] = [[q_t]] + V[[q_x]], \tag{10.2.72}$$

while from the formulas (10.2.3) and (10.2.5) we obtain

$$[[q_x]] = -\epsilon[[e_t]] - [[\vartheta_t]]. \tag{10.2.73}$$

Furthermore, taking the jump in eqn (10.2.6), we find

$$[[q]] + t_0[[q_t]] = -[[\vartheta_x]]. \tag{10.2.74}$$

Eliminating the amplitudes $[[q_t]]$ and $[[q_x]]$ from eqns (10.2.72)–(10.2.74), we get

$$\frac{\mathrm{d}}{\mathrm{d}t}[[q]] + \frac{1}{t_0}[[q]] + \epsilon V[[e_t]] = -V[[\vartheta_t]] - \frac{1}{t_0}[[\vartheta_x]]. \tag{10.2.75}$$

Also, given eqn (10.2.20)

$$\frac{\mathrm{d}}{\mathrm{d}t}[[\vartheta]] = [[\vartheta_t]] + V[[\vartheta_x]]. \tag{10.2.76}$$

This, together with the formula (10.2.47), leads to

$$[[\vartheta_x]] = \frac{1}{V}\left\{-\frac{\epsilon V^2}{V^2 - t_0^{-1}}\frac{\mathrm{d}}{\mathrm{d}t}[[e]] - [[\vartheta_t]]\right\}. \tag{10.2.77}$$

Next, on account of eqns (10.2.41) and (10.2.47), we have

$$[[q]] = -\frac{\epsilon V^2}{t_0 V(V^2 - t_0^{-1})}[[e]]. \tag{10.2.78}$$

Substituting the amplitudes $[[\vartheta_x]]$ and $[[q]]$ given by eqns (10.2.77) and (10.2.78), respectively, into eqn (10.2.75), we obtain

$$\begin{aligned}\epsilon V^2 \left\{2\frac{\mathrm{d}}{\mathrm{d}t}\,[[e]] + \frac{1}{t_0}\,[[e]]\right\} - \epsilon t_0 V^2 (V^2 - t_0^{-1})\,[[e_t]] \\ = t_0 (V^2 - t_0^{-1})^2\,[[\vartheta_t]]\,.\end{aligned} \tag{10.2.79}$$

Let us next consider the relation (recall eqn (10.2.20))

$$\frac{\mathrm{d}}{\mathrm{d}t}\,[[S]] = [[S_t]] + V\,[[S_x]]\,, \tag{10.2.80}$$

which, in light of eqns (10.2.4) and (10.2.9), becomes

$$\frac{\mathrm{d}}{\mathrm{d}t}\,[[S]] = [[e_t]] - [[\vartheta_t]] + V\,[[v_t]]\,. \tag{10.2.81}$$

Recalling eqn (10.2.28), we have

$$[[S]] = -V\,[[v]]\,, \tag{10.2.82}$$

so that

$$\frac{\mathrm{d}}{\mathrm{d}t}\,[[S]] = -V\,\{[[v_t]] + V\,[[v_x]]\}\,, \tag{10.2.83}$$

or, in view of eqns (10.2.1) and (10.2.8),

$$\frac{\mathrm{d}}{\mathrm{d}t}\,[[S]] = -V\,\{[[v_t]] + V\,[[e_t]]\}\,. \tag{10.2.84}$$

Adding eqns (10.2.81) and (10.2.84), we find

$$[[\vartheta_t]] = -2\frac{\mathrm{d}}{\mathrm{d}t}\,[[S]] - (V^2 - 1)\,[[e_t]]\,. \tag{10.2.85}$$

Let us also note from eqn (10.2.45) that

$$[[e]] = V^{-2}\,[[S]]\,. \tag{10.2.86}$$

Substituting $[[\vartheta_t]]$ and $[[e]]$ given by eqns (10.2.85) and (10.2.86), respectively, into eqn (10.2.79), we obtain

$$\begin{aligned}\epsilon \left\{2\frac{\mathrm{d}}{\mathrm{d}t}\,[[S]] + \frac{1}{t_0}\,[[S]]\right\} - \epsilon t_0 V^2 (V^2 - t_0^{-1})\,[[e_t]] \\ = t_0 (V^2 - t_0^{-1})^2 \left\{-(V^2 - 1)\,[[e_t]] - 2\frac{\mathrm{d}}{\mathrm{d}t}\,[[S]]\right\}.\end{aligned} \tag{10.2.87}$$

In view of eqn (10.2.48), this equation reduces to

$$\left(\frac{\mathrm{d}}{\mathrm{d}t} + \hat{h}\right)[[S]] = 0, \tag{10.2.88}$$

where

$$\hat{h} = \frac{\epsilon}{2t_0[\epsilon + t_0(V^2 - t_0^{-1})^2]}. \tag{10.2.89}$$

On account of eqns (10.2.88) and (10.2.89), for the ith front given by the equation $t = x/\hat{v}_i$ we find

$$\left(\frac{\mathrm{d}}{\mathrm{d}t} + \hat{h}_i\right)[[S]]_i = 0, \tag{10.2.90}$$

where

$$\hat{h}_i = \frac{\epsilon}{2t_0[\epsilon + t_0(\hat{v}_i^2 - t_0^{-1})^2]}. \tag{10.2.91}$$

Employing eqns (10.2.48) and (10.2.89) one can show that

$$\hat{h} = \frac{V^2}{2t_0}\frac{(1+\epsilon - V^2)}{t_0^{-1} - V^4}. \tag{10.2.92}$$

From this and from the definitions of $\hat{v}_i$ (recall eqns (10.2.49) and (10.2.50)) and $\hat{k}_i$ (recall eqns (10.2.69)–(10.2.71)) we infer that $\hat{h}_i$ is given by the formula (10.2.68). □

Remark 10.4 It follows from Theorem 10.2 that the amplitude $[[S]]_i$ is a decreasing function of t:

$$[[S]]_i(t) = [[S]]_i(0)\exp\{-\hat{h}_i t\}. \tag{10.2.93}$$

Remark 10.5 A behavior of the function $\hat{h}_i = \hat{h}_i(t_0, \epsilon)$ with respect to t_0 (for a fixed ϵ) is determined by a behavior of $\hat{v}_i(t_0)$ and $\hat{k}_i(t_0)$, whose graphs are shown in Figs. 6.7 and 6.9, respectively. Also, it follows from eqns (10.2.49), (10.2.50) and (10.2.91) that $\hat{h}_1 \to \epsilon/2$ and $\hat{h}_2 \to +\infty$ for $t_0 \to 0$.

10.3 Propagation of a plane acceleration wave in a thermoelastic half-space with two relaxation times

Consider a homogeneous isotropic thermoelastic half-space with two relaxation times, occupying a thermoelastic half-space $x \geq 0$. Let us assume that the half-space boundary is subjected to a time-dependent stress–heat-flux loading with the first derivatives discontinuous at $t = 0$. This means that the loading is specified in terms of a pair (S, q), where S and q denote, respectively, the stress and heat-flux vector aligned with x, whereby S_t and q_t have discontinuities at $t = x = 0$[5]. Clearly, that loading gives rise to the propagation of a plane thermoelastic wave in the half-space. Adopting, just like in Section 10.2, dimensionless constitutive parameters and dimensionless independent and dependent variables, we conclude that this wave is described by a system of dimensionless

[5] See the notations (10.2.7).

functions $(u, e, S, \vartheta, \eta, q)$, defined for every $(x, t) \in [0, \infty) \times [0, \infty)$ and satisfying the 1D field equations of the G–L theory. Thus, we have:

the strain–displacement equation

$$e = u_x, \tag{10.3.1}$$

the equation of motion

$$S_x = u_{tt}, \tag{10.3.2}$$

the energy balance

$$\eta_t = -q_x, \tag{10.3.3}$$

the constitutive equations

$$S = e - \vartheta - t_1 \vartheta_t, \tag{10.3.4}$$

$$\eta = \epsilon e + \vartheta + t_0 \vartheta_t, \tag{10.3.5}$$

$$q = -\vartheta_x. \tag{10.3.6}$$

Here, (t_0, t_1) is a dimensionless pair of relaxation times in the G–L theory ($t_1 \geq t_0 > 0$), while other symbols have the same meaning as in Section 10.2.

Just like in Section 10.2, note that, if v denotes the particle velocity in the x-direction

$$v = u_t, \tag{10.3.7}$$

then, in view of eqns (10.3.1) and (10.3.2), we obtain

$$v_x = e_t, \tag{10.3.8}$$

$$S_x = v_t. \tag{10.3.9}$$

It can also be shown that, if

$$u = \phi_x, \tag{10.3.10}$$

where ϕ is a scalar field, then a pair (u, ϑ) corresponds to a process $(u, e, S, \vartheta, \eta, q)$, provided the pair (ϕ, ϑ) satisfies the 1D potential–temperature equations of the G–L theory corresponding to zero body forces and heat sources (recall eqns (4.2.7) specialized to coupled fields dependent on x and t only).

In the following we show that the boundary stress–heat-flux loading gives rise to a superposition of plane thermoelastic waves of order $n = 2$ relative to a pair (u, ϑ) (recall Definition 10.5 of Section 10.1), that is, the waves for which

$$[[u_{tt}]] \neq 0 \quad \text{and} \quad [[\vartheta_{tt}]] \neq 0, \tag{10.3.11}$$

at a front given by

$$t = \frac{x}{V}, \tag{10.3.12}$$

where $V > 0$.

It follows from the definition of a wave of order $n = 2$ relative to the pair (u, ϑ) that

$$[[e]] = 0, \qquad [[v]] = 0, \qquad [[S]] = 0, \tag{10.3.13}$$

$$[[\vartheta]] = 0, \qquad [[\eta]] = 0, \qquad [[q]] = 0, \tag{10.3.14}$$

on the front described by eqn (10.3.12).

In order to demonstrate the existence of a wavefront (10.3.12) on which there hold the relations (10.3.11), (10.3.13) and (10.3.14), we use the relation (recall (10.2.20))

$$\frac{\mathrm{d}}{\mathrm{d}t}[[f]] = [[f_t]] + V\,[[f_x]], \tag{10.3.15}$$

where $f = f(x, t)$ is an arbitrary function.

Let us note here that the relations (10.3.1), (10.3.4)–(10.3.7), (10.3.13) and (10.3.14) imply

$$[[u_x]] = 0, \qquad [[u_t]] = 0, \tag{10.3.16}$$

$$[[\vartheta_x]] = 0, \qquad [[\vartheta_t]] = 0. \tag{10.3.17}$$

Furthermore, differentiating eqns (10.3.13) and (10.3.14) with respect to time and using eqn (10.3.15), we obtain

$$[[e_t]] + V\,[[e_x]] = 0, \tag{10.3.18}$$

$$[[v_t]] + V\,[[v_x]] = 0, \tag{10.3.19}$$

$$[[S_t]] + V\,[[S_x]] = 0, \tag{10.3.20}$$

$$[[\eta_t]] + V\,[[\eta_x]] = 0, \tag{10.3.21}$$

$$[[q_t]] + V\,[[q_x]] = 0. \tag{10.3.22}$$

Of course, none of the jumps appearing in eqns (10.3.18)–(10.3.22) vanishes on the front specified by eqn (10.3.12).

If we introduce the notation

$$[[u_{tt}]] = a, \tag{10.3.23}$$

then, on account of eqn (10.3.2), we get

$$[[S_x]] = a, \tag{10.3.24}$$

and hence, given eqn (10.3.20),

$$[[S_t]] = -Va. \tag{10.3.25}$$

Next, taking note of eqns (10.3.1), (10.3.7) and (10.3.19), we obtain

$$a + V\,[[e_t]] = 0. \tag{10.3.26}$$

Hence,

$$[[e_t]] = -\frac{a}{V}. \tag{10.3.27}$$

This relation, along with eqn (10.3.18), implies

$$[[e_x]] = \frac{a}{V^2}. \tag{10.3.28}$$

Now, taking the jump in eqn (10.3.3), we find

$$[[q_x]] = -[[\eta_t]], \tag{10.3.29}$$

which, in view of eqns (10.3.21), (10.3.5), and $(10.3.17)_1$, leads to

$$[[q_x]] = V[[\eta_x]] = V\{\epsilon[[e_x]] + t_0[[\vartheta_{tx}]]\}, \tag{10.3.30}$$

or, using eqns (10.3.6) and (10.3.28), gives

$$[[q_x]] + Vt_0[[q_t]] = \frac{\epsilon}{V}a. \tag{10.3.31}$$

Combining this equation with an equation equivalent to eqn (10.3.22)

$$[[q_x]] + \frac{1}{V}[[q_t]] = 0, \tag{10.3.32}$$

and solving the system (10.3.31) and (10.3.32) with respect to $[[q_x]]$ and $[[q_t]]$, yields

$$[[q_x]] = -\frac{\epsilon a}{Vt_0(V^2 - t_0^{-1})}, \tag{10.3.33}$$

$$[[q_t]] = \frac{\epsilon a}{t_0(V^2 - t_0^{-1})}. \tag{10.3.34}$$

Furthermore, differentiating eqns $(10.3.17)_1$ and $(10.3.17)_2$ with respect to time and using eqn (10.3.15), we find

$$[[\vartheta_{xt}]] + V[[\vartheta_{xx}]] = 0, \tag{10.3.35}$$

$$[[\vartheta_{tt}]] + V[[\vartheta_{tx}]] = 0. \tag{10.3.36}$$

From this, on account of eqn (10.3.6), we get

$$[[\vartheta_{xt}]] = -[[q_t]], \tag{10.3.37}$$

$$[[\vartheta_{xx}]] = V^{-1}[[q_t]], \tag{10.3.38}$$

$$[[\vartheta_{tt}]] = V[[q_t]]. \tag{10.3.39}$$

It follows from eqns (10.3.37)–(10.3.39) and (10.3.33) and (10.3.34) that none of the jumps appearing on the left-hand sides of eqns (10.3.37)–(10.3.39) vanish so long as $a \neq 0$.

If we differentiate eqn (10.3.4) with respect to x, take the jump of the resulting equation and use eqns (10.3.24) and (10.3.17)$_1$, then we obtain

$$a = [[e_x]] - t_1 [[\vartheta_{tx}]], \tag{10.3.40}$$

from which, on account of eqns (10.3.28) and (10.3.37), we find

$$t_1 [[q_t]] = \left(1 - \frac{1}{V^2}\right) a. \tag{10.3.41}$$

Therefore, substituting the amplitude $[[q_t]]$ given by eqn (10.3.34) into eqn (10.3.41), we infer that the necessary and sufficient condition for the propagation of a plane displacement–temperature acceleration wave in the half-space is that V be a positive root of the algebraic equation

$$(V^2 - t_0^{-1})(V^2 - 1) - \epsilon \frac{t_1}{t_0} V^2 = 0. \tag{10.3.42}$$

Similarly to what was done in Section 10.2, one can show that this equation has two positive roots. Denoting these roots by $v_{1.2} = v_{1.2}(t_0, \zeta)$ and $\zeta = t_1 \epsilon$, we obtain

$$v_{1.2}^{-2} = \frac{1}{2}\left\{1 + t_0 + \zeta \pm \Delta^{1/2}\right\}, \tag{10.3.43}$$

where

$$\Delta = (1 - t_0 + \zeta)^2 + 4\zeta t_0. \tag{10.3.44}$$

Comparing the formulas (10.3.43) and (10.3.44) with eqns (6.3.1) and (6.3.2), we conclude that the speeds v_1 and v_2 are identical, respectively, with the speeds of the first and second fundamental thermoelastic disturbance in the G–L theory. Also, note that the relations (10.3.25) and (10.3.34) with $a \neq 0 \; \forall t \geq 0$ correspond to a stress–heat-flux boundary loading with the first derivatives discontinuous at $t = 0$. Thus, we are led to the following theorem.

Theorem 10.3 *A stress–heat-flux loading with the first derivatives discontinuous at $t = 0$ and applied to the boundary of a half-space $x \geq 0$ gives rise to a thermoelastic wave consisting of two plane displacement–temperature waves of order $n = 2$ propagating in the positive direction of the x-axis with speeds of the first and second fundamental thermoelastic disturbances of the G–L theory.*

Schematic graphs of functions $v_1(\zeta)$ and $v_2(\zeta)$ for a fixed $t_0 \geq 1$ and $\zeta > 0$ are shown in Figs. 6.3 and 6.4. The graphs of functions $v_1(t_0)$ and $v_2(t_0)$ for a fixed $\zeta > 0$ and $t_0 > 1.5$ are shown in Fig. 6.6.

Let us assume the boundary loading to be

$$S(0,t) = S^{(0)}(t), \qquad q(0,t) = q^{(0)}(t), \tag{10.3.45}$$

where the functions $S^{(0)}$ and $q^{(0)}$ satisfy the conditions

$$\left[\left[S^{(0)}\right]\right](0) = 0, \qquad \left[\left[q^{(0)}\right]\right](0) = 0, \tag{10.3.46}$$

$$\left[\left[S_t^{(0)}\right]\right](0) \neq 0, \qquad \left[\left[q_t^{(0)}\right]\right](0) \neq 0. \tag{10.3.47}$$

Furthermore, let us introduce the notations

$$\dot{S}_0 = \left[\left[S_t^{(0)}\right]\right](0), \qquad \dot{q}_0 = \left[\left[q_t^{(0)}\right]\right](0). \tag{10.3.48}$$

The externally applied discontinuities (10.3.48) can be treated as superpositions of jumps $(\dot{S}_1, \dot{q}_1)$ and $(\dot{S}_2, \dot{q}_2)$ that correspond to the wavefronts at speeds υ_1 and υ_2 taken at $x = t = 0$. The pair $(\dot{S}_i, \dot{q}_i)$ is defined here by the relations

$$\dot{S}_i = [[S_t]]_i\,(0) \qquad \dot{q}_i = [[q_t]]_i\,(0), \tag{10.3.49}$$

where $[[\cdot]]_i\,(t)$ is the jump at the ith wavefront $(i = 1, 2)$.

Employing eqns (10.3.25) and (10.3.41), we now obtain the system of algebraic equations, which defines the pair $(\dot{S}_i, \dot{q}_i)$ with the help of the pair $(\dot{S}_0, \dot{q}_0)$, namely

$$\begin{gathered} \dot{S}_1 + \dot{S}_2 = \dot{S}_0, \qquad \dot{q}_1 + \dot{q}_2 = \dot{q}_0, \\ \frac{\upsilon_1^2 - 1}{\upsilon_1^3}\dot{S}_1 + t_1\dot{q}_1 = 0, \qquad \frac{\upsilon_2^2 - 1}{\upsilon_2^3}\dot{S}_2 + t_1\dot{q}_2 = 0. \end{gathered} \tag{10.3.50}$$

In the case when $\dot{S}_0 = 0$ and $\dot{q}_0 \neq 0$, a solution of that system takes the form

$$\begin{aligned} \dot{S}_1 &= \frac{t_1 t_0^{-3/2}\dot{q}_0}{(\upsilon_2 - \upsilon_1)(t_0^{-1/2} + \zeta t_0^{-1} + 1)}, \\ \dot{S}_2 &= -\frac{t_1 t_0^{-3/2}\dot{q}_0}{(\upsilon_2 - \upsilon_1)(t_0^{-1/2} + \zeta t_0^{-1} + 1)}, \\ \dot{q}_1 &= \frac{\upsilon_2(\upsilon_2^2 - t_0^{-1})\dot{q}_0}{(\upsilon_2 - \upsilon_1)(t_0^{-1/2} + \zeta t_0^{-1} + 1)}, \\ \dot{q}_2 &= \frac{\upsilon_1(t_0^{-1} - \upsilon_1^2)\dot{q}_0}{(\upsilon_2 - \upsilon_1)(t_0^{-1/2} + \zeta t_0^{-1} + 1)}. \end{aligned} \tag{10.3.51}$$

Since (recall Fig. 6.6)

$$\upsilon_2 - \upsilon_1 > 0, \quad \upsilon_2^2 - t_0^{-1} > 0, \quad t_0^{-1} - \upsilon_1^2 > 0, \tag{10.3.52}$$

on account of eqn (10.3.51) for $\dot{q}_0 < 0$ we obtain

$$\begin{gathered} \dot{S}_1 < 0, \qquad \dot{S}_2 > 0, \\ \dot{q}_1 < 0, \qquad \dot{q}_2 < 0. \end{gathered} \tag{10.3.53}$$

For an arbitrary $t > 0$, in view of eqns (10.3.25) and (10.3.41), we find

$$[[S_t]]_1\,[[q_t]]_1 > 0, \qquad [[S_t]]_2\,[[q_t]]_2 < 0. \tag{10.3.54}$$

It follows from these inequalities that, on a faster wavefront ($\upsilon = \upsilon_2$), a decrease in the stress rate is accompanied by an increase in the heat-flux rate, whereas, on a slower wavefront ($\upsilon = \upsilon_1$), an increase in the stress rate is accompanied by an increase in the heat-flux rate.

We now show the following theorem concerning $[[q_t]]_i$ (or $[[S_t]]_i$):

Theorem 10.4 *For every $t \geq 0$ the amplitude $[[q_t]]_i$ satisfies the following linear first-order differential equation*

$$\left(\frac{\mathrm{d}}{\mathrm{d}t} + h_i\right)[[q_t]]_i = 0, \tag{10.3.55}$$

where[6]

$$h_i = \frac{1}{2}\upsilon_i^2 k_i \qquad (i = 1, 2). \tag{10.3.56}$$

Here, υ_i is determined by the formulas (10.3.43) and (10.3.44) and (recall eqn (6.2.9))

$$k_{1.2} = \frac{1}{2}(1 + \epsilon \mp \alpha \Delta^{1/2}), \tag{10.3.57}$$

$$\alpha = -[(1+\epsilon)(t_0 + \zeta) - (1-\epsilon)]\Delta^{-1}, \tag{10.3.58}$$

$$\Delta = (1 - t_0 + \zeta)^2 + 4\zeta t_0. \tag{10.3.59}$$

Proof. First, let us observe that the constitutive equations (10.3.4) and (10.3.5) can be rewritten in the equivalent form

$$S = e - \vartheta - t_1 \vartheta_t, \tag{10.3.60}$$

$$\eta = \left(\frac{t_0}{t_1} + \epsilon\right) e + \left(1 - \frac{t_0}{t_1}\right)\vartheta - \frac{t_0}{t_1} S. \tag{10.3.61}$$

Differentiating eqn (10.3.61) twice with respect to x, using eqn (10.3.6) and taking the jump at the wavefront of the thus-obtained equation, yields

$$[[\eta_{xx}]] = \left(\frac{t_0}{t_1} + \epsilon\right)[[e_{xx}]] - \left(1 - \frac{t_0}{t_1}\right)[[q_x]] - \frac{t_0}{t_1}[[S_{xx}]]. \tag{10.3.62}$$

On the other hand, from eqns (10.3.30) and (10.3.32) we obtain

$$[[\eta_x]] = -\frac{1}{V^2}[[q_t]]. \tag{10.3.63}$$

Differentiating this relation with respect to t and using eqn (10.3.15), gives

$$[[\eta_{xt}]] + V[[\eta_{xx}]] = -\frac{1}{V^2}\frac{\mathrm{d}}{\mathrm{d}t}[[q_t]], \tag{10.3.64}$$

[6] According to the notations of Section 6.5 [see eqn (6.5.47)], the parameter h_i stands for the reduced attentuation coefficient of the ith fundamental thermoelastic disturbance of the G–L theory.

or, on account of eqn (10.3.3),

$$-\,[[q_{xx}]] + V\,[[\eta_{xx}]] = -\frac{1}{V^2}\frac{\mathrm{d}}{\mathrm{d}t}\,[[q_t]]\,, \tag{10.3.65}$$

which leads to an alternative form of eqn (10.3.62)

$$[[\eta_{xx}]] = -\frac{1}{V^3}\frac{\mathrm{d}}{\mathrm{d}t}\,[[q_t]] + \frac{1}{V}\,[[q_{xx}]]\,. \tag{10.3.66}$$

Furthermore, from eqn (10.3.15) we find

$$\frac{\mathrm{d}}{\mathrm{d}t}\,[[q_x]] = [[q_{xt}]] + V\,[[q_{xx}]]\,. \tag{10.3.67}$$

From this, in view of eqn (10.3.32), we obtain

$$[[q_{xx}]] = -\frac{1}{V^2}\frac{\mathrm{d}}{\mathrm{d}t}\,[[q_t]] - \frac{1}{V}\,[[q_{xt}]]\,. \tag{10.3.68}$$

Substituting $[[q_{xx}]]$ from the above into eqn (10.3.66), and setting the right-hand sides of eqns (10.3.62) and (10.3.66) equal, we get

$$\begin{gathered}\frac{2}{V^3}\frac{\mathrm{d}}{\mathrm{d}t}\,[[q_t]] + \frac{1}{V}\left(1 - \frac{t_0}{t_1}\right)[[q_t]] = \\ \frac{t_0}{t_1}\,[[S_{xx}]] - \left(\frac{t_0}{t_1} + \epsilon\right)[[e_{xx}]] - \frac{1}{V^2}\,[[q_{xt}]]\,.\end{gathered} \tag{10.3.69}$$

On the other hand, differentiating eqn (10.3.60) twice with respect to x and using eqn (10.3.6) yields

$$q_{xt} = \frac{1}{t_1}\,(S_{xx} - e_{xx} - q_x)\,. \tag{10.3.70}$$

Taking the jump on both sides of the above, and using eqn (10.3.32), results in

$$[[q_{xt}]] = \frac{1}{t_1}\left([[S_{xx}]] - [[e_{xx}]] + \frac{1}{V}\,[[q_t]]\right). \tag{10.3.71}$$

Now, substituting $[[q_{xt}]]$ from the above into the right-hand side of eqn (10.3.69) gives

$$\begin{gathered}\frac{2}{V^3}\frac{\mathrm{d}}{\mathrm{d}t}\,[[q_t]] + \frac{1}{V^3 t_1}\,\left[1 + V^2\,(t_1 - t_0)\right][[q_t]] = \\ \frac{t_0}{t_1}\frac{(V^2 - t_0^{-1})}{V^2}\,[[S_{xx}]] + \frac{[1 - V^2(t_0 + \epsilon t_1)]}{V^2 t_1}\,[[e_{xx}]]\,.\end{gathered} \tag{10.3.72}$$

Furthermore, differentiating eqns (10.3.27) and (10.3.28) with respect to time t and using eqn (10.3.15), we obtain

$$[[e_{tt}]] + V\,[[e_{tx}]] = -\frac{1}{V}\dot{a}, \tag{10.3.73}$$

$$[[e_{xt}]] + V\,[[e_{xx}]] = \frac{1}{V^2}\dot{a}. \tag{10.3.74}$$

Eliminating the amplitude $[[e_{xt}]]$ from this, yields

$$[[e_{tt}]] - V^2\,[[e_{xx}]] = -\frac{2}{V}\dot{a}, \tag{10.3.75}$$

or, given the equation of motion (10.3.2) and eqn (10.3.34), gives

$$[[S_{xx}]] = V^2\,[[e_{xx}]] - \frac{2}{V}\frac{t_0(V^2 - t_0^{-1})}{\epsilon}\frac{\mathrm{d}}{\mathrm{d}t}\,[[q_t]]\,. \tag{10.3.76}$$

Finally, substituting $[[S_{xx}]]$ from the above into eqn (10.3.72), we infer that, on account of eqn (10.3.42), the coefficient in front of $[[e_{xx}]]$ vanishes and that the equation reduces to

$$\left(\frac{\mathrm{d}}{\mathrm{d}t} + h\right)[[q_t]] = 0, \tag{10.3.77}$$

in which

$$h = \frac{\epsilon}{2t_0}\frac{[1 + V^2(t_1 - t_0)]}{[\epsilon t_1 t_0^{-1} + t_0(V^2 - t_0^{-1})^2]}. \tag{10.3.78}$$

Clearly, since $t_1 \geq t_0 > 0$ and $\epsilon > 0$, it follows that $h > 0$. Also, if $t_1 = t_0$ then $h = \hat{h}$, where $\hat{h}$ is the reduced attenuation coefficient of the L–S theory (recall Theorem 10.2). On the ith wavefront given by $t = x/v_i$, on account of eqns (10.3.77) and (10.3.78), we have

$$\left(\frac{\mathrm{d}}{\mathrm{d}t} + h_i\right)[[q_t]]_i = 0, \tag{10.3.79}$$

where

$$h_i = \frac{\epsilon}{2t_0}\frac{[1 + v_i^2(t_1 - t_0)]}{[\epsilon t_1 t_0^{-1} + t_0(v_i^2 - t_0^{-1})^2]}, \tag{10.3.80}$$

with v_i $(i = 1, 2)$ determined by eqns (10.3.43) and (10.3.44).

Finally, using eqn (10.3.42), it can be shown that

$$h = \frac{1}{2t_0}\frac{V^2(1 + \epsilon - V^2)}{(t_0^{-1} - V^4)}. \tag{10.3.81}$$

From this and the definition of v_i, and k_i (recall eqns (10.3.57)–(10.3.59)) we conclude that h_i is given by eqn (10.3.56). □

Remark 10.6 It follows from Theorem 10.4 that the amplitude $[[q_t]]_i$ is a decreasing function of t:

$$[[q_t]]_i\,(t) = [[q_t]]_i\,(0)\exp\{-h_i t\}. \tag{10.3.82}$$

Remark 10.7 It can be shown that the parameters v_i and h_i are identical with the speeds and the reduced attenuation coefficients for a plane wave propagating in the half-space $x \geq 0$, which is determined by a solution of an initial-boundary value problem for small times using the Laplace transform method [see, e.g. (Chandrasekharaiah, 1981) for $t_1 > t_0 > 0$ and (Sherief and Dhaliwal, 1981) for $t_1 = t_0 > 0$]. The pair (v_i, h_i) also determines the speeds and the reduced attentuation coefficients for the spherical and cylindrical waves generated in an infinite space by a point or line heat source, respectively [recall the solution of Nowacki type in Section 7.4 and the approximate solutions given in (Sherief, 1986) and (Sherief and Anwar, 1994)].

Remark 10.8 Problems of moving discontinuity surfaces in various generalizations of hyperbolic thermoelasticity, in particular, those dealing with the influence of thermal and piezo-electric fields on propagating elastic stress jumps, have gained momentum only recently (El-Karamany and Ezzat, 2004 and 2005).

11

TIME-PERIODIC SOLUTIONS

Similar to the situation in classical linear theory of dynamical coupled thermoelasticity, a relatively large number of problems satisfactorily solved so far in the L–S and G–L theories involve periodic-in-time thermoelastic disturbances. This is due to the fact that the assumption of periodicity allows one to reduce the field equations of both theories to classical Helmholtz equations with complex wave numbers and then to use the methods of classical elastodynamics in solving particular problems. In the case of each problem, a pair of complex wave numbers, associated with the central equation of the given theory (recall eqns (6.1.2) and (6.1.7)) generates a pair of phase velocities and a pair of damping coefficients, which are fundamental characteristics of a periodic thermoelastic wave. In this chapter we analyze several problems of periodic vibrations in the framework of the G–L theory. In Sections 11.1–11.3 the focus is on plane, spherical and cylindrical waves propagating in an unbounded thermoelastic medium. Section 11.4 is devoted to an integral representation and radiation conditions for the potential–temperature solutions of the G–L theory.

11.1 Plane waves in an infinite thermoelastic body with two relaxation times

Upon elimination from eqns (10.3.1)–(10.3.6) of the strain, stress, entropy and heat flux fields, we obtain the following 1D displacement–temperature field equations with two relaxation times

$$\begin{aligned} u_{xx} - u_{tt} &= \vartheta_x + t_1 \vartheta_{tx}, \\ \vartheta_{xx} - t_0 \vartheta_{tt} - \vartheta_t &= \epsilon u_{xt}. \end{aligned} \tag{11.1.1}$$

Here, u and ϑ denote, respectively, the displacement in the x-direction and the temperature difference; t_0 and t_1 are the relaxation times of the G–L theory ($t_1 \geq t_0 > 0$), while ϵ is the parameter of thermoelastic coupling. We assume a pair (u, ϑ) to be defined for every $(x, t) \in (-\infty, \infty) \times [0, \infty)$ through the relation

$$(u, \vartheta) = (u^0, \vartheta^0) \exp[\mathrm{i}(\omega t - \eta x)] \qquad (\mathrm{i}^2 = -1), \tag{11.1.2}$$

where ω is the prescribed frequency ($\omega > 0$) and η is the wave number,[1] which is determined from the condition that the pair (u, ϑ) be a non-trivial solution of the system (11.1.1), and (u^0, ϑ^0) is a constant non-vanishing pair.

The solution (11.1.2) represents a plane periodic-in-time wave, propagating in the x-direction. Its speed is given as

$$c = \frac{\omega}{\text{Re } \eta}. \tag{11.1.3}$$

Its damping coefficient is[2]

$$q = -\text{Im } \eta. \tag{11.1.4}$$

Its wavelength is[3]

$$\lambda = \frac{2\pi}{\text{Re } \eta}, \tag{11.1.5}$$

and the period is

$$T = \frac{2\pi}{\omega}. \tag{11.1.6}$$

Here, Re η and Im η stand, respectively, for the real and imaginary parts of η. The relations (11.1.3) and (11.1.4) imply that this number may be given in the form

$$\eta = \frac{\omega}{c} - \text{i}q. \tag{11.1.7}$$

If Re $\eta > 0$ and Im $\eta > 0$, the pair (u, ϑ) represents a damped wave propagating in the x-direction with the speed $c > 0$.

Substituting eqn (11.1.2) into eqn (11.1.1), we see that the pair (11.1.2) is a nontrivial solution to eqn (11.1.1) if the pair (u^0, ϑ^0) is a non-trivial solution of the linear homogeneous system of algebraic equations

$$\begin{aligned} u^0(\omega^2 - \eta^2) + \vartheta^0(\text{i}\eta - t_1\omega\eta) &= 0, \\ u^0\omega\eta\epsilon + \vartheta^0(\eta^2 - t_0\omega^2 + \text{i}\omega) &= 0. \end{aligned} \tag{11.1.8}$$

In order for (u^0, ϑ^0) to be a non-trivial solution of the system (11.1.8) it is necessary and sufficient for the number η to satisfy the equation

$$(\omega^2 - \eta^2)(\eta^2 + \text{i}\omega - t_0\omega^2) - \epsilon\eta^2(\text{i}\omega - t_1\omega^2) = 0, \tag{11.1.9}$$

This equation can also be written in the form

$$\eta^4 - \eta^2\omega[(1 + t_0 + \epsilon t_1)\omega - (1 + \epsilon)\text{i}] + \omega^3(t_0\omega - \text{i}) = 0. \tag{11.1.10}$$

[1] The symbol η in eqn (11.1.2) must not be confused with the entropy function of Section 10.3.

[2] The symbol q in eqn (11.1.4) must not be confused with the heat-flux function of Section 10.3.

[3] The symbol λ in eqn (11.1.5) must not be confused with the coefficient of convolution of Section 6.2.

Three cases can be identified here:

(a) If $t_1 = t_0 = 0$, eqn (11.1.10) becomes a characteristic equation of the classical coupled thermoelasticity (e.g. Deresiewicz, 1957).
(b) If $t_1 = t_0 > 0$, eqn (11.1.10) reduces to a characteristic equation of the L–S theory [e.g. (Nayfeh and Nemat-Nasser, 1971; Puri, 1973; Agarwal, 1979)].
(c) If $\epsilon = 0$, eqn (11.1.10) reduces to a characteristic equation of the uncoupled generalized dynamical theory of thermal stresses.

Recall that in a range of real materials, the non-dimensional relaxation time t_0 satisfies the inequality[4]

$$1 < t_0 < 16. \tag{11.1.11}$$

We shall now demonstrate that, for $t_0 > 0$ $(t_1 \geq t_0)$ and for arbitrary $\epsilon > 0$ and $\omega > 0$, one can find two closed-form solutions of eqn (11.1.10) corresponding, respectively, to two plane thermoelastic waves propagating in the positive x-direction. To this end, let us represent eqn (11.1.10) in the form

$$(\eta^2 - \eta_1^2)(\eta^2 - \eta_2^2) = 0, \tag{11.1.12}$$

where

$$\eta_{1.2} = -\mathrm{i}s_{1.2}(\mathrm{i}\omega), \tag{11.1.13}$$

and (recall eqns (7.4.15) and (7.4.16))

$$s_{1.2}(p) = \left(\frac{p}{2}\right)^{1/2} \{(1+\epsilon) + (1 + t_0 + \epsilon t_1)p \pm \Delta^{1/2}[(p-\alpha)^2 + \beta^2]^{1/2}\}^{1/2}. \tag{11.1.14}$$

The symbols α, β and Δ, appearing in eqn (11.1.14), are defined in Theorem 6.1 on decomposition of the G–L theory (recall eqns (6.2.4) and (6.2.5)), that is

$$\begin{aligned} \alpha &= -[(1+\epsilon)(t_0 + \epsilon t_1) - (1-\epsilon)]\Delta^{-1}, \\ \beta &= 2\sqrt{\epsilon}[1 + (1+\epsilon)(t_1 - t_0)]^{1/2}\Delta^{-1}, \\ \Delta &= (1 - t_0 + \epsilon t_1)^2 + 4\epsilon t_0 t_1. \end{aligned} \tag{11.1.15}$$

It follows from the relations (11.1.15) that

$$\alpha^2 + \beta^2 = (1+\epsilon)^2\Delta^{-1}. \tag{11.1.16}$$

In order to write eqn (11.1.13) in the form of eqn (11.1.7), so as to allow the determination of the phase velocities and damping coefficients, we note the fact that, for an arbitrary pair of real numbers (a, b), such that $b \geq 0$, the following is true

$$\sqrt{a + \mathrm{i}b} = \pm\frac{1}{\sqrt{2}}(\sqrt{\rho + a} + \mathrm{i}\sqrt{\rho - a}), \tag{11.1.17}$$

[4] See Section 6.5 and (Tao and Prevost, 1984).

in which

$$\rho = \sqrt{a^2 + b^2}. \tag{11.1.18}$$

Since $t_0 > 1$, the parameter α given by eqn $(11.1.15)_1$ is negative (see footnote 5 of Section 6.5). Thus, setting

$$a = \alpha^2 + \beta^2 - \omega^2, \qquad b = -2\alpha\omega \geq 0 \tag{11.1.19}$$

in eqn (11.1.17), and restricting ourselves to the plus sign, we obtain

$$[(\mathrm{i}\omega - \alpha)^2 + \beta^2]^{1/2} = \frac{1}{\sqrt{2}}(\sqrt{\rho + a} + \mathrm{i}\sqrt{\rho - a}), \tag{11.1.20}$$

in which

$$\rho = [(\alpha^2 + \beta^2 - \omega^2)^2 + 4\alpha^2\omega^2]^{1/2}. \tag{11.1.21}$$

From this, on account of eqn (11.1.14), we find

$$\begin{aligned} s_{1.2}^2(\mathrm{i}\omega) = \frac{\mathrm{i}\omega}{2}\Big\{&(1+\epsilon) + \mathrm{i}(1 + t_0 + \epsilon t_1)\omega \\ &\pm\frac{\Delta^{1/2}}{\sqrt{2}}(\sqrt{\rho + a} + \mathrm{i}\sqrt{\rho - a})\Big\}, \end{aligned} \tag{11.1.22}$$

or

$$\begin{aligned} s_{1.2}^2(\mathrm{i}\omega) = \frac{\omega}{2}\Big\{&-\Big[(1 + t_0 + \epsilon t_1)\omega \pm \frac{\Delta^{1/2}}{\sqrt{2}}\sqrt{\rho - a}\Big] \\ &+\mathrm{i}\Big[(1+\epsilon) \pm \frac{\Delta^{1/2}}{\sqrt{2}}\sqrt{\rho + a}\Big]\Big\}. \end{aligned} \tag{11.1.23}$$

In the formula (11.1.23) the signs + and − correspond, respectively, to the numbers $s_1^2(\mathrm{i}\omega)$ and $s_2^2(\mathrm{i}\omega)$.

Of course, we have

$$\mathrm{Im}[s_1^2(\mathrm{i}\omega)] = \frac{\omega}{2}\Big[(1+\epsilon) + \frac{\Delta^{1/2}}{\sqrt{2}}\sqrt{\rho + a}\Big] \geq 0, \tag{11.1.24}$$

and we shall now show that also

$$\mathrm{Im}[s_2^2(\mathrm{i}\omega)] = \frac{\omega}{2}\Big[(1+\epsilon) - \frac{\Delta^{1/2}}{\sqrt{2}}\sqrt{\rho + a}\Big] \geq 0. \tag{11.1.25}$$

Indeed, it will suffice to show that

$$(1+\epsilon)^2 - \frac{\Delta}{2}(\rho + a) \geq 0. \tag{11.1.26}$$

Denoting the left-hand side of inequality (1.1.26) by L and using eqns (11.1.16), (11.1.19) and (11.1.21), we obtain

$$L = \frac{\Delta}{2}\{(\alpha^2+\beta^2+\omega^2) - [(\alpha^2+\beta^2-\omega^2)^2 + 4\alpha^2\omega^2]^{1/2}\}, \tag{11.1.27}$$

or

$$L = 2\Delta\omega^2\beta^2\{(\alpha^2+\beta^2+\omega^2) + [(\alpha^2+\beta^2-\omega^2)^2 + 4\alpha^2\omega^2]^{1/2}\}^{-1}. \tag{11.1.28}$$

From this we infer that $L \geq 0$ and the inequality (11.1.25) holds true.

The inequalities (11.1.24) and (11.1.25) imply that the square root of both sides of eqn (11.1.23) can be computed with the help of formulas (11.1.17) and (11.1.18). Thus, taking the square root of eqn (11.1.23) and using eqn (11.1.17) with the plus sign, we obtain

$$s_{1.2}(\mathrm{i}\omega) = \frac{\sqrt{\omega}}{2}[\sqrt{\rho_{1.2}+a_{1.2}} + \mathrm{i}\sqrt{\rho_{1.2}-a_{1.2}}\ ], \tag{11.1.29}$$

where

$$\begin{aligned}\rho_{1.2} = \{[(1+t_0+\epsilon t_1)\omega \pm \frac{\Delta^{1/2}}{\sqrt{2}}\sqrt{\rho - a}\ ]^2 \\ +[(1+\epsilon) \pm \frac{\Delta^{1/2}}{\sqrt{2}}\sqrt{\rho + a}\ ]^2\}^{1/2},\end{aligned} \tag{11.1.30}$$

and

$$a_{1.2} = -[(1+t_0+\epsilon t_1)\omega \pm \frac{\Delta^{1/2}}{\sqrt{2}}\sqrt{\rho - a}\ ]. \tag{11.1.31}$$

Here, the numbers a and ρ are defined, respectively, by eqns $(11.1.19)_1$ and (11.1.21). Finally, substituting eqn (11.1.29) into eqn (11.1.13), we obtain the sought roots of eqn (11.1.10)

$$\eta_{1.2} = \frac{\omega}{c_{1.2}} - \mathrm{i}q_{1.2}, \tag{11.1.32}$$

with the phase velocities $c_{1.2}$ and damping coefficients $q_{1.2}$ given by

$$c_{1.2} = 2\sqrt{\omega}(\rho_{1.2} - a_{1.2})^{-1/2}, \tag{11.1.33}$$

$$q_{1.2} = \frac{\sqrt{\omega}}{2}(\rho_{1.2} + a_{1.2})^{1/2}. \tag{11.1.34}$$

Passing with ϵ to zero in these relations and introducing the notations

$$\lim_{\epsilon \to 0}\ (c_i, q_i) = (c_i^0, q_i^0) \qquad (i = 1, 2), \tag{11.1.35}$$

we obtain

$$c_1^0 = \sqrt{2\omega}[(1+t_0^2\omega^2)^{1/2} + t_0\omega]^{-1/2}, \tag{11.1.36}$$

$$q_1^0 = \sqrt{\frac{\omega}{2}}[(1+t_0^2\omega^2)^{1/2} - t_0\omega]^{1/2}, \tag{11.1.37}$$

with

$$c_2^0 = 1, \qquad q_2^0 = 0. \tag{11.1.38}$$

The pairs (c_1^0, q_1^0) and (c_2^0, q_2^0) correspond, respectively, to a purely thermal and a purely elastic wave, both propagating in the positive x-direction, in accordance with the uncoupled generalized dynamical theory of thermal stresses.

In general, the relations (11.1.2) and (11.1.32) imply that, for $\epsilon > 0$, $\omega > 0$ and $t_1 \geq t_0 > 1$, a plane thermoelastic wave propagating in the positive x direction is described by the pair (u, ϑ) in which

$$u = u_1^0 \exp(-q_1 x) \exp\left[\mathrm{i}\omega\left(t - \frac{x}{c_1}\right)\right] + u_2^0 \exp(-q_2 x) \exp\left[\mathrm{i}\omega\left(t - \frac{x}{c_2}\right)\right], \tag{11.1.39}$$

and

$$\vartheta = \vartheta_1^0 \exp(-q_1 x) \exp\left[\mathrm{i}\omega\left(t - \frac{x}{c_1}\right)\right] + \vartheta_2^0 \exp(-q_2 x) \exp\left[\mathrm{i}\omega\left(t - \frac{x}{c_2}\right)\right]. \tag{11.1.40}$$

Here, $c_{1.2}$ and $q_{1.2}$ are specified by the formulas (11.1.33) and (11.1.34), respectively. Furthermore, u_1^0 and u_2^0 are arbitrary complex numbers, while ϑ_1^0 and ϑ_2^0 are related to u_1^0 and u_2^0 through (recall eqn $(11.1.8)_1$ for $u^0 = u_i^0$, $\vartheta^0 = \vartheta_i^0$, $\eta = \eta_i$, $i = 1, 2$)

$$\vartheta_1^0 = \frac{\omega^2 - \eta_1^2}{\eta_1(t_1\omega - i)} u_1^0, \qquad \vartheta_2^0 = \frac{\omega^2 - \eta_2^2}{\eta_2(t_1\omega - i)} u_2^0, \tag{11.1.41}$$

with $\eta_{1.2}$ given by eqn (11.1.32). In the case when a thermomechanical loading

$$u(0, t) = \hat{u}_0 \exp(\mathrm{i}\omega t), \qquad \vartheta(0, t) = \hat{\vartheta}_0 \exp(\mathrm{i}\omega t), \tag{11.1.42}$$

is applied to the half-space $(x \geq 0)$ at its boundary, where the pair $(\hat{u}_0, \hat{\vartheta}_0)$ is given, the constants u_1^0 and u_2^0 are uniquely specified by the following system of equations

$$u_1^0 + u_2^0 = \hat{u}_0, \qquad \vartheta_1^0 + \vartheta_2^0 = \hat{\vartheta}_0, \tag{11.1.43}$$

in which ϑ_1^0 and ϑ_2^0 are replaced by eqns $(11.1.41)_1$ and $(11.1.41)_2$, respectively.

Clearly, a plane thermoelastic wave propagating in the negative x-direction is represented by the pair (u, ϑ) in which

$$u = u_3^0 \exp(q_1 x) \exp\left[\mathrm{i}\omega\left(t + \frac{x}{c_1}\right)\right] + u_4^0 \exp(q_2 x) \exp\left[\mathrm{i}\omega\left(t + \frac{x}{c_2}\right)\right], \tag{11.1.44}$$

and

$$\vartheta = \vartheta_3^0 \exp(q_1 x) \exp\left[\mathrm{i}\omega\left(t + \frac{x}{c_1}\right)\right] + \vartheta_4^0 \exp(q_2 x) \exp\left[\mathrm{i}\omega\left(t + \frac{x}{c_2}\right)\right]. \tag{11.1.45}$$

Here, u_3^0 and u_4^0 are arbitrary constants, while

$$\vartheta_3^0 = \frac{\omega^2 - \hat{\eta}_1^2}{\hat{\eta}_1(t_1\omega - \mathrm{i})} u_3^0, \qquad \vartheta_4^0 = \frac{\omega^2 - \hat{\eta}_2^2}{\hat{\eta}_2(t_1\omega - \mathrm{i})} u_4^0, \tag{11.1.46}$$

with $\hat{\eta}_{1.2}$ given by

$$\hat{\eta}_{1.2} = -\frac{\omega}{c_{1.2}} + \mathrm{i}q_{1.2}. \tag{11.1.47}$$

Clearly, a plane thermoelastic wave propagating in the layer $|x| \leq l$ ($2l$ = non-dimensional layer thickness) due to a displacement–temperature loading applied to its boundaries $x = \pm l$ is a superposition of the pairs defined by eqns (11.1.39), (11.1.40), (11.1.44) and (11.1.45), whereby the constants u_1^0, u_2^0, u_3^0 and u_4^0 are uniquely specified through the four displacement–temperature boundary conditions at $x = \pm l$.

Approximate formulas for the phase velocities and damping coefficients

We now derive asymptotic formulas for the phase velocities c_i and damping coefficients q_i ($i = 1, 2$) in the three cases:

(i) ω is small: $\omega \ll 1$, (ii) ω is large: $\omega \gg 1$, (iii) ϵ is small: $\epsilon \ll 1$.

In the cases (i) and (ii) the relaxation times are assumed to satisfy the inequalities

$$t_1 \geq t_0 \geq 0, \tag{11.1.48}$$

which implies that we admit a passage with the relaxation times to zero.[5]

First, let us consider the case (i). The square of s_i determining η_i (recall eqn (11.1.13)) can be written as

$$s_{1.2}^2(p) = \frac{p}{2}\varphi_{1.2}(p), \tag{11.1.49}$$

where

$$\varphi_{1.2}(p) = (1+\epsilon) + (1 + t_0 + \epsilon t_1)p \pm \Delta^{1/2}[(p-\alpha)^2 + \beta^2]^{1/2}. \tag{11.1.50}$$

Here, small values of p correspond to low frequencies ω. Thus, an asymptotic expansion of the function $s_{1.2}^2(\mathrm{i}\omega)$ for small ω corresponds to the Taylor series expansion of $\varphi_{1.2}(p)$ in a small neighborhood of $p = 0$. Computing three subsequent derivatives of $\varphi_{1.2}(p)$, we obtain

$$\varphi_{1.2}'(p) = (1 + t_0 + \epsilon t_1) \pm \Delta^{1/2}(p-\alpha)[(p-\alpha)^2 + \beta^2]^{-1/2}, \tag{11.1.51}$$

$$\varphi_{1.2}''(p) = \pm\, \Delta^{1/2}\beta^2[(p-\alpha)^2 + \beta^2]^{-3/2}, \tag{11.1.52}$$

$$\varphi_{1.2}'''(p) = \mp\, 3\Delta^{1/2}\beta^2(p-\alpha)[(p-\alpha)^2 + \beta^2]^{-5/2}. \tag{11.1.53}$$

Since (recall eqn (11.1.16))

$$\alpha^2 + \beta^2 = (1+\epsilon)^2\Delta^{-1}, \tag{11.1.54}$$

[5] Such a passage is not possible in eqns (11.1.33) and (11.1.34) since those relations were obtained under the hypothesis $t_0 > 1$.

therefore letting $p = 0$ in eqns (11.1.51)–(11.1.53), we find

$$\varphi'_{1.2}(0) = (1 + t_0 + \epsilon t_1) \mp \frac{\alpha}{1+\epsilon}\Delta, \tag{11.1.55}$$

$$\varphi''_{1.2}(0) = \pm \frac{\beta^2}{(1+\epsilon)^3}\Delta^2, \tag{11.1.56}$$

$$\varphi'''_{1.2}(p) = \pm\, 3\frac{\alpha\beta^2}{(1+\epsilon)^5}\Delta^3. \tag{11.1.57}$$

Furthermore, on account of eqns (11.1.50) and (11.1.54), we obtain

$$\varphi_1(0) = 2(1+\epsilon), \qquad \varphi_2(0) = 0. \tag{11.1.58}$$

Also note that, in view of eqn $(11.1.15)_1$, the formulas (11.1.55) reduce to the forms

$$\varphi'_1(0) = 2\left(t_0 + \epsilon t_1 + \frac{\epsilon}{1+\epsilon}\right), \qquad \varphi'_2(0) = \frac{2}{1+\epsilon}. \tag{11.1.59}$$

Thus, the functions $\varphi_1(p)$ and $\varphi_2(p)$ can be approximated by

$$\begin{aligned}\varphi_1(p) = 2(1+\epsilon)\Bigg\{1 + \frac{1}{1+\epsilon}\left(t_0 + \epsilon t_1 + \frac{\epsilon}{1+\epsilon}\right)p \\ + \frac{\beta^2\Delta^2}{4(1+\epsilon)^4}p^2 + \frac{\alpha\beta^2\Delta^3}{4(1+\epsilon)^6}p^3 + O(p^4)\Bigg\},\end{aligned} \tag{11.1.60}$$

and

$$\varphi_2(p) = \frac{2p}{1+\epsilon}\left\{1 - \frac{\beta^2\Delta^2}{4(1+\epsilon)^2}p - \frac{\alpha\beta^2\Delta^3}{4(1+\epsilon)^4}p^2 + O(p^3)\right\}. \tag{11.1.61}$$

From this, given the asymptotic expansion

$$(1+x)^{1/2} \to 1 + \frac{x}{2} \quad \text{for} \quad x \to 0, \tag{11.1.62}$$

we find

$$\begin{aligned}\sqrt{\varphi_1(p)} = \sqrt{2(1+\epsilon)}\Bigg\{1 + \frac{1}{2(1+\epsilon)}\left(t_0 + \epsilon t_1 + \frac{\epsilon}{1+\epsilon}\right)p \\ + \frac{\beta^2\Delta^2}{8(1+\epsilon)^4}p^2 + O(p^3)\Bigg\},\end{aligned} \tag{11.1.63}$$

and

$$\sqrt{\varphi_2(p)} = \sqrt{\frac{2p}{1+\epsilon}}\left\{1 - \frac{\beta^2\Delta^2}{8(1+\epsilon)^2}p - \frac{\alpha\beta^2\Delta^3}{8(1+\epsilon)^4}p^2 + O(p^3)\right\}. \tag{11.1.64}$$

Thus, the functions $s_{1.2}(p)$ defined by eqn (11.1.49) have the expansions

$$s_1(p) = \sqrt{1+\epsilon}\sqrt{p}\left\{1 + \frac{1}{2(1+\epsilon)}\left(\frac{\epsilon}{1+\epsilon} + t_0 + \epsilon t_1\right)p + \frac{\beta^2\Delta^2}{8(1+\epsilon)^4}p^2 + O(p^3)\right\}, \tag{11.1.65}$$

and

$$s_2(p) = \frac{p}{\sqrt{1+\epsilon}}\left\{1 - \frac{\beta^2\Delta^2}{8(1+\epsilon)^2}p - \frac{\alpha\beta^2\Delta^3}{8(1+\epsilon)^4}p^2 + O(p^3)\right\}. \tag{11.1.66}$$

Letting $p = \mathrm{i}\omega$ in the above and using the identity

$$\sqrt{\mathrm{i}} = \frac{1}{\sqrt{2}}(1+\mathrm{i}), \tag{11.1.67}$$

we obtain

$$s_1(\mathrm{i}\omega) = \sqrt{1+\epsilon}\,\frac{\sqrt{\omega}}{\sqrt{2}}\left\{1 - \frac{1}{2(1+\epsilon)}\left(\frac{\epsilon}{1+\epsilon} + t_0 + \epsilon t_1\right)\omega\right.$$

$$\left.-\frac{\beta^2\Delta^2}{8(1+\epsilon)^4}\omega^2 + \mathrm{i}\left[1 + \frac{1}{2(1+\epsilon)}\left(\frac{\epsilon}{1+\epsilon} + t_0 + \epsilon t_1\right)\omega - \frac{\beta^2\Delta^2}{8(1+\epsilon)^4}\omega^2\right] + O(\omega^3)\right\}, \tag{11.1.68}$$

and

$$s_2(\mathrm{i}\omega) = \frac{\omega}{\sqrt{1+\epsilon}}\left\{\frac{\beta^2\Delta^2}{8(1+\epsilon)^2}\omega + \mathrm{i}\left[1 + \frac{\alpha\beta^2\Delta^3}{8(1+\epsilon)^4}\omega^2\right] + O(\omega^3)\right\}. \tag{11.1.69}$$

From this, given eqn (11.1.13), we find

$$\eta_1 = \sqrt{1+\epsilon}\,\sqrt{\frac{\omega}{2}}\left\{\left[1 + \frac{1}{2(1+\epsilon)}\left(\frac{\epsilon}{1+\epsilon} + t_0 + \epsilon t_1\right)\omega - \frac{\beta^2\Delta^2}{8(1+\epsilon)^4}\omega^2\right]\right.$$

$$\left.-\mathrm{i}\left[1 - \frac{1}{2(1+\epsilon)}\left(\frac{\epsilon}{1+\epsilon} + t_0 + \epsilon t_1\right)\omega - \frac{\beta^2\Delta^2}{8(1+\epsilon)^4}\omega^2\right] + O(\omega^3)\right\}, \tag{11.1.70}$$

and

$$\eta_2 = \frac{\omega}{\sqrt{1+\epsilon}}\left\{\left[1 + \frac{\alpha\beta^2\Delta^3}{8(1+\epsilon)^4}\omega^2\right] - \mathrm{i}\frac{\beta^2\Delta^2}{8(1+\epsilon)^2}\omega + O(\omega^3)\right\}. \tag{11.1.71}$$

Therefore, on account of eqn (11.1.7), relations (11.1.70) and (11.1.71) yield the asymptotic formulas for phase velocities and damping coefficients for $\omega \ll 1$

$$c_1 \approx \frac{\omega}{\sqrt{1+\epsilon}\sqrt{\omega/2}}\left\{1 - \left[\frac{1}{2(1+\epsilon)}\left(\frac{\epsilon}{1+\epsilon} + t_0 + \epsilon t_1\right)\omega - \frac{\beta^2\Delta^2}{8(1+\epsilon)^4}\omega^2\right]\right\}, \tag{11.1.72}$$

$$q_1 \approx \sqrt{1+\epsilon}\sqrt{\omega/2}\left\{1 - \left[\frac{1}{2(1+\epsilon)}\left(\frac{\epsilon}{1+\epsilon} + t_0 + \epsilon t_1\right)\omega + \frac{\beta^2\Delta^2}{8(1+\epsilon)^4}\omega^2\right]\right\}, \tag{11.1.73}$$

$$c_2 \approx \sqrt{1+\epsilon}\left[1 - \frac{\alpha\beta^2\Delta^2}{8(1+\epsilon)^4}\omega^2\right], \tag{11.1.74}$$

$$q_2 \approx \frac{\beta^2\Delta^2\omega^2}{8(1+\epsilon)^4\sqrt{1+\epsilon}}. \tag{11.1.75}$$

Letting $t_0 = t_1 = 0$ in these equations, and using the relations

$$\alpha = 1-\epsilon, \qquad \beta = 2\sqrt{\epsilon}, \qquad \Delta = 1, \tag{11.1.76}$$

we obtain

$$\hat{c}_1 \approx \frac{\sqrt{2\omega}}{\sqrt{1+\epsilon}}\left\{1 - \left[\frac{\epsilon\omega}{2(1+\epsilon)^2} - \frac{\epsilon\omega^2}{2(1+\epsilon)^4}\right]\right\}, \tag{11.1.77}$$

$$\hat{q}_1 \approx \sqrt{\frac{\omega(1+\epsilon)}{2}}\left\{1 - \left[\frac{\epsilon\omega}{2(1+\epsilon)^2} + \frac{\epsilon\omega^2}{2(1+\epsilon)^4}\right]\right\}, \tag{11.1.78}$$

$$\hat{c}_2 \approx \sqrt{1+\epsilon}\left[1 - \frac{\epsilon(1-\epsilon)}{2(1+\epsilon)^4}\omega^2\right], \tag{11.1.79}$$

$$\hat{q}_2 \approx \frac{\epsilon\omega^2}{2(1+\epsilon)^2\sqrt{1+\epsilon}}. \tag{11.1.80}$$

The pairs $(\hat{c}_1, \hat{q}_1)$ and $(\hat{c}_2, \hat{q}_2)$ correspond, respectively, to a quasi-thermal and a quasi-elastic wave of classical thermoelasticity ($\epsilon > 0,\ \omega \ll 1$).

We now turn to the case (ii): $\omega \gg 1$. First, let us transform eqn (11.1.49) to the form

$$s_{1.2}^2(p) = \frac{p^2}{2}\psi_{1.2}(u), \tag{11.1.81}$$

where

$$u = p^{-1}, \tag{11.1.82}$$

and

$$\psi_{1.2}(u) = (1 + t_0 + \epsilon t_1) + (1 + \epsilon)u \pm \Delta^{1/2}[(1 - \alpha u)^2 + \beta^2 u^2]^{1/2}. \qquad (11.1.83)$$

Here, large values of ω correspond to small values of u. Thus, in order to analyze the behavior of functions $s_{1.2}^2(p)$ for large p, we shall consider Taylor's expansions of $\psi_{1.2}(u)$ in a neighborhood of $u = 0$. The first derivative of $\psi_{1.2}(u)$ is

$$\psi'_{1.2}(u) = (1 + \epsilon) \pm \Delta^{1/2}[(\alpha^2 + \beta^2)u - \alpha][(1 - \alpha u)^2 + \beta^2 u^2]^{-1/2}. \qquad (11.1.84)$$

Letting $u = 0$ in eqns (11.1.83) and (11.1.84), we find

$$\psi_{1.2}(0) = (1 + t_0 + \epsilon t_1) \pm \Delta^{1/2}, \qquad (11.1.85)$$

$$\psi'_{1.2}(0) = (1 + \epsilon) \mp \alpha \Delta^{1/2}. \qquad (11.1.86)$$

Comparing these formulas with eqns $(6.2.9)_1$ and $(6.2.9)_2$, we find that

$$\psi_{1.2}(0) = 2v_{1.2}^{-2}, \qquad \psi'_{1.2}(0) = 2k_{1.2}, \qquad (11.1.87)$$

where v_i and k_i $(i = 1, 2)$ represent, respectively, the speed and damping of the ith fundamental thermoelastic disturbance in the G–L theory (recall Section 6.2).

From the above, on account of eqn (11.1.81), we obtain

$$s_{1.2}^2(p) \approx \frac{p^2}{v_{1.2}^2}\left(1 + \frac{k_{1.2}v_{1.2}^2}{p}\right) \quad \text{as} \quad p \to \infty. \qquad (11.1.88)$$

Taking the square root of the above, and using the asymptotic formula (11.1.62), gives

$$s_{1.2}(p) \approx \frac{p}{v_{1.2}}\left(1 + \frac{h_{1.2}}{p}\right) \quad \text{as} \quad p \to \infty, \qquad (11.1.89)$$

where

$$h_i = \frac{k_i v_i^2}{2} \qquad (i = 1, 2) \qquad (11.1.90)$$

is the reduced damping coefficient for the ith fundamental thermoelastic disturbance in the G–L theory (recall eqn $(6.5.47)_1$).

Finally, letting $p = \mathrm{i}\omega$ in eqn (11.1.89) and substituting the result in the right-hand side of eqn (11.1.13), we obtain

$$\eta_{1.2} \approx \frac{\omega}{v_{1.2}} - \mathrm{i}\frac{h_{1.2}}{v_{1.2}} \quad \text{for} \quad \omega \to \infty \qquad (11.1.91)$$

Thus, for high frequencies, the phase velocities and the corresponding damping coefficients are defined by the formulas

$$c_i^\infty = v_i, \qquad q_i^\infty = h_i/v_i \qquad (i = 1, 2) \qquad (11.1.92)$$

Prior to the analysis of the case (iii), where $\epsilon \ll 1$, let us note that the formulas (11.1.91) are true for each triplet of the constitutive parameters (t_0, t_1, ϵ)

satisfying the inequalities

$$t_1 \geq t_0 \geq 0, \qquad \epsilon \geq 0. \tag{11.1.93}$$

Let us also note that the point $(t_0, t_1, \epsilon) = (1, 1, 0)$ is a bifurcation point of the function $c_i^\infty = v_i$ (recall Fig. 6.4). Thus, the asymptotic expansion of the function η_i in a Taylor series with respect to ϵ in a neighborhood of $\epsilon = 0$ is correct only outside of a neighborhood of that bifurcation point. Considering the case (iii): $\epsilon \ll 1$, we assume that $t_1 \geq t_0 > 1$, i.e. we analyze such an "external" expansion only.[6] To that end, first reduce eqn (11.1.9) to the form

$$(1 - X)[1 - X(t_0 - \sigma)] = \epsilon X(t_1 - \sigma), \tag{11.1.94}$$

where

$$X = \frac{\omega^2}{\eta^2}, \qquad \sigma = \frac{i}{\omega}, \qquad \omega > 0, \qquad |\eta| > 0. \tag{11.1.95}$$

Two "external" solutions of eqn (11.1.94) are sought in the form

$$X_1^0 = 1 + a_1\epsilon + a_2\epsilon^2 + a_3\epsilon^3 + ..., \tag{11.1.96}$$

$$X_2^0 = \frac{d_1}{d_2} + b_1\epsilon + b_2\epsilon^2 + b_3\epsilon^3 + ..., \tag{11.1.97}$$

where

$$d_1 = \frac{1}{t_1 - \sigma}, \qquad d_2 = \frac{t_0 - \sigma}{t_1 - \sigma}, \tag{11.1.98}$$

and the sequences a_n and b_n $(n = 1, 2, 3, \ldots)$ are chosen so that the numbers X_1^0 and X_2^0 satisfy eqn (11.1.94).

Substituting X_1^0 into eqn (11.1.94) and setting equal the coefficients of like powers of ϵ, we obtain $a_1, a_2, a_3, \ldots$. Thus, we arrive at

$$X_1^0 = 1 + \frac{\epsilon}{d_2 - d_1} - \frac{d_1\epsilon^2}{(d_2 - d_1)^3} + O(\epsilon^3). \tag{11.1.99}$$

A similar procedure related to X_2^0 yields

$$X_2^0 = \frac{d_1}{d_2}\left[1 - \frac{\epsilon}{d_2 - d_1} - \frac{d_2\epsilon^2}{(d_2 - d_1)^3} + O(\epsilon^3)\right]. \tag{11.1.100}$$

Retaining the terms of order ϵ only, gives

$$\frac{\omega^2}{\eta_1^2} = X_1^0 = 1 + \frac{\epsilon}{d_2 - d_1}, \tag{11.1.101}$$

$$\frac{\omega^2}{\eta_2^2} = X_2^0 = \frac{d_1}{d_2}\left[1 - \frac{\epsilon}{d_2 - d_1}\right]. \tag{11.1.102}$$

[6] An "internal" expansion of the pair (c_i, q_i) with respect to ϵ is obtained in (Tao and Prevost, 1984).

From this we obtain

$$\eta_1^2 = \omega^2 \left(1 + \frac{\epsilon}{d_2 - d_1}\right)^{-1} = \omega^2 \left(1 - \frac{\epsilon}{d_2 - d_1}\right) + O(\epsilon^2), \tag{11.1.103}$$

$$\eta_2^2 = \omega^2 \frac{d_2}{d_1} \left(1 - \frac{\epsilon}{d_2 - d_1}\right)^{-1} = \omega^2 \frac{d_2}{d_1} \left(1 + \frac{\epsilon}{d_2 - d_1}\right) + O(\epsilon^2). \tag{11.1.104}$$

Substituting d_1 and d_2 from eqn (11.1.98) into eqns (11.1.103) and (11.1.104), we find

$$\eta_1^2 = \omega^2 \left(1 - \frac{\omega t_1 - \mathrm{i}}{\omega(t_0 - 1) - \mathrm{i}} \epsilon\right) + O(\epsilon^2), \tag{11.1.105}$$

$$\eta_2^2 = \omega^2 t_0 \left(1 - \mathrm{i}\omega^{-1} t_0^{-1}\right) \left(1 + \frac{\omega t_1 - \mathrm{i}}{\omega(t_0 - 1) - \mathrm{i}} \epsilon\right) + O(\epsilon^2). \tag{11.1.106}$$

Taking square roots of both sides of these relations and using the asymptotic formula (11.1.62) (for small ϵ), we obtain

$$\eta_1 = \omega \left\{1 - \frac{[\omega^2(t_0 - 1)t_1 + 1] + \mathrm{i}\omega(t_1 - t_0 + 1)}{2[\omega^2(t_0 - 1)^2 + 1]} \epsilon\right\}, \tag{11.1.107}$$

$$\eta_2 = \omega t_0^{1/2} (1 - \mathrm{i}\omega^{-1} t_0^{-1})^{1/2} \left\{1 + \frac{[\omega^2(t_0 - 1)t_1 + 1] + \mathrm{i}\omega(t_1 - t_0 + 1)}{2[\omega^2(t_0 - 1)^2 + 1]} \epsilon\right\}. \tag{11.1.108}$$

Furthermore, assuming

$$t_1 \geq t_0 \gg 1, \tag{11.1.109}$$

we have

$$t_0 t_1 - t_1 \approx t_0 t_1, \qquad t_0 - 1 \approx t_0, \tag{11.1.110}$$

whereby eqns (11.1.107) and (11.1.108) reduce to the form

$$\eta_1 = \omega \left[1 - \frac{\omega^2 t_0 t_1 + 1}{2(\omega^2 t_0^2 + 1)} \epsilon\right] - \mathrm{i}\omega^2 \frac{(t_1 - t_0 + 1)}{2(\omega^2 t_0^2 + 1)} \epsilon, \tag{11.1.111}$$

$$\eta_2 = \omega t_0^{1/2} \left(1 - \frac{i}{\omega t_0}\right)^{1/2} \left\{1 + \frac{\omega^2 t_0 t_1 + 1}{2(\omega^2 t_0^2 + 1)} \epsilon + \mathrm{i} \frac{\omega(t_1 - t_0 + 1)}{2(\omega^2 t_0^2 + 1)} \epsilon\right\}. \tag{11.1.112}$$

If also we assume $\omega \gg 1$, then

$$\left(1 - \frac{\mathrm{i}}{\omega t_0}\right)^{1/2} \approx 1 - \frac{\mathrm{i}}{2\omega t_0}, \tag{11.1.113}$$

and the formulas (11.1.111) and (11.1.112) take the form

$$\eta_1 \approx \omega\left(1 - \frac{1}{2}\frac{t_1}{t_0}\epsilon\right) - \mathrm{i}\frac{(t_1 - t_0 + 1)}{2t_0^2}\epsilon, \tag{11.1.114}$$

$$\eta_2 \approx \omega t_0^{1/2}\left(1 + \frac{1}{2}\frac{t_1}{t_0}\epsilon\right) - \mathrm{i}\frac{1}{2t_0^{1/2}}\left(1 - \frac{t_1 + 2 - 2t_0}{2t_0}\epsilon\right). \tag{11.1.115}$$

Thus, for $t_1 \geq t_0 \gg 1$, $\omega \gg 1$ and $\epsilon \ll 1$, we obtain the following formulas for the phase velocities and damping coefficients

$$c_1 \sim \left(1 + \frac{t_1}{2t_0}\epsilon\right), \tag{11.1.116}$$

$$q_1 \sim \frac{t_1 - t_0 + 1}{2t_0^2}\epsilon, \tag{11.1.117}$$

$$c_2 \sim t_0^{-1/2}\left(1 - \frac{t_1}{2t_0}\epsilon\right), \tag{11.1.118}$$

$$q_2 \sim \frac{1}{2t_0^{1/2}}\left(1 - \frac{t_1 + 2 - 2t_0}{2t_0}\epsilon\right). \tag{11.1.119}$$

These formulas imply that the pairs (c_1, q_1) and (c_2, q_2) correspond, respectively, to a quasi-elastic and a quasi-thermal wave, similarly to the case (ii) ($\omega \gg 1$). In comparison to the "purely" elastic wave, which propagates with a unit velocity and without damping, a quasi-elastic wave possesses a higher velocity and a damping. The latter depend on the parameter ϵ and both relaxation times of the G–L theory.

On the other hand, the quasi-thermal wave possesses a lower velocity and a weaker damping than the corresponding "purely" thermal wave that propagates in an undeformed solid body. A comparison of asymptotic expansions for the cases (i)–(iii) implies that only for $\omega \ll 1$ (case (i)) does the pair (c_1, q_1) correspond to a quasi-thermal wave, while the pair (c_2, q_2) correspond to a quasi-elastic wave.

In the cases (ii) and (iii) the pairs (c_1, q_1) and (c_2, q_2) represent, respectively, the quasi-elastic and quasi-thermal waves. Explicit formulas for the pair $(c_i,\ q_i)(i = 1, 2)$ [see eqns (11.1.33) and (11.1.34)], together with the asymptotic relations obtained for the cases (i)–(iii), allow graphs of the functions c_i and q_i to be drawn in the parameter space $(t_0, t_1, \epsilon; \omega)$. The graphs of theses functions, with E and T indicating, respectively, the quasi-elastic and quasi-thermal wave, are shown in Figs. 11.1 and 11.2. Figure 11.1 displays the speeds c_E and c_T as functions of $t_0^{-1/2}$ for $\epsilon = 0.073$, $\omega = 10^5$, and $t_1/t_0 = 10.0$. Figure 11.2 displays

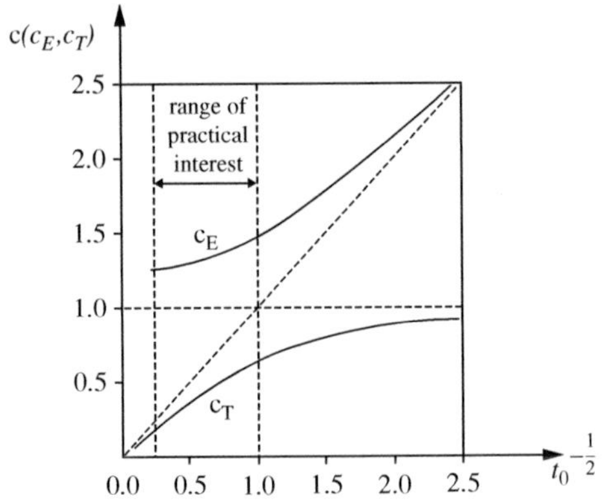

Figure 11.1 The speeds c_E and c_T as functions of $t_0^{-1/2}$ for $\epsilon = 0.073$, $\omega = 10^5$, and $t_1/t_0 = 10.0$.

the damping coefficients q_E and q_T as functions of $t_0^{-1/2}$ for the same set of input parameters.

11.2 Spherical waves produced by a concentrated source of heat in an infinite thermoelastic body with two relaxation times

Consider an unbounded homogeneous isotropic thermoelastic body with two relaxation times, referred to a spherical co-ordinate system (R, θ, φ). At the origin of the co-ordinate system there is a concentrated heat source characterized by a frequency $\omega > 0$. The source gives rise to the time-periodic

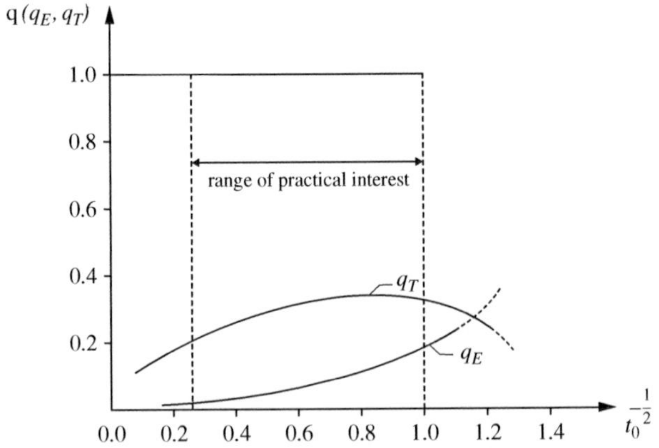

Figure 11.2 The damping coefficients q_E and q_T as functions of $t_0^{-1/2}$ for the same parameters as in Fig. 11.1.

potential–temperature spherical disturbances in terms of a pair (ϕ, ϑ) satisfying the equations[7]

$$\left.\begin{aligned} \Gamma\phi &= -\left(1 + t^0 \frac{\partial}{\partial t}\right)\delta(\mathbf{x})e^{i\omega t} \\ \left(\nabla^2 - \frac{\partial^2}{\partial t^2}\right)\phi &= \left(1 + t^0 \frac{\partial}{\partial t}\right)\vartheta \end{aligned}\right\} \quad \text{on } E^3, \tag{11.2.1}$$

and appropriate radiation conditions at infinity. Here

$$\Gamma = \left(\nabla^2 - \frac{\partial^2}{\partial t^2}\right)\left(\nabla^2 - t_0\frac{\partial^2}{\partial t^2} - \frac{\partial}{\partial t}\right) - \epsilon\nabla^2\frac{\partial}{\partial t}\left(1 + t^0\frac{\partial}{\partial t}\right) \tag{11.2.2}$$

is the *central operator* of the G–L theory (recall eqn (6.1.8)), where t_0 and t^0 are the relaxation times and ϵ is the parameter of thermoelastic coupling. Furthermore, $\delta = \delta(\mathbf{x})$ in $(11.2.1)_1$ denotes the Dirac delta in the Cartesian co-ordinate system x_i $(i = 1, 2, 3)$.

The radial displacement u_R and stresses σ_R, σ_θ and σ_φ are generated through the pair (ϕ, ϑ) according to the formulas (recall eqns (7.4.6) and (7.4.7))

$$u_R = \frac{\partial\phi}{\partial R}, \tag{11.2.3}$$

$$\begin{aligned} \sigma_R &= -\frac{2(1-2\nu)}{1-\nu}\frac{1}{R}u_R + \ddot{\phi}, \\ \sigma_\theta &= \frac{1-2\nu}{1-\nu}\left[\frac{1}{R}u_R - \left(1 + t^0\frac{\partial}{\partial t}\right)\vartheta\right] + \frac{\nu}{1-\nu}\ddot{\phi}, \\ \sigma_\varphi &= \sigma_\theta, \end{aligned} \tag{11.2.4}$$

where ν is the Poisson ratio. Given the spherical symmetry of these waves, the pair (ϕ, ϑ) depends on R and t only, while the operator ∇^2 appearing in eqns (11.2.1) and (11.2.2) takes the form

$$\nabla^2 = \frac{\partial^2}{\partial R^2} + \frac{2}{R}\frac{\partial}{\partial R}. \tag{11.2.5}$$

In the following, we assume $t_0 > 1$ (recall the inequality (11.1.11)).

It can now be shown that with the above assumptions the only solution to eqns (11.2.1) vanishing as $R \to \infty$ and representing a wave propagating from the

[7] See eqns (6.1.10) in which we let $t_1 = t^0$, $r = \delta(x)e^{i\omega t}$.

origin $\mathbf{x} = \mathbf{0}$ to infinity is the pair defined by

$$\phi(R,t) = \frac{e^{i\omega t}}{4\pi R} \frac{(1 + i\omega t^0)}{s_1^2(i\omega) - s_2^2(i\omega)} \{\exp[-Rs_1(i\omega)] - \exp[-Rs_2(i\omega)]\}, \qquad (11.2.6)$$

$$\vartheta(R,t) = \frac{e^{i\omega t}}{4\pi R} \frac{1}{s_1^2(i\omega) - s_2^2(i\omega)} \left\{ \left[s_1^2(i\omega) - (i\omega)^2\right] \exp\left[-Rs_1(i\omega)\right] - \left[s_2^2(i\omega) - (i\omega)^2\right] \exp\left[-Rs_2(i\omega)\right] \right\}, \qquad (11.2.7)$$

where the functions $s_{1.2}(i\omega)$ are given by the formulas (recall eqns (11.1.13) and (11.1.29)–(11.1.34))

$$s_{1.2}(i\omega) = q_{1.2} + i\frac{\omega}{c_{1.2}}. \qquad (11.2.8)$$

Here,

$$q_{1.2} = \frac{\sqrt{\omega}}{2}(\rho_{1.2} + a_{1.2})^{1/2}, \qquad (11.2.9)$$

$$c_{1.2} = 2\sqrt{\omega}(\rho_{1.2} - a_{1.2})^{1/2}, \qquad (11.2.10)$$

in which

$$\rho_{1.2} = \left\{ \left[(1 + t_0 + \epsilon t^0)\omega \pm \frac{\Delta^{1/2}}{\sqrt{2}}(\rho - a)^{1/2} \right]^2 + \left[(1+\epsilon) \pm \frac{\Delta^{1/2}}{\sqrt{2}}(\rho + a)^{1/2} \right]^2 \right\}^{1/2}, \qquad (11.2.11)$$

and

$$a_{1.2} = -\left[(1 + t_0 + \epsilon t^0)\omega \pm \frac{\Delta^{1/2}}{\sqrt{2}}(\rho - a)^{1/2} \right]. \qquad (11.2.12)$$

The parameters ρ and a in eqns (11.2.11) and (11.2.12) are specified by the formulas

$$\rho = [(\alpha^2 + \beta^2 - \omega^2)^2 + 4a^2\omega^2]^{1/2}, \qquad (11.2.13)$$

and

$$a = \alpha^2 + \beta^2 - \omega^2. \qquad (11.2.14)$$

Finally, the symbols α, β and Δ in eqns (11.2.11)–(11.2.14) are defined in Theorem 6.1 [recall eqns (6.2.4) and (6.2.5)]:

$$\begin{aligned} \alpha &= -[(1+\epsilon)(t_0 + \epsilon t^0) - (1-\epsilon)]\Delta^{-1}, \\ \beta &= 2\sqrt{\epsilon}[1 + (1+\epsilon)(t^0 - t_0)]^{1/2}\Delta^{-1}, \qquad (11.2.15) \\ \Delta &= (1 - t_0 + \epsilon t^0)^2 + 4\epsilon t_0 t^0. \end{aligned}$$

It follows from the formulas (11.2.9) and (11.2.10) that

$$c_{1.2} > 0 \qquad \text{and} \qquad q_{1.2} > 0, \tag{11.2.16}$$

which, on account of eqns (11.2.6) and (11.2.8), implies that the pair (ϕ, ϑ) can be represented in the form

$$(\phi, \vartheta) = (\phi^{(1)}, \vartheta^{(1)}) + (\phi^{(2)}, \vartheta^{(2)}), \tag{11.2.17}$$

where $(\phi^{(i)}, \vartheta^{(i)})$ $(i = 1, 2)$ corresponds to a spherical thermoelastic disturbance propagating from the origin of the co-ordinate system to infinity with the velocity c_i and damping q_i. Also, note that eqns (11.2.6) and (11.2.7) can be represented in an equivalent form

$$\phi(R,t) = \frac{\mathrm{e}^{\mathrm{i}\omega t}}{4\pi R}\left(\frac{1}{\upsilon_1^2} - \frac{1}{\upsilon_2^2}\right)^{-1} \frac{1 + \mathrm{i}\omega t^0}{\mathrm{i}\omega}[\overline{M}_1(R, \mathrm{i}\omega) - \overline{M}_2(R, \mathrm{i}\omega)], \tag{11.2.18}$$

$$\vartheta(R,t) = \frac{\mathrm{e}^{\mathrm{i}\omega t}}{4\pi R}\{\overline{N}_1(R, \mathrm{i}\omega) + \overline{N}_2(R, \mathrm{i}\omega) + (\hat{\alpha} + \hat{\beta}\mathrm{i}\omega)[\overline{M}_1(R, \mathrm{i}\omega) - \overline{M}_2(R, \mathrm{i}\omega)]\}, \tag{11.2.19}$$

where the functions $\overline{M}_i(R,p)$ and $\overline{N}_i(R,p)$, as well as the parameters $\upsilon_{1.2}$, $\hat{\alpha}$ and $\hat{\beta}$, are given by (recall eqns (7.4.22)–(7.4.27))

$$\overline{M}_i(R,p) = [(p - \alpha)^2 + \beta^2]^{-1/2} \exp[-Rs_i(p)], \tag{11.2.20}$$

$$\overline{N}_i(R,p) = \exp[-Rs_i(p)] \qquad (i = 1, 2), \tag{11.2.21}$$

$$\upsilon_{1.2}^{-2} = \frac{1}{2}(1 + t_0 + \epsilon t^0 \pm \Delta^{1/2}), \tag{11.2.22}$$

$$\hat{\alpha} = (1 + \epsilon)\Delta^{-1/2}, \tag{11.2.23}$$

$$\hat{\beta} = (t_0 + \epsilon t^0 - 1)\Delta^{-1/2}. \tag{11.2.24}$$

From the above and the formula

$$[(\mathrm{i}\omega - \alpha)^2 + \beta^2]^{-1/2} = \frac{1}{\rho\sqrt{2}}[(\rho + a)^{1/2} - \mathrm{i}(\rho - a)^{1/2}], \tag{11.2.25}$$

we infer that the pairs $(\phi^{(1)}, \vartheta^{(1)})$ and $(\phi^{(2)}, \vartheta^{(2)})$ are defined by the formulas

$$\phi^{(1)}(R,t) = \frac{m}{8\pi R}(A_\phi - \mathrm{i}B_\phi)\exp\left[-q_1 R + \mathrm{i}\omega\left(t - \frac{R}{c_1}\right)\right], \quad (11.2.26)$$

$$\phi^{(2)}(R,t) = -\frac{m}{8\pi R}(A_\phi - \mathrm{i}B_\phi)\exp\left[-q_2 R + \mathrm{i}\omega\left(t - \frac{R}{c_2}\right)\right], \quad (11.2.27)$$

$$\vartheta^{(1)}(R,t) = \frac{1}{8\pi R}(1 + A_\vartheta + \mathrm{i}B_\vartheta)\exp\left[-q_1 R + \mathrm{i}\omega\left(t - \frac{R}{c_1}\right)\right], \quad (11.2.28)$$

$$\vartheta^{(2)}(R,t) = \frac{1}{8\pi R}(1 - A_\vartheta - \mathrm{i}B_\vartheta)\exp\left[-q_2 R + \mathrm{i}\omega\left(t - \frac{R}{c_2}\right)\right], \quad (11.2.29)$$

where the pairs (A_ϕ, B_ϕ) and $(A_\vartheta, B_\vartheta)$ are real valued, specified through the following relations

$$A_\phi = \frac{1}{\rho\sqrt{2}}[\omega t^0(\rho + a)^{1/2} - (\rho - a)^{1/2}],$$
$$B_\phi = \frac{1}{\rho\sqrt{2}}[(\rho + a)^{1/2} + \omega t^0(\rho - a)^{1/2}], \quad (11.2.30)$$

$$A_\vartheta = \frac{1}{\rho\sqrt{2}}[\hat{\alpha}(\rho + a)^{1/2} + \omega\hat{\beta}(\rho - a)^{1/2}],$$
$$B_\vartheta = \frac{1}{\rho\sqrt{2}}[\omega\hat{\beta}(\rho + a)^{1/2} - \hat{\alpha}(\rho - a)^{1/2}]. \quad (11.2.31)$$

Furthermore,

$$m = \frac{2}{\omega}\left(\frac{1}{v_1^2} - \frac{1}{v_2^2}\right)^{-1}. \quad (11.2.32)$$

Now, the pairs $(\phi^{(1)}, \vartheta^{(1)})$ and $(\phi^{(2)}, \vartheta^{(2)})$ allow a determination of the radial displacement u_R as well as the stresses σ_R and σ_θ from the formulas (recall eqns (11.2.3) and (11.2.4))

$$u_R = u_R^{(1)} + u_R^{(2)},$$
$$\sigma_R = \sigma_R^{(1)} + \sigma_R^{(2)}, \quad (11.2.33)$$
$$\sigma_\theta = \sigma_\theta^{(1)} + \sigma_\theta^{(2)},$$

where

$$u_R^{(i)} = \frac{\partial \phi^{(i)}}{\partial R} \qquad (i = 1, 2), \tag{11.2.34}$$

$$\sigma_R^{(i)} = -\frac{4\varkappa}{R} u^{(i)} + \ddot{\phi}^{(i)} \qquad (i = 1, 2), \tag{11.2.35}$$

$$\sigma_\theta^{(i)} = 2\varkappa \left[\frac{u^{(i)}}{R} - \left(1 + t^0 \frac{\partial}{\partial t} \right) \vartheta^{(i)} \right] + (1 - 2\varkappa) \ddot{\phi}^{(i)} \qquad (i = 1, 2). \tag{11.2.36}$$

In the above

$$\varkappa = \frac{1 - 2\nu}{2 - 2\nu}. \tag{11.2.37}$$

The expressions (11.2.17), (11.2.26)–(11.2.29) and (11.2.33)–(11.2.37) represent a complex-valued solution corresponding to a complex-valued concentrated heat source of the form $\delta(\mathbf{x}) \exp(\mathrm{i}\omega t)$. If at the origin of the co-ordinate system acts a real-valued heat source of the form $\delta(\mathbf{x}) \cos(\omega t)$, it will give rise to a thermoelastic disturbance described by real parts of these formulas. Introducing the notations

$$\hat{\phi} = \mathrm{Re}\phi, \qquad \hat{\vartheta} = \mathrm{Re}\vartheta, \tag{11.2.38}$$

$$\hat{u}_R = \mathrm{Re} u_R, \tag{11.2.39}$$

$$\hat{\sigma}_R = \mathrm{Re}\sigma_R, \qquad \hat{\sigma}_\theta = \mathrm{Re}\sigma_\theta, \tag{11.2.40}$$

by virtue of eqns (11.2.17), (11.2.26)–(11.2.29) and (11.2.33)–(11.2.36), we obtain

$$\hat{\phi} = \hat{\phi}_1 + \hat{\phi}_2, \qquad \hat{\vartheta} = \hat{\vartheta}_1 + \hat{\vartheta}_2, \tag{11.2.41}$$

where

$$\hat{\phi}_1(R, t) = \frac{m}{8\pi R} \mathrm{e}^{-q_1 R} \left[A_\phi \cos\omega \left(t - \frac{R}{c_1} \right) + B_\phi \sin\omega \left(t - \frac{R}{c_1} \right) \right], \tag{11.2.42}$$

$$\hat{\phi}_2(R, t) = -\frac{m}{8\pi R} \mathrm{e}^{-q_2 R} \left[A_\phi \cos\omega \left(t - \frac{R}{c_2} \right) + B_\phi \sin\omega \left(t - \frac{R}{c_2} \right) \right], \tag{11.2.43}$$

$$\hat{\vartheta}_1(R, t) = \frac{1}{8\pi R} \mathrm{e}^{-q_1 R} \left[(1 + A_\vartheta) \cos\omega \left(t - \frac{R}{c_1} \right) - B_\vartheta \sin\omega \left(t - \frac{R}{c_1} \right) \right], \tag{11.2.44}$$

$$\hat{\vartheta}_2(R, t) = \frac{1}{8\pi R} \mathrm{e}^{-q_2 R} \left[(1 - A_\vartheta) \cos\omega \left(t - \frac{R}{c_2} \right) + B_\vartheta \sin\omega \left(t - \frac{R}{c_2} \right) \right], \tag{11.2.45}$$

and

$$\hat{u}_R = \hat{u}_R^{(1)} + \hat{u}_R^{(2)}, \tag{11.2.46}$$

in which

$$\hat{u}_R^{(1)}(R,t) = -\frac{m}{8\pi R^2}\mathrm{e}^{-q_1 R}\left\{\left[(1+q_1R)A_\phi + \frac{R\omega}{c_1}B_\phi\right]\cos\omega\left(t-\frac{R}{c_1}\right)\right.$$
$$\left. + \left[(1+q_1R)B_\phi - \frac{R\omega}{c_1}A_\phi\right]\sin\omega\left(t-\frac{R}{c_1}\right)\right\}, \tag{11.2.47}$$

$$\hat{u}_R^{(2)}(R,t) = \frac{m}{8\pi R^2}\mathrm{e}^{-q_2 R}\left\{\left[(1+q_2R)A_\phi + \frac{R\omega}{c_2}B_\phi\right]\cos\omega\left(t-\frac{R}{c_2}\right)\right.$$
$$\left. + \left[(1+q_2R)B_\phi - \frac{R\omega}{c_2}A_\phi\right]\sin\omega\left(t-\frac{R}{c_2}\right)\right\}. \tag{11.2.48}$$

For the radial stress $\hat{\sigma}_R$ and hoop stress $\hat{\sigma}_\theta$, respectively, we obtain

$$\hat{\sigma}_R = \hat{\sigma}_R^{(1)} + \hat{\sigma}_R^{(2)}, \tag{11.2.49}$$

where

$$\hat{\sigma}_R^{(1)}(R,t) = \frac{m}{8\pi R^3}\mathrm{e}^{-q_1 R}$$
$$\times\left\{\left\langle\left[4\varkappa(1+q_1R) - \omega^2R^2\right]A_\phi + 4\varkappa\frac{R\omega}{c_1}B_\phi\right\rangle\cos\omega\left(t-\frac{R}{c_1}\right)\right. \tag{11.2.50}$$
$$\left. + \left\langle\left[4\varkappa(1+q_1R) - \omega^2R^2\right]B_\phi - 4\varkappa\frac{R\omega}{c_1}A_\phi\right\rangle\sin\omega\left(t-\frac{R}{c_1}\right)\right\},$$

$$\hat{\sigma}_R^{(2)}(R,t) = -\frac{m}{8\pi R^3}\mathrm{e}^{-q_2 R}$$
$$\times\left\{\left\langle\left[4\varkappa(1+q_2R) - \omega^2R^2\right]A_\phi + 4\varkappa\frac{R\omega}{c_2}B_\phi\right\rangle\cos\omega\left(t-\frac{R}{c_2}\right)\right. \tag{11.2.51}$$
$$\left. + \left\langle\left[4\varkappa(1+q_2R) - \omega^2R^2\right]B_\phi - 4\varkappa\frac{R\omega}{c_2}A_\phi\right\rangle\sin\omega\left(t-\frac{R}{c_2}\right)\right\},$$

and

$$\hat{\sigma}_\theta = \hat{\sigma}_\theta^{(1)} + \hat{\sigma}_\theta^{(2)}, \tag{11.2.52}$$

where

$$\hat{\sigma}_\theta^{(1)}(R,t) = -\frac{m}{8\pi R^3}\mathrm{e}^{-q_1 R}\Bigg\{\Bigg\langle \left[2\varkappa(1+q_1R) + (1-2\varkappa)\omega^2R^2\right]A_\phi$$

$$+2\varkappa\frac{R\omega}{c_1}B_\phi + 2\varkappa\frac{R^2}{m}\left(1 + A_\vartheta - \omega t^0 B_\vartheta\right)\Bigg\rangle \cos\omega\left(t - \frac{R}{c_1}\right)$$

$$+\Bigg\langle \left[2\varkappa(1+q_1R) + (1-2\varkappa)\omega^2R^2\right]B_\phi \tag{11.2.53}$$

$$-2\varkappa\frac{R\omega}{c_1}A_\phi - 2\varkappa\frac{R^2}{m}(\omega t^0 + \omega t^0 A_\vartheta + B_\vartheta)\Bigg\rangle \sin\omega\left(t - \frac{R}{c_1}\right)\Bigg\},$$

$$\hat{\sigma}_\theta^{(2)}(R,t) = \frac{m}{8\pi R^3}\mathrm{e}^{-q_2 R}\Bigg\{\Bigg\langle \left[2\varkappa(1+q_2R) + (1-2\varkappa)\omega^2R^2\right]A_\phi$$

$$+2\varkappa\frac{R\omega}{c_2}B_\phi - 2\varkappa\frac{R^2}{m}\left(1 - A_\vartheta + \omega t^0 B_\vartheta\right)\Bigg\rangle \cos\omega\left(t - \frac{R}{c_2}\right)$$

$$+\Bigg\langle \left[2\varkappa(1+q_2R) + (1-2\varkappa)\omega^2R^2\right]B_\phi \tag{11.2.54}$$

$$-2\varkappa\frac{R\omega}{c_2}A_\phi + 2\varkappa\frac{R^2}{m}(\omega t^0 - \omega t^0 A_\vartheta - B_\vartheta)\Bigg\rangle \sin\omega\left(t - \frac{R}{c_2}\right)\Bigg\}.$$

Closed-form expressions of $\hat{\phi}$, $\hat{\vartheta}$, $\hat{u}_R$, $\hat{\sigma}_R$ and $\hat{\sigma}_\theta$ given, respectively, by eqns $(11.2.41)_1$, $(11.2.41)_2$, (11.2.46), (11.2.49) and (11.2.52) allow a complete (qualitative and quantitative) analysis of propagation of a spherical thermoelastic wave due to a concentrated heat source of the form $\delta(\mathbf{x})\cos(\omega t)$ in an infinite body. In particular, these formulas indicate that the wave treated as a function of R for $R > 0$ is a superposition of two spherical oscillating waves corresponding to the pairs (c_1, q_1) and (c_2, q_2) and vanishing at $R \to \infty$. This conclusion is also true for an arbitrary choice of parameters $(t, \omega; t_0, t^0, \epsilon)$ satisfying the inequalities

$$t > 0, \qquad \omega > 0, \qquad t^0 \geq t_0 > 1, \qquad \epsilon > 0. \tag{11.2.55}$$

In three special cases (i) $\omega \ll 1$, (ii) $\omega \gg 1$ and (iii) $\epsilon \ll 1$ the formulas (11.2.41)–(11.2.54) can substantially be simplified by using the asymptotic formulas from Section 11.1 for the pair (c_i, q_i) $(i = 1, 2)$.

11.3 Cylindrical waves produced by a line heat source in an infinite thermoelastic body with two relaxation times

Let us now refer an unbounded thermoelastic body to the cylindrical co-ordinate system (r, φ, z) and assume a line heat source to act along the entire axis z, having a unit intensity, being periodic in time, and independent of φ. This source gives rise to cylindrical thermoelastic waves that propagate under conditions of a plane strain, parallel to the $z = 0$ plane and independent of φ. These waves are generated through a pair (ϕ, ϑ) satisfying the equations (recall eqns (11.2.1) and (11.2.2))

$$\left.\begin{aligned} \Gamma\phi &= -\frac{\delta(r)}{2\pi r}\left(1 + t^0\frac{\partial}{\partial t}\right)\mathrm{e}^{\mathrm{i}\omega t} \\ \left(\nabla^2 - \frac{\partial^2}{\partial t^2}\right)\phi &= \left(1 + t^0\frac{\partial}{\partial t}\right)\vartheta \end{aligned}\right\} \quad \text{on} \quad \mathrm{E}^2, \tag{11.3.1}$$

and appropriate radiation conditions at infinity. Here

$$\Gamma = \left(\nabla^2 - \frac{\partial^2}{\partial t^2}\right)\left(\nabla^2 - t_0\frac{\partial^2}{\partial t^2} - \frac{\partial}{\partial t}\right) - \epsilon\nabla^2\frac{\partial}{\partial t}\left(1 + t^0\frac{\partial}{\partial t}\right), \tag{11.3.2}$$

and

$$\nabla^2 = \frac{\partial^2}{\partial r^2} + \frac{1}{r}\frac{\partial}{\partial r}. \tag{11.3.3}$$

Furthermore, $\delta = \delta(r)$ is the Dirac delta, while the remaining symbols in eqns (11.3.1) and (11.3.2) are the same as in Section 11.2.

The radial displacement u_r as well as the stresses σ_r, σ_φ, and σ_z are generated through the pair (ϕ, ϑ) according to the formulas (recall eqns (6.1.18) and (6.1.19), specialized to cylindrical co-ordinates for the case of plane strain, independent of φ)

$$u_r = \frac{\partial\phi}{\partial r}, \tag{11.3.4}$$

$$\left.\begin{aligned} \sigma_r &= -2\varkappa\frac{u_r}{r} + \ddot{\phi}, \\ \sigma_\varphi &= 2\varkappa\left[\frac{u_r}{r} - \left(1 + t^0\frac{\partial}{\partial t}\right)\vartheta\right] + (1 - 2\varkappa)\ddot{\phi}, \\ \sigma_z &= -2\varkappa\left(1 + t^0\frac{\partial}{\partial t}\right)\vartheta + (1 - 2\varkappa)\ddot{\phi}, \end{aligned}\right\} \tag{11.3.5}$$

in which (recall eqn (11.2.37))

$$\varkappa = \frac{1 - 2\nu}{2 - 2\nu}, \tag{11.3.6}$$

ν being Poisson's ratio. Similarly to what was done in Section 11.2, we assume that $t_0 > 1$.

Now, comparing the system of eqns (11.2.1) with eqn (11.3.1), we observe that latter can be obtained from the former upon integration along the axis z from $-\infty$ to ∞. This observation allows us to find a solution (ϕ, ϑ) to the system (11.3.1) by a direct integration along z of the relations (11.2.18) and (11.2.19) in which

$$R = \sqrt{r^2 + z^2}. \tag{11.3.7}$$

Thus, using the relation[8]

$$\int_{-\infty}^{\infty} \frac{\mathrm{e}^{-kR}}{R} \mathrm{d}z = 2K_0(kr) \quad \text{for} \quad \operatorname{Re} k > 0, \tag{11.3.8}$$

where $K_0 = K_0(x)$ is the modified Bessel function of the second kind and order zero and integrating eqns (11.2.18) and (11.2.19), we obtain

$$\phi(r,t) = \phi^{(1)}(r,t) + \phi^{(2)}(r,t), \tag{11.3.9}$$

$$\vartheta(r,t) = \vartheta^{(1)}(r,t) + \vartheta^{(2)}(r,t), \tag{11.3.10}$$

where

$$\phi^{(1)}(r,t) = \frac{m}{4\pi}\,\mathrm{e}^{\mathrm{i}\omega t}(A_\phi - \mathrm{i}B_\phi)K_0\left[s_1\left(\mathrm{i}\omega\right)r\right], \tag{11.3.11}$$

$$\phi^{(2)}(r,t) = -\frac{m}{4\pi}\,\mathrm{e}^{\mathrm{i}\omega t}(A_\phi - iB_\phi)K_0\left[s_2\left(\mathrm{i}\omega\right)r\right], \tag{11.3.12}$$

$$\vartheta^{(1)}(r,t) = \frac{\mathrm{e}^{\mathrm{i}\omega t}}{4\pi}(1 + A_\vartheta + iB_\vartheta)K_0\left[s_1\left(\mathrm{i}\omega\right)r\right], \tag{11.3.13}$$

$$\vartheta^{(2)}(r,t) = \frac{\mathrm{e}^{\mathrm{i}\omega t}}{4\pi}(1 - A_\vartheta - iB_\vartheta)K_0\left[s_2\left(\mathrm{i}\omega\right)r\right]. \tag{11.3.14}$$

Here (recall eqn (11.2.8))

$$s_{1.2}\left(\mathrm{i}\omega\right) = q_{1.2} + i\frac{\omega}{c_{1.2}}, \tag{11.3.15}$$

and the real-valued pairs $(A_\phi,\ B_\phi)$ and $(A_\vartheta,\ B_\vartheta)$ as well as the constant m are specified by the formulas (11.2.30)–(11.2.32).

On account of eqns (11.3.4), (11.3.9), (11.3.11) and (11.3.12), the radial displacement is given by

$$u_r(r,t) = u_r^{(1)}(r,t) + u_r^{(2)}(r,t), \tag{11.3.16}$$

[8] See formula (7.201) in (Stakgold, 1968).

where

$$u_r^{(1)}(r,t) = -\frac{m}{4\pi} e^{i\omega t}(A_\phi - iB_\phi) s_1(i\omega) K_1[s_1(i\omega) r], \qquad (11.3.17)$$

$$u_r^{(2)}(r,t) = \frac{m}{4\pi} e^{i\omega t}(A_\phi - iB_\phi) s_2(i\omega) K_1[s_2(i\omega) r]. \qquad (11.3.18)$$

The knowledge of $(u_r^{(i)}, \vartheta^{(i)})$ $(i = 1, 2)$ together with the function $\phi^{(i)}(i = 1, 2)$ allows the determination of stresses

$$\begin{aligned} \sigma_r &= \sigma_r^{(1)} + \sigma_r^{(2)}, \\ \sigma_\varphi &= \sigma_\varphi^{(1)} + \sigma_\varphi^{(2)}, \\ \sigma_z &= \sigma_z^{(1)} + \sigma_z^{(2)}, \end{aligned} \qquad (11.3.19)$$

in which, for $i = 1, 2$,

$$\begin{aligned} \sigma_r^{(i)} &= -2\varkappa \frac{u_r^{(i)}}{r} + \ddot{\phi}^{(i)}, \\ \sigma_\varphi^{(i)} &= 2\varkappa \left[\frac{u_r^{(i)}}{r} - \left(1 + t^0 \frac{\partial}{\partial t}\right) \vartheta^{(i)} \right] + (1 - 2\varkappa)\ddot{\phi}^{(i)}, \\ \sigma_z^{(i)} &= -2\varkappa \left(1 + t^0 \frac{\partial}{\partial t}\right) \vartheta^{(i)} + (1 - 2\varkappa)\ddot{\phi}^{(i)}. \end{aligned} \qquad (11.3.20)$$

The relations (11.3.9)–(11.3.20) describe cylindrical thermoelastic waves propagating in the body from the z-axis to infinity. Given the complicated form of these formulas, in the following we shall analyze the asymptotic form of these waves for $r \to \infty$. To this end, let us recall the asymptotic formulas

$$\begin{aligned} K_0(sr) &\approx \sqrt{\frac{\pi}{2}} (sr)^{-1/2} \exp(-sr), \\ K_1(sr) &\approx \sqrt{\frac{\pi}{2}} (sr)^{-1/2} \exp(-sr), \end{aligned} \qquad (11.3.21)$$

which hold for any complex number s such that $\mathrm{Re}s > 0$ and for $r \to \infty$. First, note that

$$[s_1(i\omega)]^{1/2} = \left(q_1 + i\frac{\omega}{c_1}\right)^{1/2} = C^{(1)} + iD^{(1)}, \qquad (11.3.22)$$

$$[s_2(i\omega)]^{1/2} = \left(q_2 + i\frac{\omega}{c_2}\right)^{1/2} = C^{(2)} + iD^{(2)}, \qquad (11.3.23)$$

where

$$C^{(1)} = \frac{1}{\sqrt{2}} \left[\left(q_1^2 + \frac{\omega^2}{c_1^2} \right)^{1/2} + q_1 \right]^{1/2}, \tag{11.3.24}$$

$$D^{(1)} = \frac{1}{\sqrt{2}} \left[\left(q_1^2 + \frac{\omega^2}{c_1^2} \right)^{1/2} - q_1 \right]^{1/2}, \tag{11.3.25}$$

$$C^{(2)} = \frac{1}{\sqrt{2}} \left[\left(q_2^2 + \frac{\omega^2}{c_2^2} \right)^{1/2} + q_2 \right]^{1/2}, \tag{11.3.26}$$

$$D^{(2)} = \frac{1}{\sqrt{2}} \left[\left(q_2^2 + \frac{\omega^2}{c_2^2} \right)^{1/2} - q_2 \right]^{1/2}. \tag{11.3.27}$$

From the formulas (11.3.22)–(11.3.27) we get

$$[s_1 (\mathrm{i}\omega)]^{-1/2} = \left(q_1^2 + \frac{\omega^2}{c_1^2} \right)^{-1/2} \left(C^{(1)} - \mathrm{i} D^{(1)} \right), \tag{11.3.28}$$

and

$$[s_2 (\mathrm{i}\omega)]^{-1/2} = \left(q_2^2 + \frac{\omega^2}{c_2^2} \right)^{-1/2} \left(C^{(2)} - \mathrm{i} D^{(2)} \right). \tag{11.3.29}$$

Substituting eqn $(11.3.21)_1$ into eqns (11.3.11)–(11.3.14) and eqn $(11.3.21)_2$ into eqns (11.3.17) and (11.3.18), and taking note of eqns (11.3.22)–(11.3.29), we obtain asymptotic formulas for the functions $\phi^{(i)}$, $\vartheta^{(i)}$ and $u_r^{(i)}$ valid in a neighborhood of $r = \infty$:

$$\begin{aligned} \phi^{(1)} &= \frac{m}{4} \frac{1}{\sqrt{2\pi r}} \left(A_\phi - iB_\phi \right) \left(C^{(1)} - \mathrm{i} D^{(1)} \right) \\ &\times \left(q_1^2 + \frac{\omega^2}{c_1^2} \right)^{-1/2} \exp \left[-q_1 r + \mathrm{i}\omega \left(t - \frac{r}{c_1} \right) \right], \end{aligned} \tag{11.3.30}$$

$$\begin{aligned} \phi^{(2)} &= -\frac{m}{4} \frac{1}{\sqrt{2\pi r}} \left(A_\phi - iB_\phi \right) \left(C^{(2)} - \mathrm{i} D^{(2)} \right) \\ &\times \left(q_2^2 + \frac{\omega^2}{c_2^2} \right)^{-1/2} \exp \left[-q_2 r + \mathrm{i}\omega \left(t - \frac{r}{c_2} \right) \right], \end{aligned} \tag{11.3.31}$$

$$\begin{aligned} \vartheta^{(1)} &= \frac{1}{4\sqrt{2\pi r}} \left(1 + A_\vartheta + iB_\vartheta \right) \left(C^{(1)} - \mathrm{i} D^{(1)} \right) \\ &\times \left(q_1^2 + \frac{\omega^2}{c_1^2} \right)^{-1/2} \exp \left[-q_1 r + \mathrm{i}\omega \left(t - \frac{r}{c_1} \right) \right], \end{aligned} \tag{11.3.32}$$

$$\vartheta^{(2)} = \frac{1}{4\sqrt{2\pi r}}\left(1 - A_\vartheta - \mathrm{i}B_\vartheta\right)\left(C^{(2)} - \mathrm{i}D^{(2)}\right) \times \left(q_2^2 + \frac{\omega^2}{c_2^2}\right)^{-1/2} \exp\left[-q_2 r + \mathrm{i}\omega\left(t - \frac{r}{c_2}\right)\right], \tag{11.3.33}$$

and

$$u_r^{(1)} = -\frac{m}{4\sqrt{2\pi r}}\left(A_\phi - \mathrm{i}B_\phi\right)\left(C^{(1)} + \mathrm{i}D^{(1)}\right)\exp\left[-q_1 r + \mathrm{i}\omega\left(t - \frac{r}{c_1}\right)\right], \tag{11.3.34}$$

$$u_r^{(2)} = \frac{m}{4\sqrt{2\pi r}}\left(A_\phi - \mathrm{i}B_\phi\right)\left(C^{(2)} + \mathrm{i}D^{(2)}\right)\exp\left[-q_2 r + \mathrm{i}\omega\left(t - \frac{r}{c_2}\right)\right]. \tag{11.3.35}$$

With the above, we can now determine $\sigma_r^{(i)}$, $\sigma_\varphi^{(i)}$ and $\sigma_z^{(i)}$ from the formulas

$$\begin{aligned}
\left(\sigma_r^{(\mathrm{i})} + \sigma_\varphi^{(\mathrm{i})}\right)/2 &= -\varkappa\left(1 + t^0\frac{\partial}{\partial t}\right)\vartheta^{(\mathrm{i})} + (1-\varkappa)\ddot{\phi}^{(\mathrm{i})}, \\
\left(\sigma_r^{(\mathrm{i})} - \sigma_\varphi^{(\mathrm{i})}\right)/2 &= -2\varkappa r^{-1}u_r^{(\mathrm{i})} + \varkappa\left(1 + t^0\frac{\partial}{\partial t}\right)\vartheta^{(\mathrm{i})} + \varkappa\ddot{\phi}^{(\mathrm{i})}, \\
\sigma_z^{(\mathrm{i})} &= -2\varkappa\left(1 + t^0\frac{\partial}{\partial t}\right)\vartheta^{(\mathrm{i})} + (1-2\varkappa)\ddot{\phi}^{(\mathrm{i})}.
\end{aligned} \tag{11.3.36}$$

The relations (11.3.30)–(11.3.36) describe the asymptotic form of complex thermoelastic disturbances propagating in the medium far away from the line heat source $r = 0$. If a real-valued heat source of the form $[\delta(r)/2\pi r]\cos(\omega t)$ acts along that line, the asymptotic form of disturbances is obtained by taking their real parts. Thus, denoting $\hat{f} = \mathrm{Re}f$ for any complex function f, we obtain

$$\hat{\phi}^{(1)}(r,t) = \frac{m e^{-q_1 r}}{4\sqrt{2\pi r}}\left(q_1^2 + \frac{\omega^2}{c_1^2}\right)^{-1/2} \times \left\{\left(A_\phi C^{(1)} - B_\phi D^{(1)}\right)\cos\omega\left(t - \frac{r}{c_1}\right) + \left(A_\phi D^{(1)} + B_\phi C^{(1)}\right)\sin\omega\left(t - \frac{r}{c_1}\right)\right\}, \tag{11.3.37}$$

$$\hat{\phi}^{(2)}(r,t) = -\frac{m e^{-q_2 r}}{4\sqrt{2\pi r}}\left(q_2^2 + \frac{\omega^2}{c_2^2}\right)^{-1/2} \times \left\{\left(A_\phi C^{(2)} - B_\phi D^{(2)}\right)\cos\omega\left(t - \frac{r}{c_2}\right) + \left(A_\phi D^{(2)} + B_\phi C^{(2)}\right)\sin\omega\left(t - \frac{r}{c_2}\right)\right\}, \tag{11.3.38}$$

$$\hat{\vartheta}^{(1)}(r,t) = \frac{\mathrm{e}^{-q_1 r}}{4\sqrt{2\pi r}}\left(q_1^2 + \frac{\omega^2}{c_1^2}\right)^{-1/2}$$
$$\times\left\{\left[(1+A_\vartheta)\,C^{(1)} + B_\vartheta D^{(1)}\right]\cos\omega\left(t-\frac{r}{c_1}\right)\right. \tag{11.3.39}$$
$$\left.+\left[(1+A_\vartheta)\,D^{(1)} - B_\vartheta C^{(1)}\right]\sin\omega\left(t-\frac{r}{c_1}\right)\right\},$$

$$\hat{\vartheta}^{(2)}(r,t) = \frac{\mathrm{e}^{-q_2 r}}{4\sqrt{2\pi r}}\left(q_2^2 + \frac{\omega^2}{c_2^2}\right)^{-1/2}$$
$$\times\left\{\left[(1-A_\vartheta)\,C^{(2)} - B_\vartheta D^{(2)}\right]\cos\omega\left(t-\frac{r}{c_2}\right)\right. \tag{11.3.40}$$
$$\left.+\left[(1-A_\vartheta)\,D^{(2)} + B_\vartheta C^{(2)}\right]\sin\omega\left(t-\frac{r}{c_2}\right)\right\},$$

and

$$\hat{u}_r^{(1)}(r,t) = -\frac{m\mathrm{e}^{-q_1 r}}{4\sqrt{2\pi r}}$$
$$\times\left[\left(A_\phi C^{(1)} + B_\phi D^{(1)}\right)\cos\omega\left(t-\frac{r}{c_1}\right) - \left(A_\phi D^{(1)} - B_\phi C^{(1)}\right)\sin\omega\left(t-\frac{r}{c_1}\right)\right], \tag{11.3.41}$$

$$\hat{u}_r^{(2)}(r,t) = \frac{m\mathrm{e}^{-q_2 r}}{4\sqrt{2\pi r}}$$
$$\times\left[\left(A_\phi C^{(2)} + B_\phi D^{(2)}\right)\cos\omega\left(t-\frac{r}{c_2}\right) - \left(A_\phi D^{(2)} - B_\phi C^{(2)}\right)\sin\omega\left(t-\frac{r}{c_2}\right)\right]. \tag{11.3.42}$$

Moreover, from eqn (11.3.36) we obtain

$$\left(\hat{\sigma}_r^{(1)} + \hat{\sigma}_\varphi^{(1)}\right)/2 = -\frac{\mathrm{e}^{-q_1 r}}{4\sqrt{2\pi r}}\left(q_1^2 + \frac{\omega^2}{c_1^2}\right)^{-1/2}$$
$$\times\left[\alpha_1\cos\omega\left(t-\frac{r}{c_1}\right) + \beta_1\sin\omega\left(t-\frac{r}{c_1}\right)\right], \tag{11.3.43}$$

where

$$\alpha_1 = \varkappa\left[\left(1 + A_\vartheta - \omega t^0 B_\vartheta\right)C^{(1)} + (B_\vartheta + \omega t^0(1+A_\vartheta))D^{(1)}\right]$$
$$+m(1-\varkappa)\omega^2(A_\phi C^{(1)} - B_\phi D^{(1)}), \tag{11.3.44}$$

$$\beta_1 = \varkappa\left[\left(1 + A_\vartheta - \omega t^0 B_\vartheta\right)D^{(1)} - (B_\vartheta + \omega t^0(1+A_\vartheta))C^{(1)}\right]$$
$$+m(1-\varkappa)\omega^2(A_\phi D^{(1)} + B_\phi C^{(1)}), \tag{11.3.45}$$

and

$$\left(\hat{\sigma}_r^{(2)} + \hat{\sigma}_\varphi^{(2)}\right)/2 = -\frac{\mathrm{e}^{-q_2 r}}{4\sqrt{2\pi r}}\left(q_2^2 + \frac{\omega^2}{c_2^2}\right)^{-1/2} \times \left[\alpha_2 \cos\omega\left(t - \frac{r}{c_2}\right) + \beta_2 \sin\omega\left(t - \frac{r}{c_2}\right)\right], \tag{11.3.46}$$

where

$$\alpha_2 = \varkappa\left[\left(1 - A_\vartheta + \omega t^0 B_\vartheta\right) C^{(2)} - (B_\vartheta - \omega t^0(1 - A_\vartheta))D^{(2)}\right] -m(1-\varkappa)\omega^2(A_\phi C^{(2)} - B_\phi D^{(2)}), \tag{11.3.47}$$

$$\beta_2 = \varkappa\left[\left(1 - A_\vartheta + \omega t^0 B_\vartheta\right) D^{(2)} + (B_\vartheta - \omega t^0(1 - A_\vartheta))C^{(2)}\right] -m(1-\varkappa)\omega^2(A_\phi D^{(2)} + B_\phi C^{(2)}). \tag{11.3.48}$$

We also have

$$\left(\hat{\sigma}_r^{(1)} - \hat{\sigma}_\varphi^{(1)}\right)/2 = \frac{\varkappa \mathrm{e}^{-q_1 r}}{4\sqrt{2\pi r}}\left(q_1^2 + \frac{\omega^2}{c_1^2}\right)^{-1/2} \times \left\{\gamma_1 \cos\omega\left(t - \frac{r}{c_1}\right) + \delta_1 \sin\omega\left(t - \frac{r}{c_1}\right) + \frac{2}{r}m\left(q_1^2 + \frac{\omega^2}{c_1^2}\right)^{1/2}\left[(A_\phi C^{(1)} + B_\phi D^{(1)})\cos\omega\left(t - \frac{r}{c_1}\right) -(A_\phi D^{(1)} - B_\phi C^{(1)})\sin\omega\left(t - \frac{r}{c_1}\right)\right]\right\}, \tag{11.3.49}$$

where

$$\gamma_1 = \left(1 + A_\vartheta - \omega t^0 B_\vartheta\right) C^{(1)} + (B_\vartheta + \omega t^0(1 + A_\vartheta))D^{(1)} -m\omega^2(A_\phi C^{(1)} - B_\phi D^{(1)}), \tag{11.3.50}$$

$$\delta_1 = \left(1 + A_\vartheta - \omega t^0 B_\vartheta\right) D^{(1)} - (B_\vartheta + \omega t^0(1 + A_\vartheta))C^{(1)} -m\omega^2(A_\phi D^{(1)} + B_\phi C^{(1)}), \tag{11.3.51}$$

and

$$\left(\hat{\sigma}_r^{(2)} - \hat{\sigma}_\varphi^{(2)}\right)/2 = \frac{\varkappa e^{-q_2 r}}{4\sqrt{2\pi r}}\left(q_2^2 + \frac{\omega^2}{c_2^2}\right)^{-1/2}$$
$$\times \left\{\gamma_2 \cos\omega\left(t - \frac{r}{c_2}\right) + \delta_2 \sin\omega\left(t - \frac{r}{c_2}\right)\right.$$
$$-\frac{2}{r}m\left(q_2^2 + \frac{\omega^2}{c_2^2}\right)^{1/2}\left[(A_\phi C^{(2)} + B_\phi D^{(2)})\cos\omega\left(t - \frac{r}{c_2}\right)\right.$$
$$\left.\left. -(A_\phi D^{(2)} - B_\phi C^{(2)})\sin\omega\left(t - \frac{r}{c_2}\right)\right]\right\}, \tag{11.3.52}$$

where

$$\gamma_2 = \left(1 - A_\vartheta + \omega t^0 B_\vartheta\right) C^{(2)} - (B_\vartheta - \omega t^0(1 - A_\vartheta))D^{(1)}$$
$$+ m\omega^2(A_\phi C^{(2)} - B_\phi D^{(2)}), \tag{11.3.53}$$

$$\delta_2 = \left(1 - A_\vartheta + \omega t^0 B_\vartheta\right) D^{(2)} + (B_\vartheta - \omega t^0(1 - A_\vartheta))C^{(2)}$$
$$+ m\omega^2(A_\phi D^{(2)} + B_\phi C^{(2)}). \tag{11.3.54}$$

Finally, for the axial stresses $\hat{\sigma}_z^{(i)}$ $(i = 1, 2)$ we obtain

$$\hat{\sigma}_z^{(1)} = -\frac{e^{-q_1 r}}{4\sqrt{2\pi r}}\left(q_1^2 + \frac{\omega^2}{c_1^2}\right)^{-1/2}$$
$$\times \left[\mu_1 \cos\omega\left(t - \frac{r}{c_1}\right) + \nu_1 \sin\omega\left(t - \frac{r}{c_1}\right)\right], \tag{11.3.55}$$

where

$$\mu_1 = 2\varkappa\left[\left(1 + A_\vartheta - \omega t^0 B_\vartheta\right) C^{(1)} + (B_\vartheta + \omega t^0(1 + A_\vartheta))D^{(1)}\right]$$
$$+ m(1 - 2\varkappa)\omega^2(A_\phi C^{(1)} - B_\phi D^{(1)}), \tag{11.3.56}$$

$$\nu_1 = 2\varkappa\left[\left(1 + A_\vartheta - \omega t^0 B_\vartheta\right) D^{(1)} - (B_\vartheta + \omega t^0(1 + A_\vartheta))C^{(1)}\right]$$
$$+ m(1 - 2\varkappa)\omega^2(A_\phi D^{(1)} + B_\phi C^{(1)}), \tag{11.3.57}$$

and

$$\hat{\sigma}_z^{(2)} = -\frac{e^{-q_2 r}}{4\sqrt{2\pi r}}\left(q_2^2 + \frac{\omega^2}{c_2^2}\right)^{-1/2}$$
$$\times \left[\mu_2 \cos\omega\left(t - \frac{r}{c_2}\right) + \nu_2 \sin\omega\left(t - \frac{r}{c_2}\right)\right], \tag{11.3.58}$$

where

$$\mu_2 = 2\varkappa \left[\left(1 - A_\vartheta + \omega t^0 B_\vartheta\right) C^{(2)} - (B_\vartheta - \omega t^0 (1 - A_\vartheta)) D^{(2)} \right] \\ -m(1-2\varkappa)\omega^2 (A_\phi C^{(2)} - B_\phi D^{(2)}), \tag{11.3.59}$$

$$\nu_2 = 2\varkappa \left[\left(1 - A_\vartheta + \omega t^0 B_\vartheta\right) D^{(2)} + (B_\vartheta - \omega t^0 (1 - A_\vartheta)) C^{(2)} \right] \\ -m(1-2\varkappa)\omega^2 (A_\phi D^{(2)} + B_\phi C^{(2)}). \tag{11.3.60}$$

The closed forms of functions (11.3.37)–(11.3.60) allow an asymptotic analysis of the cylindrical thermoelastic waves as $r \to \infty$ and for the parameters $(t, \omega; t_0, t^0, \epsilon)$ satisfying the inequalities

$$t > 0, \qquad \omega > 0, \qquad t^0 \geq t_0 > 1, \qquad \epsilon > 0. \tag{11.3.61}$$

The above solution is one of the simplest axially symmetric ones in the G–L theory. A more complex axially symmetric problem of time-periodic longitudinal waves in an infinite cylinder of a circular cross-section was solved in (Erbay and Şuhubi, 1986).

11.4 Integral representation of solutions and radiation conditions in the Green–Lindsay theory

11.4.1 *Integral representations and radiation conditions for the fundamental solution in the Green–Lindsay theory*

Let B denote a bounded domain in a 3D Euclidean space E^3. According to the definition in Section 7.1, the kth fundamental solution in the G–L theory satisfies the relations (recall eqns (7.1.4)–(7.1.6))

$$\hat{L}_k \phi_k = 0 \quad \text{on} \quad B \times [0, \infty), \tag{11.4.1}$$

$$\phi_k (\cdot, 0) = \dot{\phi}_k (\cdot, 0) = 0 \quad \text{on} \quad B, \tag{11.4.2}$$

$$\phi_k = f_k \quad \text{on} \quad \partial B \times [0, \infty). \tag{11.4.3}$$

Here, $\hat{L}_k$ $(k = 1, 2)$ is the wave-type operator (recall formulas (7.1.2) and (7.1.3)), while f_k is a function given on $\partial B \times [0, \infty)$. Let us now assume f_k to be periodic in time, that is

$$f_k(\mathbf{x}, t) = f_k^*(\mathbf{x}) e^{i\omega t}, \tag{11.4.4}$$

where $f_k^*(x)$ is a given function for every $x \in \partial B$ and $\omega > 0$.

In order to find a solution of eqns (11.4.1)–(11.4.3), with f_k given by eqn (11.4.4), we set

$$\phi_k(\mathbf{x}, t) = \varphi_k(\mathbf{x}, t) + \phi_k^*(\mathbf{x}) e^{i\omega t}, \tag{11.4.5}$$

where φ_k represents a "transient" part of ϕ_k, while ϕ_k^* is an "amplitude" of the harmonic part of ϕ_k. Substituting eqn (11.4.5) into eqns (11.4.1)–(11.4.3), we conclude that ϕ_k is a fundamental solution of the G–L theory if φ_k satisfies the relations

$$\hat{L}_k \varphi_k = 0 \quad \text{on} \quad \mathrm{B} \times (0, \infty), \tag{11.4.6}$$

$$\left.\begin{aligned} \varphi_k(\cdot, 0) &= -\phi_k^*(\cdot) \\ \dot{\varphi}_k(\cdot, 0) &= -\mathrm{i}\omega\phi_k^*(\cdot) \end{aligned}\right\} \quad \text{on} \quad \mathrm{B}, \tag{11.4.7}$$

$$\varphi_k = 0 \quad \text{on} \quad \partial\mathrm{B} \times (0, \infty), \tag{11.4.8}$$

with the function $\phi_k^*(\mathbf{x})$ being such that

$$\Box_k^2 \phi_k^* = 0 \quad \text{on} \quad \mathrm{B}, \tag{11.4.9}$$

$$\phi_k^* = f_k^* \quad \text{on} \quad \partial\mathrm{B}. \tag{11.4.10}$$

The operator $\Box_k^2$ appearing in eqn (11.4.9) is defined by the formula

$$\Box_k^2 = \nabla^2 + \eta_k^2, \tag{11.4.11}$$

where (recall eqns (11.1.13) and (11.1.32))

$$\eta_k = \frac{\omega}{c_k} - \mathrm{i}q_k \qquad (k = 1, 2), \tag{11.4.12}$$

while c_k and q_k denote, respectively, the phase velocity and the damping coefficient of the kth fundamental solution, all specified by the relations (11.1.33) and (11.1.34).

If we note the relation between η_k and $s_k(\mathrm{i}\omega)$ (recall eqn (11.1.13))

$$s_k(\mathrm{i}\omega) = \mathrm{i}\eta_k \qquad (k = 1, 2), \tag{11.4.13}$$

we find that a fundamental solution of the G–L theory, corresponding to a periodic-in-time boundary condition, is obtained by solving a Dirichlet problem for the Helmholtz equation with a complex wave number of the form (11.4.12).

In the following we will drop the asterisk in eqns (11.4.9) and (11.4.10), that is, we will work with

$$\Box_k^2 \phi_k = 0 \quad \text{on} \quad \mathrm{B}, \tag{11.4.14}$$

$$\phi_k = f_k \quad \text{on} \quad \partial\mathrm{B}. \tag{11.4.15}$$

In order to obtain an integral representation of the function $\phi_k(\mathbf{x})$ satisfying eqn (11.4.14), consider now the systems of equations (recall eqns (8.1.7) and (8.1.8))

$$\Box_k^2 \phi_k = 0 \quad \text{on} \quad \mathrm{B}, \tag{11.4.16}$$

$$\Box_k^2 K_k = -\delta(\mathbf{x} - \mathbf{y}) \quad \text{on} \quad \mathrm{E}^3, \tag{11.4.17}$$

where $\delta = \delta(\mathbf{x})$ is the Dirac delta and $\mathbf{y}$ is a fixed point in E^3. Multiplying eqn (11.4.16) through by K_k and eqn (11.4.17) by ϕ_k, subtracting one from another and integrating over B, we obtain (recall eqns (8.1.10) and (8.1.11))

$$\epsilon(\mathbf{y})\phi_k(\mathbf{y}) = \int\limits_{\partial \mathrm{B}} [K_k(\mathbf{x},\mathbf{y})\frac{\partial}{\partial n}\phi_k(\mathbf{x}) - \phi_k(\mathbf{x})\frac{\partial}{\partial n}K_k(\mathbf{x},\mathbf{y})]\mathrm{d}a(\mathbf{x}), \qquad (11.4.18)$$

where

$$\epsilon(\mathbf{y}) = \begin{cases} 1 \text{ for } & \mathbf{y} \in \mathrm{B}, \\ 0 \text{ for } \mathbf{y} \in \mathrm{E}^3 - \bar{\mathrm{B}}. \end{cases} \qquad (11.4.19)$$

One can show that the only solution of eqn (11.4.17) satisfying a radiation condition, that is representing a wave propagating from $\mathbf{y}$ to infinity, is the function (recall eqn (8.1.12))

$$K_k(\mathbf{x},\mathbf{y}) = \frac{1}{4\pi R}\exp(-\mathrm{i}\eta_k R), \qquad (11.4.20)$$

in which

$$R = |\mathbf{x} - \mathbf{y}|\,. \qquad (11.4.21)$$

Thus, we conclude that the following theorem is true.

Theorem 11.2 *(Helmholtz-type theorem for the fundamental solutions in the G–L theory) An arbitrary solution of eqn (11.4.16) in a bounded domain B can be represented by the surface integral*

$$\phi_k(\mathbf{y}) = \frac{1}{4\pi}\int\limits_{\partial B}\left[\left(\frac{\mathrm{e}^{-\mathrm{i}\eta_k R}}{R}\right)\frac{\partial}{\partial n}\phi_k(\mathbf{x}) - \phi_k(\mathbf{x})\frac{\partial}{\partial n}\left(\frac{\mathrm{e}^{-\mathrm{i}\eta_k R}}{R}\right)\right] da(\mathbf{x}) \quad \forall \mathbf{y} \in B. \qquad (11.4.22)$$

If we seek the kth fundamental solution of the G–L theory in the exterior of B, that is in $\mathrm{B}_e = \mathrm{E}^3 - \mathrm{B}$, under the boundary condition (11.4.4), in order to find a harmonic part of that solution corresponding to a wave propagating from the boundary ∂B to infinity, we postulate the so-called *radiation conditions of Sommerfeld type.*

Let $\mathbf{x} \in \mathrm{E}^3 - \bar{\mathrm{B}}$ and let B_r denote a ball centered at $\mathbf{x}$ and having a radius sufficiently large that B is contained in B_r. Next, let V_r denote a domain bounded by the surfaces ∂B and $\partial \mathrm{B}_r$, and consider the equations

$$\Box_k^2\phi_k = 0 \quad \text{on} \quad \mathrm{V}_r, \qquad (11.4.23)$$

$$\Box_k^2 K_k = -\delta(\mathbf{x} - \mathbf{y}) \quad \text{on} \quad \mathrm{E}^3, \qquad (11.4.24)$$

with $\mathbf{y}$ being a fixed point of E^3. Combining these equations with one another, and integrating similarly to what was done in the case of formula (11.4.22), we obtain

$$\phi_k(\mathbf{y}) = \frac{1}{4\pi} \int\limits_{\partial B \cup \partial B_r} \left[\left(\frac{e^{-i\eta_k R}}{R} \right) \frac{\partial}{\partial n} \phi_k(\mathbf{x}) - \phi_k(\mathbf{x}) \frac{\partial}{\partial n} \left(\frac{e^{-i\eta_k R}}{R} \right) \right] da(\mathbf{x}) \quad \forall \mathbf{y} \in V_r. \tag{11.4.25}$$

On the ball's surface ∂B_r : $R = r$ and $\partial/\partial n = \partial/\partial r$, where r is a component of the spherical co-ordinate system (r, θ, φ) centered at $\mathbf{y}$. Thus,

$$\int\limits_{\partial B_r} [\cdot] da = \int\limits_{\partial B_r} \frac{e^{-i\eta_k r}}{r} \left(\frac{\partial}{\partial r} + i\eta_k \right) \phi_k \, d\Sigma + \int\limits_{\partial B_r} \frac{e^{-i\eta_k r}}{r^2} \phi_k \, d\Sigma, \tag{11.4.26}$$

where

$$d\Sigma = r^2 \sin\theta d\theta d\varphi, \qquad 0 \le \theta \le \pi, \qquad 0 \le \varphi \le 2\pi. \tag{11.4.27}$$

Clearly, the integral (11.4.26) converges to zero for $r \to \infty$ provided the function ϕ_k, referred to the system (r, θ, φ), satisfies the radiation conditions[9]

$$r e^{-i\eta_k r} \left(\frac{\partial}{\partial r} + i\eta_k \right) \phi_k = o(1), \qquad e^{-i\eta_k r} \phi_k = o(1), \tag{11.4.28}$$

for every θ and φ satisfying the inequalities (11.4.27).[10] Therefore, passing in eqn (11.4.25) with r to infinity and given that $V_r \to E^3 - B$ for $r \to \infty$, we conclude that the following theorem is true.

Theorem 11.2 *An arbitrary solution of eqn (11.4.23) outside a bounded domain B, satisfying the radiation conditions (11.4.28), can be represented by the surface integral*

$$\phi_k(\mathbf{y}) = \frac{1}{4\pi} \int\limits_{\partial B} \left[\left(\frac{e^{-i\eta_k R}}{R} \right) \frac{\partial}{\partial n} \phi_k(\mathbf{x}) - \phi_k(\mathbf{x}) \frac{\partial}{\partial n} \left(\frac{e^{-i\eta_k R}}{R} \right) \right] da(\mathbf{x}) \quad \forall \mathbf{y} \in E^3 - B. \tag{11.4.29}$$

This theorem implies that, in the presence of a periodic boundary condition (11.4.4), the harmonic part of the kth fundamental solution of the G–L theory, corresponding to a wave propagating from the boundary ∂B to infinity, has the form of a product $\phi_k(y) \exp(i\omega t)$, where $\phi_k(y)$ is given by eqn (11.4.29).

[9] The relation $f(r) = o(1)$ means that $f(r) \to 0$ as $r \to \infty$.

[10] It is easy to show that $\phi_k = r^{-1} \exp(-i\eta_k r)$ satisfies the conditions (11.4.28), so that the function $\phi_k \exp(i\omega t)$, $\omega > 0$, represents a wave propagating from the point $y = 0$ to infinity.

11.4.2 *Integral representations and radiation conditions for the potential–temperature solution in the Green–Lindsay theory*

According to the definition introduced in Section 4.2, the potential–temperature problem in the G–L theory consists in finding a pair (ϕ, ϑ) on $\mathrm{B} \times [0, \infty)$ that satisfies the equations

$$\left.\begin{aligned} &\Gamma\phi = 0 \\ &\left(\nabla^2 - \frac{\partial^2}{\partial t^2}\right)\phi = \left(1 + t^0\frac{\partial}{\partial t}\right)\vartheta \end{aligned}\right\} \quad \text{on } \mathrm{B} \times [0, \infty), \tag{11.4.30}$$

$$\left.\begin{aligned} \phi(\cdot, 0) = \dot{\phi}(\cdot, 0) = 0 \\ \vartheta(\cdot, 0) = \dot{\vartheta}(\cdot, 0) = 0 \end{aligned}\right\} \quad \text{on } \mathrm{B}, \tag{11.4.31}$$

$$\phi = f \quad \vartheta = g \quad \text{on } \partial\mathrm{B} \times [0, \infty). \tag{11.4.32}$$

Here, Γ is the central operator of the G–L theory (recall eqn (6.1.8)) while f and g are known functions. Assume these functions to have the form

$$f(\mathbf{x}, t) = f^*(\mathbf{x})\mathrm{e}^{\mathrm{i}\omega t}, \quad g(\mathbf{x}, t) = g^*(\mathbf{x})\mathrm{e}^{\mathrm{i}\omega t}, \tag{11.4.33}$$

with $f^*(\mathbf{x})$ and $g^*(\mathbf{x})$ given on $\partial\mathrm{B}$.

In order to find a solution (ϕ, ϑ) of the system (11.4.30)–(11.4.33), we proceed similarly to what was done in Section 11.4.1. Thus, we assume the pair (ϕ, ϑ) to be specified by the formulas

$$\phi(\mathbf{x}, t) = \varphi(\mathbf{x}, t) + \phi^*(\mathbf{x})\mathrm{e}^{\mathrm{i}\omega t}, \tag{11.4.34}$$

$$\vartheta(\mathbf{x}, t) = \theta(\mathbf{x}, t) + \vartheta^*(\mathbf{x})\mathrm{e}^{\mathrm{i}\omega t}, \tag{11.4.35}$$

where the pair (φ, θ) represents a "transient" part of the sought solution, while (ϕ^*, ϑ^*) is a harmonic "amplitude" of that solution. Inserting eqns (11.4.34) and (11.4.35) into eqns (11.4.30)–(11.4.32) and using eqn (11.4.33), we infer that the pair (ϕ, ϑ), given by eqns (11.4.34) and (11.4.35), satisfies eqns (11.4.30)–(11.4.32) provided the pairs (φ, θ) and (ϕ^*, ϑ^*) satisfy the relations

$$\left.\begin{aligned} &\Gamma\varphi = 0 \\ &\left(\nabla^2 - \frac{\partial^2}{\partial t^2}\right)\varphi = \left(1 + t^0\frac{\partial}{\partial t}\right)\theta \end{aligned}\right\} \quad \text{on } \mathrm{B} \times [0, \infty), \tag{11.4.36}$$

$$\left.\begin{aligned} \varphi(\cdot, 0) = -\phi^*(\cdot), \quad \dot{\varphi}(\cdot, 0) = -\mathrm{i}\omega\phi^*(\cdot) \\ \theta(\cdot, 0) = -\vartheta^*(\cdot), \quad \dot{\theta}(\cdot, 0) = -\mathrm{i}\omega\vartheta^*(\cdot) \end{aligned}\right\} \quad \text{on } \mathrm{B}, \tag{11.4.37}$$

$$\varphi = \theta = 0 \quad \text{on } \partial\mathrm{B} \times (0, \infty), \tag{11.4.38}$$

and

$$\left.\begin{aligned} &\Box_1^2\Box_2^2\phi^* = 0 \\ &(\nabla^2 + \omega^2)\phi^* = (1 + \mathrm{i}\omega t^0)\vartheta^* \end{aligned}\right\} \quad \text{on } \mathrm{B}, \tag{11.4.39}$$

$$\phi^* = f^*, \quad \vartheta^* = g^* \quad \text{on } \partial\mathrm{B}. \tag{11.4.40}$$

It is clear that a solution of the original problem stated in terms of eqns (11.4.30)–(11.4.33) will depend on first solving the problem (11.4.39) and (11.4.40) and determining the initial conditions of the problem (11.4.63)–(11.4.38). The knowledge of the pair (ϕ^*, ϑ^*) will allow us, in turn, to find the pair (φ, θ) using the expansion into a series of eigenfunctions via the method already proposed in Section 7.2.

In the following we shall consider an integral representation of the pair (ϕ^*, ϑ^*), whereby, for convenience, we drop the asterisks. Thus, we have

$$\left.\begin{aligned} \Box_1^2\Box_2^2\phi &= 0 \\ (\nabla^2+\omega^2)\phi &= (1+\mathrm{i}\omega t^0)\vartheta \end{aligned}\right\} \quad \text{on B,} \tag{11.4.41}$$

$$\phi = f, \qquad \vartheta = g \qquad \text{on } \partial\mathrm{B}, \tag{11.4.42}$$

wherein the operators $\Box_k^2$ $(k = 1, 2)$ are defined by eqn (11.4.11). Together with eqns (11.4.41) and (11.4.42), let us consider the system

$$\left.\begin{aligned} \Box_1^2\Box_2^2\phi_N &= -(1+\mathrm{i}\omega t^0)\delta(\mathbf{x}-\mathbf{y}) \\ (\nabla^2+\omega^2)\phi_N &= (1+\mathrm{i}\omega t^0)\vartheta_N \end{aligned}\right\} \quad \text{on } \mathrm{E}^3, \tag{11.4.43}$$

where $\mathbf{y}$ is a fixed point in E^3, while the pair (ϕ_N, ϑ_N) satisfies suitable radiation conditions at infinity. It is easy to verify that the systems (11.4.41) and (11.4.43) can be obtained by setting $p = \mathrm{i}\omega$, respectively, in the systems (8.3.10) and (8.3.12) and identifying the pair (ϕ, ϑ) with $(\bar{\phi}, \bar{\vartheta})$, and the pair (ϕ_N, ϑ_N) with $(\bar{\phi}_N, \bar{\vartheta}_N)$ of Section 8.3. This observation leads us to conclude that the following theorem is true.

Theorem 11.3 *(Helmholtz-type theorem for the pair (ϕ, ϑ)) An arbitrary solution of eqn (11.4.41) for a bounded domain B can be represented by the surface integrals*

$$\begin{aligned} \phi(\mathbf{y}) = &\int\limits_{\partial B} \left[K_N(\mathbf{x},\mathbf{y})\frac{\partial}{\partial n}\phi(\mathbf{x}) - \phi(\mathbf{x})\frac{\partial}{\partial n}K_N(\mathbf{x},\mathbf{y})\right] da(\mathbf{x}) \\ &+ \int\limits_{\partial B} \left[\phi_N(\mathbf{x},\mathbf{y})\frac{\partial}{\partial n}\vartheta(\mathbf{x}) - \vartheta(\mathbf{x})\frac{\partial}{\partial n}\phi_N(\mathbf{x},\mathbf{y})\right] da(\mathbf{x}), \end{aligned} \tag{11.4.44}$$

$$\begin{aligned} \vartheta(\mathbf{y}) = &\int\limits_{\partial B} \left[\vartheta_N(\mathbf{x},\mathbf{y})\frac{\partial}{\partial n}\vartheta(\mathbf{x}) - \vartheta(\mathbf{x})\frac{\partial}{\partial n}\vartheta_N(\mathbf{x},\mathbf{y})\right] da(\mathbf{x}) \\ &+ \int\limits_{\partial B} \left[H_N(\mathbf{x},\mathbf{y})\frac{\partial}{\partial n}\phi(\mathbf{x}) - \phi(\mathbf{x})\frac{\partial}{\partial n}H_N(\mathbf{x},\mathbf{y})\right] da(\mathbf{x}) \end{aligned} \tag{11.4.45}$$

for every point $\mathbf{y} \in B$.

In the above, the kernels ϕ_N, ϑ_N, K_N and H_N are given by the formulas (recall eqns (8.3.25), (8.3.26), (8.3.29) and (8.3.37))

$$\phi_N = -\frac{(1 + \mathrm{i}\omega t^0)}{4\pi R} \frac{\mathrm{e}^{-\mathrm{i}\eta_1 R} - \mathrm{e}^{-\mathrm{i}\eta_2 R}}{\eta_1^2 - \eta_2^2}, \tag{11.4.46}$$

$$\vartheta_N = \frac{(\eta_1^2 - \omega^2)\mathrm{e}^{-\mathrm{i}\eta_1 R} - (\eta_2^2 - \omega^2)\mathrm{e}^{-\mathrm{i}\eta_2 R}}{4\pi R(\eta_1^2 - \eta_2^2)}, \tag{11.4.47}$$

$$K_N = -\frac{(\eta_2^2 - \omega^2)\mathrm{e}^{-\mathrm{i}\eta_1 R} - (\eta_1^2 - \omega^2)\mathrm{e}^{-\mathrm{i}\eta_2 R}}{4\pi R(\eta_1^2 - \eta_2^2)}, \tag{11.4.48}$$

$$H_N = \frac{i\epsilon\omega^3}{4\pi R} \frac{\mathrm{e}^{-\mathrm{i}\eta_1 R} - \mathrm{e}^{-\mathrm{i}\eta_2 R}}{\eta_1^2 - \eta_2^2}, \tag{11.4.49}$$

where (recall eqn (11.4.21))

$$R = |\mathbf{x} - \mathbf{y}|, \tag{11.4.50}$$

and η_k $(k = 1, 2)$ are complex numbers of the form (recall eqn (11.4.12))

$$\eta_k = \frac{\omega}{c_k} - iq_k \qquad (c_k > 0, \quad q_k > 0). \tag{11.4.51}$$

One can show that a thermoelastic wave generated through the pair (ϕ_N, ϑ_N) propagates from $\mathbf{y}$ to infinity, that is, it satisfies suitable Sommerfeld radiation conditions in E^3.

Now, in connection with the construction of an integral representation of Helmholtz type for the pair (ϕ, ϑ) in the exterior of the domain B, we shall derive an explicit form of the said conditions. This construction will be analogous to that for the pair (ϕ, ϑ) defined by eqns (11.4.44) and (11.4.45). First, one can show that, if V_r is the domain introduced in Section 11.4.1 and $\mathbf{y} \in \mathrm{V}_r$, then

$$\begin{aligned} \phi(\mathbf{y}) = &\int\limits_{\partial\mathrm{B}\cup\partial\mathrm{B}_r} \left[K_N(\mathbf{x},\mathbf{y})\frac{\partial}{\partial n}\phi(\mathbf{x}) - \phi(\mathbf{x})\frac{\partial}{\partial n}K_N(\mathbf{x},\mathbf{y})\right] da(\mathbf{x}) \\ &+ \int\limits_{\partial\mathrm{B}\cup\partial\mathrm{B}_r} \left[\phi_N(\mathbf{x},\mathbf{y})\frac{\partial}{\partial n}\vartheta(\mathbf{x}) - \vartheta(\mathbf{x})\frac{\partial}{\partial n}\phi_N(\mathbf{x},\mathbf{y})\right] da(\mathbf{x}), \end{aligned} \tag{11.4.52}$$

$$\begin{aligned} \vartheta(\mathbf{y}) = &\int\limits_{\partial\mathrm{B}\cup\partial\mathrm{B}_r} \left[\vartheta_N(\mathbf{x},\mathbf{y})\frac{\partial}{\partial n}\vartheta(\mathbf{x}) - \vartheta(\mathbf{x})\frac{\partial}{\partial n}\vartheta_N(\mathbf{x},\mathbf{y})\right] da(\mathbf{x}) \\ &+ \int\limits_{\partial\mathrm{B}\cup\partial\mathrm{B}_r} \left[H_N(\mathbf{x},\mathbf{y})\frac{\partial}{\partial n}\phi(\mathbf{x}) - \phi(\mathbf{x})\frac{\partial}{\partial n}H_N(\mathbf{x},\mathbf{y})\right] da(\mathbf{x}), \end{aligned} \tag{11.4.53}$$

where $\partial\mathrm{B}_r$ is the surface of a ball introduced in Section 11.4.1.

In order to examine the right-hand sides of these relations for $r \to \infty$ we note that, on account of eqns (11.4.46) and (11.4.48), the surface integral over $\partial \mathrm{B}_r$ appearing in eqn (11.4.52), up to a constant, equals

$$
\begin{aligned}
I_1 = \int_{\partial \mathrm{B}_r} \Bigg\{ & \left[\frac{(\eta_2^2 - \omega^2)\mathrm{e}^{-\mathrm{i}\eta_1 r} - (\eta_1^2 - \omega^2)\mathrm{e}^{-\mathrm{i}\eta_2 r}}{r} \right] \frac{\partial}{\partial r}\phi \\
& - \phi \frac{\partial}{\partial r} \left[\frac{(\eta_2^2 - \omega^2)\mathrm{e}^{-\mathrm{i}\eta_1 r} - (\eta_1^2 - \omega^2)\mathrm{e}^{-\mathrm{i}\eta_2 r}}{r} \right] \\
& + (1 + \mathrm{i}\omega t^0) \left[\left(\frac{\mathrm{e}^{-\mathrm{i}\eta_1 r} - \mathrm{e}^{-\mathrm{i}\eta_2 r}}{r} \right) \frac{\partial}{\partial r}\vartheta \right. \\
& \left. - \vartheta \frac{\partial}{\partial r} \left(\frac{\mathrm{e}^{-\mathrm{i}\eta_1 r} - \mathrm{e}^{-\mathrm{i}\eta_2 r}}{r} \right) \right] \Bigg\} \mathrm{d}\Sigma.
\end{aligned} \tag{11.4.54}
$$

This integral can be rewritten as

$$
\begin{aligned}
I_1 = \int_{\partial \mathrm{B}_r} \Bigg\{ & (\eta_2^2 - \omega^2) \left[\frac{\mathrm{e}^{-\mathrm{i}\eta_1 r}}{r} \left(\frac{\partial}{\partial r} + \mathrm{i}\eta_1 \right) \phi + \frac{\mathrm{e}^{-\mathrm{i}\eta_1 r}}{r^2} \phi \right] \\
& - (\eta_1^2 - \omega^2) \left[\frac{\mathrm{e}^{-\mathrm{i}\eta_2 r}}{r} \left(\frac{\partial}{\partial r} + \mathrm{i}\eta_2 \right) \phi + \frac{\mathrm{e}^{-\mathrm{i}\eta_2 r}}{r^2} \phi \right] \\
& + (1 + \mathrm{i}\omega t^0) \left[\frac{\mathrm{e}^{-\mathrm{i}\eta_1 r}}{r} \left(\frac{\partial}{\partial r} + \mathrm{i}\eta_1 \right) \vartheta + \frac{\mathrm{e}^{-\mathrm{i}\eta_2 r}}{r^2} \vartheta \right. \\
& \left. - \frac{\mathrm{e}^{-\mathrm{i}\eta_2 r}}{r} \left(\frac{\partial}{\partial r} + \mathrm{i}\eta_2 \right) \vartheta - \frac{\mathrm{e}^{-\mathrm{i}\eta_2 r}}{r^2} \vartheta \right] \Bigg\} \mathrm{d}\Sigma.
\end{aligned} \tag{11.4.55}
$$

Given that $\mathrm{d}\Sigma = r^2 \sin\theta \mathrm{d}\theta \mathrm{d}\varphi$ [(recall eqn (11.4.27)], the integral I_1 tends to zero as $r \to \infty$, so long as the pair (ϕ, ϑ) satisfies the radiation conditions

$$
\begin{aligned}
& r\mathrm{e}^{-\mathrm{i}\eta_1 r} \left(\frac{\partial}{\partial r} + \mathrm{i}\eta_1 \right) \left[(\eta_2^2 - \omega^2)\phi + (1 + \mathrm{i}\omega t^0)\vartheta \right] \\
& - r\mathrm{e}^{-\mathrm{i}\eta_2 r} \left(\frac{\partial}{\partial r} + \mathrm{i}\eta_2 \right) \left[(\eta_1^2 - \omega^2)\phi + (1 + \mathrm{i}\omega t^0)\vartheta \right] = o(1),
\end{aligned} \tag{11.4.56}
$$

$$
\begin{aligned}
& \mathrm{e}^{-\mathrm{i}\eta_1 r} \left[(\eta_2^2 - \omega^2)\phi + (1 + \mathrm{i}\omega t^0)\vartheta \right] \\
& - \mathrm{e}^{-\mathrm{i}\eta_2 r} \left[(\eta_1^2 - \omega^2)\phi + (1 + \mathrm{i}\omega t^0)\vartheta \right] = o(1).
\end{aligned} \tag{11.4.57}
$$

Let us also note that, on account of eqns (11.4.47) and (11.4.49), the integral over $\partial \mathrm{B}_r$ appearing in eqn (11.4.53), up to a constant, equals

$$
\begin{aligned}
I_2 = \int\limits_{\partial \mathrm{B}_r} \Bigg\{ & \left[\frac{(\eta_1^2-\omega^2)\mathrm{e}^{-\mathrm{i}\eta_1 r} - (\eta_2^2-\omega^2)\mathrm{e}^{-\mathrm{i}\eta_2 r}}{r} \right] \frac{\partial}{\partial r}\vartheta \\
& -\vartheta \frac{\partial}{\partial r} \left[\frac{(\eta_1^2-\omega^2)\mathrm{e}^{-\mathrm{i}\eta_1 r} - (\eta_2^2-\omega^2)\mathrm{e}^{-\mathrm{i}\eta_2 r}}{r} \right] \\
& + i\epsilon\omega^3 \left[\left(\frac{\mathrm{e}^{-\mathrm{i}\eta_1 r} - \mathrm{e}^{-\mathrm{i}\eta_2 r}}{r} \right) \frac{\partial}{\partial r}\phi - \phi \frac{\partial}{\partial r} \left(\frac{\mathrm{e}^{-\mathrm{i}\eta_1 r} - \mathrm{e}^{-\mathrm{i}\eta_2 r}}{r} \right) \right] \Bigg\} \mathrm{d}\Sigma.
\end{aligned}
\tag{11.4.58}
$$

The above integral can be rewritten as

$$
\begin{aligned}
I_2 = \int\limits_{\partial \mathrm{B}_r} \Bigg\{ & (\eta_1^2-\omega^2) \left[\frac{\mathrm{e}^{-\mathrm{i}\eta_1 r}}{r} \left(\frac{\partial}{\partial r} + \mathrm{i}\eta_1 \right) \vartheta + \frac{\mathrm{e}^{-\mathrm{i}\eta_1 r}}{r^2}\vartheta \right] \\
& -(\eta_2^2-\omega^2) \left[\frac{\mathrm{e}^{-\mathrm{i}\eta_2 r}}{r} \left(\frac{\partial}{\partial r} + \mathrm{i}\eta_2 \right) \vartheta + \frac{\mathrm{e}^{-\mathrm{i}\eta_2 r}}{r^2}\vartheta \right] \\
& + i\epsilon\omega^3 \left[\frac{\mathrm{e}^{-\mathrm{i}\eta_1 r}}{r} \left(\frac{\partial}{\partial r} + \mathrm{i}\eta_1 \right) \phi + \frac{\mathrm{e}^{-\mathrm{i}\eta_1 r}}{r^2}\phi \right. \\
& \left. - \frac{\mathrm{e}^{-\mathrm{i}\eta_2 r}}{r} \left(\frac{\partial}{\partial r} + \mathrm{i}\eta_2 \right) \phi - \frac{\mathrm{e}^{-\mathrm{i}\eta_2 r}}{r^2}\phi \right] \Bigg\} \mathrm{d}\Sigma.
\end{aligned}
\tag{11.4.59}
$$

A sufficient condition for the integral I_2 to tend to zero as $r \to \infty$ is that

$$
\begin{aligned}
& r\mathrm{e}^{-\mathrm{i}\eta_1 r} \left(\frac{\partial}{\partial r} + \mathrm{i}\eta_1 \right) \left[(\eta_1^2-\omega^2)\vartheta + i\epsilon\omega^3\phi \right] \\
& -r\mathrm{e}^{-\mathrm{i}\eta_2 r} \left(\frac{\partial}{\partial r} + \mathrm{i}\eta_2 \right) \left[(\eta_2^2-\omega^2)\vartheta + i\epsilon\omega^3\phi \right] = o(1),
\end{aligned}
\tag{11.4.60}
$$

$$
\begin{aligned}
& \mathrm{e}^{-\mathrm{i}\eta_1 r} \left[(\eta_1^2-\omega^2)\vartheta + i\epsilon\omega^3\phi \right] \\
& -\mathrm{e}^{-\mathrm{i}\eta_2 r} \left[(\eta_2^2-\omega^2)\vartheta + i\epsilon\omega^3\phi \right] = o(1).
\end{aligned}
\tag{11.4.61}
$$

The relations (11.4.56), (11.4.57), (11.4.60) and (11.4.61) are called the *radiation conditions of Sommerfeld type for the pair (ϕ, ϑ) in the G–L theory.* If a regular solution of the system of eqns (11.4.41) on $\mathrm{E}^3 - \mathrm{B}$ is defined as a pair (ϕ, ϑ) that satisfies eqns (11.4.41) on $\mathrm{E}^3 - \mathrm{B}$ subject to the radiation conditions (11.4.56), (11.4.57), (11.4.60) and (11.4.61), then, upon passing to the limit in the relations (11.4.52) and (11.4.53) as $r \to \infty$, we obtain the following theorem.

Theorem 11.4 *(Helmholtz-type theorem for a pair (ϕ, ϑ)) An arbitrary solution (ϕ, ϑ) of the system (11.4.41), which is regular in $E^3 - B$, can be represented* *by the surface integrals*

$$\begin{aligned}\phi(\mathbf{y}) &= \int_{\partial B} \left[K_N(\mathbf{x},\mathbf{y}) \frac{\partial}{\partial n}\phi(\mathbf{x}) - \phi(\mathbf{x}) \frac{\partial}{\partial n} K_N(\mathbf{x},\mathbf{y}) \right] da(\mathbf{x}) \\ &+ \int_{\partial B} \left[\phi_N(\mathbf{x},\mathbf{y}) \frac{\partial}{\partial n}\vartheta(\mathbf{x}) - \vartheta(\mathbf{x}) \frac{\partial}{\partial n} \phi_N(\mathbf{x},\mathbf{y}) \right] da(\mathbf{x}),\end{aligned} \tag{11.4.62}$$

$$\begin{aligned}\vartheta(\mathbf{y}) &= \int_{\partial B} \left[\vartheta_N(\mathbf{x},\mathbf{y}) \frac{\partial}{\partial n}\vartheta(\mathbf{x}) - \vartheta(\mathbf{x}) \frac{\partial}{\partial n} \vartheta_N(\mathbf{x},\mathbf{y}) \right] da(\mathbf{x}) \\ &+ \int_{\partial B} \left[H_N(\mathbf{x},\mathbf{y}) \frac{\partial}{\partial n}\phi(\mathbf{x}) - \phi(\mathbf{x}) \frac{\partial}{\partial n} H_N(\mathbf{x},\mathbf{y}) \right] da(\mathbf{x}),\end{aligned} \tag{11.4.63}$$

where the kernels ϕ_N, ϑ_N, K_N and H_N are defined, respectively, by the formulas (11.4.46), (11.4.47), (11.4.48) and (11.4.49).

By analyzing a relation between the radiation conditions (11.4.28) for the kth fundamental solution of the G–L theory and the conditions (11.4.56), (11.4.57), (11.4.60) and (11.4.61), one can show that the following theorem is also true.

Theorem 11.5 *Let $\phi = \phi_1 + \phi_2$ and $\vartheta = \vartheta_1 + \vartheta_2$, where ϕ_k is the kth fundamental solution of the G–L theory in $E^3 - B$ and ϑ_k is the temperature associated with ϕ_k. Then, the pair (ϕ, ϑ) is a regular solution of the system (11.4.41) in $E^3 - B$.*

This theorem clearly shows that by looking for a pair (ϕ, ϑ) in $\mathrm{E}^3 - \mathrm{B}$ as a sum of the fundamental pairs (ϕ_k, ϑ_k) subject to the radiation conditions (11.4.28), we also satisfy the coupled radiation conditions (11.4.56), (11.4.57), (11.4.60) and (11.4.61).[11]

Remark 11.1 In this chapter we have analyzed plane, spherical and cylindrical harmonic waves propagating in an unbounded domain as well as the integral representation of a harmonic solution to a potential–temperature boundary value problem of the G–L theory. A comparison of plane harmonic waves propagating in an infinite body for the three thermoelastic theories: classical, L–S, and G–L, was presented in (Haddow and Wegner, 1996). Stability of plane harmonic generalized thermoelastic waves was discussed in (Leslie and Scott, 2000, 2004). Harmonic

[11] The coupled radiation conditions (11.4.56), (11.4.57), (11.4.60) and (11.4.61) are less restrictive than the analogous radiation conditions of asymmetric isothermal elastodynamics proposed in (Ignaczak, 1970). Also, note that the radiation conditions for the kth fundamental solution of classical coupled thermoelasticity are given in (Ignaczak and Nowacki, 1962), while the coupled radiation conditions for a pair (ϕ, ϑ) in classical coupled thermoelasticity and in L–S theory are given in (Rożnowski, 1971, 1983). Finally, note that a general form of the radiation conditions for classical coupled thermoelasticity is proposed in the monograph (Kupradze *et al.*, 1979).

thermoelastic vibrations of an infinite body with a spherical cavity within the G–L theory were discussed in (Erbay, Erbay, and Dost, 1991). Free harmonic vibrations of a solid thermoelastic sphere within the G–L theory were studied in (Soyucok, 1991). Propagation of harmonic waves in an unbounded thermoelastic domain with a thermal memory was discussed in (Shashkov and Yanovsky, 1994). Reflection and refraction of harmonic thermoelastic waves at an interface of two semi-infinite media with two relaxation times were analyzed in (Sinha and Elsibai, 1997). Axially symmetric thermoelastic waves produced by a time-harmonic boundary normal point load and thermal source in a half-space within the L–S and G–L theories, were discussed in (Sharma, Chauhan, and Kumar, 2000). A harmonic fundamental solution in the theory of micropolar thermoelasticity without energy dissipation was obtained in (Svanadze, Tibullo, and Zampoli, 2006). Thermoelastic damping in nanomechanical resonators with finite wave speeds was studied in (Khisaeva and Ostoja-Starzewski, 2006); see Section 12.4. Finally, the time-periodic, 2D cylindrical electro-magneto-thermoelastic waves in an annular region, within the thermoelasticity without energy dissipation, were studied in (Allam, Elsibai and Abouelregal, 2007).

12

PHYSICAL ASPECTS AND APPLICATIONS OF HYPERBOLIC THERMOELASTICITY

12.1 Heat conduction

12.1.1 *Physics viewpoint and other theories*

The theories with one and two relaxation times discussed in this book are special cases of numerous theories proposed to deal with the wave-like transport of heat beginning with the work of Maxwell (1871). The activity in this direction was renewed following the experiments on *second sound* in liquid helium by Peshkov (1944), the term originating from the wave motion of heat being similar to the propagation of sound in air. This phenomenon – quantum mechanical in nature – involves heat transfer occurring by wave-like motion, rather than by diffusion. As a result, helium II has the highest thermal conductivity of any known material (several hundred times higher than copper). Peshkov then suggested that second sound might take place in materials that have a phonon gas. The constant t_0 of the L–S theory may be interpreted as the time required to establish the steady state of heat conduction in a volume element suddenly subjected to a temperature gradient. Chester (1963) quantitatively estimated t_0 in terms of measurable macroscopic parameters to be

$$t_0 = \frac{3k}{\upsilon^2 \rho c}, \tag{12.1.1}$$

where υ is the phonon velocity, k the thermal conductivity, ρ the mass density, and c the heat capacity per unit mass. It is important to mention that eqn (12.1.1) can only be used for a medium where the transport of heat occurs via the phonon gas, which is usually the case for micro- and nano-electromechanical systems (MEMS/NEMS). To the first approximation, υ can be replaced by the elastic wave velocity (Lifshitz and Roukes, 2000).

It must be noted that we still do not possess a full understanding of classical Fourier-type heat conduction in terms of microscopic physics. This was observed by Peierls (1979) and still remains a challenge, see e.g. (Klages, 2007). However, there are many known cases where the phenomenon of second sound has been experimentally observed; this includes solid He3 and He4, bismuth, lithium fluoride, sodium fluoride, sodium iodide and crystals of quartz and sapphire; Caviglia *et al.* (1992) and Dreyer and Struchtrup (1993). Not surprisingly, thermal wave propagation is a low-temperature phenomenon, typically in the range 1–20 K. As pointed out by Caviglia *et al.* (1992), the thermal wave speed decreases appreciably as the temperature increases and tends to an asymptotic value as absolute zero is approached. Also, Caviglia *et al.* (1992) discuss the experiments of Jackson *et al.* (1970) and McNelly *et al.* (1970) involving a pure crystal of NaF where three distinct waves are observed: a longitudinal quasi-elastic wave travelling fastest, followed by a transverse quasi-elastic wave and then a longitudinal quasi-temperature wave. It appears that the quasi-temperature and transverse waves travel at the same speed below 8 K.

Much work was stimulated by the early papers of physicists and mechanicians, and several reviews have appeared in the past two decades: (Joseph and Preziosi, 1989, 1990) focusing on heat conduction with finite speeds in a rigid body and (Ignaczak, 1980, 1981, 1987; Chandrasekharaiah, 1986, 1998; Hetnarski and Ignaczak, 1999, 2000) focusing on generalized thermoelasticity. Very recently, Miller and Haber (2008) have used space-time finite elements to study continuous and discontinuous thermal waves in 2D problems with rapid and localized heating of the conducting medium governed by the Maxwell–Cattaneo equation.

In a series of papers Green and Naghdi (1991, 1992, 1993) developed the following three thermoelasticity theories:

- Their (1991) paper developed a theory for describing the behavior of a thermoelastic body, which relies on an entropy balance law rather than on an entropy inequality. The novel quantity is a *thermal displacement variable*

$$\alpha = \int_{t_0}^{t} T(\mathbf{X}, \tau)\mathrm{d}\tau + \alpha_0,$$

  in which T is the "empirical" temperature, with $\mathbf{X}$ being the spatial coordinate in the reference configuration of the body.
- Their (1993) article developed another theory called *thermoelasticity of type II* allowing for heat transmission at finite speed without any energy dissipation. The displacement–temperature equations of that theory are

$$\begin{aligned} \rho\ddot{u}_i &= (C_{ijkh}u_{k,h})_{,j} - (a_{ij}\theta)_{,j} + \rho f_i \,, \\ c\ddot{\theta} &= -a_{ij}\ddot{u}_{i,j} + (k_{ik}\theta_{,k})_{,i} + \rho r \,, \end{aligned} \tag{12.1.2}$$

  with the elasticity tensor C_{ijkh} assumed positive-definite; $c > 0$. Other authors who further pursued this theory are Chandrasekharaiah (1996a;b), Ieşan (1998), Nappa (1998) and Quintanilla (1999). See also Section 12.5.1.
- Their (1992) paper developed yet another thermoelasticity theory, based on their 1991 work, called *thermoelasticity of type III* allowing for heat transport at a finite speed but with dissipation taking place. The displacement–temperature equations of this theory are

$$\begin{aligned} \rho\ddot{u}_i &= (C_{ijkh}u_{k,h})_{,j} - (a_{ij}\theta)_{,j} + \rho f_i \,, \\ c\ddot{\theta} &= -a_{ij}\ddot{u}_{i,j} + (k_{ik}\theta_{,k})_{,i} + (b_{ik}\dot{\theta}_{,k})_{,i} + \rho r \,, \end{aligned} \tag{12.1.3}$$

  with the elasticity tensor C_{ijkh} assumed positive-definite.

Quintanilla and Straughan (2000) provided new results (uniqueness, growth) for anisotropic linearized versions of thermoelasticity of type II and type III without requiring any definiteness whatsoever of the elasticity tensor, but only assuming $C_{ijkh} = C_{khij}$.

This section would not be complete without noting a paper by Scott (2008) who studied classical and generalized anisotropic thermoelasticity (G–L model) from the standpoint of Whitham's wave hierarchies; see also (Leslie and Scott, 2004). He casts the standard form of the displacement–temperature partial differential equations in four unknowns to a single equation in one variable, in terms of isothermal and isentropic wave operators.

Now, let us note two interesting consequences, one stemming from classical physics considerations and another from modern continuum thermodynamics.

12.1.2 *Consequence of Galilean invariance*

Let us begin with the Maxwell–Cattaneo equation, introduced in this book as (R1.1.8)

$$\left(1 + t_0 \frac{\partial}{\partial t}\right) q_i = -k_{ij} \frac{\partial \theta}{\partial x_j}. \tag{12.1.4}$$

As observed recently by (Christov and Jordan, 2005), the presence of a partial derivative $\partial/\partial t$ in that equation leads to a violation of the Galilean invariance, and the problem is resolved by using, instead, a material derivative $\mathcal{D}/\mathcal{D}t$. In the following, we retrace the arguments of these authors.

First, restrict the attention to an isotropic rigid heat conductor ($k_{ij} = k\delta_{ij}$) and consider a process of heat conduction in one dimension (x). In that case, $q_i = (q, 0, 0)$ and eqn (12.1.4) takes the form

$$\left(1 + t_0 \frac{\partial}{\partial t}\right) q = -k \frac{\partial \theta}{\partial x}. \tag{12.1.5}$$

The balance equation for the internal energy is postulated in the form involving the material derivative $\mathcal{D}/\mathcal{D}t$

$$\rho c_p \frac{\mathcal{D}\theta}{\mathcal{D}t} = -\frac{\partial q}{\partial x}. \tag{12.1.6}$$

where

$$\frac{\mathcal{D}}{\mathcal{D}t} = \frac{\partial}{\partial t} + v \frac{\partial}{\partial x} \tag{12.1.7}$$

and $v = v(x, t)$ is the velocity of a material point in the x direction. By eliminating q from eqns (12.1.5) and (12.1.6), we obtain the hyperbolic equation for θ

$$t_0 \left[\frac{\partial^2 \theta}{\partial t^2} + \frac{\partial}{\partial t}\left(v \frac{\partial \theta}{\partial x}\right)\right] + \frac{\partial \theta}{\partial t} + v \frac{\partial \theta}{\partial x} = \kappa \frac{\partial^2 \theta}{\partial x^2}, \tag{12.1.8}$$

where $\kappa = k/\rho c_p$. Clearly, in the case of a continuum at rest ($v = 0$), the above reduces to a damped wave equation

$$t_0 \frac{\partial^2 \theta}{\partial t^2} + \frac{\partial \theta}{\partial t} = \kappa \frac{\partial^2 \theta}{\partial x^2}. \tag{12.1.9}$$

Next, introduce the dimensionless quantities

$$x' = x/l, \quad t' = t/t_0, \quad \theta' = \theta/\theta_0, \quad v' = v/\sqrt{\kappa/t_0}, \tag{12.1.10}$$

where $l = \sqrt{\kappa t_0}$ and $\theta_0 > 0$ denote the characteristic length and reference temperature, respectively. Upon substitution into eqn (12.1.8), and omitting all the primes for simplicity of notation, we obtain

$$\frac{\partial^2 \theta}{\partial t^2} + \frac{\partial}{\partial t}\left(v \frac{\partial \theta}{\partial x}\right) + \frac{\partial \theta}{\partial t} + v \frac{\partial \theta}{\partial x} = \frac{\partial^2 \theta}{\partial x^2}. \tag{12.1.11}$$

Let us now assume a rigid motion of a half-space ($x \in [0, \infty)$) occupied by the rigid heat conductor: translation along the x-axis with constant velocity $v(x,t) = U$. Then, eqn (12.1.11) leads to

$$\frac{\partial^2 \theta}{\partial t^2} + U \frac{\partial^2 \theta}{\partial t \partial x} + \frac{\partial \theta}{\partial t} + U \frac{\partial \theta}{\partial x} = \frac{\partial^2 \theta}{\partial x^2}, \tag{12.1.12}$$

which exhibits a paradoxical property of the thermal wave speeds being nonlinear functions of U:

$$c_{1,2} = \frac{1}{2}\left(U \pm \sqrt{U^2 + 4}\right), \quad c_1 > \max(U, 0), \quad c_2 < \min(U, 0). \tag{12.1.13}$$

Furthermore, consider an initial-boundary value problem for eqn (12.1.12) with $U = 0$ in a half-space $x \geq 0$ subject to a heat source at the boundary $x = 0$. Also, consider an analogous initial-boundary value problem for a half-space subject to a heat source at the boundary $x = 0$ but moving with the velocity U. According to Galileo's principle of classical relativity, the propagation of second sound in the moving half-space should be exactly the same as that in the half-space at rest. Since the governing equations for these two problems are different, their solutions are also different, and Galileo's principle is violated. To resolve the paradox, eqn (12.1.5) is replaced by

$$\left(1 + t_0 \frac{\mathcal{D}}{\mathcal{D}t}\right) q = -k \frac{\partial \theta}{\partial x}. \tag{12.1.14}$$

Then, by eliminating q from eqns (12.1.6) and (12.1.14) we obtain

$$t_0 \frac{\partial^2 \theta}{\partial t^2} + t_0 \frac{\partial}{\partial t}\left(v \frac{\partial \theta}{\partial x}\right) + t_0 v \frac{\partial}{\partial t}\left(\frac{\partial \theta}{\partial x}\right) + t_0 v \frac{\partial}{\partial x}\left(v \frac{\partial \theta}{\partial x}\right) + \frac{\partial \theta}{\partial t} + v \frac{\partial \theta}{\partial x} = \kappa \frac{\partial^2 \theta}{\partial x^2} \tag{12.1.15}$$

and, for the case of a constant velocity $v(x,t) = U$, the corresponding dimensionless equation is

$$\frac{\partial^2\theta}{\partial t^2} + 2U\frac{\partial^2\theta}{\partial t\partial x} + \frac{\partial\theta}{\partial t} + U\frac{\partial\theta}{\partial x} = (1 - U^2)\frac{\partial^2\theta}{\partial x^2}. \tag{12.1.16}$$

In contradistinction to eqn (12.1.13), the wave speeds are

$$c_{1,2} = U \pm 1, \tag{12.1.17}$$

i.e. the sum or difference of the dimensionless frame velocity and thermal wave speed, as should be the case for a body moving with a velocity U.

The argument can be extended to a 3D case, where, in place of eqn (12.1.4), we generally have

$$q_i + t_0\left(\frac{\partial q_i}{\partial t} + v_j\frac{\partial q_i}{\partial x_j}\right) \equiv \left(1 + t_0\frac{\mathcal{D}}{\mathcal{D}t}\right)q_i = -k_{ij}\frac{\partial\theta}{\partial x_j}, \tag{12.1.18}$$

where v_j is the velocity vector of a material particle. Clearly, in the case of a linear (small strains and temperature changes) thermoelasticity the convective term in the material derivative may be neglected, and this justifies the approach adopted throughout this book.

12.1.3 *Consequence of continuum thermodynamics*

The Maxwell–Cattaneo equation, involving the material derivative postulated in Section 12.1.2, can be derived directly from thermodynamics (Ostoja-Starzewski, 2009c). In the following we adopt the thermodynamics with internal variables (TIV), see (Ziegler and Wehrli, 1987; Maugin, 1999). To this end, one needs to assume that the (specific, per unit mass) internal energy e is a function of the strains E_{ij}, entropy η and heat flux q_i

$$e = e(E_{ij}, \eta, q_i). \tag{12.1.19}$$

In other words, the state space needs to be extended to include q_i (Maugin, 2008). Furthermore, the (specific, per unit mass) dissipation ϕ needs to be taken as a function of the strain rate, the heat flux and its rate:

$$\phi = \phi(\dot{E}_{ij}, q_i, \dot{q}_i). \tag{12.1.20}$$

We are focusing on thermoelasticity, so in the above we do not need to admit other fluxes. An overdot in this section denotes a material derivative, alternately written as $\mathcal{D}/\mathcal{D}t$ in Section 12.1.2.

Now, the specific power of deformation is defined by

$$l = S_{ij}\dot{E}_{ij}, \tag{12.1.21}$$

while the classical relation holds

$$\psi = e - \theta\eta, \tag{12.1.22}$$

whereby, in view of eqn (12.1.19), we recognize that

$$\psi = \psi(E_{ij}, \theta, q_i) \quad \text{and} \quad \eta = \eta(E_{ij}, \theta, q_i). \tag{12.1.23}$$

The first fundamental law takes the form

$$\dot{e} = S_{ij}\dot{E}_{ij} - q_{i,i}\,, \tag{12.1.24}$$

where $\dot{E}_{ij}$ is the deformation rate, while the second fundamental law is written in terms of the reversible ($\eta^{*(r)}$) and irreversible ($\eta^{*(i)}$) parts of the entropy production rate ($\dot{\eta}$)

$$\dot{\eta} = \eta^{*(r)} + \eta^{*(i)} \quad \text{with} \quad \eta^{*(r)} = -\left(\frac{q_i}{\theta}\right)_{,i} \quad \text{and} \quad \eta^{*(i)} \geq 0. \tag{12.1.25}$$

Proceeding just like in TIV, from eqns (12.1.24) and (12.1.25) we obtain

$$\dot{e} - \theta\dot{\eta} = S_{ij}\dot{E}_{ij} - \frac{q_i\theta_{,i}}{\theta} - \theta\eta^{*(i)}, \tag{12.1.26}$$

or

$$\begin{aligned} S_{ij}\dot{E}_{ij} &= \left(\frac{\partial e}{\partial E_{ij}} - \theta\frac{\partial \eta}{\partial E_{ij}}\right)\dot{E}_{ij} + \left(\frac{\partial e}{\partial \theta} - \theta\frac{\partial \eta}{\partial \theta}\right)\dot{\theta} \\ &\quad + \left(\frac{\partial e}{\partial q_i} - \theta\frac{\partial \eta}{\partial q_i}\right)\dot{q}_i + \frac{q_i\theta_{,i}}{\theta} + \theta\eta^{*(i)}. \end{aligned} \tag{12.1.27}$$

If we consider the case of a rigid heat conductor, from eqn (12.1.27) at zero net flow of heat, we get

$$\frac{\partial e}{\partial \theta} = \theta\frac{\partial \eta}{\partial \theta}, \tag{12.1.28}$$

and since both sides are state functions (and hence independent of the particular process), this equation must be valid in general. Hence, in view of eqn (12.1.28), eqn (12.1.27) becomes

$$\begin{aligned} S_{ij}\dot{E}_{ij} &= \left(\frac{\partial e}{\partial E_{ij}} - \theta\frac{\partial \eta}{\partial E_{ij}}\right)\dot{E}_{ij} + \left(\frac{\partial e}{\partial q_i} - \theta\frac{\partial \eta}{\partial q_i}\right)\dot{q}_i + \frac{q_i\theta_{,i}}{\theta} + \theta\eta^{*(i)} \\ &= \frac{\partial \psi}{\partial E_{ij}}\dot{E}_{ij} + \frac{\partial \psi}{\partial q_i}\dot{q}_i + \frac{q_i\theta_{,i}}{\theta} + \theta\eta^{*(i)}, \end{aligned} \tag{12.1.29}$$

where eqn (12.1.22) has been employed.

At this point we split the stress tensor into the quasi-conservative ($S_{ij}^{(q)}$) and dissipative ($S_{ij}^{(d)}$) parts

$$S_{ij} = S_{ij}^{(q)} + S_{ij}^{(d)}, \quad \text{where} \quad S_{ij}^{(q)} = \frac{\partial \psi}{\partial E_{ij}}, \tag{12.1.30}$$

and, just like in all other thermodynamic theories, we let

$$\eta = -\frac{\partial \psi}{\partial \theta}. \tag{12.1.31}$$

As a result, eqns (12.1.29) and (12.1.25)$_3$ lead to the Clausius–Duhem inequality

$$S_{ij}^{(d)} \dot{E}_{ij} - \frac{q_i \theta_{,i}}{\theta} - \frac{\partial \psi}{\partial q_i} \dot{q}_i = \theta \eta^{*(i)} \geq 0, \tag{12.1.32}$$

or, in the direct notation where $\nabla_{\mathbf{q}}$ stands for the gradient in the space of fluxes,

$$\mathbf{S}^{(d)} \cdot \dot{\mathbf{E}} - \frac{\mathbf{q} \cdot \nabla \theta}{\theta} - \nabla_{\mathbf{q}} \psi \cdot \dot{\mathbf{q}} = \theta \eta^{*(i)} \geq 0. \tag{12.1.33}$$

Thus, we have a dissipation inequality

$$\mathbf{A}^{(d)} \cdot \mathbf{v} \geq 0, \tag{12.1.34}$$

where

$$\mathbf{A}^{(d)} = \left(\mathbf{S}^{(d)}, -\frac{\nabla \theta}{\theta}, -\nabla_{\mathbf{q}} \psi \right) \tag{12.1.35}$$

is a vector of dissipative thermodynamic forces, and

$$\mathbf{v} = \left(\dot{\mathbf{E}}, \mathbf{q}, \dot{\mathbf{q}} \right) \tag{12.1.36}$$

is a vector of conjugate thermodynamic velocities.

A general procedure based on the representation theory due to Edelen (1973, 1974) allows a derivation of the most general form of the constitutive relation subject to inequality (12.1.34). The following steps are involved: Assume $\mathbf{A}^{(d)}$ to be a function of $\mathbf{v}$, and determine it as

$$\mathbf{A}^{(d)} = \nabla_{\mathbf{v}} \phi + \mathbf{U}, \quad \text{or} \quad A_i^{(d)} = \frac{\partial \phi}{\partial v_i} + U_i, \tag{12.1.37}$$

where the vector $\mathbf{U} = (\mathbf{u}_1, \mathbf{u}_2)$ does not contribute to the entropy production

$$\mathbf{U} \cdot \mathbf{v} = 0, \tag{12.1.38}$$

while the dissipation function is

$$\phi = \int_0^1 \mathbf{v} \cdot \mathbf{A}^{(d)}(\tau \mathbf{v}) \mathrm{d}\tau, \tag{12.1.39}$$

and $\mathbf{U}$ is uniquely determined, for given $\mathbf{A}^{(d)}$, by

$$U_i = \int_0^1 \tau v_j \left[\frac{\partial A_i^{(d)}(\tau \mathbf{v})}{\partial v_j} - \frac{\partial A_j^{(d)}(\tau \mathbf{v})}{\partial v_i} \right] \mathrm{d}\tau. \tag{12.1.40}$$

Also, the symmetry relations

$$\frac{\partial\left[A_i^{(d)}(\tau\mathbf{v})-U_i\right]}{\partial v_j}=\frac{\partial\left[A_j^{(d)}(\tau\mathbf{v})-U_j\right]}{\partial v_i}$$

must hold, and these reduce to the classical Onsager reciprocity conditions

$$\frac{\partial A_i^{(d)}(\tau\mathbf{v})}{\partial v_j}=\frac{\partial A_j^{(d)}(\tau\mathbf{v})}{\partial v_i} \tag{12.1.41}$$

if and only if $\mathbf{U}=\mathbf{0}$.

In the following, we focus on a rigid heat conductor, i.e. $\dot{E}_{ij}\equiv 0$ everywhere, and hence $S_{ij}^{(d)}=0$. Also, we first consider a 1D situation (along the x-axis), so the vectors $\mathbf{A}^{(d)}$ and $\mathbf{v}$ simplify, respectively, to

$$\mathbf{A}^{(d)}=\left(-\frac{\theta_x}{\theta},-\frac{\partial\psi}{\partial q}\right),\qquad \mathbf{v}=(q,\dot{q})\,. \tag{12.1.42}$$

The simplest expressions for $\mathbf{U}$ satisfying the above relations are

$$U_1=\frac{\lambda t_0}{\theta}\dot{q}\,,\qquad U_2=-\frac{\lambda t_0}{\theta}q, \tag{12.1.43}$$

whereby the dissipation function is a quadratic form

$$\phi(\mathbf{v})\equiv\phi(q,\dot{q})=\frac{1}{2\theta}\lambda q^2+\frac{1}{2}G\dot{q}^2. \tag{12.1.44}$$

On account of eqn (12.1.37), we obtain

$$\begin{aligned}-\frac{\theta_x}{\theta}&\equiv A_1^{(d)}=\frac{\lambda q}{\theta}+U_1=\frac{\lambda}{\theta}q+\frac{\lambda t_0}{\theta}\dot{q},\\ -\frac{\partial\psi}{\partial q}&\equiv A_{2i}^{(d)}=G\dot{q}+U_2=G\dot{q}-\frac{\lambda t_0}{\theta}q.\end{aligned} \tag{12.1.45}$$

Let us now make these observations:

(i) Equation $(12.1.45)_1$ immediately yields the Maxwell–Cattaneo law $-k\theta_x=q+t_0\dot{q}$, with $k=1/\lambda$; see eqn (12.1.14).

(ii) Equation $(12.1.45)_2$ is satisfied identically provided a quadratic form of the free energy $\psi(q)=\frac{t_0}{2\theta}\lambda q^2$, just like in eqn (R1.1.5), is adopted with θ being set equal to θ_0, along with $G=0$. Regarding the latter, note that $\dot{q}$ must still be kept as an argument of $\phi(q,\dot{q})$ in eqn (12.1.20). Interestingly, the assumption $\theta\sim\theta_0$ is involved in eqn $(12.1.45)_2$ but not in eqn $(12.1.45)_1$.

(iii) Other expressions for $\mathbf{U}$ and $\phi(q,\dot{q})$ – but then leading to non-linear heat conduction laws, and therefore to non-linear field equations – can be explored henceforth. At this point, one may consider various forms of homogeneous and quasi-homogeneous functions in place of $\lambda q^2/\theta$ (Ziegler and Wehrli, 1987). Regarding the quadratic form above, note that the factor

1/2 could be incorporated into G in eqn (12.1.44) because ϕ is not required to equal the entropy production rate.

(iv) Another derivation implying the Maxwell–Cattaneo law can be found in Edelen (1993), although that approach involves a dissipation function as a functional of forces $[\phi = \phi(\mathbf{A}^{(d)})]$ along with a split of thermodynamic velocities $\mathbf{v} = \nabla_{\mathbf{A}^{(d)}}\phi + \mathbf{U}$ with the condition $\mathbf{U} \cdot \mathbf{A}^{(d)} = 0$. Edelen, looking for $\mathbf{v}$ as a function of $\mathbf{A}^{(d)}$, adopts a dissipation function involving $\theta_{,i}/\theta$ and $\dot{\theta}$, and derives a non-linear partial differential equation for heat conduction which in a third-order approximation becomes a telegraph equation. However, our approach based on TIV and the representation theory can be applied not only to thermoelastic (e.g. thermoelasticity theories with one or two relaxation times) but also to thermo-inelastic solids with second-sound effects. Also, the same vectors as in eqn (12.1.42) follow by using a rational thermodynamics instead of a TIV approach.

(v) The derivation above may be generalized to 3D, where

$$\mathbf{v} = (\mathbf{q}, \dot{\mathbf{q}}), \qquad \mathbf{A}^{(d)} = \left(-\frac{\nabla\theta}{\theta}, -\nabla_{\mathbf{q}}\psi\right), \qquad \mathbf{U} = (\mathbf{U}_1, \mathbf{U}_2). \tag{12.1.46}$$

Now,

$$U_{1i} = \frac{\lambda_{ij}t_0}{\theta}\dot{q}_j, \qquad u_{2i} = -\frac{\lambda_{ij}t_0}{\theta}q_j, \tag{12.1.47}$$

and the dissipation function is

$$\phi(q_i, \dot{q}_i) = \frac{1}{2\theta_0}\lambda_{ij}q_iq_j + \frac{1}{2}G_{ij}\dot{q}_i\dot{q}_j, \tag{12.1.48}$$

which jointly lead to eqn (12.1.18).

12.2 Thermoelastic helices and chiral media

12.2.1 *Homogeneous case*

Several years ago, an isothermal elastodynamic helix model has been generalized to account for coupled thermo-elastodynamic effects (Ostoja-Starzewski, 2003, 2008). The constitutive equations of such a helix are

$$\begin{aligned} F &= A_1u_x + A_2\varphi_x, \\ M &= A_3u_x + A_4\varphi_x \ . \end{aligned} \tag{12.2.1}$$

Here, F is the axial force in the helix (normalized by the cross-sectional area A and the Young's modulus E), M is the torque carried by the helix (normalized by the cube of radius R and the Young modulus E), u_x is the axial strain, φ_x is the angle of twist per unit length, Fig. 12.1(a). [Recall that the notation $f_x = \partial f/\partial x$ and $f_t = \partial f/\partial t$ has already been introduced in Chapter 10.] The strain energy

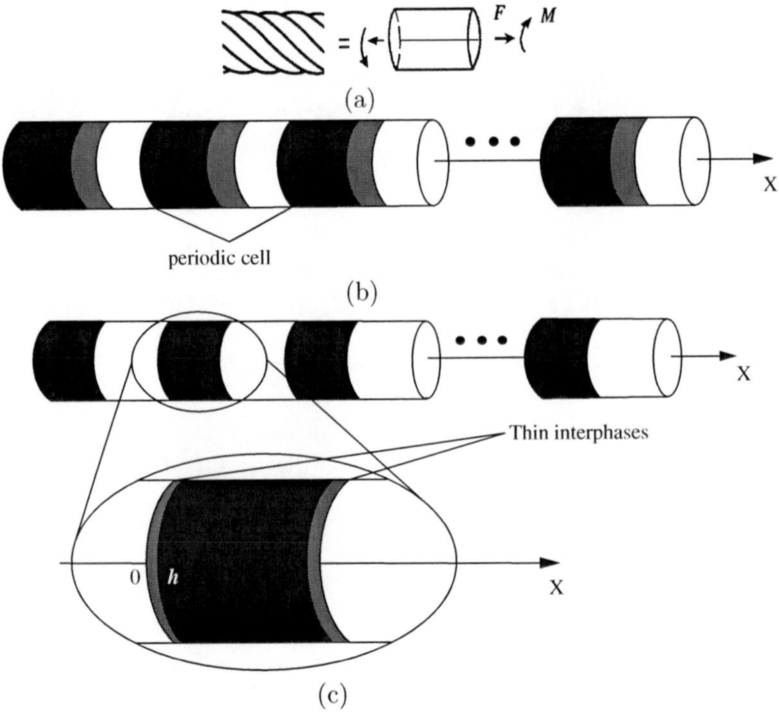

Figure 12.1 (a) A wire-rope made of several strands, each of the same helix angle. (b) A heterogeneous rod with a three-phase microperiodic structure (Vivar-Pèrez *et al.*, 2006)). (c) A composite helix and its periodic cell, also showing a thin interphase of thickness h between the black and white phases (Vivar-Pèrez *et al.*, 2008).

density $W = W(u_x, \varphi_x)$ is defined by

$$W(u_x, \varphi_x) = \frac{1}{2}u_x A_{11} u_x + \frac{1}{2}u_x(A_{12} + A_{21})\varphi_x + \frac{1}{2}\varphi_x A_{22}\varphi_x, \tag{12.2.2}$$

and $W(u_x, \varphi_x)$ is positive-definite iff

$$A_{11} > 0, \qquad A_{22} > 0, \qquad A_{12} = A_{21}, \qquad A_{11}A_{22} - A_{12}A_{21} > 0. \tag{12.2.3}$$

Equations (12.2.1), together with the equations of motion, lead to a system of two coupled wave equations for the pair (u, φ)

$$\begin{aligned} \rho A\ddot{u} &= A_1 u_{xx} + A_2 \varphi_{xx}, \\ \rho J\ddot{\varphi} &= A_3 u_{xx} + A_4 \varphi_{xx}, \end{aligned} \tag{12.2.4}$$

where ρ is the mass density and J is the mass polar moment of inertia.

Accounting for the thermal expansion in the axial direction and using the basic model of Fourier conductivity along the strand's axis, leads to the three coupled equations governing a triplet (u, φ, θ)

$$\begin{aligned} \rho A \ddot{u} &= A_1 u_{xx} + A_2 \varphi_{xx} - \alpha E \theta_x, \\ \rho J \ddot{\varphi} &= A_3 u_{xx} + A_4 \varphi_{xx}, \\ \rho c_v \dot{\theta} &= K \theta_{xx} - \alpha E \theta_0 \dot{u}_x. \end{aligned} \tag{12.2.5}$$

By letting

$$(u, \varphi, \theta) = (\hat{u}_0, \hat{\varphi}_0, \hat{\theta}_0) \exp[\mathrm{i}k(x - ct)], \tag{12.2.6}$$

where $(\hat{u}_0, \hat{\varphi}_0, \hat{\theta}_0)$ is a constant triplet, and substituting eqn (12.2.6) into eqns (12.2.5), we obtain the characteristic equation for the velocities c in terms of the wave number k and all the material parameters

$$\begin{vmatrix} A\rho c^2 - A_1 & -A_2 & -\mathrm{i}E\alpha/k \\ -A_3 & J\rho c^2 - A_4 & 0 \\ \dfrac{c\alpha E\theta_0}{\rho c_v} & 0 & \dfrac{K}{\rho c_v} - \mathrm{i}\dfrac{c}{k} \end{vmatrix} = 0. \tag{12.2.7}$$

This equation shows that the wave motion is not only dispersive [as in the mechanical model described by eqn (12.2.4)], but also damped. More specifically, eqn (12.2.7) leads to a fifth-order algebraic equation for the roots c, which can be solved numerically so as to assess the velocity c as a function of the wave number k, in the presence of weak thermal effects expressed by the dimensionless thermoelastic coupling constant (ε) and the thermal diffusivity at constant deformation (k_v)

$$\varepsilon = \frac{\alpha^2 E\theta_0}{\rho c_v}, \qquad k_v = \frac{K}{\rho c_v}. \tag{12.2.8}$$

As a particular case, considering the values pertaining to an oceanographic steel cable (Samras *et al.*, 1974), it was found that, with ε increasing from zero up, while keeping $k_v = 0$, there are two velocities, c_1 and c_2, that increase linearly. With ε in metals taking values up to 0.1, those speeds may easily go up by a few per cent. On the other hand, increasing k_v from zero up, while keeping $\varepsilon = 0$, has no effect on c_1 and c_2.

Employing the Maxwell–Cattaneo model for heat conduction, with the relaxation time $t_0 > 0$, we replace eqns (12.2.5) by

$$\begin{aligned} \rho A \ddot{u} &= A_1 u_{xx} + A_2 \varphi_{xx} - \alpha E \theta_x, \\ \rho J \ddot{\varphi} &= A_3 u_{xx} + A_4 \varphi_{xx}, \\ \theta_{xx} - \frac{\rho c_v}{K}(\dot{\theta} + t_0 \ddot{\theta}) &= \frac{\alpha E \theta_0}{K}\left(\dot{u}_x + t_0 \ddot{u}_x\right). \end{aligned} \tag{12.2.9}$$

While this leads to a sixth-order algebraic equation for the roots c, the ensuing numerical analysis has revealed that they are only weakly affected by t_0. We return to the setup of a 1D, chiral counterpart of the L–S theory in Section 12.2.3.

12.2.2 *Heterogeneous case and homogenization*

The thermo-elastodynamics of a helix with a periodic structure has been studied in (Vivar-Pèrez *et al.*, 2006). More precisely, the attention was focused on a bar composed of two, or more, locally homogeneous, thermoelastic materials distributed in periodic cells, and arranged sequentially as shown in Fig. 12.1(b). The length of this periodic cell is assumed to be very small as compared with the unit length taken, and this is expressed by a parameter $\varepsilon \ll 1$. Thus, at each point x and for every $t \geq 0$, we consider three field variables: the displacement $u(x,t)$, the twist angle $\varphi(x,t)$, and the temperature $\theta(x,t)$, as well as three flux variables: the axial and torsional stresses $F(x,t)$ and $M(x,t)$, and the negative of the heat flux through the cross-sectional area: $-q(x,t)$. The constitutive equations governing these triplets are

$$\begin{aligned} F &= A_1 u_x + A_2 \varphi_x - \alpha E(\theta - \theta_0), \\ M &= A_3 u_x + A_4 \varphi_x, \\ q + t_0 q_t &= -K\theta_x \ , \end{aligned} \tag{12.2.10}$$

and the balance equations read

$$\begin{aligned} F_x &= \rho A u_{tt}, \\ M_x &= \rho J \varphi_{tt}, \\ -q_x &= \alpha E \theta_0 u_{tx} + \rho c_v \theta_t, \end{aligned} \tag{12.2.11}$$

where the coefficients A_1, A_2, A_3, A_4, ρ, J, c_v, α, E, and K are periodic functions of x since we are dealing with a periodic helix, and t_0, θ_0 are positive constants. Upon setting $t_0 = 0$, we have the case of a parabolic heat conduction; otherwise, the equations describe a hyperbolic thermoelastic helix.

By introducing the notations

$$v = \begin{bmatrix} u \\ \varphi \\ \theta \end{bmatrix}, \qquad f = \begin{bmatrix} F \\ M \\ -q \end{bmatrix},$$

$$Q_0 = \begin{bmatrix} A_1 & A_2 & 0 \\ A_3 & A_4 & 0 \\ 0 & 0 & K \end{bmatrix}, \quad Q_1 = \begin{bmatrix} 0 & 0 & -\alpha E \\ 0 & 0 & 0 \\ 0 & 0 & 0 \end{bmatrix}, \quad T = \begin{bmatrix} 0 & 0 & 0 \\ 0 & 0 & 0 \\ 0 & 0 & t_0 \end{bmatrix}, \tag{12.2.12}$$

$$P_0 = \begin{bmatrix} 0 & 0 & 0 \\ 0 & 0 & 0 \\ \alpha E \theta_0 & 0 & 0 \end{bmatrix}, \quad P_1 = \begin{bmatrix} \rho A & 0 & 0 \\ 0 & \rho J & 0 \\ 0 & 0 & 0 \end{bmatrix}, \quad P_2 = \begin{bmatrix} 0 & 0 & 0 \\ 0 & 0 & 0 \\ 0 & 0 & \rho c_v \end{bmatrix},$$

eqns (12.2.10) and (12.2.11) are both cast in the following compact form

$$\begin{aligned} f + T\frac{\partial f}{\partial t} &= Q_0 \frac{\partial v}{\partial x} + Q_1 v, \\ \frac{\partial f}{\partial t} &= P_0 \frac{\partial^2 v}{\partial t \partial x} + P_1 \frac{\partial^2 v}{\partial t^2} + P_2 \frac{\partial v}{\partial t}. \end{aligned} \tag{12.2.13}$$

With ε being a very small parameter, the asymptotic homogenization method (Bensoussan *et al.*, 1978; Bakhvalov and Panasenko, 1989) can be used to obtain the homogenized rod equations

$$\begin{aligned} \Phi &= \bar{A}_1 v_x + \bar{A}_2 \psi_x - \overline{\alpha E} \gamma, \\ M &= \bar{A}_3 u_x + \bar{A}_4 \psi_x, \\ -\Theta - \bar{t}_0 \Theta_t &= \bar{K} \gamma_x, \end{aligned} \tag{12.2.14}$$

and

$$\begin{aligned} \Phi_x &= \overline{\rho A}\ v_{tt}, \\ M_x &= \overline{\rho J}\ \psi_{tt}, \\ -\Theta_x &= \overline{\alpha E \theta_0}\ v_{tx} + \overline{\rho c_v}\ \gamma_t. \end{aligned} \tag{12.2.15}$$

Here, Φ, M, and Θ are the average fields of F, M, and q, respectively; v and ψ are the average fields of u and φ, respectively. Also, γ is the average temperature field, while the effective coefficients $\bar{A}_1$, $\bar{A}_2$, ... $\bar{K}$ are given explicitly in terms of the properties of constituent phases (Vivar-Pèrez *et al.*, 2006). These results are valid in the case of wavelengths much longer than the length of the unit cell. However, formulas for shorter wavelengths can be derived by admitting higher-order terms in the expansion.

Vivar-Pèrez *et al.* (2008) discussed the isothermal elastodynamics of a helix made of a sequence of unit cells, each containing a thin imperfect interphase embedded within a finite number of other phases, Fig. 12.1(c). This analysis can be generalized to the case of a helix made of thermoelastic phases.

12.2.3 *Plane waves in non-centrosymmetric micropolar thermoelasticity*

Plane harmonic waves in a helical-type (i.e. non-centrosymmetric or chiral) micropolar, isotropic linear thermoelastic material were studied over two decades ago, see, e.g. Nowacki (1986). One starts from the free energy density in a quadratic form

$$F = \frac{1}{2}\gamma_{ji} a_{jikl} \gamma_{kl} + \frac{1}{2}\kappa_{ji} c_{jikl} \kappa_{kl} + \gamma_{ji} b_{jikl} \kappa_{kl} - \eta_{ji} \gamma_{ji} \vartheta - \zeta_{ji} \kappa_{ji} \vartheta - \frac{c_e}{2\theta_0} \vartheta^2, \tag{12.2.16}$$

and, using the classical relations (with S denoting the entropy, to avoid clash of notation with η_{ji})

$$\sigma_{ji} = \frac{\partial F}{\partial \gamma_{ji}}, \qquad \mu_{ji} = \frac{\partial F}{\partial \kappa_{ji}}, \qquad S = -\frac{\partial F}{\partial \vartheta}, \tag{12.2.17}$$

and assuming isotropy, finds (see also eqns (0.18) in Dyszlewicz, 2004)

$$\begin{aligned}\sigma_{ji} &= (\mu+\alpha)\,\gamma_{ji} + (\mu-\alpha)\,\gamma_{ij} + (\lambda\gamma_{kk} - \eta\vartheta)\,\delta_{ji} + (\chi+\nu)\,\kappa_{ji} \\ &+ (\chi-\nu)\,\kappa_{ij} + \kappa\delta_{ij}\kappa_{kk}, \\ \mu_{ji} &= (\chi+\nu)\,\gamma_{ji} + (\chi-\nu)\,\gamma_{ij} + \kappa\gamma_{kk}\delta_{ij} + (\gamma+\varepsilon)\,\kappa_{ji} \\ &+ (\gamma-\varepsilon)\,\kappa_{ij} + (\beta\kappa_{kk} - \zeta\vartheta)\,\delta_{ji}, \\ S &= \eta\gamma_{kk} + \zeta\kappa_{kk} + \frac{c_e}{\theta_0}\vartheta.\end{aligned} \tag{12.2.18}$$

This shows that, besides μ, λ, α, β, ε, γ for the non-centrosymmetric micropolar material, there are three additional (chiral) constants: χ, ν, and κ.

Remark 12.1 (i) In the centrosymmetric case, α, β, ε, γ are four micropolar constants with ε here being, of course, different from the thermoelastic coupling constant ε used earlier. (ii) κ is just a material constant, as opposed to κ_{ij}, which is the torsion-curvature tensor.

Upon substituting the constitutive equations $(12.2.18)_{1,2}$ into the linear and angular momentum balance equations

$$\sigma_{ji,j} = \rho\ddot{u}_i, \qquad e_{ijk}\sigma_{kj} + \mu_{ji,j} = J\ddot{\varphi}_i\ , \tag{12.2.19}$$

and using the kinematic relations

$$\varepsilon_{ji} = u_{j,i} - e_{kij}\varphi_k, \qquad \kappa_{ij} = \varphi_{j,i}\ , \tag{12.2.20}$$

we obtain the equations of motion in displacements $\mathbf{u}$, rotations $\boldsymbol{\varphi}$, and temperature ϑ (where $\partial_t u \equiv \dot{u}$)

$$\begin{aligned} &\Box_2\mathbf{u} + (\lambda+\mu-\alpha)\,grad\,(div\mathbf{u}) + 2\alpha\,(rot\boldsymbol{\varphi}) \\ +(\chi+\nu)\,\nabla^2\boldsymbol{\varphi} &+ (\chi-\nu+\kappa)\,grad\,(div\boldsymbol{\varphi}) = \eta\,(grad\vartheta)\,, \\ &\Box_4\boldsymbol{\varphi} + (\beta+\gamma-\varepsilon)\,grad\,(div\boldsymbol{\varphi}) + 2\alpha\,(rot\mathbf{u}) \\ +4v\,(rot\boldsymbol{\varphi}) &+ (\chi+\nu)\,\nabla^2\mathbf{u} + (\chi-\nu+\kappa)\,grad\,(div\mathbf{u}) = \zeta\,(grad\vartheta)\,. \end{aligned} \tag{12.2.21}$$

Here,

$$\Box_2 = (\mu+\alpha)\,\nabla^2 - \rho\partial_t^2 \quad \text{and} \quad \Box_4 = (\gamma+\varepsilon)\,\nabla^2 - 4\alpha - J\partial_t^2 \tag{12.2.22}$$

denote the d'Alembert and Klein–Gordon operators, respectively.

Regarding the heat conduction, we begin with Fourier's law for an isotropic material

$$q_i = -K\vartheta_{,i}, \tag{12.2.23}$$

where K denotes thermal conductivity, so as to avoid a clash of notation with the wave number k in eqn (12.2.26) below. Also, $\vartheta = \theta - \theta_0$, with θ_0 being the reference temperature. Using the expression for S, and recalling the energy balance $\theta_0\dot{S} = -q_{i,i}$, we arrive at the energy equation in terms of u_i, φ_i, and ϑ

$$\left(K\nabla^2 - c_e\partial_t\right)\vartheta - \theta_0\eta\dot{u}_{i,i} - \theta_0\zeta\dot{\varphi}_{i,i} = 0, \tag{12.2.24}$$

where, interestingly, the term $\zeta\dot{\varphi}_{i,i}$ arises due to the non-centrosymmetry of the material, i.e. its helical microstructure. Here, $c_e = \rho c_v$, with c_v being the heat capacity.

Assuming $\mathbf{u}$ and φ to depend on x_1 and t only, we arrive at two disjoint systems of equations. The first of these involves wave motion of the triplet $(u_1, \varphi_1, \vartheta)$, and is written below

$$\begin{aligned} \left[(\lambda+2\mu)\,\partial_1^2 - \rho\partial_t^2\right] u_1 + (\kappa+2\chi)\,\partial_t^2\varphi_1 - \eta\partial_1\vartheta &= 0, \\ (\kappa+2\chi)\,\partial_1^2 u_1 + \left[(\beta+2\gamma)\,\partial_1^2 - 4\alpha - J\partial_t^2\right]\varphi_1 - \zeta\partial_1\vartheta &= 0, \\ -\eta\partial_1\partial_t u_1 - \zeta\partial_1\partial_t\varphi_1 + \theta_0^{-1}\left(K\,\partial_1^2 - c_e\,\partial_t\right)\vartheta &= 0. \end{aligned} \tag{12.2.25}$$

The second system involves transverse and rotational wave motions, uncoupled from the temperature field, and we shall not concern ourselves with them.

Upon substituting the time- and space-harmonic wave forms

$$\begin{aligned} u_1(x_1,t) &= A\exp\left[\mathrm{i}k(x_1-ct)\right], \\ \varphi_1(x_1,t) &= B\exp\left[\mathrm{i}k(x_1-ct)\right], \\ \vartheta &= C\exp\left[\mathrm{i}k(x_1-ct)\right], \end{aligned} \tag{12.2.26}$$

where A, B, and C are constants, into eqn (12.2.25), we find the system of three homogeneous algebraic equations

$$\begin{aligned} \left[\rho c^2k^2 - (\lambda+2\mu)\,k^2\right]A + \left[-(\kappa+2\chi)\,k^2\right]B - (\mathrm{i}k\eta)C &= 0, \\ \left[-(\kappa+2\chi)\,k^2\right]A + \left[-(\beta+2\gamma)\,k^2 - 4\alpha + \mathrm{i}Jc^2k^2\right]B - (\mathrm{i}\zeta k)C &= 0, \\ -c\eta k^2 A - c\zeta k^2 B + \left(-\frac{K}{\theta_0}k^2 + \mathrm{i}\frac{c_e}{\theta_0}kc\right)C &= 0, \end{aligned} \tag{12.2.27}$$

which leads to a characteristic determinant/equation for the roots c in terms of k and all the material parameters (upon division by k^2 above)

$$\begin{vmatrix} \rho c^2 - (\lambda+2\mu) & -(\kappa+2\chi) & -\mathrm{i}\eta/k \\ -(\kappa+2\chi) & Jc^2 - (\beta+2\gamma) - 4\alpha/k^2 & -\mathrm{i}\zeta/k \\ c\eta & c\zeta & \dfrac{K}{\theta_0} - \mathrm{i}\dfrac{c_e}{\theta_0}c/k \end{vmatrix} = 0. \tag{12.2.28}$$

In essence, the wave motions described by $(u_1, \varphi_1, \vartheta)$ are dispersive and damped. Additionally, by comparison to the case of the 1D helix studied in Section 12.2.1, we note these principal differences:

(i) In the determinant (12.2.28) in which the elements are denoted by C_{ij} $(i,j=1,2,3)$, the terms C_{23} and C_{32} are non-zero.
(ii) The term C_{22} contains $4\alpha/k^2$, which, already in the simpler case of absence of thermal effects, leads to dispersive effects.

A solution of eqn (12.2.28) has not been attempted due to the present difficulty of obtaining explicit values of material parameters χ, ν, κ, and ζ. Recently, progress has been made on derivations of centrosymmetric micropolar parameters of composite materials from a microstructure (e.g. Ostoja-Starzewski, 2008a), but further work still needs to be done on the non-centrosymmetric case. Thus, at present, the analysis of the 1D helix in Section 12.2.1 provides guidance about plane waves in a 3D hemitropic thermoelastic continua.

12.3 Surface waves

The relaxation times of a generalized dynamic coupled thermoelasticity, as opposed to the classical thermoelasticity, give rise to interesting effects on the propagation of surface waves. In the following, we review the results of Wojnar on stationary thermoelastic surface waves in an isotropic homogeneous half-space within the G–L theory (1986, 1988a, 1988b). The starting point is given by the field equations (2.1.11) of the mixed displacement–temperature problem in Chapter 2, which now become

$$\begin{aligned} \mu u_{i,jj} + (\lambda+\mu) u_{j,ji} - \gamma(\vartheta + t_1\dot{\vartheta})_{,j} &= \rho \ddot{u}_i \\ k\vartheta_{,ii} &= C_E(\dot{\vartheta} + t_0\ddot{\vartheta}) + \gamma\theta_0 \dot{u}_{j,j} \end{aligned} \quad \text{on} \quad \mathrm{B}\times[0,\infty), \tag{12.3.1}$$

where B stands for the domain: $|x_1| < \infty,\ x_2 \geq 0,\ |x_3| < \infty$, while

$$\gamma = (3\lambda + 2\mu)\alpha_0, \tag{12.3.2}$$

with α_0 denoting the coefficient of linear thermal expansion. The boundary conditions at $x_2 = 0$ involve (i) $S_{i2}(x_1, 0, x_3, t) = 0$ and (ii) a free heat transfer. Thus, in view of eqn (1.3.45) restricted to an isotropic body and assuming the fields independent of x_3, and $u_3 = 0$, at the boundary we have

$$\begin{aligned} &u_{1,2} + u_{2,1} = 0, \qquad \lambda u_{1,1} + (\lambda + 2\mu)u_{2,2} - \gamma(\vartheta + t_1\dot{\vartheta}) = 0, \\ &\eta_1\vartheta + \eta_2\vartheta_{,2} = 0 \quad \text{for the constants } \eta_1,\ \eta_2 \text{ such that} \quad 1 \geq \eta_1 \geq 0, \quad \eta_2 \geq 0. \end{aligned} \tag{12.3.3}$$

To solve the generalized thermoelastic surface-wave problem, we let

$$A_j = \alpha_j c_1/\varkappa, \tag{12.3.4}$$

and substitute

$$(u_1, u_2, \vartheta) = \sum_{j=1}^{3} (U^{(j)}, V^{(j)}, \vartheta^{(j)}) \exp[-A_j x_2 + \mathrm{i}(kx_1 - \Omega t)] \tag{12.3.5}$$

into eqns (13.3.3). This leads to three linear homogeneous algebraic equations for $(U^{(1)}, U^{(2)}, U^{(3)})$, whose non-zero solution is assured by the following condition

$$\begin{aligned} &[qG\eta_2(\alpha_2^2 + \alpha_3^2 + \omega^2 - q^2 + \alpha_2\alpha_3) + \eta_0(q^2 - \omega^2 + \alpha_2\alpha_3)]^2 = \\ &\qquad (qG\eta_0 + \eta_2\alpha_2\alpha_3)^2(\alpha_2^2 + \alpha_3^2 + \alpha_2\alpha_3), \end{aligned} \tag{12.3.6}$$

where

$$\eta_0 = \eta_1 \varkappa / c_1, \qquad G = \frac{q}{\alpha_1}\left(\frac{\alpha_1^2 + q^2}{2q^2}\right)^2, \qquad \varkappa = K/C_E$$
$$\omega = \Omega/\Omega^*, \qquad \Omega^* = c_1^2/\varkappa, \quad c_1^2 = (\lambda + 2\mu)/\rho, \qquad q = kc_1/\Omega^*, \tag{12.3.7}$$

and $\alpha_i > 0$ $(i = 1, 2, 3)$ satisfy the relations

$$\alpha_1^2 = q^2 - \beta^2\omega^2, \qquad \alpha_2^2 + \alpha_3^2 = -P - Q - R, \qquad \alpha_2^2\alpha_3^2 = PQ - q^2R, \tag{12.3.8}$$

where

$$P = \omega^2 - q^2, \qquad Q = -q^2 + \omega^2\tau_0 + \mathrm{i}\omega, \qquad R = \epsilon\omega(\mathrm{i} + \omega\tau_1). \tag{12.3.9}$$

In eqns (12.3.8) and (12.3.9) the material parameters β, τ_0, τ_1, and ϵ are defined by

$$\beta = c_1/c_2, \qquad c_2^2 = \mu/\rho, \qquad \tau_0 = t_0\Omega^*, \qquad \tau_1 = t_1\Omega^*,$$
$$\varepsilon = mh\varkappa, \qquad h = \gamma\theta_0/K, \qquad m = \gamma/(\lambda + 2\mu), \tag{12.3.10}$$

For the limiting value $\varepsilon = 0$, eqn (12.3.6) splits into the equations

$$G^2 = 1 - \omega^2/q^2, \qquad q^2 = \omega^2\tau_0 + \mathrm{i}\omega, \tag{12.3.11}$$

of which the first one is the classical equation of a Rayleigh wave, while the second one is the dispersion equation of a thermal wave.

Another special case occurs for $\eta_1 = 0$ and $\eta_2 = 1$, whereby eqn (12.3.6) reduces to

$$G^2[q^2 - \omega^2(\tau_0 + \varepsilon\tau_1) - \mathrm{i}\omega(1 + \varepsilon) + \alpha_2\alpha_3]^2 =$$
$$(\alpha_2^2\alpha_3^2/q^2)[2q^2 - \omega(1 + \tau_0 + \varepsilon\tau_1) - \mathrm{i}\omega(1 + \varepsilon) + 2\alpha_2\alpha_3]. \tag{12.3.12}$$

From the above, for $\omega \to 0$ and $q \to 0$ as $\omega/q \to$ *const*, and for $\tau_1 \geq \tau_0 > 0$, we recover the result of Lockett (1958)

$$G^2 = (1 + \varepsilon - \omega^2/q^2)/(1 + \varepsilon). \tag{12.3.13}$$

With the notations

$$X = \omega^2/q^2, \qquad T_0 = \tau_0 + \frac{\mathrm{i}}{\omega}, \qquad T_1 = \tau_1 + \frac{\mathrm{i}}{\omega},$$
$$\mathcal{G} \equiv G^2 = \left(1 - \frac{\beta^2}{2}X\right)^4 / \left(1 - \beta^2 X\right), \tag{12.3.14}$$
$$A^2 \equiv (\alpha_2\alpha_3/q^2)^2 = (1 - X)(1 - XT_0) - \varepsilon X T_1,$$

eqn (12.3.12) takes the form

$$\mathcal{G}[1 - X(T_0 + \varepsilon T_1) + A]^2 = A^2[2 - X - X(T_0 + \varepsilon T_1) + 2A]. \tag{12.3.15}$$

On the other hand, if the relaxation times are equal, $\tau_1 = \tau_0$ (that is, if $T_0 = T_1$), there holds the dispersion equation due to Nayfeh and Nemat-Nasser (1971) for the L–S theory

$$\mathcal{G}[1-(1+\varepsilon)XT_0+A]^2 = A^2[2-X-(1+\varepsilon)XT_0+2A], \tag{12.3.16}$$

where

$$A^2 = (1-X)(1-XT_0) - \varepsilon XT_0. \tag{12.3.17}$$

This shows that an analogy first observed for the plane thermoelastic waves in the L–S and G–L theories (Agarwal, 1978) carries over to the surface waves. More specifically, that analogy says that, while there is no a priori correspondence between the L–S and G–L theories, the results for plane waves in the G–L theory carry over to those of the L–S theory provided $\tau_0 = \tau_1$; see also (Abd-Alla and Al-Dawy, 2001).

Wojnar solved the dispersion equation (12.3.16) and his results are reproduced here in Fig. 12.2. These graphs show the squared speed of the surface wave, $c^2 = [1/\text{Re}(1/\sqrt{X})]^2$, and its damping coefficient, $\eta = \omega\text{Im}(1/\sqrt{X})$, for four values of the ratio τ_1/τ_0, as functions of $1/\tau_0$, i.e. the reciprocal of the smaller of both relaxation times. Observe that, for decreasing values of τ_1, the solutions c^2 tend to the straight asymptotes indicated by the broken lines. The horizontal line $c_R^2 = 0.2817$ corresponds to the mechanical Rayleigh wave mode, while the linear relation $c^2 = 1/\tau_0$ corresponds to the thermal mode. As discussed at length in (Wojnar, 1988), these results are consistent with the asymptotic analysis.

The effect of coupled generalized thermoelastic material response versus the classical one has also been examined in a number of problems involving interaction of one or two half-spaces: rigid die sliding on a deformable body, surface or interface wave propagation, and dynamic thermoelastic fracture (Brock, 2005, 2006, 2007, 2008a,b). In particular, Brock (2005) found that, when a rigid die slides with friction at constant subcritical speed on a thermoelastic half-space with one relaxation time, there is a clear dependence of the speeds of body waves and a Rayleigh wave on the thermoelastic coupling and the relaxation time. Furthermore, Brock (2006) examined a dynamic steady state occurring when the half-space is debonded from a rigid insulated substrate at constant speed by moving shear and normal line loads. He obtained the field variables on the debonded surface and the still-bonded interface for the sub-Rayleigh, super-Rayleigh/subsonic, lower and upper transonic, and supersonic speed ranges.

These investigations have been extended in (Brock, 2007) to include thermoelasticity with two relaxation times (i.e. the G–L theory) in two half-spaces (1 and 2). Interestingly, it has been found that Stoneley waves can be generated by a thermal mismatch alone (i.e. without any mismatch in the elastic properties), in classical coupled thermoelasticity with the Fourier heat flow as well as in thermoelasticity with one or two relaxation times. Furthermore, an asymptotic analysis has shown that, for very long times after load application, a residual temperature for all three theories obeys the Fourier heat conduction. Also, a time step load gives rise to a propagating step in temperature for the Fourier and two relaxation time models, and to a propagating pulse for the single-relaxation

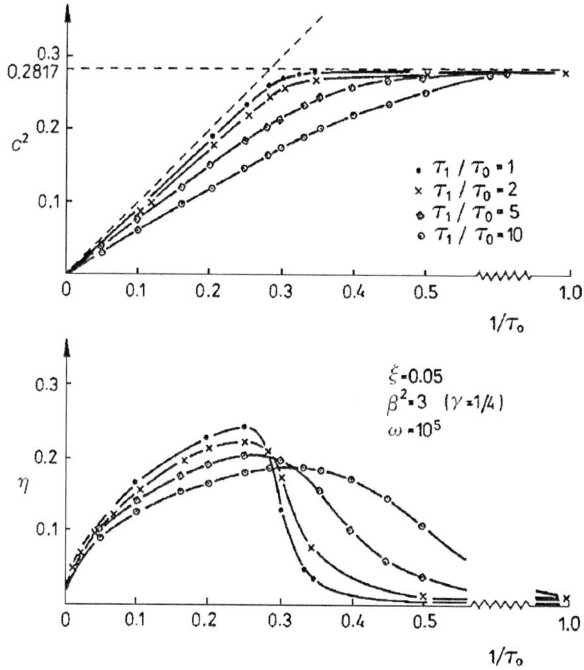

Figure 12.2 The squared speed c^2 and damping coefficient η of a generalized surface wave treated as functions of $1/\tau_0$, for the coupling constant $\varepsilon = 0.05$, (which corresponds to a medium with a Poisson ratio $\nu = 1/4$), for $\tau_1/\tau_0 = 1$, 2, 5, 10 at the frequency $\omega = 10^5$; from (Wojnar, 1988).

time model (Brock, 2008a). These studies have now been extended to a dynamic thermoelastic fracture (Brock, 2008b).

12.4 Thermoelastic damping in nanomechanical resonators

12.4.1 *Flexural vibrations of a thermoelastic Bernoulli–Euler beam*

High-frequency nano-electromechanical systems (so-called NEMS) attract widespread attention of many researchers due to a number of important applications, such as the ultrasensitive mass detection, mechanical signal processing, scanning probe microscopes, etc. The main challenges in fabrication and design of such devices include an increase of their natural frequency and a minimization of dissipation effects (Roukes, 2000). Thermoelastic damping, which results from coupling between the mechanical and temperature fields, is identified as one of the main mechanisms of internal dissipation in NEMS resonators and places a fundamental limit on their force sensitivity. While most work to date has focused on analysis of thermoelastic damping of beam resonators within the framework of the classical theory of thermoelasticity (e.g. Kinra and Milligan, 1994; Bishop and Kinra, 1997; Roukes, 2000), in a recent paper we examined it from the standpoint of the L–S theory (Khisaeva and Ostoja-Starzewski, 2006).

The energy balance equation of the L–S theory for a thermoelastic isotropic body is

$$\vartheta_{,ii} = \frac{\rho c}{k}\left(\dot{\vartheta} + t_0\ddot{\vartheta}\right) + \frac{E\alpha}{k(1-2\nu)}\theta_0\left(\dot{u}_{k,k} + t_0\ddot{u}_{k,k}\right), \tag{12.4.1}$$

where u_i is the displacement vector, ϑ is the temperature change, ρ is the mass density, c is the specific heat, E is Young's modulus of the material, ν is Poisson's ratio, α is the linear thermal expansion coefficient, θ_0 is the absolute equilibrium temperature, and k is the isotropic thermal conductivity. The entropy production within the body per unit time per unit mass, in which a heat flow takes place, can be calculated from the entropy balance equation

$$\dot{\eta} = -\frac{1}{\rho\theta^2}q_i\theta_{,i}, \tag{12.4.2}$$

where η, θ, and q_i denote the entropy, absolute temperature, and heat flux, respectively. Assuming small deviations of temperature from the equilibrium value, θ in the denominator can be replaced by θ_0. For a 1D case, in which $\vartheta = \vartheta(y,t)$ and the temperature gradient is not equal to zero, elimination of $q_i = q(y,t)\delta_{i2}$ from eqn (12.4.2) and the Maxwell–Cattaneo equation leads to

$$t_0\ddot{\eta} + \dot{\eta}\left(1 + 2t_0\frac{\dot{\vartheta}}{\theta_0} - t_0\frac{\dot{\vartheta}'}{\vartheta'}\right) = \frac{k}{\rho\theta_0^2}(\vartheta')^2, \tag{12.4.3}$$

where the prime denotes a derivative with respect to y.

The thermoelastic damping is defined as the ratio of the energy dissipated to the energy stored in the body over the same period of time; it is expressed as

$$\psi = \int_V \psi_L \mathrm{d}V = \frac{\int_V \Delta W \mathrm{d}V}{\int_V W \mathrm{d}V} = \frac{\int_V \psi_L W \mathrm{d}V}{\int_V W \mathrm{d}V}, \tag{12.4.4}$$

where $\psi_L = \Delta W/W$ is the local specific damping capacity, W is the elastic energy density stored in the body, given by $W = \frac{1}{2}S_{ij}E_{ij}$, and ΔW is the total work lost throughout the body, which can be related to the entropy generation by the following equation

$$\Delta W = \rho\theta_0\eta. \tag{12.4.5}$$

The interest being in a resonator, attention is focused on a thin homogeneous, isotropic, elastic beam of thickness h with a constant rectangular cross-section (Fig. 12.3) subjected to a steady-state displacement condition $u(x,t) \equiv u_2(x,t) = U(x)\mathrm{e}^{\mathrm{i}\omega t}$ at the neutral axis, where $U(x)$ is a prescribed function and ω is the circular frequency in radians per second.

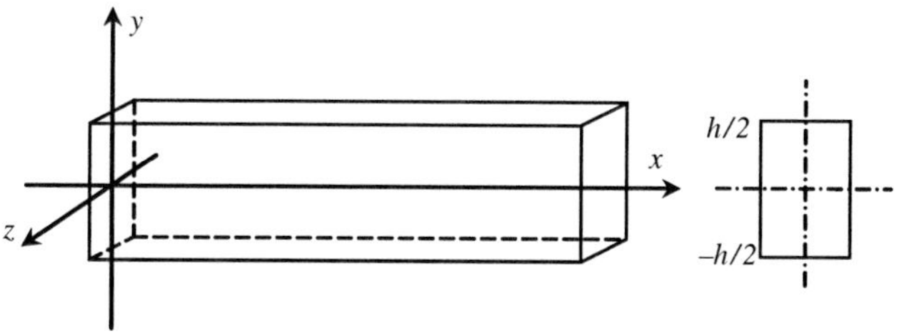

Figure 12.3 The co-ordinate system and geometry of a beam.

In the same vein, we take the dilatation and temperature fields as

$$\begin{gathered} E_{kk} = -(1-2\nu)E_{xx} = -(1-2\nu)(-yu_{xx}) = -(1-2\nu)\kappa_0 y \mathrm{e}^{\mathrm{i}\omega t}, \\ \vartheta = \theta_0(1+i) + V(y)\mathrm{e}^{\mathrm{i}\omega t}, \end{gathered} \tag{12.4.6}$$

where $\kappa_0 = 2\sigma/(hE)$, σ is the pressure on the beam's surface, and $V = V(y)$ is to be found. Since we note that the conduction of heat in the beam is much faster than the exchange of heat with the environment, the energy balance equation (12.4.1) becomes

$$\frac{\mathrm{d}^2\vartheta^*}{\mathrm{d}Y^2} - \pi^2\Omega\vartheta^*\left(\mathrm{i} - \gamma\Omega\right) = -2\pi^2\Omega Y\left(\mathrm{i} - \gamma\Omega\right). \tag{12.4.7}$$

Here, $Y = y/h$ is the non-dimensional co-ordinate, $\vartheta^* = \vartheta/\Delta T$ is the normalized temperature (ΔT being the change in temperature at the compressed upper surface of the beam by a constant stress σ under "adiabatic" conditions $\Delta T = -\frac{\alpha}{\rho c}\sigma\theta_0 = \frac{\alpha}{\rho c}\frac{h}{2}E\theta_0\kappa_0$.) and $\Omega = \omega\tau$ is the normalized frequency, where $\tau = \rho c h^2/\pi^2 k$ is the characteristic time in Zener's model of thermoelastic damping (Zener, 1948), while $\gamma = t_0/\tau$ is the normalized relaxation time.

A general solution to eqn (12.4.3) for the entropy production per unit time per unit mass, in which ϑ is generated by ϑ^* satisfying eqn (12.4.7), is

$$\begin{aligned} \frac{\partial\eta}{\partial t} = \frac{\Delta T^2 k}{\rho h^2\theta_0^2\omega(1+\omega^2 t_0^2)} \Bigg\{ & \left[B_I \sin(\omega t) - B_R\cos(\omega t)\right]^2 \\ -\omega t_0 & \left[\frac{1}{2}\sin(2\omega t)\left(B_I^2 - B_R^2\right) - B_I B_R \cos(2\omega t)\right]\Bigg\} \\ & + C_1 \mathrm{e}^{-t/t_0}\left[B_I\sin\left(\omega t\right) - B_R\cos\left(\omega t\right)\right], \end{aligned} \tag{12.4.8}$$

where B_R and B_I are, respectively the real and imaginary parts of the temperature gradient $\partial\vartheta^*/\partial Y$ and C_1 is an arbitrary constant. Defining the period as $t^* = 2\pi/\omega$, the change of entropy over the period is

$$\Delta\eta = \frac{\pi\Delta T^2 k}{\rho h^2\theta_0^2\omega(1+\omega^2 t_0^2)}\left(B_I^2 + B_R^2\right), \tag{12.4.9}$$

so that, on account of eqn (12.4.5), we obtain the averaged damping across the beam thickness

$$\psi = \frac{3\psi_0}{\pi^2\Omega(1+\gamma^2\Omega^2)} \int_{-1/2}^{1/2} (B_I^2 + B_R^2)\, \mathrm{d}Y, \tag{12.4.10}$$

where ψ_0 is a characteristic Zener damping defined by $\psi_0 = 2\pi\alpha^2 E\theta_0/\rho c$.

12.4.2 *Numerical results and discussion*

The average damping capacity of the beam resonator is plotted and analyzed in Fig. 12.4 according to the above formulas for three of the most common NEMS materials with various γ values: silicon, quartz and diamond. It is found that the temperature remains at the reference state at low frequencies and represents a line (adiabatic limit when $\Omega \to \infty$) analogously to the classical temperature distribution (Kinra and Milligan, 1994). At intermediate frequencies the L–S theory gives rise to a wave-like temperature distribution across the beam thickness with the temperature exceeding the adiabatic limit by 5 times for silicon and 20 times for diamond. Evidently, the nature of such a temperature distribution is defined by the difference in speed of propagation of mechanical and thermal waves. As in the classical case, at low frequencies, the system has enough time to relax and the temperature distribution doesn't depend on Y; at higher frequencies the system has no time to relax and the temperature curve represents a line – an adiabatic limit. When the frequency of vibration is in the intermediate range, thermal waves have time to propagate for some finite distance, which, as a result, produce a wave-like temperature distribution across the thickness of the beam.

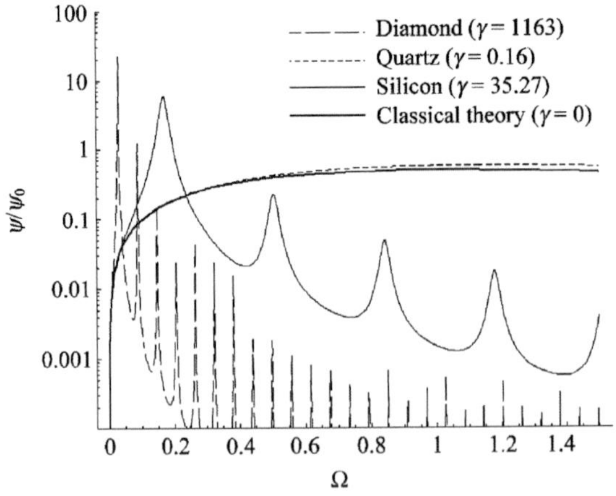

Figure 12.4 A comparison of classical thermoelastic damping with a generalized thermoelastic damping for different values of the parameter γ.

An interesting observation is made for a real component of the complex-valued temperature. One can observe cooling instead of heating in the compressed surface of the beam for specific ranges of frequency (negative values of the temperature disturbances). Furthermore, the finite speed of thermal wave propagation results in the existence of not one damping peak as in the classical theory, but many peaks, with a decreasing amplitude as the frequency tends to infinity (Fig. 12.4). The maximum value of damping peaks exceeds the classical prediction by ~ 5 (Si) to ~ 33 (diamond) times. It is important to note that in the L–S theory the maximum damping occurs at much lower frequencies than in the classical theory, which may result in high energy losses if the frequency of operation falls into this range. Conversely, a higher frequency of operation will result in much lower damping than expectations of the classical model.

The currently attainable frequencies for the fundamental flexural modes of thin nanobeams of dimension ($0.1 \times 0.01 \times 0.01\,\mu$m) are in the range of 1.9 to 12 GHz, which corresponds to $\Omega = 0.0044$–0.0279 for quartz, $\Omega = 0.0002$–0.0013 for silicon and $\Omega = 0.000017$–0.00011 for diamond. From examination of the graphical results, in light of real nanomechanical system vibrations, it can be concluded that, at frequencies attainable at present, second sound effect doesn't have any significant influence on damping capacity of the resonators and therefore a good enough approximation can be obtained using the classical theory. However, in the frequency range of 10^{12} Hz, the modified heat-conduction equation (R1.1.8) gives completely different results from the classical one and needs to be employed.

Remark 12.2 The thermoelastic damping of a beam whose material is governed by the G–L theory should next be studied.

12.5 Fractional calculus and fractals in thermoelasticity

12.5.1 *Anomalous heat conduction*

There exist many materials and physical situations where classical thermoelasticity, based on a Fourier-type heat conduction, breaks down: low-temperature regimes, amorphous media, colloids, glassy and porous materials, man-made and biological materials/polymers, transient loading, etc. In such cases one needs to use a generalized thermoelasticity (and, more generally, thermo-viscoelasticity) theory, based on an anomalous heat-conduction model involving time-fractional (non-integer order) derivatives.

A starting point of this brief review is the idea of Povstenko (2005a,b, 2008) who replaces the classical heat-conduction equation $\rho c\ \partial\vartheta/\partial t = k\nabla^2\vartheta$ for a macroscopically homogeneous isotropic medium by an *anomalous* one, i.e. by an equation with time-fractional (strictly speaking, non-integer) derivatives

$$\rho c \frac{\partial^\alpha \vartheta}{\partial t^\alpha} = k\nabla^2\vartheta, \tag{12.5.1}$$

where the real parameter $0 < \alpha \leq 2$, and ∇^2 indicates a Laplacian in physical space.

Remark 12.3 (i) Clearly, the Newtonian calculus, to which we are overwhelmingly accustomed, employs derivatives of integer order α only. (ii) The term "fractional" is a misnomer as it accommodates any real, not just a rational number.

Now, two important types of anomalous heat transport can be distinguished:

(i) the *subdiffusion regime*, characterized by $0 < \alpha < 1$;
(ii) the *superdiffusion regime*, characterized by $1 < \alpha \leq 2$.

In the language of statistical physics and stochastic processes (which, in fact, are the basis of all the diffusion phenomena), we would say that the particles of *subdiffusion* (respectively, *superdiffusion*) regime move slower (respectively, faster) than those of classical diffusion ($\alpha = 1$). The limiting case of $\alpha = 2$ is known as *ballistic diffusion* (Kimmich, 2002).

The superdiffusion case of eqn (12.5.1) has its origin in a time-non-local dependence between the heat flux and temperature gradient

$$q(t) = -k \int_0^t K(t-\tau)\nabla\vartheta(\tau)\mathrm{d}\tau, \tag{12.5.2}$$

with the kernel

$$K(t-\tau) = \frac{1}{\Gamma(\alpha-1)}(t-\tau)^{\alpha-2}, \quad 1 < \alpha < 2, \tag{12.5.3}$$

where Γ is the gamma function.

From the standpoint of individual particle motions, the anomalous diffusion is characterized by a mean-squared displacement having a power-law time dependence

$$\langle x^2 \rangle \sim t^{\alpha}, \quad \alpha \neq 1. \tag{12.5.4}$$

The case of $\alpha = 1$ corresponds to the classical Brownian process model. Several different random process models were proposed for the case of $\alpha \neq 1$.

The fractional derivative implied in eqn (12.5.1) is that of Caputo (1967, 1969):

$$\frac{\mathrm{d}^{\alpha}}{\mathrm{d}t^{\alpha}} f(t) \equiv D^{\alpha} f(t) = \begin{cases} \dfrac{1}{\Gamma(n-\alpha)} \displaystyle\int_0^t (t-\tau)^{n-\alpha-1} \dfrac{\mathrm{d}^n f(\tau)}{\mathrm{d}\tau^n} d\tau, & n-1 < \alpha < n, \\ \dfrac{\mathrm{d}^{\alpha}}{\mathrm{d}t^{\alpha}} f(t), & \alpha = n. \end{cases} \tag{12.5.5}$$

This derivative offers several well-known advantages over the Riemann–Liouville derivative

$$\frac{\mathrm{d}^\alpha}{\mathrm{d}t^\alpha} f(t) \equiv D_R^\alpha f(t) = \begin{cases} \dfrac{\mathrm{d}^n}{\mathrm{d}t^n}\left[\dfrac{1}{\Gamma(n-\alpha)}\displaystyle\int_0^t (t-\tau)^{n-\alpha-1} f(\tau)\mathrm{d}\tau\right], & n-1<\alpha<n, \\ \dfrac{\mathrm{d}^\alpha}{\mathrm{d}t^\alpha} f(t), & \alpha = n; \end{cases} \tag{12.5.6}$$

see also (Gorenflo and Mainardi, 1997, 1998).

In order to model the quasi-static thermal stresses due to the temperature field governed by eqn (12.5.1), Povstenko uses

$$\rho c \frac{\partial^\alpha \vartheta}{\partial t^\alpha} = K \nabla^2 \vartheta, \tag{12.5.7}$$

the equilibrium equation

$$\mu u_{i,kk} + (\lambda + \mu) u_{k,ki} = \beta(\lambda + 2\mu/3)\vartheta_{,i} \tag{12.5.8}$$

along with the stress–strain temperature relation of linear thermoelasticity

$$S_{ij} = 2\mu E_{ij} + [\lambda E_{kk} - \beta(\lambda + 2\mu/3)\vartheta]\delta_{ij}. \tag{12.5.9}$$

Here, to avoid a clash of notation, $\beta/3$ denotes the coefficient of linear thermal expansion (i.e. the same as α elsewhere in this book).

Remark 12.4 Since the superdiffusion case of eqn (12.5.1) interpolates the heat-conduction equation ($\alpha = 1$) and the wave equation ($\alpha = 2$), the thermoelasticity with eqns (12.5.1) and (12.5.8 and 9) interpolates the classical quasi-static thermoelasticity and the quasi-static thermoelasticity without energy dissipation due to Green and Naghdi (1993).

Several initial-boundary value problems for eqns (12.5.7)–(12.5.9) have been solved by Povstenko. To illustrate their salient features, we recall a 1D Cauchy problem (Povstenko, 2005b). In this case, eqns (12.5.7–9) are to be satisfied for $|x| < \infty$ and $t > 0$ together with the initial conditions

$$\begin{aligned} t = 0: \vartheta = p\,\delta(x) \text{ for } & \quad 0 < \alpha \le 2, \\ t = 0: \quad \frac{\partial \vartheta}{\partial t} = 0 \quad \text{for} & \quad 1 < \alpha \le 2, \end{aligned} \tag{12.5.10}$$

where p is a prescribed constant. Also, the boundary conditions ensuring vanishing at infinity are postulated

$$\left.\begin{aligned} \lim_{|x|\to\infty} u(x,t) = 0, \\ \lim_{|x|\to\infty} \vartheta(x,t) = 0, \end{aligned}\right\} \quad \forall t > 0. \tag{12.5.11}$$

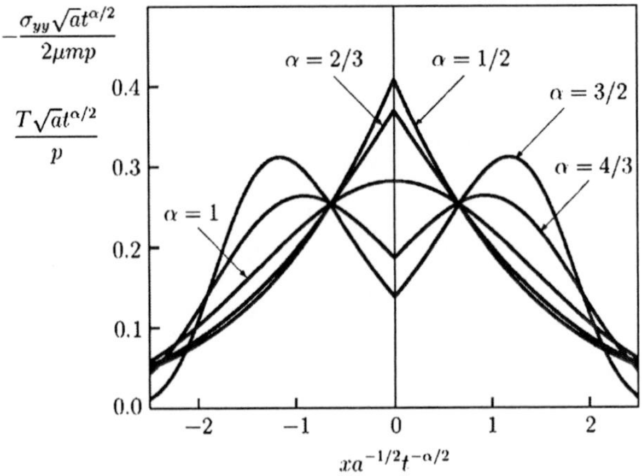

Figure 12.5 Dependence of temperature ϑ (denoted here by T) and stress σ_{yy} on the similarity variable z, after Povstenko (2005b).

The resulting temperature and stress distributions corresponding to various values of α (1/2, 2/3, 1, 4/3, 3/2) are shown in Fig. 12.5 for a similarity variable $z = xa^{-1/2}t^{-\alpha/2}$ where $a = K/\rho c$, and $m = \beta(3\lambda + 2\mu)/3(\lambda + 2\mu)$.

To summarize, only the heat transfer alone (i.e. in a rigid conductor) and its effect on the diffusive strains/stresses has been studied within the quasi-static theory, but no extension of the quasi-static solution to include the inertia terms in eqn (12.5.8) has been obtained as yet.

12.5.2 *Fractal media*

Continuum thermomechanics of fractal media

Continuum mechanics is naturally suited to deal primarily with media exhibiting spatially homogeneous properties. If the materials are heterogeneous and random, then it is hoped that the statistics are describable by the conventional Euclidean geometry and, therefore, by the conventional calculus. Needless to say, such media are ubiquitous in nature, yet they fall outside the realm of classical continuum mechanics. Another consideration is that, as already in Section 12.5.1, the constitutive responses of many materials are best described by fractional calculus. In physics, the connection between transport phenomena in fractal geometries and fractional models has been known for quite some time (Feder, 2007). Although many advances have been made, a solution of boundary value problems of fractal media in the vein of continuum mechanics is still an open issue. However, a first step in that direction has recently been taken by Tarasov (2005a,b), who developed the continuum-type equations of conservation of mass, linear momentum and energy for fractals from the standpoint of dimensional regularization, and, on that basis studied some fluid mechanics and wave problems. This allowed the development of a continuum-type expression of the second law of thermodynamics for fractal media in the setting of thermomechanics with internal variables (Ostoja-Starzewski, 2007a), from which followed a generalized energy balance equation coupling the temperature and mechanical fields (Ostoja-Starzewski, 2007b).

First, let the mass m in a fractal medium obey the power law relation

$$m(R) = kR^D, \qquad D < 3, \tag{12.5.12}$$

where R is a box size (or a sphere radius, effectively a lengthscale of measurement), D is a fractal dimension of mass, and k is a proportionality constant. It follows that the power law (12.5.12) describes the scaling of mass with R. Focusing on fractal porous media, that law is rewritten as

$$m_D(R) = m_0 \left(\frac{R}{R_{\mathrm{p}}} \right)^D, \tag{12.5.13}$$

where R_{p} is the average pore radius, and m_0 is the mass at $R_p = R$; this is a reference case. With eqn (12.5.13) we use D to denote the fractal dimension of mass in a domain W, while the boundary ∂W of W has dimension d. In general, d equals neither 2 nor $D - 1$.

At this point, the conventional equation giving mass in a 3D region W

$$m(W) = \int_W \rho(\mathbf{r})\, \mathrm{d}^3\mathbf{r} \tag{12.5.14}$$

has to be generalized to

$$m_D(W) = \frac{2^{3-D}\Gamma(3/2)}{\Gamma(D/2)} \int_W \rho(\mathbf{r}) \left|\mathbf{r} - \mathbf{r}_0\right|^{D-3} \mathrm{d}^3\mathbf{r}, \tag{12.5.15}$$

where $\mathbf{r}_0$ is a fixed point of W. Assuming the fractal medium to be spatially homogeneous

$$\rho(\mathbf{r}) = \rho_0 = const, \tag{12.5.16}$$

eqn (12.5.14) is replaced by the fractional integral

$$m_D(W) = \frac{\pi 2^{5-D}\Gamma(3/2)}{\Gamma(D/2)} \rho_0 \int_W R^{D-1} \mathrm{d}R = \frac{\pi 2^{5-D}\Gamma(3/2)}{D\Gamma(D/2)} \rho_0 R^D, \tag{12.5.17}$$

where $\mathbf{R} = \mathbf{r} - \mathbf{r}_0$ and $|\mathbf{R}| = R$. That is, the fractal medium with a non-integer mass dimension D is described using a fractional integral of order D. This allows an interpretation of the fractal (intrinsically discontinuous) medium as a continuum. In particular, the next step is Tarasov's reformulation of the Green–Gauss Theorem

$$\int_{\partial W} f v_k n_k \mathrm{d}A_{\mathrm{d}} = \int_W c_3^{-1}(D, R)\, div\left(c_2(d, R)\, f\mathbf{v}\right) \mathrm{d}V_D, \tag{12.5.18}$$

where f is an arbitrary function, $\mathbf{v}$ is the velocity, and the infinitesimal surface ($\mathrm{d}A_{\mathrm{d}}$) and volume ($\mathrm{d}V_D$) elements of the fractal body are related to the conventional infinitesimal elements $\mathrm{d}A_2$ ($\equiv \mathrm{d}A$) and $\mathrm{d}V_3$ ($\equiv \mathrm{d}V$) by

$$\mathrm{d}A_{\mathrm{d}} = c_2\,(d,R)\,\mathrm{d}A_2, \qquad \mathrm{d}V_D = c_3\,(D,R)\,\mathrm{d}V_3. \tag{12.5.19}$$

On account of eqns (12.5.19), the left-hand side in eqn (12.5.18) is a fractional integral, equal to a conventional integral $\int_{\partial W} c_2\,(d,R)\,f\mathbf{v}\cdot\mathbf{n}\mathrm{d}A_2$. Similarly, the right-hand side in eqn (12.5.18) is a fractional integral equal to a conventional integral $\int_W div\,(c_2\,(d,R)\,f\mathbf{v})\,\mathrm{d}V_3$.

Tarasov (2005a,b) gave expressions for c_2 and c_3 based on a Riesz measure. These are very well suited for isotropic fractal media, and, as a result, the equations governing problems in 1D cannot be consistently obtained from the equations governing problems in 3D. For example, the one-dimensional fractal wave equation is not equivalent to that of a plane wave in three dimensions. That drawback can be removed by introducing a product measure instead, whereby a possible anisotropy is also incorporated, further ensuring that the mechanical approach to continuum mechanics is consistent with the energetic approach (Li and Ostoja-Starzewski, 2009). To this end, note that, while the mass distribution in conventional continuum mechanics is

$$\mathrm{d}\mu(\mathbf{x}) = \rho(\mathbf{x})\mathrm{d}V_3, \tag{12.5.20}$$

where $\rho(x)$ is mass density and $\mathrm{d}V_3$ is the Lebesgue measure in $\mathbb{R}^3$, the product measure we now introduce is

$$\mathrm{d}\mu_k(x_k) = \rho(\mathbf{x})c_1(\alpha_k, x_k)\mathrm{d}x_k, \quad k = 1,2,3. \tag{12.5.21}$$

Thus, while eqn (12.5.20) applies to a non-fractal mass distribution $M \sim x_1x_2x_3$, eqn (12.5.21) applies to a fractal mass distribution $M \sim x_1^{\alpha_1}x_2^{\alpha_2}x_3^{\alpha_3}$, the total fractal dimension being $D = \alpha_1 + \alpha_2 + \alpha_3$. For simplicity, here we adopt a form based on a Riemann–Liouville integral

$$c_1^{(k)} = \frac{|x_k|^{\alpha_k - 1}}{\Gamma(\alpha_k)}, \quad k = 1,2,3, \tag{12.5.22}$$

so as to give

$$\begin{aligned} c_2^{(k)} &= c_1^{(i)}c_1^{(j)} = \frac{|x_i|^{\alpha_i-1}|x_j|^{\alpha_j-1}}{\Gamma(\alpha_i)\Gamma(\alpha_j)}, \quad i,j \neq k, \\ c_3 &= c_1^{(1)}c_1^{(2)}c_1^{(3)} = \frac{|x_1|^{\alpha_1-1}|x_2|^{\alpha_2-1}|x_3|^{\alpha_3-1}}{\Gamma(\alpha_1)\Gamma(\alpha_2)\Gamma(\alpha_3)}. \end{aligned} \tag{12.5.23}$$

One can show that eqns (12.5.23) are consistent with eqns (12.5.19). The ensuing calculus of tensor fields, relying on eqn (12.5.18), involves the following operator (or, generalized derivative) of spatial gradient

$$\begin{aligned} \nabla_k^D f &= c_3^{-1}(D,R)\frac{\partial}{\partial x_k}\left[c_2(d,R) f\right] \\ &\equiv c_3^{-1}(D,R)\nabla_k\left[c_2(d,R) f\right] = \left[c_1^{(k)}\right]^{-1}\nabla_k f, \end{aligned} \tag{12.5.24}$$

where the last equality follows from the relations (12.5.22) and (12.5.23).

On the other hand, a derivation of the Reynolds transport theorem for fractal media undergoing finite motions (Ostoja-Starzewski, 2009a)

$$\frac{\mathrm{d}}{\mathrm{d}t}\int_W f(x,t)\,\mathrm{d}V_D = \int_W \left[\frac{\partial}{\partial t} f + (f v_k)_{,k}\right]\mathrm{d}V_D \tag{12.5.25}$$

dictates the conventional material derivative

$$\frac{\mathcal{D}f}{\mathcal{D}t} = \frac{\partial f}{\partial t} + v_k\frac{\partial f}{\partial x_k}. \tag{12.5.26}$$

This is in contrast to $(\mathcal{D}/\mathcal{D}t)_D = \partial f/\partial t + c(D,d,R)v_k f_{,k}$, with $c(D,d,R) = c_3^{-1}(D,R)\,c_2(d,R)$, of Tarasov's formulation, which was adopted rather intuitively. In consequence, in all previous results for mechanics of fractal media $(\mathcal{D}/\mathcal{D}t)_D$ is to be simply replaced by the conventional material derivative $\mathcal{D}/\mathcal{D}t$, thereby leading to certain simplifications. In particular, the balance equations of fractal media become:

- the fractional equation of continuity:

$$\frac{\mathcal{D}\rho}{\mathcal{D}t} = -\rho\nabla_k^D v_k, \tag{12.5.27}$$

- the fractional equation of balance of linear momentum density:

$$\rho\frac{\mathcal{D}v_k}{\mathcal{D}t} = \rho f_k + \nabla_l^D S_{kl}, \tag{12.5.28}$$

- the fractional equation of balance of internal energy density:

$$\rho\frac{\mathcal{D}u}{\mathcal{D}t} = S_{kl}v_{k,l} - \nabla_k^D q_k, \tag{12.5.29}$$

- the second law of thermodynamics (unchanged from TIV):

$$S_{ij}^{(d)} d_{ij} + \beta_{ij}^{(d)}\dot{\alpha}_{ij} - \frac{\theta_{,k}\, q_k}{\theta} \geq 0. \tag{12.5.30}$$

In the above, S_{kl} is the Cauchy stress (symmetric according to the balance of angular momentum, employed just like in non-fractal media), although a generalization to fractal micropolar media is possible. As usual, d_{ij} is the deformation rate, α_{ij} is the internal parameter tensor, and $\beta_{ij}^{(d)}$ is the dissipative internal stress tensor. Note that, in a non-fractal medium ($D = 3,\ d = 2$) $c_2^{(k)} = 1$ and $c_3 = 1$, whereby one recovers conventional forms of local relations of continuum mechanics.

Thermodynamics and Thermoelasticity

Constitutive laws of fractal media now follow from the above, and a number of key relations of Ziegler and Wehrli (1987) – such as laws governing complex and compound processes and the associated Onsager reciprocity conditions and Legendre transformations – carry over to fractal media. In particular, the energy balance equation of classical non-linear thermoelastodynamics, coupling the thermal and mechanical fields, is generalized to

$$\rho c_p \frac{\theta}{\theta_0} \dot{\theta} = -(3\lambda + 2\mu)\alpha\theta \dot{E}_{(1)} - \nabla_i^D q_i, \tag{12.5.31}$$

where $\dot{E}_{(1)}$ is the first invariant of the strain rate tensor, c_p is the specific heat at constant pressure, while $\theta_0 > 0$ is a reference temperature. In the special case of $\dot{E}_{(1)} = 0$ we have the equation describing heat conduction in a fractal rigid conductor

$$\rho c_p \frac{\theta}{\theta_0} \dot{\theta} = -\nabla_i^D q_i \ . \tag{12.5.32}$$

Assuming a Fourier-type heat flow everywhere in the fractal medium, a linearization of eqn (12.5.31) leads to the generalization of energy balance equation

$$\rho c_p \dot{\theta} = -(3\lambda + 2\mu)\alpha\theta_0 \dot{E}_{(1)} + \nabla_i^D \left(k \frac{\partial \theta}{\partial x_i} \right), \tag{12.5.33}$$

while eqn (12.5.32) becomes

$$\rho c_p \dot{\theta} = \nabla_i^D \left(k \frac{\partial \theta}{\partial x_i} \right), \tag{12.5.34}$$

where k is the thermal conductivity of the material.

Turning to the second sound in a rigid conductor with fractal geometry, following Sections 12.1.2 and 12.1.3 we first adopt the Maxwell–Cattaneo equation (12.1.14). Given the remarks following eqns (12.5.24) and (12.5.26), this leads to a telegraph equation for fractal materials (a generalization of eqn (12.1.15))

$$t_0 \rho c_p \ddot{\theta} + \rho c_p \dot{\theta} = \left[c_1^{(k)} \right]^{-1} k \frac{\partial^2 \theta}{\partial x^2}, \tag{12.5.35}$$

with the coefficient $c_1^{(k)}$ accounting for the fractal structure in one dimension, recall eqn (12.5.22). In this vein, the entire approach may be extended to hyperbolic thermoelasticity in fractal deformable materials, which, however, is a subject matter presently outside this book. On the other hand, some progress has already been made on other topics in mechanics of fractal media: transport equations, extremum and variational principles in elastic and inelastic materials, turbulence and fracture mechanics (Ostoja-Starzewski, 2008b, 2009a,b; Ostoja-Starzewski and Li, 2009).

13

NON-LINEAR HYPERBOLIC RIGID HEAT CONDUCTOR OF THE COLEMAN TYPE

In Chapters 1–11 an emphasis on the mathematical development of the linear hyperbolic theory of thermoelasticity was made, while Chapter 12 focused on several physical aspects and micromechanical applications of the theory. In the present chapter a rigid but non-linear hyperbolic heat conductor; that obeys the law of conservation of energy, the dissipation inequality, Cattaneo's equation, and a generalized energy–entropy relation with a parabolic variation of the energy and entropy along the heat flux axis (Coleman *et al.*, 1982, 1983, 1986); is analyzed. In Section 13.1 the basic field equations for a 1D case are recalled, while in Section 13.2 a number of closed-form solutions to the non-linear governing equations are obtained. Finally, in Section 13.3 a method of weakly non-linear geometric optics is applied to obtain an asymptotic solution to the Cauchy problem with a weakly perturbed initial condition associated with the non-linear model (Ignaczak and Domański, 2008).

13.1 Basic field equations for a 1D case

A 1D non-linear homogeneous isotropic rigid heat conductor proposed by Coleman *et al.* (1982, 1986) obeys the following field equations.

The law of conservation of energy

$$\frac{\partial e}{\partial t} = -\frac{\partial q}{\partial x} + r. \tag{13.1.1}$$

The dissipation inequality

$$\frac{\partial \sigma}{\partial t} \equiv \frac{\partial s}{\partial t} + \frac{\partial}{\partial x}\left(\frac{q}{T}\right) - \frac{r}{T} \geq 0. \tag{13.1.2}$$

Cattaneo's equation

$$q + \tau\frac{\partial q}{\partial t} = -k\frac{\partial T}{\partial x}. \tag{13.1.3}$$

The energy–entropy relation

$$\frac{\mathrm{d}e}{T} = \mathrm{d}s + \frac{\tau}{k}\frac{q}{T^2}\mathrm{d}q, \tag{13.1.4}$$

where

$$e = e(T, q) = e_0(T) + \frac{\tau}{k}\frac{1}{T}q^2, \tag{13.1.5}$$

$$s = s(T, q) = s_0(T) + \frac{1}{2}\frac{\tau}{k}\frac{1}{T^2}q^2, \tag{13.1.6}$$

and

$$\frac{\mathrm{d}e_0}{T} = \mathrm{d}s_0, \qquad \frac{\mathrm{d}e_0}{\mathrm{d}T} = \rho c. \tag{13.1.7}$$

In eqns (13.1.1)–(13.1.7), $T = T(x,t)$ and $q = q(x,t)$ represent the absolute temperature and heat flux, respectively; $e = e(x,t)$ and $s = s(x,t)$ denote the internal energy and entropy, respectively; $\sigma = \sigma(x,t)$ and $r = r(x,t)$ are the entropy production and external heat source, respectively; τ and k are the relaxation time and thermal conductivity, respectively; while ρ and c represent the density and specific heat, respectively.

The material parameters obey the inequalities

$$\tau > 0, \qquad k > 0, \qquad \rho > 0, \qquad c > 0. \tag{13.1.8}$$

Note that an equivalent form of eqn (13.1.4) reads

$$\frac{1}{T}\left(\frac{\partial e}{\partial T}\mathrm{d}T + \frac{\partial e}{\partial q}\mathrm{d}q\right) = \frac{\partial s}{\partial T}\mathrm{d}T + \frac{\partial s}{\partial q}\mathrm{d}q + \frac{\tau}{k}\frac{q}{T^2}\mathrm{d}q, \tag{13.1.9}$$

or

$$\frac{1}{T}\frac{\partial e}{\partial T} = \frac{\partial s}{\partial T}, \tag{13.1.10}$$

and

$$\frac{1}{T}\frac{\partial e}{\partial q} = \frac{\partial s}{\partial q} + \frac{\tau}{k}\frac{q}{T^2}. \tag{13.1.11}$$

By substituting e and s from eqns (13.1.5) and (13.1.6), respectively, into eqns (13.1.10) and (13.1.11), and using eqn (13.1.7) we find that eqns (13.1.10) and (13.1.11) are identically satisfied. Therefore, the energy–entropy relation (13.1.4) is identically satisfied.

Also, note that the dissipation inequality (13.1.2) is satisfied in the following sense. By dividing eqn (13.1.1) by $T > 0$ and using eqn (13.1.3) we obtain

$$\begin{aligned}\frac{1}{T}\frac{\partial e}{\partial t} &= -\frac{1}{T}\frac{\partial q}{\partial x} + \frac{r}{T} = -\frac{\partial}{\partial x}\left(\frac{q}{T}\right) - \frac{1}{T^2}\frac{\partial T}{\partial x}q + \frac{r}{T} \\ &= -\left[\frac{\partial}{\partial x}\left(\frac{q}{T}\right) - \frac{r}{T}\right] + \frac{q}{kT^2}\left(q + \tau\frac{\partial q}{\partial t}\right).\end{aligned} \tag{13.1.12}$$

Next, it follows from eqns (13.1.5)–(13.1.7) that

$$\frac{1}{T}\frac{\partial e}{\partial t} = \frac{\partial s}{\partial t} + \frac{\tau}{k}\frac{q}{T^2}\frac{\partial q}{\partial t}. \tag{13.1.13}$$

Hence, substituting eqn (13.1.13) into eqn (13.1.12) we obtain

$$\frac{\partial \sigma}{\partial t} \equiv \frac{\partial s}{\partial t} + \frac{\partial}{\partial x}\left(\frac{q}{T}\right) - \frac{r}{T} = \frac{q^2}{kT^2} \geq 0. \tag{13.1.14}$$

Note that if the second term on the RHS of eqn (13.1.4) as well as the second terms on the RHSs of eqns (13.1.5) and (13.1.6) are ignored, we arrive at a classical rigid heat conductor of Cattaneo's type for which the dissipation inequality (13.1.2) is satisfied in an approximate form [see p. 290 in (Ignaczak, 1989a)].

Finally, note that by substituting eqn (13.1.5) into eqn (13.1.1) and combining the result with eqn (13.1.3), the following non-linear field equations in terms of a pair (T, q) are obtained

$$\begin{aligned} &\left(\rho c - \frac{\tau}{k}\frac{q^2}{T^2}\right)\frac{\partial T}{\partial t} + 2\frac{\tau}{k}\frac{q}{T}\frac{\partial q}{\partial t} + \frac{\partial q}{\partial x} = r, \\ &\frac{1}{k}\left(q + \tau\frac{\partial q}{\partial t}\right) + \frac{\partial T}{\partial x} = 0, \end{aligned} \tag{13.1.15}$$

or

$$\begin{aligned} &\frac{\partial}{\partial t}\left[\rho c T + \frac{\tau}{k}\left(\frac{q^2}{T}\right)\right] + \frac{\partial q}{\partial x} = r, \\ &\frac{1}{k}\left(q + \tau\frac{\partial q}{\partial t}\right) + \frac{\partial T}{\partial x} = 0. \end{aligned} \tag{13.1.16}$$

If we introduce the dimensionless fields

$$\theta = \frac{T}{T_0}, \qquad Q = \frac{q}{q_0}, \qquad R = \frac{r}{r_0}, \tag{13.1.17}$$

and the dimensionless variables

$$\bar{x} = \frac{x}{x_0}, \qquad \bar{t} = \frac{t}{t_0}, \tag{13.1.18}$$

where

$$x_0 = \frac{2\kappa}{v\nu_0}, \qquad t_0 = \frac{2\kappa}{\nu_0^2}, \tag{13.1.19}$$

$$\kappa = \frac{k}{\rho c}, \qquad \nu_0 = \sqrt{\frac{k}{\tau}}, \tag{13.1.20}$$

$$q_0 = \frac{k\nu_0 T_o}{\kappa}, \qquad r_0 = \frac{\rho c T_0}{t_0}, \tag{13.1.21}$$

and T_0 is a reference temperature, and omit the bars over the variables, we arrive at the dimensionless form of eqns (13.1.16)

$$\begin{aligned} \frac{\partial}{\partial t}\left(\theta + \frac{Q^2}{\theta}\right) + \frac{\partial Q}{\partial x} &= R, \\ \frac{\partial Q}{\partial t} + 2Q + \frac{\partial \theta}{\partial x} &= 0. \end{aligned} \tag{13.1.22}$$

If $R = 0$, eqns (13.1.22) reduce to eqns (21) and (22) from (Bai and Lavine, 1995).

Also, note that ν_0 in eqn (13.1.20) represents a "thermal propagation speed," and $\nu_0 \to \infty$ as $\tau \to 0+0$. Therefore, a dimensionless heat-conduction process (θ, Q) corresponds to a physical heat-conduction process (T, q) for which $x_0 \to 0, t_0 \to 0, q_0 \to 0$ and $r_0 \to 0$ as $\tau \to 0+0$. In addition, $\nu_0 = x_0/t_0 \to \infty$ as $\tau \to 0+0$. Since the non-linear model is expected to transmit thermal waves with finite speeds when τ is small, a thermal wave obeying eqns (13.1.22) is to propagate with a dimensionless speed $\nu < \nu_0$.

13.2 Closed-form solutions

13.2.1 *Closed-form solution to a time-dependent heat-conduction Cauchy problem*

We let

$$R(x,t) \equiv R(t) = R_0 \exp(-\alpha t), \tag{13.2.1}$$

where

$$R_0 > 0 \quad \text{and} \quad \alpha > 0, \tag{13.2.2}$$

and formulate the following Cauchy problem. Find a pair $[\theta(t), Q(t)]$ that satisfies

$$\left(\frac{\mathrm{d}}{\mathrm{d}t} + 2\right) Q = 0 \quad \text{for} \quad t > 0, \tag{13.2.3}$$

$$\frac{\mathrm{d}}{\mathrm{d}t}\left(\theta + \frac{Q^2}{\theta}\right) = R \quad \text{for} \quad t > 0, \tag{13.2.4}$$

subject to the initial conditions

$$\theta(0) = \theta_0, \quad Q(0) = Q_0, \tag{13.2.5}$$

where θ_0 and Q_0 are prescribed dimensionless constants such that

$$\theta_0 > Q_0 > 0. \tag{13.2.6}$$

We are to show that there is a unique solution to the non-linear initial problem (13.2.3)–(13.2.6), and the solution is expressed in terms of elementary functions. First, we note that a solution $Q = Q(t)$ of eqn (13.2.3) subject to the initial condition $(13.2.5)_2$ takes the form

$$Q(t) = Q_0 \exp(-2t) \quad \text{for} \quad t \geq 0, \tag{13.2.7}$$

Next, integrating eqn (13.2.4) over the interval $[0, t]$ and using eqns (13.2.1) and $(13.2.5)_1$, we obtain

$$\theta + \frac{Q^2}{\theta} = \theta_0 + \frac{Q_0^2}{\theta_0} + \frac{R_0}{\alpha}[1 - \exp(-\alpha t)] \quad \text{for} \quad t \geq 0. \tag{13.2.8}$$

Finally, substituting $Q = Q(t)$ from eqn (13.2.7) into the LHS of eqn (13.2.8), and solving the resulting quadratic equation in θ, we obtain

$$\theta(t) = \frac{1}{2\theta_0}\left[\theta_0^2 + Q_0^2 + \frac{R_0\theta_0}{\alpha}[1 - \exp(-\alpha t)]\right]$$
$$+\frac{1}{2\theta_0}\left\{\left[\theta_0^2 + Q_0^2 + \frac{R_0\theta_0}{\alpha}[1 - \exp(-\alpha t)]\right]^2 - 4\theta_0^2 Q_0^2 \exp(-4t)\right\}^{1/2}. \tag{13.2.9}$$

The second solution of the quadratic equation in θ with minus in front of the square root has to be rejected because the solution vanishes as $t \to \infty$, and only positive temperatures are admissible for every $t \geq 0$. To show that $\theta = \theta(t)$ given by eqn (13.2.9) is a solution of eqn (13.2.8), we note that an equivalent form of eqn (13.2.8) reads

$$(\theta - \theta^*)^2 = (\theta^*)^2 - Q_0^2 \exp(-4t), \tag{13.2.10}$$

where

$$\theta^* = \theta^*(t) = \frac{1}{2\theta_0}\left[\theta_0^2 + Q_0^2 + \frac{R_0\theta_0}{\alpha}[1 - \exp(-\alpha t)]\right]. \tag{13.2.11}$$

Also, note that

$$\varphi(t) \equiv (\theta^*)^2 - Q_0^2 \exp(-4t) > 0 \quad \text{for} \quad t \geq 0, \tag{13.2.12}$$

since

$$\varphi(0) = \frac{1}{\theta_0^2}(\theta_0^2 - Q_0^2)^2 > 0 \tag{13.2.13}$$

and

$$\varphi'(t) = R_0\theta^*(t)\exp(-\alpha t) + 4Q_0^2 \exp(-4t) > 0 \quad \text{for} \quad t \geq 0. \tag{13.2.14}$$

Therefore, the only solution $\theta = \theta(t)$ of eqn (13.2.10) that is positive for $t \geq 0$ takes the form

$$\theta(t) = \theta^*(t) + \sqrt{[\theta^*(t)]^2 - Q_0^2 \exp(-4t)} \tag{13.2.15}$$

that is identical to eqn (13.2.9). This completes the proof that $\theta = \theta(t)$ given by eqn (13.2.9) satisfies eqn (13.2.8).

Since

$$\theta^*(0) = \frac{1}{2\theta_0}(\theta_0^2 + Q_0^2), \tag{13.2.16}$$

therefore, letting $t = 0$ in eqn (13.2.15), we obtain

$$\theta(0) = \frac{1}{2\theta_0}\left[\theta_0^2 + Q_0^2 + |\theta_0^2 - Q_0^2|\right]. \tag{13.2.17}$$

This, together with the inequalities (13.2.6), implies that $\theta = \theta(t)$ also satisfies the initial condition $(13.2.5)_1$. As a result, the pair $[\theta(t), Q(t)]$ in which $Q(t)$ and $\theta(t)$ are given by eqns (13.2.7) and (13.2.9), respectively, represents a solution to the Cauchy problem.

Note that the solution possesses the properties

$$[\theta(t), Q(t)] \to (\theta_0, Q_0) \quad \text{as} \quad t \to 0 \tag{13.2.18}$$

and

$$[\theta(t), Q(t)] \to (\theta_\infty, 0) \quad \text{as} \quad t \to \infty, \tag{13.2.19}$$

where

$$\theta_\infty = \frac{1}{\theta_0}\left(\theta_0^2 + Q_0^2 + \frac{R_0\theta_0}{\alpha}\right) > \theta_0 > 0. \tag{13.2.20}$$

Since

$$\theta'(t) > 0 \quad \text{for} \quad t \geq 0, \tag{13.2.21}$$

therefore the temperature $\theta = \theta(t)$ is a monotonically increasing function of time. On the other hand, by eqn (13.2.7)

$$Q'(t) < 0 \quad \text{for} \quad t \geq 0, \tag{13.2.22}$$

and, therefore, $Q = Q(t)$ is a monotonically decreasing function of time. Plots of the functions $Q = Q(t)$ and $\theta = \theta(t)$ are shown in Fig. 13.1.

The following corollary, in which a new physical property of the non-linear heat-conduction model of Coleman's type is revealed, holds true.

Corollary 13.1 *In the non-linear theory of a rigid heat conductor proposed by Coleman et al. (1982, 1983, 1986) there is a closed-form time-dependent heat-conduction process $[\theta(t), Q(t)]$ that represents a homogeneous heating with a zero spatial temperature gradient during which the entropy production rate is positive for every $t \geq 0$.*

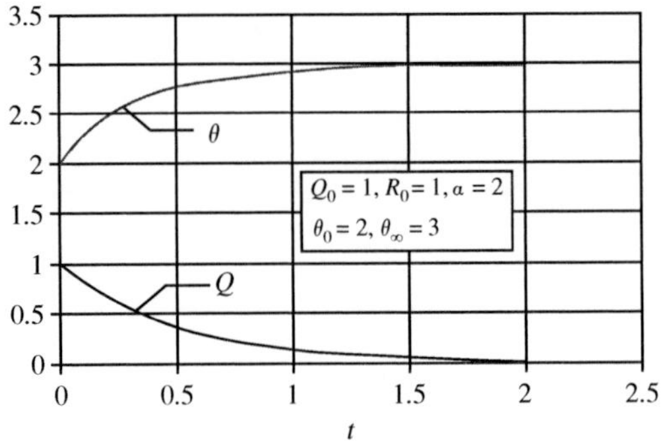

Figure 13.1 Plot of the functions $Q = Q(t)$ and $\theta = \theta(t)$ over the time interval $[0, \infty)$.

13.2.2 *Travelling-wave solutions*

The travelling-wave solutions are related to eqns (13.1.22) with $R = 0$:

$$\begin{aligned} \frac{\partial}{\partial t}\left(\theta + \frac{Q^2}{\theta}\right) + \frac{\partial Q}{\partial x} &= 0, \\ \frac{\partial Q}{\partial t} + 2Q + \frac{\partial \theta}{\partial x} &= 0. \end{aligned} \tag{13.2.23}$$

By letting

$$\theta = \theta(\xi), \qquad Q = Q(\xi), \tag{13.2.24}$$

where

$$\xi = x - vt, \qquad |x| \leq \infty, \qquad t \geq 0, \tag{13.2.25}$$

and $v > 0$ is a dimensionless velocity, and substituting eqns (13.2.24) into eqns (13.2.23), the following non-linear ordinary differential equations are obtained

$$\begin{aligned} \dot{\theta}[v(Q^2\theta^{-2} - 1)] + \dot{Q}(1 - 2vQ\theta^{-1}) &= 0, \\ \dot{\theta} - v\dot{Q} &= -2Q, \end{aligned} \tag{13.2.26}$$

or, equivalently,

$$\begin{aligned} \dot{\theta} &= \frac{2vQ\left(2\dfrac{Q}{\theta} - \dfrac{1}{v}\right)}{v^2\left(\dfrac{Q}{\theta} - \dfrac{1}{v} - 1\right)\left(\dfrac{Q}{\theta} - \dfrac{1}{v} + 1\right)}, \\ \dot{Q} &= \frac{2vQ\left(\dfrac{Q^2}{\theta^2} - \dfrac{1}{v}\right)}{v^2\left(\dfrac{Q}{\theta} - \dfrac{1}{v} - 1\right)\left(\dfrac{Q}{\theta} - \dfrac{1}{v} + 1\right)}, \end{aligned} \tag{13.2.27}$$

provided

$$\left(\frac{Q}{\theta} - \frac{1}{v}\right)^2 - 1 \neq 0. \tag{13.2.28}$$

In eqns (13.2.26) and (13.2.27) the superimposed dot represents the derivative with respect to ξ ($\cdot = \mathrm{d}/\mathrm{d}\xi$); and the straight lines of the (θ, Q)-plane

$$\frac{Q}{\theta} = \frac{1}{v} + 1, \quad \frac{Q}{\theta} = \frac{1}{v} - 1 \tag{13.2.29}$$

represent the characteristics of eqns (13.2.27) at which $\dot{\theta}$ and $\dot{Q}$ become unbounded.

Also note that, if θ is treated as a function of Q, that means if $\theta = \theta(Q)$, then it follows from eqns (13.2.27) that

$$\frac{\mathrm{d}\theta}{\mathrm{d}Q} = 2\frac{\dfrac{Q}{\theta}\left(1 - \dfrac{1}{2v}\dfrac{\theta}{Q}\right)}{\left(\dfrac{Q}{\theta}\right)^2 - 1}, \tag{13.2.30}$$

provided

$$\frac{Q^2}{\theta^2} - 1 \neq 0. \tag{13.2.31}$$

Therefore, the straight lines of the (θ, Q)-plane

$$\frac{Q}{\theta} = 1, \quad \frac{Q}{\theta} = -1 \tag{13.2.32}$$

are the characteristic lines for eqn (13.2.30) at which $\mathrm{d}\theta/\mathrm{d}Q$ becomes unbounded.

Moreover, if

$$0 < v < \frac{1}{2}, \tag{13.2.33}$$

then, eqn (13.2.30) is satisfied by the straight lines

$$\theta_+ = \left[\frac{1}{2v} + \sqrt{\left(\frac{1}{2v}\right)^2 - 1}\right] Q_+ \tag{13.2.34}$$

and

$$\theta_- = \left[\frac{1}{2v} - \sqrt{\left(\frac{1}{2v}\right)^2 - 1}\right] Q_-. \tag{13.2.35}$$

Figure 13.2 shows the solutions (13.2.34) and (13.2.35) in the (θ, Q)-plane.

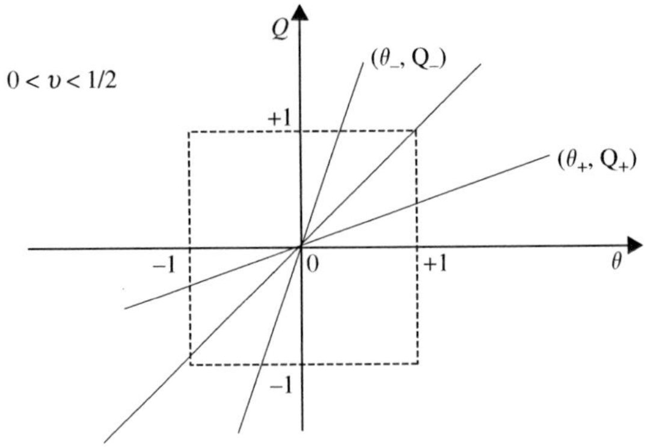

Figure 13.2 Locus of points in the (θ, Q) plane obeying eqns (13.2.34) and (13.2.35).

To obtain another solution to eqn (13.2.30), we note that eqn (13.2.30) can be written as

$$\frac{\mathrm{d}\theta}{\mathrm{d}Q} = f\left(\frac{\theta}{Q}\right), \tag{13.2.36}$$

where

$$f\left(\frac{\theta}{Q}\right) = 2\frac{\frac{\theta}{Q}\left(1 - \frac{1}{2\upsilon}\frac{\theta}{Q}\right)}{1 - \left(\frac{\theta}{Q}\right)^2}. \tag{13.2.37}$$

By letting

$$U = \frac{\theta}{Q}, \tag{13.2.38}$$

in which $U = U(Q)$ and $\theta = \theta(Q)$, or equivalently,

$$QU(Q) = \theta(Q), \tag{13.2.39}$$

and taking the total derivative of eqn (13.2.39) with respect to θ, we obtain

$$\mathrm{d}\theta = U\mathrm{d}Q + Q\mathrm{d}U, \tag{13.2.40}$$

or

$$\frac{\mathrm{d}\theta}{\mathrm{d}Q} = U + \frac{Q}{dQ}\mathrm{d}U. \tag{13.2.41}$$

Substituting eqn (13.2.41) into the LHS of eqn (13.2.36) and using the notation (13.2.38) we obtain

$$\frac{\mathrm{d}Q}{Q} = \frac{\mathrm{d}U}{f(U) - U}, \tag{13.2.42}$$

where

$$f(U) = 2\frac{U\left(1 - \frac{1}{2v}U\right)}{1 - U^2}. \tag{13.2.43}$$

By restricting the parameter v to the interval defined by the inequality (13.2.33)[1] we find that

$$\frac{1}{f(U) - U} = \frac{1}{U} - \frac{1}{U - U_1} - \frac{1}{U - U_2}, \tag{13.2.44}$$

where

$$U_1 = \frac{1}{2v} + \sqrt{\left(\frac{1}{2v}\right)^2 - 1} \tag{13.2.45}$$

and

$$U_2 = \frac{1}{2v} - \sqrt{\left(\frac{1}{2v}\right)^2 - 1}. \tag{13.2.46}$$

Therefore, integrating eqn (13.2.42) we obtain

$$\int_{Q_0}^{Q} \frac{\mathrm{d}q}{q} = \int_{U_0}^{U} \frac{\mathrm{d}u}{f(u) - u}, \tag{13.2.47}$$

where $Q_0 > 0$ is an arbitrary constant and

$$U_0 = U(Q_0) = \theta(Q_0)/Q_0 > 0. \tag{13.2.48}$$

Finally, substituting eqn (13.2.44) into the RHS of eqn (13.2.47), and computing the integrals we obtain

$$|Q| = Q_0 \left| \frac{U(U_0 - U_1)(U_0 - U_2)}{U_0(U - U_1)(U - U_2)} \right|. \tag{13.2.49}$$

In the following we let

$$Q > 0. \tag{13.2.50}$$

[1] One can prove that, for $v \geq 1/2$, a travelling wave propagates with an infinite velocity as $\tau \to 0 + 0$.

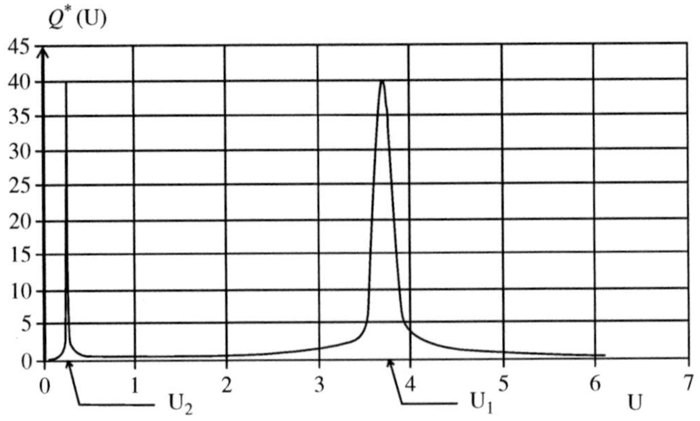

Figure 13.3 Plot of the function $Q^*(U) = Q(U)/Q_0^*$ for $U > 0$.

Then, it follows from the definition of U [see eqn (13.2.38)] that $U > 0$, and from eqn (13.2.49) we obtain

$$Q = Q_0^* \frac{U}{|(U - U_1)(U - U_2)|}, \tag{13.2.51}$$

where

$$Q_0^* = \frac{Q_0}{U_0} |(U_0 - U_1)(U_0 - U_2)| \,. \tag{13.2.52}$$

A plot of function $Q^* = Q^*(U) = Q(U)/Q_0^*$ for $U > 0$ is shown in Fig. 13.3.

Note that, by selecting U_0 in the form

$$U_0 = \frac{1}{2}\left(\frac{1}{v} + \frac{1}{Q_0}\right) + \sqrt{\frac{1}{4}\left(\frac{1}{v} + \frac{1}{Q_0}\right)^2 - 1} \quad \text{for} \quad Q_0 > 0, \tag{13.2.53}$$

we obtain

$$Q_0^* = 1. \tag{13.2.54}$$

Also, note that

$$U_0 = U_0(Q_0) \tag{13.2.55}$$

and

$$U_0 \to U_1 \quad \text{as} \quad Q_0 \to \infty, \tag{13.2.56}$$

$$U_0 \to \infty \quad \text{as} \quad Q_0 \to 0. \tag{13.2.57}$$

From now on, we assume that the condition (13.2.54) holds true. It follows then from Fig. 13.3 that $Q^* = Q(U)$ is unbounded at $U = U_1$ and $U = U_2$; and it is invertible on each of the four intervals: (i) $0 < U < U_2$, (ii) $U_2 < U \leq 1$,

(iii) $1 \leq U < U_1$, and (iv) $U > U_1$. Hence, on each of the four intervals

$$U = U(Q), \tag{13.2.58}$$

or, equivalently,

$$\theta = QU(Q). \tag{13.2.59}$$

Case (i) : $0 < U < U_2$. In this case we obtain

$$U = \frac{1}{2}\left(\frac{1}{v} + \frac{1}{Q}\right) - \sqrt{\frac{1}{4}\left(\frac{1}{v} + \frac{1}{Q}\right)^2 - 1} \quad \text{for} \quad Q > 0, \tag{13.2.60}$$

Case (ii) : $U_2 < U \leq 1$. In this case

$$U = \frac{1}{2}\left(\frac{1}{v} - \frac{1}{Q}\right) - \sqrt{\frac{1}{4}\left(\frac{1}{v} - \frac{1}{Q}\right)^2 - 1} \quad \text{for} \quad Q \geq \frac{1}{1/v - 2}, \tag{13.2.61}$$

and

$$U \to U_2 \quad \text{as} \quad Q \to \infty, \tag{13.2.62}$$

$$U \to 1 \quad \text{as} \quad Q \to \frac{1}{1/\nu - 2}. \tag{13.2.63}$$

Case (iii) : $1 \leq U < U_1$. In this case

$$U = \frac{1}{2}\left(\frac{1}{v} - \frac{1}{Q}\right) + \sqrt{\frac{1}{4}\left(\frac{1}{v} - \frac{1}{Q}\right)^2 - 1} \quad \text{for} \quad Q \geq \frac{1}{1/v - 2}, \tag{13.2.64}$$

and

$$U \to 1 \quad \text{as} \quad Q \to \frac{1}{1/v - 2}, \tag{13.2.65}$$

$$U \to U_1 \quad \text{as} \quad Q \to \infty. \tag{13.2.66}$$

Case (iv) : $U > U_1$. In this case

$$U = \frac{1}{2}\left(\frac{1}{v} + \frac{1}{Q}\right) + \sqrt{\frac{1}{4}\left(\frac{1}{v} + \frac{1}{Q}\right)^2 - 1} \quad \text{for} \quad Q > 0, \tag{13.2.67}$$

and

$$U \to U_1 \quad \text{as} \quad Q \to \infty, \tag{13.2.68}$$

$$U \to \infty \quad \text{as} \quad Q \to 0. \tag{13.2.69}$$

The temperature fields corresponding to the cases (i), (ii), (iii), and (iv), respectively, are represented by the formulas: For $0 < \theta < U_2 Q;\ Q > 0$

$$\theta = \frac{1}{2}\left(1 + \frac{Q}{v}\right) - \sqrt{\frac{1}{4}\left(1 + \frac{Q}{v}\right)^2 - Q^2}. \tag{13.2.70}$$

For $U_2 Q \leq \theta \leq Q;\ Q \geq 1/(1/v - 2)$

$$\theta = \frac{1}{2}\left(\frac{Q}{v} - 1\right) - \sqrt{\frac{1}{4}\left(\frac{Q}{v} - 1\right)^2 - Q^2}. \tag{13.2.71}$$

For $Q \leq \theta \leq U_1 Q;\ Q \geq 1/(1/v - 2)$

$$\theta = \frac{1}{2}\left(\frac{Q}{v} - 1\right) + \sqrt{\frac{1}{4}\left(\frac{Q}{v} - 1\right)^2 - Q^2}. \tag{13.2.72}$$

For $\theta \geq U_1 Q;\ Q > 0$

$$\theta = \frac{1}{2}\left(1 + \frac{Q}{v}\right) + \sqrt{\frac{1}{4}\left(1 + \frac{Q}{v}\right)^2 - Q^2}. \tag{13.2.73}$$

The formulas (13.2.70)–(13.2.73) may be used to construct travelling-wave solutions. In the following we are to obtain a travelling-wave solution associated with the temperature (13.2.73). To this end we transform eqn $(13.2.26)_2$ to the form

$$\mathrm{d}\xi = -\frac{1}{2Q}\left(\frac{\mathrm{d}\theta}{\mathrm{d}Q} - v\right)\mathrm{d}Q. \tag{13.2.74}$$

Substituting $\theta = \theta(Q)$ from eqn (13.2.73) into eqn (13.2.74) we obtain

$$\mathrm{d}\xi = -\frac{1}{4v}\left\{\frac{1 - 2v^2}{Q} + \frac{2v}{Q}\frac{\mathrm{d}}{\mathrm{d}Q}\sqrt{\frac{1}{4}\left(1 + \frac{Q}{v}\right)^2 - Q^2}\right\}\mathrm{d}Q. \tag{13.2.75}$$

Integrating this equation we arrive at the travelling-wave solution represented by $Q = Q(\xi)$:

$$\xi - \xi_0 = -\frac{1}{4v}\int_{Q(\xi_0)}^{Q(\xi)}\left\{\frac{1 - 2v^2}{q} + \frac{2v}{q}\frac{\mathrm{d}}{\mathrm{d}q}\sqrt{\frac{1}{4}\left(1 + \frac{q}{v}\right)^2 - q^2}\right\}\mathrm{d}q, \tag{13.2.76}$$

where ξ_0 is a fixed point of the ξ-axis and $Q(\xi_0) > 0$.

Clearly, eqn (13.2.76) defines $Q = Q(\xi)$ in an implicit form, and together with eqn (13.2.73) represents a travelling-wave solution $[\theta(\xi), Q(\xi)]$ on the interval $\xi_1 \leq \xi \leq \xi_2$ over which there is a unique function $Q = Q(\xi)$ that satisfies eqn (13.2.76).

Also, note that an alternative form of eqn (13.2.76) is obtained by computing the integrals on the RHS of eqn (13.2.76) in a closed form. To this end, note that by integrating by parts we obtain

$$\int_{Q(\xi_0)}^{Q(\xi)} \left\{ \frac{1}{q} \frac{\mathrm{d}}{\mathrm{d}q} \sqrt{\frac{1}{4}\left(1+\frac{q}{v}\right)^2 - q^2} \right\} \mathrm{d}q =$$

$$\frac{1}{Q(\xi)}\sqrt{\frac{1}{4}\left(1+\frac{Q(\xi)}{v}\right)^2 - Q^2(\xi)} - \frac{1}{Q(\xi_0)}\sqrt{\frac{1}{4}\left(1+\frac{Q(\xi_0)}{v}\right)^2 - Q^2(\xi_0)}$$
$$+ \int_{Q(\xi_0)}^{Q(\xi)} \left\{ \frac{1}{q^2} \sqrt{\frac{1}{4}\left(1+\frac{q}{v}\right)^2 - q^2} \right\} \mathrm{d}q. \tag{13.2.77}$$

To compute the integral on RHS of this equation we observe that

$$\sqrt{\frac{1}{4}\left(1+\frac{q}{v}\right)^2 - q^2} = \frac{1}{2v}\sqrt{1-4v^2}\sqrt{(q+q_1)(q+q_2)}, \tag{13.2.78}$$

where

$$q_1 = \frac{1}{1/v - 2}, \qquad q_2 = \frac{1}{1/v + 2} \tag{13.2.79}$$

and

$$\int \frac{1}{q^2}\sqrt{(q+q_1)(q+q_2)} dq = -\frac{1}{q}\sqrt{(q+q_1)(q+q_2)}$$
$$+ \ln\left[\sqrt{(q+q_1)(q+q_2)} + q + (q_1+q_2)/2\right]$$
$$+ \frac{q_1+q_2}{2\sqrt{q_1 q_2}} \ln\left\{ \frac{q}{\sqrt{(q+q_1)(q+q_2)} + q\left[(q_1+q_2)/2\sqrt{q_1q_2}\right] + \sqrt{q_1 q_2}} \right\}. \tag{13.2.80}$$

Therefore,

$$2v \int_{Q(\xi_0)}^{Q(\xi)} \left\{ \frac{1}{q^2} \sqrt{\frac{1}{4}\left(1+\frac{q}{v}\right)^2 - q^2} \right\} dq = -2v \left\{ \frac{1}{q} \sqrt{\frac{1}{4}\left(1+\frac{q}{v}\right)^2 - q^2} \right\}_{q=Q(\xi_0)}^{q=Q(\xi)}$$
$$+ v\sqrt{(1/v)^2 - 4}$$
$$\times \left\{ \ln\left[\sqrt{\frac{1}{4}\left(1+\frac{q}{v}\right)^2 - q^2} + \frac{q}{2}\sqrt{(1/v)^2 - 4} + \frac{1}{2v\sqrt{(1/v)^2-4}} \right] \right\}_{q=Q(\xi_0)}^{q=Q(\xi)}$$
$$+ \left\{ \ln q - \ln\left[\sqrt{\frac{1}{4}\left(1+\frac{q}{v}\right)^2 - q^2} + \frac{q}{2v} + \frac{1}{2} \right] \right\}_{q=Q(\xi_0)}^{q=Q(\xi)}, \tag{13.2.81}$$

and eqn (13.2.76) takes the form

$$\psi(Q) = \exp(-4v\xi), \tag{13.2.82}$$

where

$$\psi(Q) = Q^{2(1-v^2)}[M(Q)]^{v\sqrt{(1/v)^2-4}}[N(Q)]^{-1} \tag{13.2.83}$$

and

$$M(Q) = \sqrt{\frac{1}{4}\left(1+\frac{Q}{v}\right)^2 - Q^2} + \frac{Q}{2}\sqrt{(1/v)^2-4} + \frac{1}{2v\sqrt{(1/v)^2-4}}, \tag{13.2.84}$$

$$N(Q) = \sqrt{\frac{1}{4}\left(1+\frac{Q}{v}\right)^2 - Q^2} + \frac{Q}{2v} + \frac{1}{2}. \tag{13.2.85}$$

Therefore, a travelling wave in the non-linear hyperbolic rigid heat conductor is represented by a pair $[\theta(\xi), Q(\xi)]$ in which $\theta = \theta(\xi)$ is expressed in terms of $Q = Q(\xi)$ by eqn (13.2.73) and $Q = Q(\xi) = \psi^{-1}[\exp(-4v\xi)]$; and the inequalities $\theta \geq U_1 Q,\ Q > 0$ hold true.

13.3 Asymptotic method of weakly non-linear geometric optics applied to the Coleman heat conductor

In this section a Cauchy problem with a weakly perturbed initial condition for the non-linear hyperbolic heat conduction model is analyzed. The field equations are taken in the form [see eqns (13.2.23)]

$$\begin{aligned} &\frac{\partial}{\partial t}\left(\theta + \frac{Q^2}{\theta}\right) + \frac{\partial Q}{\partial x} = 0, \\ &\frac{\partial Q}{\partial t} + 2Q + \frac{\partial \theta}{\partial x} = 0, \end{aligned} \tag{13.3.1}$$

where

$$\theta = \theta(t,x), \qquad Q = Q(t,x), \qquad t \geq 0, \qquad |x| < \infty. \tag{13.3.2}$$

An alternative form of eqns (13.3.1) reads

$$\begin{aligned} &\left(1 - \frac{Q^2}{\theta^2}\right)\frac{\partial \theta}{\partial t} + 2\frac{Q}{\theta}\frac{\partial Q}{\partial t} + \frac{\partial Q}{\partial x} = 0, \\ &\frac{\partial Q}{\partial t} = -\left(2Q + \frac{\partial \theta}{\partial x}\right). \end{aligned} \tag{13.3.3}$$

Therefore, substituting $\partial Q/\partial t$ from eqn $(13.3.3)_2$ into eqn $(13.3.3)_1$, and dividing the resulting equation by $(1 - Q^2/\theta^2)$, we transform eqns (13.3.3) to the non-homogeneous quasi-linear matrix partial differential equation of the first order for an unknown vector $\mathbf{u} = (\theta,\ Q)^T$

$$\frac{\partial}{\partial t}\mathbf{u} + \mathbf{A}(\mathbf{u})\frac{\partial}{\partial x}\mathbf{u} = \mathbf{f}(\mathbf{u}), \tag{13.3.4}$$

where

$$\mathbf{u} = \begin{pmatrix} \theta \\ Q \end{pmatrix}, \quad \mathbf{f}(\mathbf{u}) = 2Q \begin{pmatrix} 2U/(1-U^2) \\ -1 \end{pmatrix}, \quad U = \frac{Q}{\theta}, \tag{13.3.5}$$

and

$$\mathbf{A}(\mathbf{u}) = \begin{pmatrix} -2U/(1-U^2) & 1/(1-U^2) \\ 1 & 0 \end{pmatrix}. \tag{13.3.6}$$

We are going to study the following initial value problem for eqn (13.3.4): Find a solution $\mathbf{u} = \mathbf{u}(t, x)$ to the equation

$$\frac{\partial}{\partial t}\mathbf{u} + \mathbf{A}(\mathbf{u})\frac{\partial}{\partial x}\mathbf{u} = \mathbf{f}(\mathbf{u}), \qquad t \geq 0, \qquad |x| < \infty, \tag{13.3.7}$$

subject to the initial condition

$$\mathbf{u}(0, x) = \mathbf{u}_0 + \varepsilon \mathbf{u}^*(x, x/\varepsilon), \qquad |x| < \infty, \tag{13.3.8}$$

where ε is a small positive number, $\mathbf{u}^* = \mathbf{u}^*(x, y)$ is a prescribed function on the $x - y$ plane, and $\mathbf{u}_0$ is a constant vector defined in terms of a dimensionless constant temperature $\theta_0 > 0$:

$$\mathbf{u}_0 = \begin{pmatrix} \theta_0 \\ 0 \end{pmatrix}. \tag{13.3.9}$$

It follows from eqns (13.3.4)–(13.3.6) that the problem (13.3.7)–(13.3.9) is non-linear and highly singular as the functions $f_1 = f_1(\mathbf{u})$, $A_{11} = A_{11}(\mathbf{u})$ and $A_{12} = A_{12}(\mathbf{u})$ are non-linear and become unbounded for $|U| = 1$. This is a reason why a smooth solution to the problem should be sought in a region of the (θ, Q)-plane excluding the lines $\theta = |Q| > 0$.

The following Lemmas hold true:

Lemma 13.1 *The matrix equation (13.3.4) is strictly hyperbolic and genuinely non-linear in the sense of Lax provided* $|U| \neq 1$.

Proof. One can show that the matrix $\mathbf{A}(\mathbf{u})$ has two real eigenvalues $\lambda_1(\mathbf{u})$ and $\lambda_2(\mathbf{u})$ corresponding to the right eigenvectors $\mathbf{r}_1(\mathbf{u})$ and $\mathbf{r}_2(\mathbf{u})$, respectively, and they are given by

$$\lambda_1(\mathbf{u}) = \frac{1}{1+U}, \qquad \mathbf{r}_1(\mathbf{u}) = \frac{1}{\sqrt{1+(1+U)^2}} \begin{pmatrix} 1 \\ 1+U \end{pmatrix}, \tag{13.3.10}$$

$$\lambda_2(\mathbf{u}) = -\frac{1}{1-U}, \qquad \mathbf{r}_2(\mathbf{u}) = \frac{1}{\sqrt{1+(1-U)^2}} \begin{pmatrix} 1 \\ -(1-U) \end{pmatrix}. \tag{13.3.11}$$

This implies that the matrix equation (13.3.4) is strictly hyperbolic for $|U| \neq 1$ (Lax, 1957). To show that the matrix equation (13.3.4) is genuinely nonlinear, we compute the gradient of $\lambda_i(\mathbf{u})$ with respect to $\mathbf{u}(i = 1, 2)$ and obtain

$$\nabla_u \lambda_1(\mathbf{u}) = \begin{pmatrix} \partial\lambda_1/\partial u_1 \\ \partial\lambda_1/\partial u_2 \end{pmatrix} = \frac{1}{u_1}\frac{1}{(1+U)^2}\begin{pmatrix} U \\ -1 \end{pmatrix} \tag{13.3.12}$$

and

$$\nabla_u \lambda_2(\mathbf{u}) = \begin{pmatrix} \partial\lambda_2/\partial u_1 \\ \partial\lambda_2/\partial u_2 \end{pmatrix} = -\frac{1}{u_1}\frac{1}{(1-U)^2}\begin{pmatrix} -U \\ 1 \end{pmatrix}. \tag{13.3.13}$$

Therefore, by virtue of eqns (13.3.10)–(13.3.13), we obtain

$$[\nabla_u \lambda_1(\mathbf{u})] \cdot \mathbf{r}_1(\mathbf{u}) = -\frac{1}{u_1}\frac{1}{\sqrt{1+(1+U)^2}}\frac{1}{(1+U)^2} \tag{13.3.14}$$

and

$$[\nabla_u \lambda_2(\mathbf{u})] \cdot \mathbf{r}_2(\mathbf{u}) = \frac{1}{u_1}\frac{1}{\sqrt{1+(1-U)^2}}\frac{1}{(1-U)^2}. \tag{13.3.15}$$

Since

$$[\nabla_u \lambda_1(\mathbf{u})] \cdot \mathbf{r}_1(\mathbf{u}) \neq 0 \quad \text{and} \quad [\nabla_u \lambda_2(\mathbf{u})] \cdot \mathbf{r}_2(\mathbf{u}) \neq 0 \quad \text{for} \quad |U| \neq 1, \tag{13.3.16}$$

the matrix equation (13.3.4) is genuinely non-linear for $|U| \neq 1$ (Lax, 1957). □

Lemma 13.2 *The matrix* $\mathbf{A}(\mathbf{u})$ *has two real eigenvalues* $\lambda_1(\mathbf{u})$ *and* $\lambda_2(\mathbf{u})$ *corresponding to the left eigenvectors* $l_1(\mathbf{u})$ *and* $l_2(\mathbf{u})$*, respectively, and they are given by*

$$\lambda_1(\mathbf{u}) = \frac{1}{1+U}, \ \boldsymbol{l}_1(\mathbf{u}) = \frac{1}{2}\sqrt{1+(1+U)^2}\begin{pmatrix} 1-U \\ 1 \end{pmatrix}, \tag{13.3.17}$$

$$\lambda_2(\mathbf{u}) = -\frac{1}{1-U}, \ \boldsymbol{l}_2(\mathbf{u}) = \frac{1}{2}\sqrt{1+(1-U)^2}\begin{pmatrix} 1+U \\ -1 \end{pmatrix}. \tag{13.3.18}$$

In addition

$$\boldsymbol{l}_\alpha \cdot \mathbf{r}_\beta = \delta_{\alpha\beta} \qquad \alpha, \beta = 1, 2. \tag{13.3.19}$$

Proof. It is easy to show that

$$\boldsymbol{l}_\alpha(\mathbf{u})\,[\mathbf{A}(\mathbf{u}) - \lambda_\alpha \mathbf{I}] = \mathbf{0} \qquad \alpha = 1, 2, \tag{13.3.20}$$

where $\mathbf{I}$ is a unit tensor, and the pairs $[\lambda_1(\mathbf{u}), \boldsymbol{l}_1(\mathbf{u})]$ and $[\lambda_2(\mathbf{u}), \boldsymbol{l}_2(\mathbf{u})]$ are given by eqns (13.3.17) and (13.3.18), respectively. Also, using eqns (13.3.10), (13.3.11), (13.3.17), and (13.3.18) we check that the orthogonality conditions (13.3.19) hold true. □

Using the methods of weakly non-linear geometric optics (WNGO) one can prove the following theorem (Hunter and Keller, 1983; DiPerna and Majda, 1985; Domański, 2000).

Theorem 13.1 *An asymptotic solution to the problem (13.3.7)–(13.3.9) that represents a non-linear hyperbolic small-amplitude and high-frequency heat wave propagating along the x-axis takes the form*

$$\mathbf{u}(t,x) = \mathbf{u}_0 + \varepsilon \sum_{\alpha=1}^{2} \sigma_\alpha \left(t, x, \frac{x-\lambda_\alpha t}{\varepsilon}\right) \mathbf{r}_\alpha(\mathbf{u}_0) + O(\varepsilon^3), \tag{13.3.21}$$

where

$$\lambda_1 = 1, \quad \lambda_2 = -1, \quad \mathbf{r}_1(\mathbf{u}_0) = \frac{1}{\sqrt{2}}\begin{pmatrix}1\\1\end{pmatrix}, \quad \mathbf{r}_2(\mathbf{u}_0) = \frac{1}{\sqrt{2}}\begin{pmatrix}1\\-1\end{pmatrix} \tag{13.3.22}$$

and the functions $\sigma_\alpha = \sigma_\alpha(t,x,\eta)$ *satisfy the transport equations*

$$\frac{\partial \sigma_1}{\partial t} + \frac{\partial \sigma_1}{\partial x} + \sigma_1 - \frac{1}{2}\frac{\partial \sigma_1^2}{\partial \eta} = 0, \tag{13.3.23}$$

$$\frac{\partial \sigma_2}{\partial t} - \frac{\partial \sigma_2}{\partial x} + \sigma_2 + \frac{1}{2}\frac{\partial \sigma_2^2}{\partial \eta} = 0, \tag{13.3.24}$$

and, without any loss of generality, we let $\theta_0 = 1/\sqrt{2}$ *in eqn (13.3.9).*

Proof. We are to discuss the following asymptotic initial-value problem. Find a solution $\mathbf{u}^\varepsilon = \mathbf{u}^\varepsilon(t,x)$ to the equation

$$\frac{\partial}{\partial t}\mathbf{u}^\varepsilon + \mathbf{A}(\mathbf{u}^\varepsilon)\frac{\partial}{\partial x}\mathbf{u}^\varepsilon = \mathbf{f}(\mathbf{u}^\varepsilon), \qquad t \geq 0, \qquad |x| \leq \infty, \tag{13.3.25}$$

subject to the initial condition

$$\mathbf{u}^\varepsilon(0,x) = \mathbf{u}_0 + \varepsilon \mathbf{u}^*(x, x/\varepsilon), \qquad |x| \leq \infty. \tag{13.3.26}$$

By looking for the solution in the form

$$\mathbf{u}^\varepsilon(t,x) = \mathbf{u}_0 + \varepsilon[\mathbf{u}_1(t,x;\eta) + \varepsilon \mathbf{u}_2(t,x;\eta)] + O(\varepsilon^3), \tag{13.3.27}$$

where

$$\eta = \frac{x-\lambda_\alpha t}{\varepsilon}, \quad \lambda_\alpha = \lambda_\alpha(\mathbf{u}_0), \quad \text{and} \quad \alpha = 1 \quad \text{or} \quad \alpha = 2, \tag{13.3.28}$$

and letting

$$\tilde{\mathbf{u}}(t,x;\eta) = \mathbf{u}_1(t,x;\eta) + \varepsilon \mathbf{u}_2(t,x;\eta), \tag{13.3.29}$$

and using Taylor's expansions, we obtain

$$\mathbf{f}(\mathbf{u}^\varepsilon) = \mathbf{f}(\mathbf{u}_0 + \varepsilon\tilde{\mathbf{u}}) = \mathbf{f}(\mathbf{u}_0) + \varepsilon[\nabla_u \mathbf{f}(\mathbf{u}_0)]\tilde{\mathbf{u}} + ... \tag{13.3.30}$$

and

$$\begin{aligned}\mathbf{f}(\mathbf{u}^\varepsilon) &= \mathbf{f}(\mathbf{u}_0) + \varepsilon[\nabla_u \mathbf{f}(\mathbf{u}_0)](\mathbf{u}_1 + \varepsilon \mathbf{u}_2 + ...) \\ &= \mathbf{f}(\mathbf{u}_0) + \varepsilon[\nabla_u \mathbf{f}(\mathbf{u}_0)]\mathbf{u}_1 + \varepsilon^2[\nabla_u \mathbf{f}(\mathbf{u}_0)]\mathbf{u}_2 ....\end{aligned} \tag{13.3.31}$$

In a similar way, for the matrix $\mathbf{A}(\mathbf{u}^\varepsilon)$ Taylor's expansion reads

$$\mathbf{A}(\mathbf{u}^\varepsilon) = \mathbf{A}(\mathbf{u}_0) + \varepsilon[\nabla_u \mathbf{A}(\mathbf{u}_0)](\mathbf{u}_1 + \varepsilon \mathbf{u}_2 + \ldots). \tag{13.3.32}$$

Hence, we obtain

$$\begin{aligned}
&\mathbf{A}(\mathbf{u}^\varepsilon)\frac{\partial}{\partial x}\mathbf{u}^\varepsilon \\
&= \mathbf{A}(\mathbf{u}_0)\left[\varepsilon\frac{\partial}{\partial x}(\mathbf{u}_1 + \varepsilon\mathbf{u}_2)\right] + \varepsilon[\nabla_u \mathbf{A}(\mathbf{u}_0)](\mathbf{u}_1 + \varepsilon\mathbf{u}_2)\left[\varepsilon\frac{\partial}{\partial x}(\mathbf{u}_1 + \varepsilon\mathbf{u}_2)\right] + \ldots \\
&= \mathbf{A}(\mathbf{u}_0)\frac{\partial}{\partial \eta}\mathbf{u}_1 + \varepsilon\mathbf{A}(\mathbf{u}_0)\left(\frac{\partial}{\partial x}\mathbf{u}_1 + \frac{\partial}{\partial \eta}\mathbf{u}_2\right) + \varepsilon[\nabla_u \mathbf{A}(\mathbf{u}_0)]\mathbf{u}_1\frac{\partial}{\partial \eta}\mathbf{u}_1 + O(\varepsilon^2)
\end{aligned} \tag{13.3.33}$$

and

$$\frac{\partial}{\partial t}\mathbf{u}^\varepsilon = -\lambda_\alpha\frac{\partial}{\partial \eta}\mathbf{u}_1 + \varepsilon\left(\frac{\partial}{\partial t}\mathbf{u}_1 - \lambda_\alpha\frac{\partial}{\partial \eta}\mathbf{u}_2\right) + \varepsilon^2\frac{\partial}{\partial t}\mathbf{u}_2 + \ldots. \tag{13.3.34}$$

Substituting eqns (13.3.31), (13.3.33), and (13.3.34) into eqn (13.3.25), and dividing both sides by ε, we obtain

$$\begin{aligned}
&\varepsilon^{-1}[\mathbf{A}(\mathbf{u}^\varepsilon) - \lambda_\alpha\mathbf{I}]\frac{\partial}{\partial \eta}\mathbf{u}_1 \\
&+\varepsilon^0\left\{[\mathbf{A}(\mathbf{u}_0) - \lambda_\alpha\mathbf{I}]\frac{\partial}{\partial \eta}\mathbf{u}_2 + [\nabla_u \mathbf{A}(\mathbf{u}_0)]\mathbf{u}_1\frac{\partial}{\partial \eta}\mathbf{u}_1 + \mathbf{A}(\mathbf{u}_0)\frac{\partial}{\partial x}\mathbf{u}_1 + \frac{\partial}{\partial t}\mathbf{u}_1\right\} \\
&+O(\varepsilon) = \varepsilon^{-1}\mathbf{f}(\mathbf{u}_0) + [\nabla_u \mathbf{f}(\mathbf{u}_0)]\mathbf{u}_1 + O(\varepsilon^2).
\end{aligned} \tag{13.3.35}$$

By equating the coefficients of ε^{-1} and ε^0 in eqn (13.3.35), we obtain, respectively,

$$[\mathbf{A}(\mathbf{u}_0) - \lambda_\alpha\mathbf{I}]\frac{\partial}{\partial \eta}\mathbf{u}_1 = \mathbf{f}(\mathbf{u}_0), \tag{13.3.36}$$

and

$$[\mathbf{A}(\mathbf{u}_0) - \lambda_\alpha\mathbf{I}]\frac{\partial}{\partial \eta}\mathbf{u}_2 + \mathbf{g}_\alpha = \mathbf{h}_\alpha, \tag{13.3.37}$$

where

$$\mathbf{g}_\alpha = [\nabla_u \mathbf{A}(\mathbf{u}_0)]\mathbf{u}_1\frac{\partial}{\partial \eta}\mathbf{u}_1 + \mathbf{A}(\mathbf{u}_0)\frac{\partial}{\partial x}\mathbf{u}_1 + \frac{\partial}{\partial t}\mathbf{u}_1, \tag{13.3.38}$$

and

$$\mathbf{h}_\alpha = [\nabla_u \mathbf{f}(\mathbf{u}_0)]\mathbf{u}_1. \tag{13.3.39}$$

To obtain a solution $\partial\mathbf{u}_1/\partial\eta$ of eqn (13.3.36) the following solvability condition must be satisfied

$$\boldsymbol{l}_\alpha(\mathbf{u}_0)\cdot\mathbf{f}(\mathbf{u}_0) = 0 \quad \alpha = 1, 2. \tag{13.3.40}$$

Similarly, there is a solution $\partial \mathbf{u}_2/\partial\eta$ of eqn (13.3.37) provided

$$\boldsymbol{l}_\alpha(\mathbf{u}_0)\cdot(\mathbf{g}_\alpha-\mathbf{h}_\alpha)=0 \qquad \alpha=1,2. \tag{13.3.41}$$

Since by virtue of eqns (13.3.5) and (13.3.9) $\mathbf{f}(\mathbf{u}_0)=\mathbf{0}$, therefore eqn (13.3.40) is identically satisfied, and eqn (13.3.36) reduces to

$$[\mathbf{A}(\mathbf{u}_0)-\lambda_\alpha\mathbf{I}]\frac{\partial}{\partial\eta}\mathbf{u}_1=\mathbf{0}. \tag{13.3.42}$$

A solution to eqn (13.3.42) can be taken in the form

$$\mathbf{u}_1(t,x;\eta)=\sigma_\alpha(t,x;\eta)\mathbf{r}_\alpha(\mathbf{u}_0). \tag{13.3.43}$$

Therefore, substituting eqn (13.3.43) into eqns (13.3.38) and (13.3.39), respectively, we obtain

$$\begin{aligned}\mathbf{g}_\alpha &= [\nabla_u\mathbf{A}(\mathbf{u}_0)]\mathbf{r}_\alpha(\mathbf{u}_0)\mathbf{r}_\alpha(\mathbf{u}_0)\sigma_\alpha\frac{\partial}{\partial\eta}\sigma_\alpha\\ &+\mathbf{A}(\mathbf{u}_0)\mathbf{r}_\alpha(\mathbf{u}_0)\frac{\partial}{\partial x}\sigma_\alpha+\mathbf{r}_\alpha(\mathbf{u}_0)\frac{\partial}{\partial t}\sigma_\alpha\end{aligned} \tag{13.3.44}$$

and

$$\mathbf{h}_\alpha=[\nabla_u\mathbf{f}(\mathbf{u}_0)]\mathbf{r}_\alpha(\mathbf{u}_0)\sigma_\alpha. \tag{13.3.45}$$

In components, eqns (13.3.44) and (13.3.45), respectively, take the form

$$\begin{aligned}g_\delta^{(\alpha)} &= A_{\delta\beta,\gamma}(\mathbf{u}_0)r_\gamma^{(\alpha)}(\mathbf{u}_0)r_\beta^{(\alpha)}(\mathbf{u}_0)\sigma_\alpha\frac{\partial}{\partial\eta}\sigma_\alpha\\ &+A_{\delta\beta}(\mathbf{u}_0)r_\beta^{(\alpha)}(\mathbf{u}_0)\frac{\partial}{\partial x}\sigma_\alpha+r_\delta^{(\alpha)}(\mathbf{u}_0)\frac{\partial}{\partial x}\sigma_\alpha\end{aligned} \tag{13.3.46}$$

and

$$h_\delta^{(\alpha)}=f_{\delta,\beta}(\mathbf{u}_0)r_\beta^{(\alpha)}(\mathbf{u}_0)\sigma_\alpha, \tag{13.3.47}$$

where

$$f_{\delta,\beta}(\mathbf{u}_0)=\frac{\partial f_\delta}{\partial u_\beta}(\mathbf{u}_0),\qquad A_{\delta\beta,\gamma}(\mathbf{u}_0)=\frac{\partial A_{\delta\beta}}{\partial u_\gamma}(\mathbf{u}_0),\qquad \alpha,\beta,\gamma,\delta=1,2 \tag{13.3.48}$$

and for an arbitrary vector $\mathbf{a}_\alpha$ the following notation is used

$$\mathbf{a}_\alpha\equiv\mathbf{a}^{(\alpha)}=\begin{pmatrix}a_1^{(\alpha)}\\ a_2^{(\alpha)}\end{pmatrix} \tag{13.3.49}$$

and in eqns (13.3.46) and (13.3.47) the summation convention over the indices β and γ is observed.

Also, note that

$$\mathbf{r}_1(\mathbf{u}_0) = \frac{1}{\sqrt{2}}\begin{pmatrix}1\\1\end{pmatrix}, \quad \mathbf{r}_2(\mathbf{u}_0) = \frac{1}{\sqrt{2}}\begin{pmatrix}1\\-1\end{pmatrix},$$
$$\boldsymbol{l}_1(\mathbf{u}_0) = \frac{1}{\sqrt{2}}\begin{pmatrix}1\\1\end{pmatrix}, \quad \boldsymbol{l}_2(\mathbf{u}_0) = \frac{1}{\sqrt{2}}\begin{pmatrix}1\\-1\end{pmatrix}, \tag{13.3.50}$$

and

$$f_{1,1}(\mathbf{u}_0) = 0, \quad f_{1,2}(\mathbf{u}_0) = 0, \quad f_{2,1}(\mathbf{u}_0) = 0, \quad f_{2,2}(\mathbf{u}_0) = -2. \tag{13.3.51}$$

Therefore, by virtue of eqns (13.3.39), (13.3.43), and (13.3.51) we obtain

$$\mathbf{h}_\alpha = \begin{pmatrix}0 & 0\\0 & -2\end{pmatrix}\begin{pmatrix}u_1^{(1)}\\u_2^{(1)}\end{pmatrix} = \begin{pmatrix}0\\-2u_2^{(1)}\end{pmatrix} = -2\sigma_\alpha\begin{pmatrix}0\\r_2^{(\alpha)}\end{pmatrix} \tag{13.3.52}$$

and

$$\boldsymbol{l}_\alpha(\mathbf{u}_0)\cdot\mathbf{h}_\alpha = l_2^{(\alpha)}h_2^{(\alpha)} = -2\sigma_\alpha\, l_2^{(\alpha)}r_2^{(\alpha)} = -\sigma_\alpha. \tag{13.3.53}$$

Also, it follows from eqn (13.3.50) that

$$\boldsymbol{l}_\alpha(\mathbf{u}_0)\cdot\mathbf{r}_\alpha(\mathbf{u}_0) = 1, \qquad \boldsymbol{l}_\alpha(\mathbf{u}_0)\cdot[\mathbf{A}(\mathbf{u}_0)\,\mathbf{r}_\alpha(\mathbf{u}_0)] = \lambda_\alpha(\mathbf{u}_0) \tag{13.3.54}$$

and

$$\Gamma^{(\alpha)} = \boldsymbol{l}_\alpha\cdot[\nabla_u A(\mathbf{u}_0)]\; r_\alpha(\mathbf{u}_0)\; r_\alpha(\mathbf{u}_0)$$
$$= l_\mu^{(\alpha)}(\mathbf{u}_0)A_{\mu\beta,\gamma}(\mathbf{u}_0)r_\gamma^{(\alpha)}(\mathbf{u}_0)r_\beta^{(\alpha)}(\mathbf{u}_0) = -\frac{\lambda_\alpha(\mathbf{u}_0)}{\theta_0\sqrt{2}}. \tag{13.3.55}$$

Finally, using eqns (13.3.53)–(13.3.55) we reduce eqns (13.3.41) to the transport equations

$$\frac{\partial\sigma_1}{\partial t} + \frac{\partial\sigma_1}{\partial x} + \sigma_1 - \frac{1}{2\sqrt{2}\,\theta_0}\frac{\partial\sigma_1^2}{\partial\eta} = 0, \tag{13.3.56}$$

$$\frac{\partial\sigma_2}{\partial t} - \frac{\partial\sigma_2}{\partial x} + \sigma_2 + \frac{1}{2\sqrt{2}\,\theta_0}\frac{\partial\sigma_2^2}{\partial\eta} = 0. \tag{13.3.57}$$

Hence, if we let $\theta_0 = 1/\sqrt{2}$, we arrive at eqns (13.3.23) and (13.3.24). □

In the following, we are to prove that there is a unique closed-form asymptotic solution to the problem (13.3.7)–(13.3.9) if the initial data are suitably restricted. To this end we introduce the notation

$$y = \frac{x}{\varepsilon}, \quad |y| \le \infty, \tag{13.3.58}$$

and assume that $\mathbf{u}^* = \mathbf{u}^*(x, x/\varepsilon)$ in eqn (13.3.8) takes the form

$$\mathbf{u}^* = \mathbf{u}^*(x, x/\varepsilon) = \mathbf{u}^*(x/\varepsilon) \equiv \mathbf{u}^*(y), \qquad |y| < \infty, \tag{13.3.59}$$

where $\mathbf{u}^* = \mathbf{u}^*(y)$ is a prescribed function. The following theorem holds true.

Theorem 13.2 *Suppose that the function* $\mathbf{u}^* = \mathbf{u}^*(y)$ *satisfies the conditions*

$$e < \mathbf{r}_1(\mathbf{u}_0) \cdot \mathbf{u}^*(y) \leq \mathbf{r}_2(\mathbf{u}_0) \cdot \mathbf{u}^*(y) < \infty, \qquad |y| < \infty, \tag{13.3.60}$$

where e *is the base of natural logarithms* ($e = 2.7182$). *Then there is a unique closed-form asymptotic solution to the problem (13.3.7)–(13.3.9) in which the initial condition (13.3.8) is replaced by the condition*

$$\mathbf{u}(0, x) = \mathbf{u}_0 + \varepsilon\, \mathbf{u}^*(y) \tag{13.3.61}$$

and this solution is represented by

$$\mathbf{u}(t, x) = \mathbf{u}_0 + \varepsilon \sum_{\alpha=1}^{2} \sigma_\alpha(t, \eta_\alpha)\, \mathbf{r}_\alpha(\mathbf{u}_0) + O(\varepsilon^2), \tag{13.3.62}$$

where

$$\eta_\alpha = \frac{x - \lambda_\alpha t}{\varepsilon}, \quad \lambda_1 = 1, \quad \lambda_2 = -1 \tag{13.3.63}$$

$$\mathbf{r}_1(\mathbf{u}_0) = \frac{1}{\sqrt{2}} \begin{pmatrix} 1 \\ 1 \end{pmatrix}, \quad \mathbf{r}_2(\mathbf{u}_0) = \frac{1}{\sqrt{2}} \begin{pmatrix} 1 \\ -1 \end{pmatrix} \tag{13.3.64}$$

and the functions $\sigma_\alpha = \sigma_\alpha(t, \eta)$ $(\alpha = 1, 2)$ *are determined implicitly by the formulas*

$$\sigma_1(t, \eta) = Y_1[t + \ln \sigma_1(t, \eta)] + \eta - y \qquad 0 \leq t < \tau_1, \quad |\eta| < \infty, \quad |y| < \infty, \tag{13.3.65}$$

$$\sigma_2(t, \eta) = Y_2[t + \ln \sigma_2(t, \eta)] - \eta + y \qquad 0 \leq t < \tau_2, \quad |\eta| < \infty, \quad |y| < \infty, \tag{13.3.66}$$

in which the functions $Y_\alpha = Y_\alpha(y)$ *are defined by*

$$Y_\alpha = \frac{\mathbf{r}_\alpha(\mathbf{u}_0) \cdot \mathbf{u}^*(y)}{\ln[\mathbf{r}_\alpha(\mathbf{u}_0) \cdot \mathbf{u}^*(y)]}, \tag{13.3.67}$$

and τ_α *is a blow-up time of the amplitude* $\sigma_\alpha = \sigma_\alpha(t, \eta_\alpha)$ *given by*

$$\tau_\alpha = Y_\alpha \frac{\ln Y_\alpha - 1}{\varepsilon^{-1} - Y_\alpha}, \qquad e < Y_1 \leq Y_2 < \varepsilon^{-1}. \tag{13.3.68}$$

Moreover,

$$0 < \tau_1 \leq \tau_2 < \infty. \tag{13.3.69}$$

Proof. It follows from Theorem 13.1 that $\mathbf{u}(t, x)$ given by eqn (13.3.62) is an asymptotic solution to eqn (13.3.7) provided the functions $\sigma_\alpha = \sigma_\alpha(t, \eta)$ $(\alpha =$

1, 2) satisfy the transport equations

$$\frac{\partial \sigma_1}{\partial t} + \sigma_1 - \frac{1}{2}\frac{\partial \sigma_1^2}{\partial \eta} = 0, \tag{13.3.70}$$

$$\frac{\partial \sigma_2}{\partial t} + \sigma_2 + \frac{1}{2}\frac{\partial \sigma_2^2}{\partial \eta} = 0. \tag{13.3.71}$$

Also, it is easy to check that $\sigma_1 = \sigma_1(t, \eta)$ and $\sigma_2 = \sigma_2(t, \eta)$ satisfy eqns (13.3.70) and (13.3.71), respectively. Hence, $\mathbf{u}(t, x)$ given by eqns (13.3.62)–(13.3.67) is an asymptotic solution to eqn (13.3.7); and it satisfies the initial condition (13.3.61) provided

$$\sum_{\alpha=1}^{2} \sigma_\alpha(0, y)\,\mathbf{r}_\alpha(\mathbf{u}_0) = \mathbf{u}^*(y), \tag{13.3.72}$$

or, equivalently,

$$\sigma_\alpha(0, y) = \mathbf{r}_\alpha(\mathbf{u}_0) \cdot \mathbf{u}^*(y). \tag{13.3.73}$$

By letting $t = 0$ and $\eta = y$ in eqns (13.3.65) and (13.3.66) and using eqn (13.3.67) we find that eqn (13.3.73) is identically satisfied. As a result, $\mathbf{u}(t, x)$ is an asymptotic solution to eqn (13.3.7) subject to the initial condition (13.3.61).

To show that τ_1 is a blow-up time of the amplitude $\sigma_1 = \sigma_1(t, \eta_1)$ we differentiate eqn (13.3.65) with respect to t and η, respectively, and obtain

$$\frac{\partial \sigma_1}{\partial t}(t, \eta) = Y_1\, \frac{\sigma_1(t, \eta)}{\sigma_1(t, \eta) - Y_1} \tag{13.3.74}$$

and

$$\frac{\partial \sigma_1}{\partial \eta}(t, \eta) = \frac{\sigma_1(t, \eta)}{\sigma_1(t, \eta) - Y_1}. \tag{13.3.75}$$

By letting $\eta = \eta_1 = (x - t)/\varepsilon$ in eqns (13.3.74) and (13.3.75) we find that the first partial derivatives $[\partial \sigma_1/\partial t](t, \eta_1)$ and $[\partial \sigma_1/\partial \eta](t, \eta_1)$ become unbounded if $t \to \tau_1$, where τ_1 satisfies the equation

$$\sigma_1\left(\tau_1, \frac{x - \tau_1}{\varepsilon}\right) = Y_1. \tag{13.3.76}$$

Let τ_1 be a solution of eqn (13.3.76). Substituting $t = \tau_1,\ \eta = (x - \tau_1)/\varepsilon$ in eqn (13.3.65) we obtain

$$\sigma_1\left(\tau_1, \frac{x - \tau_1}{\varepsilon}\right) = Y_1 \left[\tau_1 + \ln \sigma_1\left(\tau_1, \frac{x - \tau_1}{\varepsilon}\right)\right] + \frac{x - \tau_1}{\varepsilon} - \frac{x}{\varepsilon}. \tag{13.3.77}$$

Hence, by eliminating $\sigma_1[\tau_1, (x-\tau_1)/\varepsilon]$ from eqns (13.3.76) and (13.3.77) we obtain

$$\tau_1 = \varphi(Y_1), \tag{13.3.78}$$

where the function $\varphi = \varphi(Y)$ is defined by

$$\varphi(Y) = Y \frac{\ln Y - 1}{\varepsilon^{-1} - Y}, \qquad e < Y < \varepsilon^{-1}. \tag{13.3.79}$$

Similarly, to show that τ_2 is a blow-up time of $\sigma_2 = \sigma_2(t, \eta_2)$ we differentiate eqn (13.3.66) with respect to t and η, respectively, and obtain

$$\frac{\partial \sigma_2}{\partial t}(t, \eta) = Y_2 \, \frac{\sigma_2(t, \eta)}{\sigma_2(t, \eta) - Y_2} \tag{13.3.80}$$

and

$$\frac{\partial \sigma_2}{\partial \eta}(t, \eta) = -\frac{\sigma_2(t, \eta)}{\sigma_2(t, \eta) - Y_2}. \tag{13.3.81}$$

Therefore, $[\partial \sigma_2/\partial t](t, \eta_2)$ and $[\partial \sigma_2/\partial \eta](t, \eta_2)$ become unbounded if $t \to \tau_2$, where τ_2 satisfies the equations

$$\sigma_2\left(\tau_2, \frac{x + \tau_2}{\varepsilon}\right) = Y_2, \tag{13.3.82}$$

$$\sigma_2\left(\tau_2, \frac{x + \tau_2}{\varepsilon}\right) = Y_2 \left[\tau_2 + \ln\, \sigma_2\left(\tau_2, \frac{x + \tau_2}{\varepsilon}\right)\right] - \frac{x + \tau_2}{\varepsilon} + \frac{x}{\varepsilon}. \tag{13.3.83}$$

By eliminating $\sigma_2[\tau_2, (x + \tau_2)/\varepsilon]$ from eqns (13.3.82) and (13.3.83), we obtain

$$\tau_2 = \varphi(Y_2), \tag{13.3.84}$$

where the function $\varphi = \varphi(Y)$ is defined by eqn (13.3.79) for every $Y \in (e,\ \varepsilon^{-1})$.

To show the inequalities (13.3.68) and (13.3.69), consider the function $\psi = \psi(z)$ defined by

$$\psi(z) = \frac{z}{\ln z}, \qquad e < z < \infty. \tag{13.3.85}$$

By differentiating with respect to z we obtain

$$\psi'(z) = \frac{1}{(\ln\, z)^2}(\ln\, z - 1). \tag{13.3.86}$$

Hence,

$$\psi'(z) > 0\,, \qquad e < z < \infty, \tag{13.3.87}$$

which means that $\psi = \psi(z)$ is an increasing function of z for $z \in (e, \infty)$. Therefore,

$$e < \psi(z_1) \leq \psi(z_2) < \infty \quad \text{for} \quad e < z_1 \leq z_2 < \infty. \tag{13.3.88}$$

By letting $z_1 = \mathbf{r}_1(\mathbf{u}_0) \cdot \mathbf{u}^*(y)$ and $z_2 = \mathbf{r}_2(\mathbf{u}_0) \cdot \mathbf{u}^*(y)$ in eqn (13.3.88), and using the notations (13.3.67) and (13.3.85), as well as the hypothesis (13.3.60), we arrive at the inequalities

$$e < Y_1 \leq Y_2 < \varepsilon^{-1} \quad (\varepsilon \to 0). \tag{13.3.89}$$

This completes the proof of inequality (13.3.68).

To show that inequalities (13.3.69) hold true, recall eqn (13.3.79) in the form

$$\varphi(Y) = Y\frac{\ln Y - 1}{\varepsilon^{-1} - Y}, \qquad e < Y < \varepsilon^{-1}. \tag{13.3.90}$$

By differentiating eqn (13.3.90) with respect to Y we obtain

$$\varphi'(Y) = (\varepsilon^{-1} - Y)^{-2}(\varepsilon^{-1}\ln Y - Y). \tag{13.3.91}$$

Since

$$-Y > -\frac{1}{\varepsilon}, \tag{13.3.92}$$

therefore

$$\frac{1}{\varepsilon}\ln Y - Y > \frac{1}{\varepsilon}(\ln Y - 1) > 0 \quad \text{as} \quad Y > e. \tag{13.3.93}$$

Hence,

$$\varphi'(Y) > 0 \quad \text{for} \quad e < Y < \varepsilon^{-1}, \tag{13.3.94}$$

which means that

$$0 < \varphi(Y_1) \leq \varphi(Y_2) < \infty \quad \text{for} \quad e < Y_1 \leq Y_2 < \varepsilon^{-1} \quad (\varepsilon \to \infty). \tag{13.3.95}$$

Therefore, by virtue of eqns (13.3.78), (13.3.84), and (13.3.68), we arrive at the inequality (13.3.69). □

Remark 13.1 The hypothesis (13.3.60) is satisfied provided the vector field $\mathbf{u}^* = \mathbf{u}^*(y)$ satisfies the inequalities

$$e\sqrt{2} < u_1^*(y) + u_2^*(y) \leq u_1^*(y) \quad \text{for} \quad |y| < \infty. \tag{13.3.96}$$

Therefore, a pair of shock heat waves, occurring in the non-linear initial value problem is due to a small but high-frequency initial temperature and a small but high-frequency initial negative heat flux that are imposed on an isothermal state of the rigid heat conductor.

The inequalities (13.3.69) imply that the asymptotic solution accommodates a shock heat wave in which a steepening of the σ_1 profile is followed by a steepening of the σ_2 profile.

Theorem 13.2 reveals the typical feature of a non-linear hyperbolic wave motion that even when we start with smooth initial data we may end up with a shock wave. The following result shows that for an even function $\mathbf{u}^* = \mathbf{u}^*(y)$ there is a single shock thermal wave for the Cauchy problem (13.3.7)–(13.3.9) in which the initial condition (13.3.8) is replaced by eqn (13.3.61).

Theorem 13.3 *Suppose that the function* $\mathbf{u}^* = \mathbf{u}^*(y)$ *satisfies the conditions*

$$\mathbf{u}^*(y) = [\vartheta(y),\ 0]^T, \quad |y| < \infty, \tag{13.3.97}$$

$$\vartheta(y) = \vartheta(-y) \geq 0, \quad |y| < \infty, \tag{13.3.98}$$

$$\exists\ y_0 > 0: \quad \vartheta'(y_0) > \sqrt{2}, \tag{13.3.99}$$

where $\vartheta = \vartheta(y)$ *is a prescribed function. Then, an asymptotic solution to the problem (13.3.7)–(13.3.9) in which the initial condition (13.3.8) is replaced by eqn (13.3.61) takes the form (13.3.62)–(13.3.64), where* $\sigma_\alpha = \sigma_\alpha(\tau, \zeta)$, $\alpha, \beta = 1, 2$ *are given by*

$$\begin{aligned} \sigma_1(\tau,\zeta) &= \exp(-\tau)\sigma(\tau,\zeta), \ 0 \leq \tau < \tau_0, \quad |\zeta| < \infty, \\ \sigma_2(\tau,\zeta) &= \exp(-\tau)\sigma(\tau,-\zeta), 0 \leq \tau < \tau_0, \quad |\zeta| < \infty, \end{aligned} \tag{13.3.100}$$

in which

$$\sigma(\tau,\zeta) = \frac{1}{\sqrt{2}}\vartheta(\zeta + \sigma - \sigma\exp(-\tau)), \quad 0 \leq \tau < \tau_0, \ |\zeta| < \infty, \tag{13.3.101}$$

and $\tau_0 > 0$ *is a blow-up time such that*

$$\begin{aligned} \left|\frac{\partial \sigma_\alpha}{\partial \tau}(\tau,\zeta)\right| &\to \infty \quad \text{as} \quad \tau \to \tau_0 - 0, \quad |\zeta| < \infty, \\ \left|\frac{\partial \sigma_\alpha}{\partial \zeta}(\tau,\zeta)\right| &\to \infty \quad \text{as} \quad \tau \to \tau_0 - 0, \quad |\zeta| < \infty. \end{aligned} \tag{13.3.102}$$

Proof. First, we check that $\sigma_1 = \sigma_1(\tau,\zeta)$ and $\sigma_2 = \sigma_2(\tau,\zeta)$, respectively, satisfy the equations

$$\frac{\partial \sigma_1}{\partial \tau} + \sigma_1 - \sigma_1\frac{\partial \sigma_1}{\partial \zeta} = \exp(-\tau)\left[\frac{\partial \sigma}{\partial \tau} - \sigma\frac{\partial \sigma}{\partial \zeta}\exp(-\tau)\right](\tau,\zeta) \tag{13.3.103}$$

and

$$\frac{\partial \sigma_2}{\partial \tau} + \sigma_2 + \sigma_2\frac{\partial \sigma_2}{\partial \zeta} = \exp(-\tau)\left[\frac{\partial \sigma}{\partial \tau} - \sigma\frac{\partial \sigma}{\partial \zeta}\exp(-\tau)\right](\tau,-\zeta). \tag{13.3.104}$$

Next, by differentiating eqn (13.3.101) we obtain

$$\left[\frac{1-\exp(-\tau)}{\sqrt{2}}\vartheta'(\cdot) - 1\right]\left[\frac{\partial \sigma}{\partial \tau} - \sigma\frac{\partial \sigma}{\partial \zeta}\exp(-\tau)\right] = 0 \tag{13.3.105}$$

for any point (τ,ζ) of the region

$$0 \leq \tau < \tau^*, \quad |\zeta| < \infty \tag{13.3.106}$$

over which the LHS of eqn (13.3.105) makes sense.

Therefore, $\sigma_1 = \sigma_1(\tau, \zeta)$ and $\sigma_2 = \sigma_2(\tau, \zeta)$ satisfy eqns (13.3.70) and (13.3.71), respectively, provided

$$\left[\frac{1-\exp(-\tau)}{\sqrt{2}}\vartheta'(\cdot) - 1\right] \neq 0, \quad 0 \leq \tau < \tau^*, \ |\zeta| < \infty \tag{13.3.107}$$

and $\sigma = \sigma(\tau, \zeta)$ satisfies the equation

$$\frac{\partial \sigma}{\partial \tau} - \sigma \frac{\partial \sigma}{\partial \zeta} \exp(-\tau) = 0, \quad 0 \leq \tau < \tau^*, \ |\zeta| < \infty. \tag{13.3.108}$$

By letting

$$y_0 = \zeta_0 + [1 - \exp(-\tau_0)]\sigma(\tau_0, \zeta_0), \quad \tau_0 > 0, \ |\zeta_0| < \infty, \tag{13.3.109}$$

we find that the condition

$$\frac{1-\exp(-\tau_0)}{\sqrt{2}}\vartheta'(y_0) - 1 = 0 \tag{13.3.110}$$

is equivalent to the existence of $\tau_0 > 0$ of the form

$$\tau_0 = \ln \frac{\vartheta'(y_0)}{\vartheta'(y_0) - \sqrt{2}}. \tag{13.3.111}$$

Therefore, the asymptotic solution to the Cauchy problem takes the form (13.3.62)–(13.3.64), where $\sigma_\alpha = \sigma_\alpha(\tau, \zeta)$, $0 \leq \tau < \tau_0$, $|\zeta| < \infty$, $\alpha, \beta = 1, 2$, are given by eqns (13.3.100) and (13.3.101), provided the initial condition is satisfied

$$\sigma_\alpha(0, \zeta) = \mathbf{r}_\alpha(\mathbf{u}_0) \cdot \mathbf{u}^*(\zeta), \quad |\zeta| < \infty, \tag{13.3.112}$$

or, equivalently,

$$\begin{aligned} \sigma_1(0, y) = \sigma(0, y) &= \frac{1}{\sqrt{2}}\vartheta(y), \qquad\qquad |y| < \infty, \\ \sigma_2(0, y) = \sigma(0, -y) &= \frac{1}{\sqrt{2}}\vartheta(-y) = \frac{1}{\sqrt{2}}\vartheta(y), \ |y| < \infty. \end{aligned} \tag{13.3.113}$$

It is easy to show that the hypothesis (13.3.97)–(13.3.99) together with the definition of the eigenvector $\mathbf{r}_\alpha(\mathbf{u}_0)$ imply eqn (13.3.112).

Finally, to show that the asymptotic solution blows up as $\tau \to \tau_0 - 0$, we calculate the first partial derivatives of $\sigma_\alpha = \sigma_\alpha(\tau, \zeta)$ given by eqns (13.3.100) and (13.3.101) and obtain

$$\frac{\partial \sigma_1}{\partial \tau} = \left[\frac{1-\exp(-\tau)}{\sqrt{2}}\,\vartheta'(\cdot) - 1\right]^{-1}\left[1 - \frac{1}{\sqrt{2}}\,\vartheta'(\cdot)\right]\sigma_1, \tag{13.3.114}$$

$$\frac{\partial \sigma_1}{\partial \zeta} = -\left[\frac{1-\exp(-\tau)}{\sqrt{2}}\,\vartheta'(\cdot) - 1\right]^{-1}\frac{1}{\sqrt{2}}\,\vartheta'(\cdot)\,\exp(-\tau), \tag{13.3.115}$$

and

$$\frac{\partial \sigma_2}{\partial \tau} = \left[\frac{1 - \exp(-\tau)}{\sqrt{2}} \, \vartheta'(\cdot) - 1 \right]^{-1} \left[1 - \frac{1}{\sqrt{2}} \, \vartheta'(\cdot) \right] \sigma_2, \qquad (13.3.116)$$

$$\frac{\partial \sigma_2}{\partial \zeta} = - \left[\frac{1 - \exp(-\tau)}{\sqrt{2}} \, \vartheta'(\cdot) - 1 \right]^{-1} \frac{1}{\sqrt{2}} \, \vartheta'(\cdot) \, \exp(-\tau). \qquad (13.3.117)$$

By letting $\tau \to \tau_0 - 0$ in eqns (13.3.114)–(13.3.117) we find that the limit relations (13.3.102) hold true, and as a result the asymptotic solution (13.3.62)–(13.3.64) blows up as $\tau \to \tau_0 - 0$. □

Clearly, it follows from Theorem 13.3 that the asymptotic solution is unique if there is a unique implicit function $\sigma = \sigma(t, \eta)$ defined by eqn (13.3.101). One can prove the following theorem.

Theorem 13.4 *Suppose that the function $\vartheta = \vartheta(y)$ of Theorem 13.3 takes the form*

$$\vartheta(y) = \frac{1}{\sqrt{2}} y^2, \quad |y| < \infty. \qquad (13.3.118)$$

Then there is the unique explicit function $\sigma = \sigma(\tau, \zeta)$ that complies with eqns (13.3.101) and (13.3.118)

$$\sigma(\tau, \zeta) = \frac{1}{[1 - \exp(-\tau)]^2} \left\{ 1 - \zeta[1 - \exp(-\tau)] - \{1 - 2\zeta[1 - \exp(-\tau)]\}^{1/2} \right\} \qquad (13.3.119)$$

for

$$0 \le \tau < \tau_0^*, \quad \frac{1}{2} < \zeta < \frac{1}{2} \, \frac{1}{1 - \exp(-\tau)}, \qquad (13.3.120)$$

where τ_0^* is the blow-up time of $\sigma = \sigma(\tau, \zeta)$

$$\tau_0^* = - \ln \left(1 - \frac{1}{2\zeta} \right) \quad \textit{for} \quad \zeta > \frac{1}{2}, \qquad (13.3.121)$$

that is

$$\left| \frac{\partial \sigma}{\partial \tau}(\tau, \zeta) \right| \to +\infty \quad \textit{as} \quad \tau \to \tau_0^* - 0, \qquad (13.3.122)$$

$$\left| \frac{\partial \sigma}{\partial \zeta}(\tau, \zeta) \right| \to +\infty \quad \textit{as} \quad \tau \to \tau_0^* - 0. \qquad (13.3.123)$$

Proof. The function $\vartheta = \vartheta(y)$ is an even function of y for every $|y| < \infty$

$$\vartheta'(y) = y\sqrt{2}, \quad \vartheta''(y) = \sqrt{2}, \quad |y| < \infty \qquad (13.3.124)$$

and

$$\frac{\vartheta'(y)}{\sqrt{2}} > 1 \quad \text{for} \quad y > 1, \tag{13.3.125}$$

that means the condition (13.3.99) of Theorem 13.3 is satisfied for any point $y > 1$.

By letting

$$y(\tau,\zeta) = \zeta + [1 - \exp(-\tau)]\sigma(\tau,\zeta) \tag{13.3.126}$$

in eqn (13.3.101) we obtain

$$\frac{y-\zeta}{1-\exp(-\tau)} = \frac{1}{\sqrt{2}}\vartheta(y). \tag{13.3.127}$$

Substituting $\vartheta = \vartheta(y)$ from eqn (13.3.118) into eqn (13.3.127) we arrive at the quadratic equation for y

$$\left(y - \frac{1}{1-\exp(-\tau)}\right)^2 = \frac{1}{[1-\exp(-\tau)]^2}\left\{1 - 2\zeta[1-\exp(-\tau)]\right\}. \tag{13.3.128}$$

Note that there are two real-valued solutions to this equation

$$y^*_{1.2} = \frac{1}{1-\exp(-\tau)} \pm \frac{1}{1-\exp(-\tau)}\left\{1 - 2\zeta[1-\exp(-\tau)]\right\}^{1/2}, \tag{13.3.129}$$

provided

$$1 - 2\zeta[1-\exp(-\tau)] > 0, \quad \textit{and} \quad 1-\exp(-\tau) > 0 \tag{13.3.130}$$

and, by eqn (13.3.126), there are the two functions σ^*_1 and σ^*_2 corresponding to y^*_1 and y^*_2, respectively

$$\sigma^*_{1.2} = \frac{1}{[1-\exp(-\tau)]^2}\left\{1 - \zeta[1-\exp(-\tau)] \pm \{1 - 2\zeta[1-\exp(-\tau)]\}^{1/2}\right\}. \tag{13.3.131}$$

Also, note that of the two functions σ^*_1 and σ^*_2 only $\sigma^*_2 \equiv \sigma(\tau,\zeta)$ with "minus" in front of the square root is finite at $\tau = 0$, since

$$\sigma^*_2(\tau,\zeta) \to \frac{1}{2}\zeta^2 \quad \text{as} \quad \tau \to 0+0. \tag{13.3.132}$$

Therefore, the function $\sigma = \sigma(\tau,\zeta)$ takes the form (13.3.119)–(13.3.123) where τ^*_0 is the only root of the equation

$$1 - 2\zeta[1-\exp(-\tau^*_0)] = 0 \quad \text{for} \quad \zeta > 1/2. \tag{13.3.133}$$

□

Theorems 13.3 and 13.4 imply the following result.

Theorem 13.5 *An asymptotic solution to the Cauchy problem (13.3.7)–(13.3.9) in which the initial condition (13.3.8) is replaced by eqn (13.3.61) and* $\mathbf{u}^* = \mathbf{u}^*(y)$ *satisfies the conditions (13.3.97) and (13.3.118) takes the unique explicit closed-form (13.3.62)–(13.3.64), where* $\sigma_\alpha = \sigma_\alpha(t, \eta_\alpha)$ *are defined by the formulas*

$$\sigma_1(t, \eta_1) = \sigma(t, \eta_1) \exp(-t) \tag{13.3.134}$$

and

$$\sigma_2(t, \eta_2) = \sigma(t, -\eta_2) \exp(-t). \tag{13.3.135}$$

Here,

$$\begin{aligned} \sigma(t, \eta_1) = {} & \frac{1}{(1-e^{-t})^2} \left\{ 1 - \eta_1(1-e^{-t}) - [1 - 2\eta_1(1-e^{-t})]^{1/2} \right\} \\ & \textit{for} \quad 0 \le t < \tau_1^*, \quad \frac{1}{2} < \eta_1 \le \frac{1}{2} \frac{1}{1-e^{-t}}, \end{aligned} \tag{13.3.136}$$

where

$$\tau_1^* = -\ln\left(1 - \frac{1}{2\eta_1}\right) \tag{13.3.137}$$

and

$$\begin{aligned} \sigma(t, -\eta_2) = {} & \frac{1}{(1-e^{-t})^2} \left\{ 1 + \eta_2(1-e^{-t}) - [1 + 2\eta_2(1-e^{-t})]^{1/2} \right\} \\ & \textit{for} \quad 0 \le t < \tau_2^*, \quad -\frac{1}{2} \frac{1}{1-e^{-t}} < \eta_2 \le \frac{1}{2}, \end{aligned} \tag{13.3.138}$$

where

$$\tau_2^* = -\ln\left(1 + \frac{1}{2\eta_2}\right). \tag{13.3.139}$$

As a result, the asymptotic solution of Theorem 13.5 experiences the two blows-up at the times $t = \tau_1^*$ and $t = \tau_2^*$. Since $\tau_1^* > \tau_2^*$, the asymptotic solution represents a heat shock wave in which σ_2 blows up earlier than σ_1.

A plot of the function $\sigma = \sigma(t, \eta) \equiv \Sigma(\psi, \eta)$ for fixed values of η, and $\psi = \psi(t)$ defined by

$$t = -\ln\left(1 - \frac{1}{\psi}\right), \tag{13.3.140}$$

$$\frac{1}{2} < \eta < \frac{1}{2}\psi < \infty, \tag{13.3.141}$$

is shown on Fig. 13.4. A blow-up of the function $\Sigma = \Sigma(\psi, \eta)$ at a fixed point of the ψ-axis is clearly seen from this figure.

Remark 13.2 A method of WNGO is used in (Oncu et al., 1994) to study the 1D non-linear hyperbolic thermoelastic waves in a semi-space subject to a

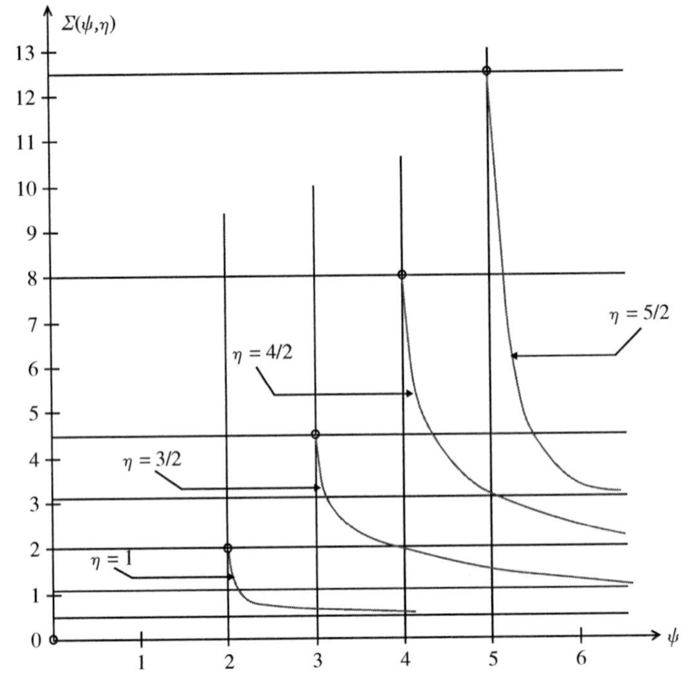

Figure 13.4 Plot of the function $\sigma(t, \eta) \equiv \Sigma(\psi, \eta)$ for fixed values of η.

small strain-temperature load on its boundary. Also, a number of interesting results on non-linear hyperbolic thermoelastic waves include those based on a thermoelasticity with internal variables (see: Kosiński, 1974 and 1975; Hetnarski and Ignaczak, 1996) as well as those dealing with well-posedness of the governing equations (Tarabek et al., 1992; Chrzęszczyk et al., 1993a,b).

REFERENCES

Abd-Alla, A.N. and Al-Dawy, A.A.S. (2001). Thermal relaxation times effect on Rayleigh waves in generalized thermoelastic media. *J. Thermal Stresses* **24**(4), 367–382.

Abramowitz, M. and Stegun, I. (1965). *Handbook of Mathematical Functions.* Dover Publ.

Achenbach, J.D. (1968). The influence of heat conduction on propagating stress jumps. *J. Mech. Phys. Solids* **16**, 273–283.

Agarwal, V.K. (1979). On plane waves in generalized thermoelasticity. *Acta Mech.* **31**, 185–198.

Al-Qahtani, H.M. Datta, S.K. and Mukdadi, O. M. (2005). Laser- generated thermoelastic waves in an anisotropic infinite plate: FEM analysis. *J. Thermal Stresses* **28**, 1099–1122.

——— (2008). Laser-generated thermoelastic waves in an anisotropic infinite plate: Exact analysis. *J. Thermal Stresses* **31**, 569–583.

Bai, C. and Lavine, A.S. (1995). On hyperbolic heat conduction and the second law of thermodynamics. *ASME J. Heat Transfer* **117**, 256–263.

Bakhvalov, N. and Panasenko, G. (1989). *Homogenisation: Averaging Processes in Periodic Media.* Kluwer, Dordrecht.

Bem, Z. (1982a). Existence of a generalized solution in thermoelasticity with one relaxation time. *J. Thermal Stresses* **5**(2), 195–206.

—— (1982b). Existence of a generalized solution in thermoelasticity with two relaxation times. Part I. *J. Thermal Stresses* **5**(3–4), 395–404.

—— (1983). Existence of a generalized solution in thermoelasticity with two relaxation times. Part II. *J. Thermal Stresses* **6**(2–4), 281–299.

Bensoussan, A. Papanicolaou, G.C. and Lions, J.L. (1978). *Asymptotic Analysis for Periodic Structures.* North-Holland Publishing Company, Amsterdam.

Biały, J. (1983). *Domain of Influence in Thermoelasticity with Finite Wave Speeds.* Ph.D. Thesis (in Polish), Kielce University of Technology, Kielce, Poland.

Bishop, J.E. and Kinra, V.K. (1997). Elastothermodynamic damping in laminated composites. *Int. J. Solids Struct.* **34**, 1075.

Boley, B.A. (1955). The Applications of Saint-Venant's principle in dynamical problems. *J. Appl. Mech.* **22**(2), 204–206.

—— (1972). Approximate analysis of thermally induced vibrations of beams and plates. *ASME J. Appl. Mech.* **39**, 212–216.

—— and Barber, A.D. (1957). Dynamical response of beams and plates subject to rapid heating. *ASME J. Appl. Mech.* **24**, 413–416.

—— and Hetnarski, R.B. (1968). Propagation of discontinuities in coupled thermoelastic problems. *ASME J. Appl. Mech.* **35**, 489–494.

—— and Weiner, J.H. (1960). *Theory of Thermal Stresses*, Dover Pub., New York.

Brock, L. M. (2005). Thermal relaxation effects in rapid sliding contact with friction. *Acta Mech.* **176**, 185–196.

Brock, L. M. (2006). Debonding of a thermoelastic material from a rigid substrate at any constant speed: thermal relaxation effects. *Acta Mech.* **184**, 185–196.

——(2007). Stoneley waves generation in joined materials with and without thermal relaxation due to thermal mismatch. *ASME J. Appl. Mech.* **74**, 1019–1025.

——(2008a). Stoneley signals in perfectly bonded dissimilar thermoelastic half-spaces with and without thermal relaxation. *J. Mech. Mater. Struct.*, in press.

——(2008b). Dynamic crack extension along the interface of materials that differ in thermal properties: Convection and thermal relaxation. *ASME J. Appl. Mech.* **75**, 021018-1-7.

Brun. L. (1975). L'onde simple thérmoélastique linéaire. *J. Méc.* **14**(5), 863–885.

Carlson, D.E. (1972). Linear Thermoelasticity, in *Encyclopedia of Physics* **VIa/2**, *Mechanics of Solids* **II**, Springer-Verlag.

Caputo, M. (1967). Linear models of dissipation whose Q is almost frequency independent. Part II. *Geoph. J. R. Astron. Soc.* **13**, 529–539.

——(1969). *Elasticità e Dissipazione.* Zanichelli, Bologna.

Cattaneo, C. (1948). Sulla condizione del calore. *Atti. del Seminario Mat. e Fis. Univ. Modena* **3**, 83–101.

Caviglia, G., Morro, A. and Straughan, B. (1992). Thermoelasticity at cryogenic temperatures. *Int. J. Non-Lin. Mech.* **27**, 251–263.

Chadwick, P. (1960). *Thermoelasticity, The Dynamical Theory, Progress in Solid Mechanics* **1** (ed. I.N. Sneddon and R. Hill), 263–328, North-Holland.

Chandrasekharaiah, D.S. (1981). Wave propagation in a thermoelastic half-space. *Ind. J. Pure Appl. Math.* **12**(2), 226–241.

——(1984). A uniqueness theorem in generalized thermoelasticity. *J. Tech. Phys.* **25** (3–4), 345–350.

——(1986). Thermoelasticity with second sound: A review. *Appl. Mech. Rev.* **39**(3), 355–376.

——(1998). Hyperbolic thermoelasticity: A review of recent literature. *Appl. Mech. Rev.* **51**(12), 705–729.

Chester, M. (1963). Second sound in solids. *Phys. Rev.* **131**(5), 2013–2015.

Chirita, S. (1995). Saint-Venant's principle in linear thermoelasticity. *J. Thermal Stresses* **18**, 485–496.

——(2007). Spatial behavior of the strongly elliptic anisotropic thermoelastic materials. *J. Thermal Stresses* **30**, 859–873.

Christov, C.I. and Jordan, P.M. (2005). Heat conduction paradox involving second-sound propagation in moving media. *Phys. Rev. Lett.* **94**, 154301-1-4.

Chrzęszczyk, A. (1993a), Initial-boundary value problems for nonlinear equations of generalized thermoelasticity and elasticity, *Math. Meth. Appl. Sci.* **16**, 49–60.

——(1993b), Well-posedness of the equations of generalized thermoelasticity, *J. Elast.* **16**, 69–80.

Coleman, B.D., Fabrizio, M. and Owen, D.R. (1982). On the thermodynamics of second sound in dielectric crystals. *Arch. Rat. Mech. Anal.* **80**(2), 135–158.

——————(1986). Thermodynamics and the constitutive relations for second sound in crystals, in: *New Perspectives in Thermodynamics.* J. Serrin, ed., 171–185, Berlin.

Coleman, B.D. and Owen, D.R. (1983). On the nonequilibrium behavior of solids that transport heat by second sound. *Comp. Math. Appls.* **9**(3), 527–546.

Dai, W., Wang, H., Jordan, P.M., Mickens, R.E. and Bejan, A. (2008). A mathematical model for skin burn injury induced by radiation heating. *Int. J. Heat Mass Transfer* **51**(23–24), 5497–5510.

Danilovskaya, V.I. (1950). Thermal stresses in an elastic half-space due to sudden heating of its boundary (in Russian). *Prikl. Mat. Mech.* **14**(3).

Day, W.A. (1985). *Heat Conduction within Linear Thermoelasticity.* Springer Tracts in Natural Philosophy **30**, Springer-Verlag, New York.

Deresiewicz, H. (1957). Plane waves in a thermoelastic solid. *J. Acoust. Soc. Am.* **29**(2), 204–209.

Dhaliwal, R.S. and Singh, A. (1980). *Dynamic Coupled Thermoelasticity*, Hindustan Publ. Corp. Delhi, India.

DiPerna, R. and Majda, A. (1985). The validity of geometric optics for weak solutions of conservation laws. *Comm. Math. Phys.* **98**, 313–347.

Domański, W. (2000). Weakly nonlinear elastic plane waves in a cubic crystal. *Contemporary Mathematics*, Am. Math. Soc., **255**, pp. 45–61.

Dreyer, W. and Struchtrup, H. (1993). Heat pulse experiments revisited. *Continuum Mech. Thermodyn.* **5**, 3–50.

Edelen, D.G.B. (1973). On the existence of symmetry relations and dissipative potentials. *Arch. Rat. Mech. Anal.* **51**, 218–227.

—— (1974*a*). Primitive thermodynamics: A new look at the Clausius-Duhem inequality. *Int. J. Eng. Sci.* **12**, 121–141.

—— (1974*b*). On the characterization of fluxes in nonlinear reversible thermodynamics. *Int. J. Eng. Sci.* **12**, 397–411.

—— (1993). *The College Station Lectures on Thermodynamics*, a collection of the lectures presented at Texas A&M University, sponsored by the Department of Aerospace Engineering.

El-Karamany, A.S. and Ezzat, M.A. (2004). Discontinuities in generalized thermoviscoelasticity under four theories. *J. Thermal Stresses* **27**, 1187–1212.

—— —— (2005). Propagation of discontinuities in thermopiezoelectric rod. *J. Thermal Stresses* **28**, 997–1030.

Erbay, S. and Şuhubi, E.S. (1986). Longitudinal wave propagation in a generalized thermoelastic cylinder. *J. Thermal Stresses* **9**, 279–295.

Eringen, A.C. and Şuhubi, E.S. (1975). *Elastodynamics, II Linear Theory.* Academic Press.

Feder, J. (2007). *Fractals (Physics of Solids and Liquids).* Springer.

Fichera, G. (1997). A boundary value problem connected with response of semi-space to a short laser pulse. *Rend. Mat. Acc. Lincei* **8**(9), 197–228.

Francis, P.H. (1972). Thermo-mechanical effects in elastic wave propagation, A survey. *J. Sound Vib.* **21**(2), 181–192.

Gładysz, J. (1982). *Convolutional Variational Principles in Thermoelasticity with Finite Wave Speeds* (in Polish). Ph.D. Thesis, Wrocław University of Technology, Wrocław, Poland.

Gorenflo, R. and Mainardi, F. (1997). Fractional calculus: Integral and differential equations of fractional order. *Arch. Mech.* **50**, 377–388.

—— —— (1998). Fractional calculus and stable probability distributions. in *Fractals and Fractional Calculus in Continuuum Mechanics*, eds. A. Carpinteri and F. Mainardi, 223–290.

Green, A.E. (1972). A note on linear thermoelasticity. *Mathematika* **19**, 69–75.

Green, A.E. and Laws, N. (1972). On the entropy production inequality. *Arch. Rational Mech. Anal.* **45**, 47–53.

——and Lindsay, K.A. (1972). Thermoelasticity. *J. Elast.* **2**(1), 1–7.

——and Naghdi, P.M. (1991). A re-examination of the basic postulates of thermomechanics. *Proc. R. Soc. Lond. A* **432**, 171–194.

————(1992). On undamped heat waves in an elastic solid. *J. Thermal Stresses* **15**, 253–264.

————(1993). Thermoelasticity without energy dissipation. *J. Elast.* **31**, 189–208.

Gurtin, M.E. (1972). The linear theory of elasticity, in *Encyclopedia of Physics* **VIa/2**, *Mechanics of Solids* **II**, Springer-Verlag.

Hetnarski, R.B. and Ignaczak, J. (1993). Generalized thermoelasticity: Closed form solutions. *J. Thermal Stresses* **16**(4), 473–498.

————(1999). Generalized thermoelasticity. *J. Thermal Stresses* **22**, 451–476.

————(2000). Nonclassical dynamical thermoelasticity. *Int. J. Solids Struct.* **37** (1–2), 215–224.

————(2004). *Mathematical Theory of Elasticity.* Taylor and Francis, New York.

Hunter, J.K. and Keller, J. (1983). Weakly nonlinear high frequency waves, *Comm. Pure Appl. Math.* **36**, 547–569.

Ieşan, D. (1966). Principes variationnels dans la théorie de la thérmoélasticité couplee. *Ann. Stiint. Univ. A.I. Cuza, Iasi, Sect. I, Matematica* **12**, 439–456.

——(1998). On the theory of thermoelasticity without energy dissipation. *J. Thermal Stresses* **21**, 295–307.

——(2004). *Thermoelastic Models of Continua.* Kluwer Academic Pub., Dordrecht.

Ignaczak, J. (1970). Radiation conditions of Sommerfeld type for elastic materials with microstructure. *Bull. Acad. Polon. Sci., Sér. Sci. Tech.* **18**(6), 251–257.

——(1974). Domain of influence theorem for stress equations of motion of linear elastodynamics. *Bull. Acad. Polon. Sci., Sér. Sci. Tech.* **22**(9).

——(1976). Thermoelastic counterpart to Boggio's theorem of linear elastodynamics. *Bull. Acad. Polon. Sci., Sér. Sci. Tech.* **24**(3), 129–137.

——(1978a). Decomposition theorem for thermoelasticity with finite wave speeds. *J. Thermal Stresses* **1**(1), 41–52.

——(1978b). Domain of influence theorem in linear thermoelasticity. *Int. J. Eng. Sci.* **16**, 139–145.

——(1978c). A uniqueness theorem for stress-temperature equations of dynamic thermoelasticity. *J. Thermal Stresses* **1**(2), 163–170.

——(1979). Uniqueness in generalized thermoelasticity. *J. Thermal Stresses* **2**(2), 171–175.

——(1980a). A variational description of stress and heat flux in dynamic thermoelasticity. *Collected Articles.* Kielce College of Education, Kielce, Poland.

——(1980b). Thermoelasticity with finite wave speeds – a survey, in *Thermal Stresses in Severe Environments*, ed. D.P.H. Hasselman and R.A. Heller. Plenum Press, New York and London, 15–30.

——(1981a). On a three-dimensional solution of dynamic thermoelasticity with two relaxation times. *J. Thermal Stresses* **4**(3–4), 357–385.

——(1981b). Linear dynamic thermoelasticity – A survey. *Shock Vib. Dig.* **13**(9), 3–8.

——(1982). A note on uniqueness in thermoelasticity with one relaxation time. *J. Thermal Stresses* **5**(3–4), 257–263.

——(1983). Thermoelastic polynomials. *J. Thermal Stresses* **6**(1), 45–55.

——(1985). A strong discontinuity wave in thermoelasticity with relaxation times. *J. Thermal Stresses* **8**(1), 25–40.

——(1987). Linear dynamic thermoelasticity – A survey: 1981–1984. *Shock Vib. Dig.* **19**(6), 11–17.

Ignaczak, J. (1989a). Generalized thermoelasticity and its applications, in *Thermal Stresses* **III** (ed. R.B. Hetnarski), 279–354, North-Holland, Amsterdam.

——(1989b). *Thermoelasticity with Finite Wave Speeds.* Lecture Notes (in Polish). Ossolineum, Wrocław-Warszawa, Kraków, Gdańsk, Łódź, 1–104.

——(1991). Domain of influence results in generalized thermoelasticity – a survey. *Appl. Mech. Rev.* **44**(9), 375–382.

——(1998). Saint-Venant type decay estimates for transient heat conduction in a composite rigid semi-space. *J. Thermal Stresses* **21**, 185–204.

——(2001). Dual-phase-lag model of rigid heat conductor revisted. in *Proc. Fourth Int. Cong. Thermal Stresses - TS2001*, Osaka Institute of Technology, Osaka, Japan, 511–514.

——(2002). Saint-Venant's principle for a microperiodic composite thermoelastic semi-space: The dynamical refined averaged theory. *J. Thermal Stresses* **25**(12), 1065–1079.

——(2006). Nonlinear hyperbolic heat conduction problem: Closed-form solutions. *J. Thermal Stresses* **29**(11), 999–1018.

——and Biały, J. (1980). Domain of influence in thermoelasticity with one relaxation time. *J. Thermal Stresses* **3**(3), 391–399.

——Carbonaro, B. and Russo, R. (1986). A domain of influence theorem in thermoelasticity with one relaxation time. *J. Thermal Stresses* **9**(1), 79–91.

——and Domański, W. (2008). Nonlinear hyperbolic rigid heat conductor of the Coleman type. *J. Thermal Stresses* **31**(5), 416–437.

——and Nowacki, W. (1962). The Sommerfeld radiation conditions for coupled problems of thermoelasticity. *Arch. Mech. Stos.* **14**(1), 3–14.

Jackson, H. E., Walker, C. T. and McNelly, T. F. (1970). Second sound in NaF. *Phys. Rev. Lett.* **25**, 26–29.

Jakubowska, M. (1982). *Kirchhoff's Formula for a Thermoelastic Body*, Ph.D. Thesis (in Polish), Inst. Fund. Tech. Res., Polish Academy of Sciences, Warsaw, Poland.

Kaliski, S. (1965). Wave equations in thermoelasticity. *Bull. Acad. Polon. Sci., Sér. Sci. Tech.* **13**, 253–260.

Khisaeva, Z.F. and Ostoja-Starzewski, M. (2006). Thermoelastic damping in nanomechanical resonators with finite wave speeds. *J. Thermal Stresses* **29**, 201–216.

Kimmich, R. (2002). Strange kinetics, porous media, and NMR. *Chem. Phys.* **284**, 253–285.

King, A.C., Needham, D.J. and Scott, N.H. (1989). The effects of weak hyperbolicity on the diffusion of heat. *Proc. R. Soc. Lond. A* **454**, 1659–1679.

Kinra, V.K. and Milligan, K.B. (1994). A second law analysis of thermoelastic damping. *J. Appl. Mech.* **61**, p. 71.

Klages, R. (2007). *Microscopic Chaos, Fractals and Transport in Nonequilibrium Statistical Mechanics.* World Scientific, Singapore.

Kosiński, W. (1974). Behaviour of the acceleration and shock waves in materials with internal state variables. *Int. J. Non-Linear Mech.* **9**, 481–499.

Kosiński, W. (1975). One-dimensional shock waves in solids with internal state variables. *Arch. Mech.* **27**, 445–458.

Kupradze, V.D., Gegelia, T.G., Basheleishvili, M.O. and Burchuladze, T.V. (1979). *Three-Dimensional Problems of Mathematical Theory of Elasticity and Thermoelasticity.* North-Holland, Amsterdam.

Landau, L. (1941). *J. Phys.* **5**, 71.

Lax, P.D. (1957). Hyperbolic systems of conservation laws. *Comm. Pure Appl. Math.* **10**, 537–566.

Lebon, G. (1982). A generalized theory of thermoelasticity. *J. Tech. Physics* **23**, 37–46.

Leslie, D.J. and Scott, N.H. (2004). Wave stability for constrained materials in anisotropic generalized thermoelasticity. *Math. Mech. Solids* **9**, 513–542.

Li, J. and Ostoja-Starzewski, M. (2009). Fractal solids, product measures, and continuum mechanics, *Proc. R. Soc. Lond. A* **465**, 2521–2536.

Lifshitz, R. and Roukes, M.L. (2000). Thermoelastic damping in micro- and nanomechanical systems. *Phys. Rev. B* **61**, 5600–5609.

Lord, H.W. and Shulman, Y. (1967). A generalized dynamical theory of thermoelasticity, *J. Mech. Phys. Solids* **15**, 299–309.

Maugin, G.A. (1999). *The Thermomechanics of Nonlinear Irreversible Behaviors – an Introduction*, World Scientific, Singapore.

Maugin, G.A. (2008). Private Communication.

Maxwell, J.C. (1867). On the dynamical theory of gases. *Phil. Trans. Roy. Soc. Lond.* **157**, 49–88.

McNelly, T.F., Rogers, S.J., Channin, D.J., Rollefson, R.J., Goubau, W.M., Schmidt, G.E., Krumhansl, J.A. and Pohl, R.O. (1970). Heat pulses in NaF: Onset of second sound. *Phys. Rev. Lett.* **24**, 100–102.

Mikusiński, J. (1957). *Operational Calculus*, PWN-Polish Scientific Publishers. English translation (1967) Pergamon Press.

Miller, S.T. and Haber, R.B. (2008). A spacetime discontinuous Galerkin method for hyperbolic heat conduction. *Comp. Meth. Appl. Mech. Eng.* **198**, 194–209.

Müller, I. (1971). The coldness, a universal function in thermoelastic bodies. *Arch. Rational Mech. Anal.* **41**, 319–331.

——and Ruggeri, T. (1993). *Extended Thermodynamics*, Springer-Verlag.

————(1998). *Rational Extended Thermodynamics*, Springer-Verlag.

Nappa, L. (1998). Spatial decay estimates for the evolution equations of linear thermoelasticity without energy dissipation. *J. Thermal Stresses* **21**, 581–592.

Nayfeh, A. and Nemat-Nasser, S. (1971). Thermoelastic waves in solids with thermal relaxation. *Acta Mech.* **12**, 53–69.

Nickell, R.E. and Sackman, J.L. (1968). Variational principles for linear coupled thermoelasticity. *Quart. Appl. Math.* **26**, 11–26.

Nowacki, Witold (1957). A dynamical problem of thermoelasticity. *Arch. Mech. Stos.* **9**(3), 325–334.

——(1962). *Thermoelasticity.* Pergamon Press.

——(1975). *Dynamic Problems of Thermoelasticity.* Noordhoff; translated from the Polish edition (1966), PWN-Polish Scientific Publishers.

——(1986). *Theory of Asymmetric Elasticity*, Pergamon Press. Oxford/PWN-Polish Scientific Publishers, Warsaw.

Ostoja-Starzewski, M. (2003). Thermoelastic waves in a helix with parabolic or hyperbolic heat conduction. *J. Thermal Stresses* **26**, 1205–1219.

——(2007a). Towards thermomechanics of fractal media. *J. Appl. Math. Phys. (ZAMP)* **58**(6), 1085–1096.

——(2007b). Towards thermoelasticity of fractal media. *J. Thermal Stresses* **30**, 889–896.

Ostoja-Starzewski, M. (2008a). *Microstructural Randomness and Scaling in Mechanics of Materials*, Chapman & Hall/CRC Press/Taylor & Francis.

——(2008b). On turbulence in fractal porous media. *J. Appl. Math. Phys. (ZAMP)* **59**(6), 1111–1117.

——(2009a). Continuum mechanics models of fractal porous media: Integral relations and extremum principles, *J Mech. Mater. Struct.*, in press.

——(2009b). Extremum and variational principles for elastic and inelastic media with fractal geometries, *Acta Mech.* **205**, 161–170. (doi:10.1007/s00707-009-0169-0)

——(2009c). A derivation of the Maxwell-Cattaneo equation from the free energy and dissipation potentials. *Int. J. Eng. Sci.* **47**, 807–810. (doi:10.1016/j.ijengsci.2009.03.002)

——and Li, J. (2009d). Fractal materials, beams and fracture mechanics, *ZAMP*, in press. (doi:10.1007/s00033-009-8120-8)

Parkus, H. (1959). *Instationäre Wärmespannuungen.* Springer, 96–97.

Pao, Y.-H. and Banerjee, D.K. (1978). A theory of anisotropic thermoelasticity at low reference temperature. *J. Thermal Stresses* **1**(1), 99–112.

Peierls, R. (1979). *Surprises in Theoretical Physics.* Princeton Series in Physics. Princeton University Press, Princeton, NJ.

Peshkov, V. (1944). "Second sound" in helium II. *J. Phys. U.S.S.R.* **8**, 131.

Podstrigach, Y.S. and Kolano, Y.M. (1976). *Generalized Thermomechanics* (in Russian), Izdat. Nauk, Dumka, Kiev.

Povstenko, Y.Z. (2005a). Fractional heat conduction equation and associated thermal stress. *J. Thermal Stresses* **28**, 83–102.

——(2005b). Stresses exerted by a source of diffusion in a case of a non-parabolic diffusion equation. *Int. J. Eng. Sci.* **43**, 977–991.

——(2008). Two-dimensional axisymmetric stresses exerted by instantaneous pulses and sources of diffusion in an infinite space in a case of time-fractional diffusion equation. *Int. J. Solids Struct.* **44**, 2324–2348.

Prevost, J.H. and Tao, D. (1983). Finite element analysis of dynamic coupled thermoelasticity problems with relaxation times. *J. Appl. Mech.* **50**, 817–822.

Puri, P. (1973). Plane waves in generalized thermoelasticity. *Int. J. Eng. Sci.* **11**, 735–744; Errata *ibid.* **12**, 339–340.

Quintanilla, R. (1999). On the spatial behaviour in thermoelasticity without energy dissipation. *J. Thermal Stresses* **22**, 213–224.

——(2001). End effects in thermoelasticity. *Math. Meth. Appl. Sci.* **24**(2), 93–102.

——and Straughan, B. (2000). Growth and uniqueness in thermoelasticity. *Proc. R. Soc. Lond. A* **456**, 1419–1429.

Roukes, M.L. (2000). Nanoelectromechanical Systems, *Technical Digest of the 2000 Solid-State Sensor and Actuator Workshop*, p. 1.

Rożnowski, T. (1971). Radiation conditions of Sommerfeld type for coupled thermoelasticity. *Bull. Acad. Polon. Sci., Sér. Sci. Tech.* **19**(7–8), 287–290.

——(1983). Integral representation and uniqueness theorems in generalized thermoelasticity. *Arch. Mech. Stos.* **35**(1), 17–36.

Samras, R.K., Skop, R.A. and Millburn, D.A. (1974). An analysis of coupled extensional-torsional oscillations in wire rope. *ASME J. Eng. Ind.* **96**, 1130–1135.

Scott, N.H. (2008). Thermoelasticity and generalized thermoelasticity viewed as wave hierarchy. *IMA J. Appl. Math.* **73**, 123–136.

Sherief, H.H. (1986). Fundamental solution of the generalized thermoelastic problem for short times. *J. Thermal Stresses* **9**, 151–164.

——and Anwar, M.N. (1994). Two-dimensional generalized thermoelasticity problem for an infinitely long cylinder. *J. Thermal Stresses* **17**(2), 213–227.

——and Dhaliwal, R.S. (1981). Generalized one-dimensional thermal shock problem for small times. *J. Thermal Stresses* **4**, 407–420.

Stakgold, I. (1968). *Boundary Value Problems of Mathematical Physics* **II**. McMillan Co.

Sternberg, E. (1954). On Saint-Venant's Principle. *Quart. Appl. Math.* **11**(4), 393–402.

Strikverda, J.C. and Scott, A.M. (1984). Thermoelastic response to a short laser pulse. *J. Thermal Stresses* **7**(1), 1–17.

Şuhubi, E.S. (1975). Thermoelastic solids, in *Continuum Physics* **2**(2) (ed. Eringen, A.C.), Academic Press.

——(1982). A generalized theory of simple thermomechanical materials. *Int. J. Eng. Sci.* **20**, 365–371.

Tao, D. and Prevost, J.H. (1984). Relaxation effects on generalized thermoelastic waves. *J. Thermal Stresses* **7**(1), 78–89.

Tarabek, M.A. (1992). On the existence of smooth solutions in one-dimensional nonlinear thermoelasticity with second sound, *Quart. Appl. Math.* **50**, 727–742.

Tarasov, V.E. (2005a). Continuous medium model for fractal media. *Phys. Lett. A* **336**, 167–174.

——(2005b). Fractional hydrodynamic equations for fractal media. *Ann. Phys.* **318**(2), 286–307.

Tzou, D.Y. (1997). *Macro – To Microscale Heat Transfer; The Lagging Behavior.* Taylor & Francis, Washington, DC.

Vernotte, P. (1958). Les paradoxes de la théorie continue de l'équation de la chaleur. *C.R. Acad. Sci.* **246** (22) 3154–3155.

Vivar-Pèrez, J.M., Bravo-Castillero, J., Rodriguez-Ramos, R. and Ostoja-Starzewski, M. (2006). Homogenization of a micro-periodic helix with parabolic or hyperbolic heat conduction. *J. Thermal Stresses* **29**, 467–483.

————————(2008). The effect of imperfect contact on the homogenization of a micro-periodic helix. *Mathematics and Mechanics of Solids*, in press.

Watson, G.N. (1958). *A Treatise on the Theory of Bessel Functions.* Cambridge University Press.

Wilmański, K. (1998). *Thermomechanics of Continua*, Springer-Verlag.

Wojnar, R. (1984). The plane stress state in thermoelasticity with relaxation times (in Polish). *Biuletyn WAT* **33**(1), 105–113.

——(1985a). Uniqueness of displacement-heat flux and stress-temperature problems in thermoelasticity with one relaxation time. *J. Thermal Stresses* **8**, 351–364.

——(1985b). Uniqueness of the displacement-heat flux problem in thermoelasticity with two relaxation times. *Bull. Acad. Polon. Sci., Sér. Sci. Tech.* **33**(5–6), 217–227.

——(1985c). Uflyand-Mindlin's plate equations in thermoelasticity with one relaxation time. *Bull. Acad. Polon. Sci., Sér. Sci. Tech.* **33**(7–8), 325–349.

——(1985d). Uflyand-Mindlin's plate equations in thermoelasticity with relaxation times. *J. Tech. Physics* **26**(3–4), 377–403.

——(1986). Rayleigh waves in thermoelasticity with relaxation times. *2nd Int. Conf. on Surface Waves in Plasmas and Solids*, Yugoslavia (1985), World Scientific, Singapore, 682–685.

Wojnar, R. (1988a). Surface waves in thermoelasticity with relaxation times, in *Recent Developments in Surface Acoustic Waves, Euromech* **226** (ed. G.F. Parker and G.A. Maugin), Springer-Verlag, 335–341.

——(1988b). Surface waves in thermoelasticity with relaxation times (in Polish). *Theor. Appl. Mech.* **1**, 55–72.

Yourgrau, W., *Treatise on Irreversible and Statistical Thermophysics: an Introduction to Nonclassical Thermodynamics*, New York, MacMillan, 1966.

Zaremba, S. (1915). Sopra un theorema d'unicita relativo alla equazione delle onde sferiche. *Atti. Reale Accad. Lincei., Cl. Sci. Fis. e Mat. Natur. [5]* **24**, 904.

Zener, C. (1948). *Elasticity and Anelasticity of Metals*, University of Chicago Press, Chicago.

Ziegler, H. and Wehrli, C. (1987). The derivation of constitutive relations from the free energy and the dissipation functions. *Adv. Appl. Mech.* **25** (ed. T.Y. Wu and J.W. Hutchinson), 183–238, Academic Press, New York.

ADDITIONAL REFERENCES

Abbas, I. (2008). Finite element method of thermal shock problem in a non-homogeneous isotropic hollow cylinder with two relaxation times. *Eng. Res.* **72**(2), 101–110.

Abd-Alla, A.N. and Abbas, I. (2002). A problem of generalized magnetothermoelasticity for an infinitely long, perfectly conducting cylinder. *J. Thermal Stresses* **25**, 1009–1025.

Achenbach, J.D. (2005). The thermoelasticity of laser-based ultrasonics. *J. Thermal Stresses* **28**(6–7), 713–727.

Agarwal, V.K. (1979). On plane electromagneto-thermoelastic plane waves. *Acta Mech.* **34**, 181–191.

Al-Huniti, N.S. and Al-Nimr, M.A. (2000). Behavior of thermal stresses in a rapidly heated thin plate. *J. Thermal Stresses* **23**, 293–307.

——— (2004). Thermoelastic behavior of a composite slab under a rapid dual-phase-lag heating. *J. Thermal Stresses* **27**, 607–623.

——— and Megdad, M.M. (2003). Thermal induced vibrations in a thin plate under wave heat conduction model. *J. Thermal Stresses* **26**, 943–962.

Allam, M.N., Elsibai, K.A. and Abouelregal, A.E. (2002). Thermal stresses in a harmonic field for an infinite body with a circular cylindrical hole without energy dissipation. *J. Thermal Stresses* **25**, 57–67.

Allam, M.N.M., Elsibai, K.A. and Abouelregal, A.E. (2007). Electromagneto-thermoelastic plane waves without energy dissipation for an infinitely long annular cylinder in a harmonic field. *J. Thermal Stresses* **30**, 195–210.

Al-Nimr, M.A. and Al-Huniti, N.S. (2000). Transient thermal stresses in a thin elastic plate due to a rapid dual-phase-lag heating. *J. Thermal Stresses* **23**, 731–746.

Anderson, C.V.D.R. and Tamma, K.K. (2006). Novel heat conduction model for bridging different space and time scales. *Phys. Rev. Lett.* **96**, 184301-1-4.

Anwar, M.N. (1991). Problem in generalized thermoelasticity for a half-space subject to smooth heating of its boundary. *J. Thermal Stresses* **14**, 241–254.

—— and Sherief, H.H. (1988). State space approach to generalized thermoelasticity. *J. Thermal Stresses* **11**, 353–365.

——— (1994a). Boundary integral equation formulation for thermoelasticity with two relaxation times. *J. Thermal Stresses* **17**(2), 257–270.

——— (1994b). State-space approach to two-dimensional generalized thermoelasticity problems. *J. Thermal Stresses* **17**, 567–590.

Aouadi, M. (2006). Thermomechanical interactions in a generalized thermomicrostretch elastic half-space. *J. Thermal Stresses* **29**, 511–528.

—— (2007). Eigenvalue approach to linear micropolar thermoelasticity under distributed loading. *J. Thermal Stresses* **30**, 421–440.

—— (2007). Uniqueness and reciprocity theorems in the theory of generalized thermoelastic diffusion. *J. Thermal Stresses* **30**, 665–678.

Aouadi, M. (2008). Generalized theory of thermoelastic diffusion for anisotropic media. *J. Thermal Stresses* **31**, 270–285.

——and El-Karamany, A.S. (2003). Plane waves in generalized thermoviscoelastic material with relaxation time and temperature dependent properties. *J. Thermal Stresses* **26**, 197–222.

Arcisz, M. and Kosiński, W. (1996). Hugoniot relations for different heat conduction laws. *Int. J. Non-Linear Mech.* **19**, 17–38.

Bagri, A. and Eslami, M.R. (2004). Generalized coupled thermoelasticity of disks based on the Lord-Shulman model. *J. Thermal Stresses* **27**, 691–704.

———— (2007a). A unified generalized thermoelasticity formulation: Application to thick functionally graded cylinders. *J. Thermal Stresses* **30**, 911–930.

———— (2007b). Analysis of thermoelastic waves in functionally graded hollow spheres based on the Green-Lindsay theory. *J. Thermal Stresses* **30**, 1175–1193.

———— (2007c). A unified generalized thermoelasticity; solution for cylinders and spheres, *Int. J. Mech. Sci.* **49**, 1325–1335.

——Taheri, H., Eslami, M.R. and Fariborz, S. (2006). Generalized coupled thermoelasticity of a layer. *J. Thermal Stresses* **29**, 359–370.

Baksi, A. and Bera, R.K. (2006). Relaxation effects on plane wave propagation in rotating magneto-thermo-viscoelastic medium. *J. Thermal Stresses* **29**, 753–769.

Bassiouny, E. and Youssef, H.M. (2008). Two-temperature generalized thermopiezoelectricity of finite rod subjected to different types of thermal load. *J. Thermal Stresses* **31**, 237–245.

Bem, Z. (1988). Existence theorem for the stress–heat-flux equations in thermoelasticity with one relaxation time. *Bull. Pol. Acad. Sci, Tech. Sci.* **36**(7–9), 493–499.

Brorson, S.D., Fujimoto, J.G. and Ippen, E.P. (1987). Femtosecond electronic heat transport dynamics in thin gold film. *Phys. Rev. Lett.* **59**, 1962–1965.

Brun, L. and Molinari, A. (1980). Unidimensional progressive waves in generalized thermoelasticity, *J. Thermal Stresses.* **3**, 67–76.

Burchuladze, T.V. (1979). Nonstationary initial boundary value contact problems of generalized elastothermodiffusion. *Georgian Math. J.* **4**, 1–18.

Carbonaro, B. and Ignaczak, J. (1987). Some theorems in temperature-rate dependent thermoelasticity for unbounded domains. *J. Thermal Stresses* **10**(1), 193–220.

Chandrasekharaiah, D.S. (1984). A temperature-rate-dependent theory of thermopiezoelectricity. *J. Thermal Stresses* **7**, 293–306.

——(1987). Complete solutions of a coupled system of partial differential equations arising in thermoelasticity. *Quart. Appl. Math.* **45**(3), 471–480.

——(1996a). A uniqueness theorem in the theory of thermoelasticity without energy dissipation. *J. Thermal Stresses* **19**, 267–272.

——(1996b). One-dimensional wave propagation in the linear theory of thermoelasticity without energy dissipation. *J. Thermal Stresses* **19**, 695–710.

——(1998). Variational and reciprocal principles in thermoelasticity without energy dissipation. *Proc. Ind. Acad. Sci. (Math. Sci.)* **108**, 209–215.

——(2003). Author's Response to "Comments on the articles "Hyperbolic thermoelasticity: A review of recent literature" (Chandrasekharaiah D.S. 1998, *Appl Mech Rev* 51(12), 705–729) and "Thermoelasticity with second sound: A review" (Chandrasekharaiah D.S., 1986) *Appl. Mech. Rev. 39*(3), 355–376)". *Appl. Mech. Rev.* **51**(12), 453.

Chandrasekharaiah, D.S. and Srinath, K.S. (1997). Thermoelastic plane waves without energy dissipation in a half-space due to time-dependent heating of the boundary. *J. Thermal Stresses* **20**, 659–676.

Chattopadhyay, N.C. and Biswas, M. (2007). Study of thermal stress generated in an elastic half-space in the context of generalized thermoelasticity theory. *J. Thermal Stresses* **30**, 107–129.

Chen, J.K., Beraun, J.E. and Tzou, D.Y. (2002). Thermomechanical response of metals heated by ultrashort-pulsed lasers. *J. Thermal Stresses* **24**, 537–558.

Chen, T.C. and Weng, C.J. (1989). Generalized coupled transient thermoelastic problem of a square cylinder with elliptical hole. *J. Thermal Stresses* **12**, 305–320.

Chirita, S. (1988). Some applications of the Lagrange identity in thermoelasticity with one relaxation time. *J. Thermal Stresses* **11**, 207–231.

Choudhary, S. and Deswal, S. (2008). Distributed loads in an elastic solid with generalized thermo-diffusion. *Arch. Mech.* **60**(2), 139–160.

Choudhuri, S.K. (2007). On a thermoelastic three-phase lag model. *J. Thermal Stresses* **30**, 231–238.

Christov, C. I. (2008). On the evolution of localized wave packets governed by a dissipative wave equation. *Wave Motion* **45**, 154–161.

Ciancio, V. and Quintanilla, R. (2007). Thermodynamics of materials with internal variables in the context of the Green and Naghdi theories. *Balk. J. Geom. Appl.* **12**(1), 16–31.

Ciarletta, M. (1996). Thermoelasticity of non-simple materials with thermal relaxation. *J. Thermal Stresses* **19**, 731–748.

——(1999). A theory of micropolar thermoelasticity without energy dissipation. *J. Thermal Stresses* **22**, 581–594.

——Scalia, A. and Svanadze, M. (2007). Fundamental solution in the theory of micropolar thermoelasticity for material with voids. *J. Thermal Stresses* **30**, 213–227.

Ciumasu, S.G. and Vieru, D. (1993). Uniqueness and variational theorems of linear generalized thermoelasticity with microstructure. *Bull. Inst. Polit. Iasi* **39**(1–4), 121–127.

Coleman, B.D. and Newman, D.C. (1988). Implications of nonlinearity in the theory of second sound in solids. *Phys. Rev. B.* **37**(4), 1492–1498.

——and Lai, P.H. (1993). Calculations for bismuth of consequences of nonlinearities in the theory of second sound. *Phys. Rev. B.* **190**, 247–255.

————(1994a). Nonstationary interference patterns in the theory of second sound in solids. *Physica B* **198**, 361–368.

————(1994b). Waves of discontinuity and sinusoidal waves in the theory of second sound in solids. *Arch. Rat. Mech. Anal.* **126**, 1–20.

De Cicco, S. and Diaco, M. (2002). A theory of thermoelastic materials with voids and without energy dissipation. *J. Thermal Stresses* **25**, 493–503.

——and Nappa, L. (1999). On the theory of thermomicrostretch elastic solids. *J. Thermal Stresses* **22**, 565–580.

Deng, A.S. and Liu, J. (2003). Non-Fourier heat conduction effect on prediction of temperature transients and thermal stresses in skin preservation. *J. Thermal Stresses* **26**, 779–798.

Dhaliwal, R.S. and Rokne, J.G. (1988). Green's functions in generalized micropolar thermoelasticity. *J. Thermal Stresses* **11**, 257–271.

——— (1989). One-dimensional thermal shock problem with two relaxation times. *J. Thermal Stresses* **12**, 259–279.

—— Saxena, H.S. and Rokne, J.G. (1991). Generalized magneto-thermoelastic waves in an infinite elastic solid with a cylindrical cavity. *J. Thermal Stresses* **14**, 353–369.

—— and Wang, J. (1993). Green's functions in generalized micropolar thermoelasticity. *Appl. Mech. Rev.* **46**(11, Pt. 2), S316–S326.

——— (1995). Small-time Green's function in temperature rate-dependent thermoelasticity. *J. Thermal Stresses* **18**, 13–24.

Ding, X.F., Furukawa, T. and Nakanishi, H. (2001). Analysis of the shear-focusing effect in a solid cylinder subject to instantaneous heating based on the generalized thermoelasticity. *J. Thermal Stresses* **24**, 383–394.

Dyszlewicz, J. (2004). *Micropolar Theory of Elasticity.* Springer-Verlag, Berlin.

El-Karamany, A.S. (2003). Boundary integral equation formulation in generalized thermoviscoelasticity with rheological volume and density in material having temperature dependent properties. *J. Thermal Stresses* **26**, 123–147.

—— (2007). Constitutive laws, uniqueness theorem and Hamilton's principles in linear micropolar thermopiezoelectric/piezomagnetic continuum with relaxation time. *J. Thermal Stresses* **30**, 59–80.

El-Maghraby, N.M. (2004). Two-dimensional problem in generalized thermoelasticity with heat sources. *J. Thermal Stresses* **27**, 227–239.

—— (2005). A two-dimensional problem for a thick plate with heat sources in generalized thermoelasticity. *J. Thermal Stresses* **28**, 1227–1241.

—— (2008). A two-dimensional generalized thermoelasticity problem for a half-space under the action of a body force. *J. Thermal Stresses* **31**(6), 557–568.

Erbay, H.A., Erbay, S. and Dost, S. (1991). Green's function in temperature-rate dependent thermoelasticity. *J. Thermal Stresses* **14**, 161–171.

Ezzat, M.A. (1997). Generation of generalized magneto-thermoelastic waves by thermal shock in a perfectly conducting half-space. *J. Thermal Stresses* **20**, 617–633.

—— and El-Karamany, A.S. (2002*a*). The uniqueness and reciprocity theorems for generalized thermoviscoelasticity for anisotropic media. *J. Thermal Stresses* **25**, 507–522.

——— (2002*b*). Magnetothermoelasticity with thermal relaxation in a conducting medium with variable electric and thermal conductivity. *J. Thermal Stresses* **25**, 859–875.

——— (2006). Propagation of discontinuities in magneto-thermoelastic half-space. *J. Thermal Stresses* **29**, 331–358.

——— and Samaan, A.A. (2001). State-space formulation of generalized thermoviscoelasticity with thermal relaxation. *J. Thermal Stresses* **24**, 823–846.

———— and Zakaria, M. (2003). The relaxation effects of the volume properties of viscoelastic material in generalized thermoelasticity with thermal relaxation. *J. Thermal Stresses* **26**, 671–690.

Ezzat, M.A. and Othman, M.I. (2002). State-space approach to generalized magneto-thermoelasticity with thermal relaxation in a medium of perfect conductivity. *J. Thermal Stresses* **25**, 409–429.

Ezzat, M.A. and Othman, M.I. and El-Karamany, A.S. (2001*a*). Electro-magneto-thermoelastic plane waves with thermal relaxation in a medium of perfect conductivity. *J. Thermal Stresses* **24**, 411–432.

——————(2001*b*). The dependence of the modulus of elasticity on the reference temperature in generalized thermoelasticity. *J. Thermal Stresses* **24**, 1159–1176.

——————(2002). State-space approach to two-dimensional generalized thermoviscoelasticity with one relaxation time. *J. Thermal Stresses* **25**, 295–316.

Falques, A. (1982). Thermoelasticity and heat conduction with memory effects. *J. Thermal Stresses* **5**, 145–160.

Fatori, L.H., Lueders, E. and Muñoz Rivera, J.E. (2003). Transmission problem for hyperbolic thermoelastic systems. *J. Thermal Stresses* **26**, 739–763.

Fichera, G. (1992). Is the Fourier theory of heat propagation paradoxical? *Rendiconti del Circolo Matematico di Palermo, Serie II*, **41**, 5–28.

Fox, N. (1969). Generalized thermoelasticity. *Int. J. Eng. Sci.* **7**, 437–445.

Gawinecki, J. (1987). Existence, uniqueness and regularity of the first initial-boundary value problem for hyperbolic equations of temperature-rate dependent solids. *Bull. Pol. Acad., Sci. Tech. Sec.* **35** (7–8), 411–419.

——(1987). The Cauchy problem for the linear hyperbolic equations system of the theory of thermoelasticity. *Bull. Pol. Acad. Sci., Tech. Sci.* **35**(7–8), 421–433.

Gawinecki, J.A., Sikorska, B. Nakamura, G. and Rafa, J. (2007). Mathematical and physical interpretation of the solution to the initial-boundary value problem in linear hyperbolic thermoelasticity. *ZAMM* **87** (10), 715–746.

Ghaleb, A.F. (1986). A model of continuum thermoelastic media within the frame of extended thermodynamics. *Int. J. Eng. Sci.* **24**(5), 765–771.

——and El-Deen Mohamedein, M.Sh. (1989). A heat conduction equation with three relaxation times. Particular Solutions. *Int. J. Eng. Sci.* **27**(11), 1367–1377.

Ghoneim, H. and Dalo, D.N. (1987). Thermoviscoplasticity with second sound effects. *J. Thermal Stresses* **10**, 357–366.

Gładysz, J. (1985). Convolutional variational principles for thermoelasticity with finite wave speeds. *J. Thermal Stresses* **8**, 205–226.

——(1986). Approximate one-dimensional solution in linear thermoelasticity with finite wave speeds. *J. Thermal Stresses.* **9**, 45–57.

Gorman, M.J. (1993). *Numerical Analysis of One-Dimensional Waves in Generalized Thermoelasticity.* M.Sc. Thesis, Dept. Mech. Eng., Rochester Institute of Technology, Rochester, NY.

Gurtin, M.E. and Pipkin, A.C. (1969). A general theory of heat conduction with finite wave speeds. *Arch. Rational Mech. Analysis* **31**, 113–126.

Haddow, J.B. and Wegner, J.L. (1996). Plane harmonic waves for three thermoelastic theories. *Math. Mech. Solids* **1**, 111–127.

He, T. and Li, S. (2006). A two-dimensional generalized electromagneto-thermoelastic problem for a half-space. *J. Thermal Stresses* **29**, 683–698.

Hector, L.G., Kim, W.S. and Ozisik, M.N. (1992). Hyperbolic heat conduction due to a mode locked laser pulse train. *Int. J. Eng. Sci.* **30**(12), 1731–1744.

Hetnarski, R.B. and Eslami, M.R. (2009). *Thermal Stresses – Advanced Theory and Applications.* Springer-Verlag.

——and Ignaczak, J. (1994). Generalized thermoelasticity: Response of semi-space to a short laser pulse. *J. Thermal Stresses* **17**, 377–396.

——— (1995). Soliton-like waves in a low-temperature nonlinear rigid heat conductor. *Int. J. Eng. Sci.* **33**, 1725–1741.

——— (1996). Soliton-like waves in a low-temperature nonlinear thermoelastic solid. *Int. J. Eng. Sci.* **34**, 1767–1787.

——— (1997). On soliton-like thermoelastic waves. *Appl. Anal.* **65**, 183–204.

Ieşan, D. (1983). Thermoelasticity of nonsimple materials. *J. Thermal Stresses* **6**, 167–188.

Ignaczak, J. (1989). Solitons in a nonlinear rigid heat conductor. *J. Thermal Stresses* **12**, 403–423.

—— (1990). Soliton-like solutions in a nonlinear dynamic coupled thermoelasticity. *J. Thermal Stresses* **13**, 73–98.

—— (2005). The second law of thermodynamics for a two-temperature model of heat transport in metal films. *J. Thermal Stresses* **28**, 929–942.

—— (2006). Nonlinear hyperbolic heat conduction: closed-form solutions. *J. Thermal Stresses* **29**, 999–1018.

—— and Mrówka-Matejewska, E. (1990). Green's function in temperature-rate dependent thermoelasticity. *J. Thermal Stresses* **11**, 257–271.

Iovane, G. and Passarella, F. (2004). Saint-Venant's principle in dynamic porous thermoelastic media with memory heat flux. *J. Thermal Stresses* **27**, 983–999.

Irschik, H. and Gusenbauer, M. (2007). Body force analogy for transient thermal stresses. *J. Thermal Stresses* **30**, 965–975.

Ivanov, T.P. (1987). Propagation of one-dimensional waves in generalized thermomechanics (in Russian). *Adv. Mech.* **10**(4), 131–166.

Jakubowska, M. (1982). Kirchhoff's formula for a thermoelastic solid. *J. Thermal Stresses* **5**, 127–144.

—— (1984). Kirchhoff's type formula in thermoelasticity with finite wave speeds. *J. Thermal Stresses* **7**, 259–283.

Jordan, P.M. and Puri, P. (2001). Thermal stresses in a spherical shell under three thermoelastic models. *J. Thermal Stresses* **24**, 47–70.

Jou, D., Casas-Vazquez, J. and Lebon, G. (1996). *Extended Irreversible Thermodynamics*, Springer-Verlag, Berlin.

Joseph, D.D. and Preziosi, L. (1989). Heat waves. *Rev. Mod. Phys.* **61**, 41–73.

——— (1990). Addendum to the paper "Heat waves" [*Rev. Mod. Phys.* **61**, 41–73]. *Rev. Mod. Phys.* **62**, 375–391.

Kawamura, R., Tanigawa, Y. and Kusuki, S. (2008). Fundamental thermo-elasticity equations for thermally induced flexural vibration problems for inhomogeneous plates and thermoelastic dynamical responses to a sinusoidally varying surface temperature. *J. Eng. Math.* **61**(2–4), 143–160.

Karakostas, G. and Massalas, C.V. (1991). Some basic results on the generalized theory of linear thermoelasticity proposed by Green and Lindsay. *Eur. J. Mech. A/Solids* **10**, 193–211.

Kosiński, W. and Frischmuth, K. (2001). Thermomechanical coupled waves in a nonlinear medium. *Wave Motion* **34**, 131–131–141.

Kosiński, W. and Szmit, K. (1977). On waves in elastic materials at low temperature: Part I. Hyperbolicity in thermoelasticity; Part II. Principal waves. *Bull. Acad. Pol. Sci., Ser. Tech. Sci.* **25**, 23–30.

Kumar, R. and Ailawalia, P. (2006). Time harmonic sources at micropolar thermoelastic medium possessing cubic symmetry with one relaxation time. *Europ. J. Mech. A: Solids* **25**(2), 271–282.

——and Rani, L. (2004). Deformation due to mechanical and thermal sources in generalized thermoelastic half-space with voids. *J. Thermal Stresses* **28**, 123–145.

Lakusta, K.V. and Lenyuk, M.P. (1980). Estimates of the validity range for the hyperbolic equation of heat-conduction in homogeneous symmetric continuous bodies, *J. Eng. Phys.* **39**(5), 930–935.

——and Timofeev, M.P. (1980). Estimating the range of applicability of the hyperbolic thermal conductivity equation, *J. Eng. Phys.* **37**(2), 366–370.

Lazzari, B. and Nibbi, R. (2007). On the energy decay of a linear hyperbolic thermoelastic system with dissipative boundary. *J. Thermal Stresses* **30**, 1159–1172.

Leseduarte, M.C. and Quintanilla, R. (2006). Thermal stresses in type III thermoelastic plates. *J. Thermal Stresses* **29**, 485–503.

Leslie, D.J. and Scott, N.H. (2000). Wave stability for near-incompressibility at uniform temperature or entropy in generalized isotropic thermoelasticity. *Math. Mech. Solids* **5**, 157–202.

Lockett, F.J. (1958). Effect of thermal properties of a solid on the velocity of Rayleigh waves. *J. Mech. Phys. Solids* **7**, 71–75.

Mallik, S.H. and Kanoria, M. (2008). A two-dimensional problem for a transversely isotropic generalized thermoelastic thick plate with spatially varying heat source. *Eur. J. Mech. A-Solid* **27**(4), 607–621.

Maruszewski, B.T., Drzewiecki, A. and Starosta, R. (2007). Anomalous features of the thermomagnetoelastic field in a vortex array in a superconductor: Propagation of Love waves. *J. Thermal Stresses* **30**, 1049–1065.

McDonough, J.M., Kunadin, I., Kumar, R.R. and Yang, T. (2006). An alternative discretization and solution procedure for the dual-phase-lag equation. *J. Comp. Phys.* **219**, 103–171.

Misra, J.C., Chattopadhyay, N.C. and Chakravorty, A. (2000). Study of thermoelastic wave propagation in a half-space using G-N theory. *J. Thermal Stresses* **23**, 327–351.

——and Samanta, S.C. (1982). Thermal shock in a viscoelastic half-space. *J. Thermal Stresses* **5**, 365–375.

Morro, A. and Ruggeri, T. (1987). Second sound and internal energy in solids. *Int. J. Non-Linear Mech.* **22**(1), 27–36.

Mukhopadhyay, S. (1999). Relaxation effects in thermally induced vibrations in a generalized thermoviscoelastic medium with a spherical cavity. *J. Thermal Stresses* **22**, 829–840.

——(2000). Effects of thermal relaxations on thermoviscoelastic interactions in an unbounded body with a spherical cavity subject to a periodic boundary load. *J. Thermal Stresses* **23**, 675–684.

——(2002a). Thermoelastic interactions in a transversely isotropic elastic medium with a cylindrical hole subject to ramp-type increase in boundary temperature or load. *J. Thermal Stresses* **25**, 341–362.

Mukhopadhyay, S. (2002b). Thermoelastic interactions without energy dissipation in an unbounded medium with a spherical cavity due to a thermal shock at the boundary. *J. Thermal Stresses* **25**, 877–887.

Naji, M., Al-Nimr, M.A. and Mallouh, M. (2003). Thermal stresses under the effect of the microscopic heat conduction model. *J. Thermal Stresses* **26**, 41–53.

——— and Al-Huniti, N.S. (2001). Thermal stresses in a rapidly heated plate using the parabolic two-step heat conduction equation. *J. Thermal Stresses* **24**, 399–410.

Nappa, L. (1998). Spatial decay estimates for the evolution equations of linear thermoelasticity without energy dissipation. *J. Thermal Stresses* **21**, 581–592.

Nayfeh, A. (1977). Propagation of thermoelastic disturbances in non-Fourier solids. *AIAA J.* **15**, 957–960.

Noda, N., Furukawa, T. and Ashida, F. (1989). Generalized thermoelasticity of an infinite solid with a hole. *J. Thermal Stresses* **12**, 385–402.

Oncu, T.S. and Moodie, T.B. (1991). Padé-extended ray series expansions in generalized thermoelasticity. *J. Thermal Stresses* **14**, 85–99.

——— (1994). Asymptotic analysis of a nonlinear problem in thermoelasticity. *Stud. Appl. Math.* **93**, 163–186.

Orisamolu, I.R., Singh, M.N.K. and Singh, M.C. (2002). Propagation of coupled thermomechanical waves in uniaxial inelastic solids. *J. Thermal Stresses* **25**, 927–949.

Othman, M.I.A. (2002). Lord-Shulman theory under the dependence of the modulus of elasticity on the reference temperature in two-dimensional generalized thermoelasticity. *J. Thermal Stresses* **25**, 1027–1045.

Payne, L.E. and Song, J.C. (2002). Growth and decay in generalized thermoelasticity. *Int. J. Eng. Sci.* **40**, 385–400.

Puri, P. and Jordan, P.M. (2003). Comments on the articles "Hyperbolic thermoelasticity: A review of recent literature" (Chandrasekhariah, D.S. 1998, *Appl. Mech. Rev.* **51**(12), 705–729) and "Thermoelasticity with second sound: A review" (Chandrasekharaiah, D.S., 1986) *Appl. Mech. Rev.* **39**(3), 355–376). *Appl. Mech. Rev.* **56**(4), 451–453.

Qunitanilla, R. (1999). On the spatial behavior in thermoelasticity without energy dissipation. *J. Thermal Stresses* **22**, 213–229.

—— (2002). Existence in thermoelasticity with energy dissipation. *J. Thermal Stresses* **25**, 195–202.

—— (2003). A condition on the delay parameters in the one-dimensional dual-phase-lag thermoelasticity theory. *J. Thermal Stresses* **26**, 713–721.

—— (2008a). A well-posed problem for the dual-phase-lag heat conduction. *J. Thermal Stresses* **31**, 260–269.

—— (2008b). Type II thermoelasticity. A new aspect. *J. Thermal Stresses* **31**, in press.

—— and Straughan, B. (2002). Explosive instabilities in heat transmission. *Proc. R. Soc. Lond. A* **458**, 2833–2837.

Racke, R. (2003). Asymptotic behavior of solutions in linear 2D or 3D thermoelasticity with second sound. *Quart. Appl. Math.* **61**, 315–328.

Rossikhin, Y.A. and Shitikova, M.V. (2007). Analysis of a hyperbolic system modelling the thermoelastic impact of two rods. *J. Thermal Stresses* **30**, 943–963.

Roychoudhuri, S.K. (2007). One-dimensional thermoelastic waves in elastic half-space with dual phase-lag effects. *J. Mech. Mater. Struct.* **2**(3), 488–503.

Rubin, M.B. (1972). Hyperbolic heat conduction and the second law. *Int. J. Eng. Sci.* **30**, 1665–1676.

Rychlewska, J., Szymczak, J. and Woźniak, C. (2004). On the modelling of the hyperbolic heat transfer problems in periodic lattice-type conductors. *J. Thermal Stresses* **27**, 825–841.

Saleh, H.A. (2005). Problem in generalized thermoelasticity for a half-space under the action of a body force. *J. Thermal Stresses* **28**, 253–266.

Sare, H.D.F., Rivera Muñoz, J.E. and Racke, R. (2008). Stability for a transmission problem in thermoelasticity with second sound. *J. Thermal Stresses*, in press.

Saxena, H.S., Dhaliwal, R.S. and Rokne, J.G. (1991). Half-space problem in one-dimensional generalized magneto-thermoelasticity. *J. Thermal Stresses* **14**, 65–84.

Saxton, K. and Saxton, R. (2006). Phase transmission and change of type in low-temperature heat propagation. *SIAM J. Appl. Math.* **66** (5), 1689–1702.

Sharma, J.N. (1997). An alternative state space approach to generalized thermoelasticity. *J. Thermal Stresses* **20**, 115–127.

Sharma, M.D. (2006). Wave propagation in anisotropic generalized thermoelastic media. *J. Thermal Stresses* **29**, 629–642.

——(2007). Propagation of inhomogeneous waves in a generalized thermoelastic anisotropic medium. *J. Thermal Stresses* **30**, 679–691.

Sharma, J.N. and Chauhan, R.S. (2001). Mechanical and thermal sources in a generalized thermoelastic half-space. *J. Thermal Stresses* **24**, 651–675.

————and Kumar, R. (2000). Time-harmonic sources in a generalized thermoelastic continuum. *J. Thermal Stresses* **23**, 657–674.

——and Kumar, V.(1996). On the axisymmetric problems of generalized thermoelasticity. *J. Thermal Stresses* **19**, 781–794.

————and Chand, D. (2003). Reflection of generalized thermoelastic waves from the boundary of a half-space. *J. Thermal Stresses* **26**, 925–942.

——Kumar, S. and Sharma, Y.D. (2008). Propagation of Rayleigh surface waves in microstretch thermoelastic continua under inviscid fluid loading. *J. Thermal Stresses* **31**, 18–39.

——and Patania, V. (2005). Propagation of leaky surface waves in thermoelastic solids due to inviscid fluid loadings. *J. Thermal Stresses* **28**, 485–519.

——and Sharma, P.K. (2002). Free vibration analysis of homogeneous transversely isotropic thermoelastic cylindrical panel. *J. Thermal Stresses* **25**, 169–182.

————and Gupta, S.K. (2004). Steady-state response to moving loads in thermoelastic solid media. *J. Thermal Stresses* **27**, 931–951.

——Sharma, Y.D. and Sharma, P.K. (2008). On the propagation of elasto-thermodiffusive surface waves in heat-conducting materials. *J. Sound* Vib. **315**(4–5), 927–938.

——and Singh, D. (2002). Circular crested thermoelastic waves in homogeneous isotropic plates. *J. Thermal Stresses* **25**, 1179–1193.

————and Kumar, R. (2004). Propagation of generalized viscothermoelastic Rayleigh-Lamb waves in homogeneous isotropic plates. *J. Thermal Stresses* **27**, 645–668.

——Thankur, N. and Singh, S. (2007). Propagation characteristics of elastic-thermodiffusive surface waves in a semiconductor material half-space. *J. Thermal Stresses* **30**, 357–380.

Sharma, J.N., and Walia, V. (2006). Straight and circular crested Lamb waves in generalized piezothermoelastic plates. *J. Thermal Stresses* **29**, 529–551.

Shashkov, A.G. and Yanovsky, S.Y. (1994). Propagation of harmonic thermoelastic waves in general theory of heat conduction with finite wave speeds. *J. Thermal Stresses* **17**, 101–114.

Sherief, H.H. (1987). On uniqueness and stability in generalized thermoelasticity. *Quart, Appl. Math.* **44**(4), 773–778.

——(1993). State space formulation for generalized thermoelasticity with one relaxation time including heat sources. *J. Thermal Stresses* **16**, 163–180.

——and Anwar, M.N. (1986). A problem in generalized thermoelasticity. *J. Thermal Stresses* **9**, 165–181.

————(1986). Two-dimensional problem of a moving heated punch in generalized thermoelasticity. *J. Thermal Stresses* **9**, 325–343.

————(1992). Generalized thermoelasticity problem for a plate subjected to moving heat sources on both sides. *J. Thermal Stresses* **15**, 489–505.

——and Darwish, A.A. (1998). A short time solution for a problem in thermoelasticity of an infinite medium with a spherical cavity. *J. Thermal Stresses* **21**, 811–828.

——and El-Maghraby, N.M. (2003). An internal penny-shaped crack in an infinite thermoelastic solid. *J. Thermal Stresses* **26**, 333–352.

————(2005). A mode-I crack problem for an infinite space in generalized thermoelasticity. *J. Thermal Stresses* **28**, 465–484.

——Elmisiery, A.E.M. and Elhagary, M.A. (2004). Generalized thermoelastic problem for an infinitely long hollow cylinder for short times. *J. Thermal Stresses* **27**, 885–902.

——and Ezzat, M.A. (1994). Solution of the generalized problem of thermoelasticity in the form of a series of functions. *J. Thermal Stresses* **17**, 75–95.

——and Hamza, F.A. (1994). Generalized thermoelastic problems of a thick plate under axisymmetric temperature distribution. *J. Thermal Stresses* **17**, 435–452.

————(1996). Generalized two-dimensional thermoelastic problems in spherical regions under axisymmetric distributions. *J. Thermal Stresses* **19**, 55–76.

————and El-Sayed, A.M. (2005). Theory of generalized micropolar thermoelasticity and an axisymmetric half-space problem. *J. Thermal Stresses* **28**, 409–437.

————Saleh, H.A. (2004). The theory of generalized thermoelastic diffusion. *Int. J. Eng. Sci..* **42**, 591–608.

——and Helmy, K.A. (1999). A two-dimensional generalized thermoelasticity problem for a half-space. *J. Thermal Stresses* **22**, 897–910.

————(2002). A two-dimensional problem for a half-space in magnetothermoelasticity with thermal relaxation. *Int. J. Eng. Sci.* **40**, 587–604.

——and Yosef, H.M. (2004). Short-time solution for a problem in magnetothermoelasticity with thermal relaxation. *J. Thermal Stresses* **27**, 537–559.

Singh, M.C. and Tran, D.V.D. (1994). Propagation of waves in nonlinear uniaxial coupled thermoelastic solids. *J. Thermal Stresses* **17**, 21–44.

Sinha, S.B. and Elsibai, K.A. (1996). Thermal stresses for an infinite body with a spherical cavity with two relaxation times. *J. Thermal Stresses* **19**, 267–272.

————(1996). Reflection of thermoelastic waves at a solid half-space with two relaxation times. *J. Thermal Stresses* **19**, 749–762.

Sinha, S.B. and Elsibai, K.A. (1997). Reflection and refraction of thermoelastic waves at an interface of two semi-infinite media with two relaxation times. *J. Thermal Stresses* **20**, 129–145.

Sinha, S.B. and Elsibai, K.A. (1998). Effect of the second relaxation time on the propagation of Stoneley waves. *J. Thermal Stresses* **21**, 41–53.

Soyucok, A. (1991). Free vibrations of a generalized thermoelastic sphere. *J. Thermal Stresses* **11**, 173–192.

Strunin, D.V. (2001). On characteristic times in generalized thermoelasticity. *J. Appl. Mech.* **68**, 816–817.

——Melnik, R.V.N. and Roberts, A.J. (2001). Coupled thermomechanical waves in hyperbolic thermoelasticity. *J. Thermal Stresses* **24**, 121–140.

Suh, C.S. and Burger, C.P. (1998). Thermoelastic modeling of laser-induced stress waves in plates. *J. Thermal Stresses* **21**, 829–897.

Sumi, N. (2001). Numerical solutions of thermoelastic wave problems by the method of characteristics. *J. Thermal Stresses* **24**, 509–530.

——(2007). Two-dimensional thermo-mechanical wave propagation in generalized coupled theory of thermoelasticity. *J. Thermal Stresses* **30**, 897–909.

——and Ashida, F. (2003). Solution for thermal and mechanical waves in a piezoelectric plate by the method of characteristics. *J. Thermal Stresses* **26**, 1113–1123.

Svanadze, M., Tibullo, V. and Zampoli, V. (2006). Fundamental solution in the theory of micropolar thermoelasticity without energy dissipation. *J. Thermal Stresses* **29**, 57–66.

Taheri, H., Fariborz, S. and Eslami, M.R. (2004). Thermoelasticity solution of a layer using the Green-Naghdi model. *J. Thermal Stresses* **27**, 795–809.

————————(2005). Thermoelastic analysis of an annulus using the Green-Naghdi model. *J. Thermal Stresses* **28**, 911–927.

Tamma, K.K. and Nambura, R.R. (1997). Computational approaches with applications to nonclassical and classical thermomechanical problems. *Appl. Mech. Rev.* **50**(9), 514–551.

——and Zhou, X. (1998). Macroscale and microscale thermal transport and thermomechanical interactions: Some noteworthy perspectives. *J. Thermal Stresses* **21**, 405–449.

Tao, D.D.J. (1988). A numerical technique for dynamic coupled thermoelasticity problems with relaxation times. *J. Thermal Stresses* **12**, 483–487.

Tehrani, P.H. and Eslami, M.R. (2000). Boundary element analysis of Green-Lindsay theory under thermal mechanical shock in a finite domain. *J. Thermal Stresses* **23**, 773–792.

————and Azari, S. (2006). Analysis of thermoelasticity crack problems using Green-Lindsay theory. *J. Thermal Stresses* **29**, 317–330.

Tran, D.V.D. and Singh, M.C. (2004). Nonlinear uniaxial thermoelastic waves by the method of characteristics. *J. Thermal Stresses* **27**, 741–777.

————(2007). Nonlinear uniaxial thermoelastic waves with second sound by finite difference-flux corrected transport method. *J. Thermal Stresses* **30**, 239–268.

Tzou, D.Y., Chen, J.K. and Beraun, J.E. (2005). Recent development of ultra fast thermoelasticity. *J. Thermal Stresses* **28**, 563–594.

Verma, K.L. and Hasebe, N. (2001). Wave propagation in plates of general anisotropic media in generalized thermoelasticity. *Int. J. Eng. Sci.* **39**, 1739–1764.

————(2002). On the dynamic response with or without energy dissipation in the thermoelastic rotating media. *Int. J. Appl. Mech. Eng.* **7**, 1329–1348.

Vick, B., and Ozisik, M.N. (1983). Growth and decay of a thermal pulse predicted by the hyperbolic heat conduction equation. *J. Heat Transfer* **105**, 902–907.

Wang, J. and Dhaliwal, R.S. (1993). Uniqueness in generalized thermoelasticity in unbounded domains. *J. Thermal Stresses* **16**, 1–9.

———(1993). Fundamental solutions of the generalized thermoelastic equations. *J. Thermal Stresses* **16**, 135–161.

Wang, X. and Xu, X. (2002). Thermoelastic wave in a metal induced by ultrafast laser pulses. *J. Thermal Stresses* **25**, 457–473.

Wilhelm, H.E. and Choi, S.H. (1975). Nonlinear hyperbolic theory of thermal waves in metals. *J. Chem. Phys.* **63**, 2119–2123.

Woźniak, C., Wierzbicki, E. and Woźniak, M. (2002). A macroscopic model for the heat propagation in the microperiodic composite solids. *J. Thermal Stresses* **25**, 283–293.

Yilbas, B.S. and Ageeli, N. (2006). Thermal stress development due to laser step input pulse heating. *J. Thermal Stresses* **29**, 721–751.

Youssef, H.M. (2005). Generalized thermoelasticity of an infinite body with a cylindrical cavity and variable material properties. *J. Thermal Stresses* **28**, 521–532.

Zhou, X., Tamma, K.K. and Anderson, C.V.D.R. (2001). On a new C- and F-processes heat conduction model and the associated generalized theory of dynamic thermoelasticity. *J. Thermal Stresses* **24**, 531–564.

NAME INDEX

SUBJECT INDEX